METABOLIC ECOLOGY

Cover image
Photograph of a Laysan Albatross (*Phoebastria immutabilis*) by
D. Costa. Drawings and photographs of organisms courtesy of
R. Beckett, S. R. Jennings and J. H. Nichols.

METABOLIC ECOLOGY

A SCALING APPROACH

Edited by Richard M. Sibly, James H. Brown,
and Astrid Kodric-Brown

WILEY-BLACKWELL

A John Wiley & Sons, Ltd., Publication

This edition first published 2012 © 2012 by John Wiley & Sons, Ltd.

Wiley-Blackwell is an imprint of John Wiley & Sons, formed by the merger of Wiley's global Scientific, Technical and Medical business with Blackwell Publishing.

Registered office: John Wiley & Sons, Ltd, The Atrium, Southern Gate, Chichester, West Sussex, PO19 8SQ, UK

Editorial offices: 9600 Garsington Road, Oxford, OX4 2DQ, UK
The Atrium, Southern Gate, Chichester, West Sussex, PO19 8SQ, UK
111 River Street, Hoboken, NJ 07030-5774, USA

For details of our global editorial offices, for customer services and for information about how to apply for permission to reuse the copyright material in this book please see our website at www.wiley.com/wiley-blackwell.

Library of Congress Cataloging-in-Publication Data applied for.

Hardback: 9780470671535; paperback: 9780470671528.

A catalogue record for this book is available from the British Library.

Wiley also publishes its books in a variety of electronic formats. Some content that appears in print may not be available in electronic books.

Set in 9/11 pt PhotinaMT by Toppan Best-set Premedia Limited
Printed and bound in Malaysia by Vivar Printing Sdn Bhd

1   2012

CONTENTS

v

**PART III   PRACTICAL
APPLICATIONS, 259**

COMPANION WEBSITE
This book is accompanied by a companion website:
www.wiley.com/go/sibly/metabolicecology
With figures and tables from the book for downloading,
together with updates and additional resources

NOTES ON CONTRIBUTORS

Ken H. Andersen is a professor in theoretical marine ecology at the National Institute of Aquatic Resources at the Technical University of Denmark. He studies how the marine ecosystem responds to perturbations, in particular fishing, using size-spectrum models and metabolic arguments. ttp://ken.haste.dk

Kristina J. Anderson-Teixeira received her PhD in Biology under James H. Brown and also studied under Marcy E. Litvak at the University of New Mexico. She is currently working as a postdoctoral research associate with Evan H. DeLucia at the University of Illinois. Her research focuses on understanding how climate shapes terrestrial ecosystems, quantifying the climate regulation services of terrestrial ecosystems, and applying knowledge of ecosystem–climate interactions to inform land-use decisions in an era of climate change.

Lisa Patrick Bentley is currently an NSF postdoctoral fellow at the University of Arizona. She works on extending metabolic scaling theory to account for additional aspects of plant hydraulics, xylem function, and carbon flux. Her aim is to integrate key plant physiological processes that affect ecosystem-level dynamics. Her approach addresses an increasing need for predictive models that scale from the leaf to globe in order to inform future research and government policy decisions.

Julia Blanchard is a Lecturer in the Department of Animal and Plant Sciences at the University of Sheffield and an Honorary Research Fellow of Imperial College London. Julia teaches and researches both the fundamental and applied ecology of marine populations, communities, and ecosystems. Her current research involves linking macroecology, food webs and fisheries ecology to inform marine ecosystem management. www.sheffield.ac.uk/aps/staff-and-students/acadstaff/blanchard-julia

Alison Boyer is a Research Assistant Professor in the Department of Ecology asd Evolutionary Biology at the University of Tennessee. She uses ecological informatics and the fossil record to examine community ecology and extinction risk in island birds. She is also engaged in research at broader scales to examine processes governing biological diversity. http://eeb.bio.utk.edu/boyer/index.html

James H. Brown is Distinguished Professor of Biology at the University of New Mexico, Albuquerque. He led the development of the Metabolic Theory of Ecology on which this book is largely based. He has a long history of research in biogeography and macroecology, taking a large-scale statistical approach to questions about abundance, distribution, and diversity. http://biology.unm.edu/jhbrown/index.shtml

John Bruno is a marine ecologist in the Department of Biology at the University of North Carolina at Chapel Hill. His research is focused on marine biodiversity, coral reef ecology and conservation and the impacts of climate change on marine ecosystems. John earned his PhD in Ecology and Evolutionary Biology from Brown University and was a postdoctoral fellow in disease ecology at Cornell University. www.brunolab.net

Oskar Burger is a Postdoctoral Fellow at the Max Planck Institute for Demographic Research. He has many interests in both the social and natural sciences centered around understanding large-scale evolutionary constraints on human/primate life history. His latest projects include the evolution of the post-reproductive lifespan, cross-species primate mortality patterns, life-history invariants across the tree of life, and the effects of variation in energy consumption on human demographics.

Chris Carbone works at the Institute of Zoology, Zoological Society of London, Regent's Park, and is interested in understanding drivers of species distributions and abundance. His work focuses particularly on the role of body size and consumer–resource relationships in shaping organism ecology and population processes, and he is developing a fondness for dinosaurs.

Daniel Costa is a Distinguished Professor of Ecology and Evolutionary Biology at the University of California at Santa Cruz. He research focuses on the ecology and physiology of marine mammals and seabirds. He has worked in almost every habitat from the Galapagos to the Antarctic with a broad range of animals including penguins, albatross, seals, sea lions, whales, and dolphins. http://bio.research.ucsc.edu/people/costa/

Jennifer Dunne is a Professor at the Santa Fe Institute (www.santafe.edu) and Co-Director of the Pacific Ecoinformatics and Computational Ecology Lab (www.foodwebs.org). Her research seeks to identify fundamental patterns and principles of ecological network structure, robustness, and dynamics at multiple spatial and temporal scales.

Brian Enquist is a Professor in the Department of Ecology and Evolutionary Biology at the University of Arizona (www.salvias.net/~brian). He uses physiological, theoretical, computational, and informatics approaches in order to discover general principles that shape the: (i) origin of biological scaling laws; (ii) scaling of plant functional traits to ecology and evolution; and (iii) origin and maintenance of functional and phylogenetic diversity. He addresses these questions in tropical forests and alpine ecosystems.

S. K. Morgan Ernest is an Associate Professor in the Department of Biology at Utah State University. She teaches courses on Animal Community Ecology, Macroecology, and non-majors courses in Biology. While she has been involved in research spanning from life-history theory to paleoecology, she is particularly interested in how energetic constraints impact the structure and dynamics of communities. http://ernestlab.weecology. org

Stephanie Forrest is Professor of Computer Science, University of New Mexico, and External Professor of the Santa Fe institute. Her research studies adaptive systems, including immunology, evolutionary computation, biological modeling, and computer security. http://www.cs.unm.edu/~forrest/

Jamie Gillooly is an Assistant Professor of Biology at the University of Florida. Jamie played a primary role in developing the Metabolic Theory of Ecology, and is currently working to extend these energy-based models in new directions. Current projects include the study of animal communication, animal migration, and aging and disease. In addition, Jamie has been working to foster collaboration between artists and scientists as a "scholar-in-residence" in the School of Fine Arts and Art History at the University of Florida.

John Gittleman is Dean and Professor of Ecology in the Odum School of Ecology at the University of Georgia where he teaches Macroecology and Conservation Ecology. His research interests are large-scale ecological and evolutionary problems, specifically related to speciation, extinction, and global biodiversity. http://blackbear.ecology.uga.edu/gittleman/

Marcus Hamilton is a Postdoctoral Fellow at the Santa Fe Institute and an Adjunct Professor of Anthropology at the University of New Mexico. His research focuses on general principles that drive the structure, dynamics, and diversity of human societies in the past, present, and future, integrating perspectives from across the social, biological, and physical sciences. http://www.unm.edu/~marcusj/home.htm

Jon F. Harrison is a Professor in the School of Life Sciences at Arizona State University where he teaches a variety of courses in biology and physiology. His research focuses on environmental and ecological physiology of insects. http://jharrison.faculty.asu.edu

April Hayward is a Postdoctoral Associate at the University of Florida whose research ultimately focuses on understanding how complex biological systems emerged from a prebiotic soup. Current efforts toward this end center on understanding the flow and retention of matter and energy through different levels of biological organization.

Ryan Hechinger is a research scientist at the University of California, Santa Barbara. He has three overarching and related research goals. One is to evaluate the importance of parasites in ecosystems. Another is to use parasites to test and refine general ecological and evolutionary theory. The third is to "keep it real." http://www.lifesci.ucsb.edu/~hechinge/

Nick Isaac works at the Natural Environment Research Council's Centre for Ecology and Hydrology in Wallingford, Oxfordshire, on questions about species' distribution and abundance over large scales, using data on mammals, insects, and birds. Of particular interest is the relative contribution of intrinsic biological traits and extrinsic environmental drivers in shaping biodiversity, and how these patterns change at different spatial, temporal, and taxonomic scales.

Simon Jennings is a Principal Scientist at the Centre for Environment, Fisheries and Aquaculture Science, Lowestoft, and Professor of Environmental Science at the University of East Anglia. His research focuses on assessing the sustainability of human and environmental impacts on marine populations, communities, and ecosystems as well as developing and applying tools to support marine environmental management.

Walter Jetz is Associate Professor in the Ecology and Evolutionary Biology Department at Yale University. Using mostly terrestrial vertebrates and plants as study systems, his interdisciplinary research draws on elements of biogeography, community ecology, landscape ecology, macroecology, global change ecology, evolution, comparative biology, biodiversity informatics, and conservation, aiming to integrate across scales of geography and ecological organization – from local to global assemblages. http://www.yale.edu/jetz/

William Karasov is Professor in the Department of Forest and Wildlife Ecology at University of Wisconsin-Madison, where he teaches Animal Physiological Ecology. He researches digestive physiology, nutritional ecology, animal energetics, and ecotoxicology of vertebrates. http://forestandwildlifeecology.wisc.edu/facstaff/karasov.html

Michael Kaspari is a Presidential Professor and Director of the EEB graduate program at the University of Oklahoma where he teaches Community Ecology and Introduction to Biology. He uses metabolic approaches to understand the structure of brown, or detrital, food webs and how their structure and function varies geographically. He is particularly fond of tropical forests and ants. http://faculty-staff.ou.edu/K/Michael.E.Kaspari-1/AntLab_Home.htm

Drew Kerkhoff is Associate Professor in the Departments of Biology and Mathematics at Kenyon College. He teaches Ecology, Statistics, Biogeography, and Mathematical Biology, as well as introductory and non-majors courses. His research includes experimental, field, and macroecological studies of biological scaling and biodiversity, mostly focused on plants and, more recently, insects. http://biology.kenyon.edu/kerkhoff

Astrid Kodric-Brown is a Professor of Biology at the University of New Mexico, Albuquerque. Her research interests include the behavioral ecology of freshwater fishes, especially the evolution of mate recognition systems and their role in speciation in pupfishes (*Cyprinodon*); the allometry of sexually selected traits; and community structure and conservation of desert fishes. http://biology.unm.edu/biology/kodric/

Armand Kuris is Professor of Zoology at the University of California Santa Barbara where he teaches Parasitology, Invertebrate Zoology, Higher Invertebrates and Evolutionary Medicine. With Kevin Lafferty and Ryan Hechinger, he researches the role of infectious processes in ecosystems. They focus on the flow of energy through trophic levels and the evolution of parasitism, investigating biological control of schistosomiasis and exotic marine pests. http://www.lifesci.ucsb.edu/eemb/labs/kuris/index.html

Kevin Lafferty is an ecologist with the US Geological Survey and adjunct faculty at University of California Santa Barbara. He studies the ecology of infectious disease and conservation biology, primarily in marine systems. http://www.werc.usgs.gov/person.aspx?personID=166.

Elena Litchman is an associate professor at Michigan State University. Her research interests are ecology and evolution of freshwater and marine phytoplankton and aquatic ecosystem responses to global environmental change. She uses experiments, field studies, and mathematical models to investigate how abiotic factors and biotic interactions jointly determine phytoplankton community structure. http://www.kbs.msu.edu/people/faculty/litchman

Brian McGill works at the School of Biology and Ecology & Sustainability Solutions Initiative, University of Maine, where he studies biodiversity and organism–environment interactions at large scales (large areas, long time periods). A primary question is understanding how geographic ranges will shift in response to climate change. He also uses a heavily informatic approach involving large datasets and advanced statistics to analyze them.

Melanie Moses is an Assistant Professor in the Computer Science Department, with a joint appointment in Biology, at the University of New Mexico. She uses mathematical models and computational concepts to understand complex biological systems. Her research focuses on networks, including scaling in cardiovascular networks and networks of information exchange in ant colonies and immune systems. http://cs.unm.edu/~melaniem

Mary O'Connor is an Assistant Professor in the Department of Zoology and Biodiversity Research Centre at the University of British Columbia in Vancouver, BC, Canada, where she teaches Advanced Ecology and Marine Ecology. She researches the ecosystem-level consequences of environmental temperature change, and explores the contribution of ecological theory to a stronger understanding of climate change impacts. http://www.zoology.ubc.ca/~oconnor

Jordan G. Okie is a NASA Astrobiology Institute and Arizona State University School of Earth and Space Exploration postdoctoral fellow. He is interested in biological scaling, macroecology, macroevolution, microbial ecology and biogeography, astrobiology, and the role of metabolism in ecology and evolution.

Owen Petchey is Professor of Integrative Ecology at the Institute of Evolutionary Biology and Environmental Studies, University of Zurich. His group's research about biodiversity and ecological networks aims to improve our understanding of the causes and consequences of extinctions.

Scott Shaffer is an Assistant Professor in the Department of Biological Sciences at San Jose State University. He teaches Physiological Ecology and Introductory Biology and his research focuses on the ecological energetics, functional morphology, and behavioral ecology of seabirds and marine mammals. http://www.biology.sjsu.edu/facultystaff/sshaffer/sshaffer.aspx

Richard Sibly is Professor in the School of Biological Sciences at the University of Reading where he teaches Behavioural Ecology and Population Biology. He researches metabolic ecology questions with members of Jim Brown's laboratory at the University of New Mexico and also works to promote the use of Agent-Based Models (ABMs) more widely in ecology. http://www.reading.ac.uk/biologicalsciences/about/staff/r-m-sibly.aspx

Felisa Smith is an Associate Professor of Biology at the University of New Mexico. She studies factors influencing mammalian body size across time, space, and hierarchical scales. Current projects include field investigations of the physiological and morphological trade-offs to life in an extreme environment, paleomiddens investigations of microevolutionary response to late Quaternary climate change, and macroecological studies of mammalian body size across evolutionary time and geographic space. http://biology.unm.edu/fasmith/

Patrick R. Stephens is an Assistant Research Scientist in the Odum School of Ecology at the University of Georgia. He uses phylogenetic methods to explore questions that lie at the intersection of ecology and evolutionary biology, and is particularly interested in the origins of large-scale patterns of community structure and diversity. http://www.ecology.uga.edu/facultyMember.php?Stephens-348/

David Storch is based at the Charles University in Prague, Czech Republic. His interests embrace macroecology, biogeography, and evolutionary ecology, namely patterns in species richness, null models of species abundances and distributions, and geometrical issues concerning biological diversity. He has edited the book *Scaling Biodiversity* (Cambridge University Press, 2007), and has coauthored several other books on ecology and evolution. http://www.cts.cuni.cz/~storch/

Peter Vitousek is Clifford G. Morrison Professor of Population and Resources in the Department of Biology at Stanford University, where he teaches ecology and biogeochemistry. His research focuses on the biogeochemistry of nitrogen, the ecosystems of the Hawaiian Islands, and the dynamics of Polynesian agriculture prior to European contact. http//www.stanford.edu/group/Vitousek/

Robert Walker is Assistant Professor of Anthropology at the University of Missouri. He researches the evolution of human bio-cultural variation with a focus on phylogenetic methods, kin co-residence patterns, marriage practices, and life-history variation across human populations. http://anthropology.missouri.edu/people/walker.html

James S. Waters is a Biology PhD candidate in the School of Life Sciences at Arizona State University working under the guidance and mentorship of Jon F. Harrison. His dissertation research focuses on how the functional integration of social insect colonies scales with colony size. He is also involved with research on the biomechanics and respiratory physiology of insect tracheal systems.

Ethan White is an Assistant Professor with joint appointments in the Department of Biology and the Ecology Center at Utah State University. His research addresses a broad range of questions in macroecology and quantitative ecology, and he is actively involved in the development of tools to facilitate environmental informatics. http://whitelab.weecology.org

Xiao Xiao is a graduate student in the Department of Biology and the Ecology Center at Utah State University. Her research focuses on the application of mathematics, statistics, and informatics in the study of macroecological patterns.

PREFACE

Metabolic ecology gained modern impetus from the publication in 2004 of the paper entitled "Toward a Metabolic Theory of Ecology" by J. H. Brown, A. P. Allen, V. M. Savage, and G. B. West. This paper inspired much subsequent research. Our intention in this book is to showcase the achievements of metabolic ecology, to engage with its critics, and to show how the subject is now maturing by accommodating additional processes relevant to particular areas of application.

The topic is ripe for a book that reviews recent work and highlights promising areas for future development. By inviting multiple authors with different perspectives and expertise, we hope to have produced a balanced book – and one that goes beyond some of the recent controversies to highlight the fundamental roles of biological metabolism in organism–environment interactions. Our intention was that the book be written to be accessible to upper-level undergraduates, and to contain enough meat to hold the interest of graduate students and even senior scientists. We wanted an introduction to this emerging area that inspires by illustrating how the scientific process is applied in diverse but particular cases.

If we have succeeded, the book may help to define metabolic ecology as an emerging field.

We asked the authors to succinctly address a particular topic, present the theoretical foundations and some seminal empirical studies in clear, non-jargony language, and illustrate the major points with key figures and tables presented in a clear and simple form. We asked them to focus on significant recent work, relate it to the metabolic ecology paradigm, and to address alternative explanations and still unanswered questions. Figures are in color where possible, and have been redrawn professionally to provide the book with a common style. Our overarching aim has been to provide interesting, easy-to-read reviews that can be used systematically in lecture and seminar courses and that together provide an authoritative treatment that will be widely used and will enthuse future generations to study metabolic ecology.

Because metabolic ecology draws on several areas, notably biological scaling, allometry, and comparative biochemistry and physiology in addition to all subdisciplines of ecology, it must contend with different terminologies, methodologies, and traditions. We have therefore made some attempt to standardize terminology and included a glossary. We use the term Arrhenius equation to refer to expressions previously called the Van't Hoff-Arrhenius equation or the Boltzmann factor. In graphs plotting biological traits against the inverse of absolute temperature we have added an additional temperature scale in °C. In these graphs it is generally appropriate to use natural logarithms, ln, for the biological traits, for the reasons described in Chapter 2, section 2.3. (Note that $\ln(x) = 2.303 \log_{10}(x)$.) In other cases we have generally been consistent in using $\log_{10}$ rather than the natural logarithm, ln, even though many original papers used the latter, and where appropriate we have indicated the logarithmic scaling by logarithmic placing of ticks on the axes. We regret that we have not achieved complete consistency in describing the power laws derived from log-log plots. Sometimes these are written $y = mx + c$, where

$y = \log_{10} Y$, $x = \log_{10} X$, and $c = \log_{10} C$, but in other places we have written $Y = C\ X^m$. In the former case m is referred to as the "slope," in the latter "exponent." The term population growth rate is used, and given the symbol r, except for microbes, where microbiologists have traditionally used the symbol μ. The maximum population growth rate is written r_{max}. We hope that these efforts will be repaid by making the material in the book and the original literature more accessible to general readers.

The idea for the book emerged in a discussion between Jim and Richard on the bus returning to Boston from the fourth Gordon Conference on the Metabolic Basis of Ecology held in Biddeford, Maine, in July 2010. A book outline was drafted and potential contributors were contacted in August, agreement to contribute was obtained by October, outlines of contributions by November, and first drafts by February 2011. Reviews were solicited, often from fellow contributors, and the edited manuscript was delivered to Wiley-Blackwell at the end of May. The enthusiastic response of the authors exceeded our expectations,

and allowed the editors to produce the book rapidly and efficiently.

Thanks are due to many people, not all of whom are listed below. In particular we thank A. Crowden for guiding the development of the book, C. Martinez del Rio, S. Buskirk, and A. Clarke for clarifying the history of the subject, and R. M. Alexander, A. P. Allen, A. Davidson, J. Elser, J. Hone, L. Hansen, C. Hou, M. J. Konner, J. Kozlowski, C. Martinez del Rio, R. McGarvey, R. E. Michod, V. M. Savage, E. Shock, R. L. Sinsabaugh, C. D. Takacs-Vesbach, F. B. Taub, and C. Venditti who, in addition to contributors, helped with the review process. We also thank IMPPS RCN (NSF DEB 0541625, F. A. Smith, S. K. Lyons, and S. K. M. Ernest, PIs) which provided opportunity for editors and some contributors to meet. But special thanks go to the contributors, who gave generously of their time, enthusiasm, and energy.

Richard Sibly, Jim Brown,
and Astrid Kodric-Brown
Santa Fe, May 2011

INTRODUCTION: METABOLISM AS THE BASIS FOR A THEORETICAL UNIFICATION OF ECOLOGY

James H. Brown, Richard M. Sibly, and Astrid Kodric-Brown

SUMMARY

1 Ecology is fundamentally metabolic, because metabolism is the biological processing of energy and materials and all ecological interactions involve exchanges of energy and materials. The metabolic rate is the pace of living and interacting with the environment.

2 This book is based on the premise that metabolism can provide a unified theoretical basis for ecology similar to that which genetics provides for evolution.

3 The metabolic theory of ecology (MTE) aimed to provide mathematical equations for the mechanistic underpinnings of ecology, characterizing how body size and temperature, through their effects on metabolic rate, affect the rates and times

of ecological processes, spanning levels of organization from individuals to ecosystems, and explaining variation across species with diverse phylogenetic relationships, body sizes, and body temperatures.

4 It is important to distinguish between MTE, which is based on a specific idealized model for the effects of body size and temperature on metabolic rate and may have serious limitations, and a more general metabolic framework for ecology, which seems to have broad applications.

5 The chapters of this book highlight some of the accomplishments and prospects for metabolic ecology applied to different levels of ecological organization and different kinds of organisms.

I METABOLISM IS TO ECOLOGY AS GENETICS IS TO EVOLUTION?

Ecology has much to contribute to understanding the current status and future prospects of the environment, biodiversity, and our own species. However, the extent to which ecology can meet the challenges of the twenty-first century is currently limited by its lack of a unified theoretical foundation. To be sure, mathematical and conceptual models have played a seminal role in the historical development of many subdisciplines of ecology, including population regulation, food-web structure, island biogeography, and ecosystem dynamics. However, the use of alternative currencies – e.g., individuals for populations, species for islands, and nutrient fluxes for ecosystems – has hampered making connections among theories and inhibited unification of ecology as a science.

This book explores the prospect that metabolism can serve as a basis for a unified theory of ecology in much the same way that genetics has served as a basis for a unified theory of evolution (Table 1). Evolution concerns the translation, transmission, and transformation of information within and between generations. Genetics has long provided a unified theoretical foundation for evolution, because all organisms use the same molecules of DNA and RNA and the same processes of replication, translation, and transcription to code for the maintenance, growth, and reproduction of the organism.

Ecology, by contrast, concerns interactions among organisms and between organisms and their abiotic environment. Such biotic and abiotic interactions are fundamentally metabolic: metabolism is defined as the transformation of energy and materials within an organism. Metabolic rate sets both the total demands of an organism for resources from its environment, and the total allocation of resources to all biological activities. Consequently, metabolic rate is the most fundamental biological rate; it is literally the speed of living. As in genetics, the molecular basis of metabolism is universal. All organisms use the same molecules of ATP and NAD, and basically the same metabolic reactions to transform energy and matter into usable forms, which are then allocated to maintenance, growth, and reproduction.

The prospects for developing a unified theory of ecology based on metabolism depend on the extent to which there are general principles, similar to the laws of genetics, that dictate how metabolic processes play out in organism–environment interactions across multiple levels of biological organization. The burgeoning studies exploring the linkages between metabolism and ecology are identifying such principles, codifying them in mathematical form, evaluating them empirically, and applying them to develop a unified theoretical basis for ecology. The striking parallels between the genetic theory of evolution and the metabolic theory of ecology are highlighted in Table 1. Both theories base their claims for universality on how fundamental principles of physics, chemistry, and biology, and common structures and functions at lower levels of biological organization, give rise to predictable phenomena at higher levels. Both rely on the shared features of living things that underlie their diversity and complexity.

II METABOLIC ECOLOGY IS MORE THAN SCALING AND MTE

There may be a tendency to equate metabolic ecology with recent papers on the scaling of metabolic rate with body size and temperature. That would be understandable, given the attention elicited by recent papers by West, Brown, and Enquist (e.g., West et al. 1997, 1999). Although these papers on the scaling of metabolic rate and other biological activities with body size and temperature have generated considerable interest in metabolism and its application to ecology, they have also generated considerable skepticism and controversy. In particular, the paper by Brown et al. (2004) attempted to make the case for a particular idealized metabolic theory of ecology (MTE; see Brown and Sibly, Chapter 2) and to lay some tentative foundations. This paper received a mixed response, but it also stimulated a flurry of activity, much of which is addressed in the following chapters. As these chapters also indicate, however, there is much more to metabolic ecology than MTE.

Indeed, as indicated in the title of the 2004 paper, MTE was envisioned as a tentative step "toward a metabolic theory of ecology." It attempted to do three things. The first was to highlight the fundamental role of energy metabolism. Organisms and ecosystems are highly organized systems maintained far from thermodynamic equilibrium by the uptake and transformation of energy. Organisms use photosynthesis or chemosynthesis to capture physical energy and convert it into organic compounds and ATP, the universal molecule that powers all biological activities. The rate of

Table 1 Parallels and differences in the theoretical bases, biological mechanisms, and emergent phenomena of evolution and ecology. Evolution and ecology use different universal molecules and reactions, which "scale up" to generate emergent structures at whole-organism and higher levels of organization.

Components	Evolution	Ecology
Fundamental processes	Transmission and modification of chemically encoded information	Exchange of energy and materials
Universal molecules	DNA RNAs Enzymes	ATP NAD Enzymes Co-factors
Universal reactions	Replication Transcription Repair	Glycolysis TCA cycle Light and dark reactions of photosynthesis
Expression at whole-organism level	Mendel's laws Quantitative inheritance Sex determination	Dependence of metabolic rate and life-history traits on body mass and temperature
Expression at population, community, and ecosystem levels	Hardy–Weinberg frequencies *F* statistics Fisher's fundamental theorem Sex ratio	Population growth and regulation Food-web structure and dynamics Ecosystem flux and storage Species diversity patterns

energy transformation and ATP synthesis is the metabolic rate. This is the most fundamental biological rate. It literally is the pace of life.

The second goal was to emphasize the role of metabolic processes in ecology. Interactions between organisms and their environments involve exchanges of energy, materials, and information. These fluxes are all part of metabolism in the broad sense. They are all dependent on metabolic rate, because energy powers and controls the exchanges. So, for example, the primary production of an ecosystem is the sum of the carbon fixation of all the autotrophic organisms; the growth rate of a population is the rate of incorporation of energy and materials into new individuals; and the information conveyed in birdsong is generated by the singer transforming metabolic energy into sound waves. All of these are examples of metabolic ecology.

The third goal was to use the physics, chemistry, and biology of metabolism to develop a rigorous, quantitative, predictive basis for much of ecology. It has been known for more than a century that the metabolic rates of animals, plants, and microbes are governed by three primary factors: body size, temperature, and availability of material resources, such as water and nutrients. Physiologists have quantified the effects of these factors on metabolic rate. It has also been known for decades that metabolic processes govern the role of

biota in the biogeochemical cycling of carbon, nitrogen, and other elements in the biosphere.

MTE was based on the premise that it is possible to use body size, temperature, water and nutrient availability, and a few other variables to understand how photosynthesis, respiration, and other metabolic processes at the molecular, cellular, and organismal levels give rise to emergent structures and dynamics at the levels of population, community, ecosystem, and biosphere (see Table 2). One contribution of Brown et al. (2004), and the source of much of the controversy, was their effort to combine these effects into a single mathematical model or central equation

$$B = B_0 M^\alpha e^{-E/kT} \tag{1}$$

where B, the dependent variable, is the rate of some biologically mediated ecological process, B_0 is a normalization constant that typically varies with the nature of the process, kind of organism, and environmental setting, M^α is the power-law body size term, where M is body mass and α is the scaling exponent, and $e^{-E/kT}$ is the exponential temperature dependence term, where E is an "activation energy," k is Boltzmann's constant, and T is temperature in kelvin.

MTE has come to be equated with the specific form of this equation, where α is some multiple of 1/4,

Table 2 Principles of metabolic ecology that apply to different phenomena and levels of organization. These applications are illustrated by a few selected examples, each considered at two levels of specificity: (1) general, which does not rely on specific models or make precise predictions; and (2) metabolic theory of ecology (MTE), which is based on specific equations and makes a priori quantitative predictions. Here, we present the more specific form of MTE developed in Brown et al. (2004), which assumes specific values of the allometric exponents, α, and activation energies, E, rather than the more general form that stems from equation 1. Some of these relationships, especially the MTE prediction for species diversity, are still highly speculative and need to be more rigorously developed theoretically and explored empirically.

Level/phenomenon	General	MTE
Whole-organism rates of metabolism and resource use: animals	Vary with taxon, body size, and environmental conditions	$B = B_0 M^{3/4} e^{-0.65/kT}$
Life history rates: e.g., fertility rate, mortality rate: ectothermic animals	Vary with both intrinsic biological traits (such as body size and temperature) and extrinsic environment (such as temperature, food supply, predators)	$R = R_0 M^{-1/4} e^{-0.65/kT}$ where R_0 varies predictably with taxonomic or functional group, lifestyle, and habitat
Population carrying capacity: ectothermic animals	Varies with both intrinsic biological traits (such as body size and trophic level) and extrinsic environment (such food supply)	$K = K_0 M^{-3/4} e^{0.65/kT}$ where K_0 varies predictably with rate of resource supply (which includes trophic level)
Whole-ecosystem rate of carbon turnover: terrestrial ecosystems	Varies spatially with vegetation type and temporally with temperature and precipitation	$T = T_0 [C] M^{-1/4} e^{-0.30/kT}$ where [C] is soil moisture, and M is the average size of the dominant plants
Species richness: animals in terrestrial ecosystems	Varies with taxon, body size, environmental variables (temperature and precipitation), and historical factors	$S = S_0 [C] M^{-3/4} e^{-0.65/kT}$ where S_0 varies with taxon and historic context and [C] is soil moisture

such as 3/4 for whole-organism metabolic rate, −1/4 for mass-specific metabolic rate and most ecological rates, and 1/4 for biological times, and E is approximately 0.65 eV for aerobic respiration and respiration-dependent rates and approximately 0.30 eV for hotosynthesis and photosynthesis-dependent rates (Enquist et al. 1998; Brown et al. 2004; Allen et al. 2005). On the one hand, many authors have applied this strict expression of MTE to their own systems, and found that it predicts their data well. On the other hand, most criticisms have focused on this narrow, idealized version of MTE. These have either questioned the specific values assumed for the scaling exponent, α, or the activation energy, E, or they have questioned whether the body size term is strictly power law or whether the temperature dependence term is strictly exponential and hence Arrhenius–Boltzmann.

This strict version of MTE and points of support and criticism are addressed in the following chapters. Our intention is not to produce a book that is uncritically favorable of MTE. Rather it is to produce a balanced book that explores the metabolic underpinnings of

ecology to the fullest extent possible. Our vision is based on the seemingly unassailable proposition that life is sustained by the flux, storage, and turnover of three ecological currencies: energy, matter, and information. The first currency, energy, is harnessed in a biologically useful form initially by autotrophs, mostly by plants. The energy of sunlight is captured in photosynthesis, stored as biomass in the chemical bonds of reduced organic compounds comprising biomass, and then released in respiration to fuel biological work. The second currency, matter, includes carbon, nitrogen, phosphorus, and other elements comprising biomass. The third currency, information, is encoded in the DNA that is transmitted from parent to offspring and in the physical-chemical forms that organisms use to sense their environment and communicate with each other. These three currencies are quantitatively related to each other because of their interacting roles in the maintenance of life and the integration of structure and function at molecular to whole-organism levels of organization. At higher levels of organization – individual organism, population, community, and ecosystem – all interactions between organisms and their environments involve the fluxes, transformations, and storage of these three basic currencies.

In this sense ecology is fundamentally metabolic, and we can look to metabolism for the mechanistic basis for a unifying theory of ecology. We believe that most ecologists, regardless of their views on MTE, will agree with this perspective. It should be productive to ask what is exchanged in an ecological interaction, and what variables and processes affect the exchange. By focusing on the metabolic currencies of energy, materials, and information, it should be possible to unify different subdisciplines and make connections across levels of organization and spatial and temporal scales. This promise of metabolic ecology transcends MTE, the general form of equation 1, and the more specific form that assumes certain values of α and E.

It would be unfortunate, however, to abandon prematurely the effort to provide a more detailed, quantitative, and predictive theoretical foundation for ecology based on fundamental metabolic processes. This was the intent of Brown et al. (2004) and it remains a goal of this book. To appreciate both the power and the limitations, it is necessary to understand the role of theory and general models. They are not intended to capture all the rich details of biological diversity and ecological interactions. Obviously no simple equation with just a few variables and terms can do this. They are intended, however, to capture the most important features of phenomena in a deliberately oversimplified quantitative form that can be used to understand general empirical patterns and to make a priori, quantitative, testable predictions. The validity and utility of the theory and models will ultimately rest on how well they perform in these applications.

The analogy between a metabolic theory of ecology and the genetic theory of evolution is illustrative (see also White, Xiao, Isaac, and Sibly, Chapter 1). The Hardy–Weinberg equilibrium and Fisher's fundamental theorem and sex ratio theory were never intended to account for all of the variation. Take Hardy–Weinberg as an example. The mathematical model is based explicitly on Mendelian genetics and incorporates all of the assumptions of Mendel's laws. From the frequencies of alleles, it predicts the frequency of genotypes in an idealized population in the absence of drift, selection, migration, and mutation. Empirical studies with sufficiently large sample sizes almost always reveal statistically significant deviations from the predicted frequencies. This does not mean that the theory is wrong or useless. It has become a cornerstone of evolutionary and population genetics, because it provides a precise, quantitative baseline based on simple underlying assumptions. We now recognize that factors such as differential selection on genotypes can account for deviations from the Hardy–Weinberg equilibrium, and we have extensions of the theory that can take these processes into account and provide new, revised, more accurate predictions.

Can a general body of theory that makes grossly oversimplified assumptions and is written in the language of mathematical equations make a comparable contribution to the progress and unification of ecology? This book will present much of the currently available evidence, but the jury is still out.

III ORGANIZATION OF THE BOOK

The book is intended to flesh out the prospects for a metabolic perspective on ecology by providing examples from past studies and suggesting avenues for future work. Part I lays out the foundations. It begins with a chapter by White, Xiao, Isaac, and Sibly on the methodological tools used in both mathematical models and statistical analysis of data. In the next chapter Brown and Sibly show the origin of the "master equation" of MTE and address some of its limitations in the light of recent criticisms.

The next four chapters focus on the physiological ecology of individual organisms. In Chapter 3 on stoichiometry, Kaspari considers the role of essential chemical elements and compounds in metabolism and the effects of these limiting materials on ecology. In Chapter 4, Kerkhoff addresses how organisms allocate energy and materials to ontogenetic growth and development. Sibly continues this theme in the next chapter, focusing on the way natural selection has shaped the life history so as to maximize fitness by optimally allocating the products of metabolism among maintenance, growth, and reproduction. Then Hayward, Gillooly, and Kodric-Brown highlight examples of the role of metabolism in behavior.

The next chapters expand the scope to higher levels of organization, considering how individual metabolism and metabolic rate affect populations and communities (Chapter 7, by Isaac, Carbone, and McGill), foraging ecology and food web organization (Chapter 8, by Petchey and Dunne), and ecosystem processes (Chapter 9, by Anderson-Teixeira and Vitousek). The next two chapters extend metabolic ecology even farther; Gittleman and Stephens (Chapter 10) consider the implications of metabolic rate for evolutionary rates, and Storch (Chapter 11) addresses relationships between metabolism and the geographic patterns of biodiversity.

Part II showcases applications of metabolic ecology to particular taxonomic groups, with Chapters 12 to 20, respectively, on prokaryotes and unicellular eukaryotes (by Okie), phytoplankton (by Litchman), land plants (by Enquist and Bentley), marine invertebrates (by O'Connor and Bruno), insects (by Waters and Harrison), terrestrial vertebrates (by Karasov), marine vertebrates (by Costa and Shaffer), parasites (by Hechinger, Lafferty, and Kuris), and finally, humans (by Hamilton, Burger, and Walker). The number of chapters in this part and the variety of their contents is an encouraging sign that metabolic ecology has already had broad impact.

Part III presents some of the practical implications and applications of the metabolic perspective. There are chapters on human impacts on and management of fisheries (by Jennings, Andersen, and Blanchard), conservation (by Boyer and Jetz), and climate change (by Anderson-Teixeira, Smith, and Ernest). This is followed by a chapter by Moses and Forrest on extensions of some of the conceptual framework and models of metabolic ecology to human-engineered systems, from transportation networks to computer chips.

Together, these chapters showcase the power and promise of a metabolic approach to ecology. They review some past contributions and place them in the unifying context of how metabolic processes of energy, material, and information exchange affect interactions between organisms and their abiotic and biotic environments. They also offer speculations on promising areas for future research. Although we recognize that readers will naturally be more interested in some chapters than others, we hope that most will read the entire volume, and thereby gain an appreciation of the diverse strands and their interconnections – for the unity that underlies the diversity.

IV A FINAL COMMENT

On the one hand, this is a book of blatant advocacy. It would not have been written if the authors of the chapters had not been enthusiastic about metabolic ecology. Indeed, their response to the strict timetable and the suggestions of the reviewers and editors has been a bit overwhelming. We hope that readers will appreciate not only the work that has gone into every chapter, but also the connections and integration among chapters in an effort to give a unified, synthetic view of the past contributions and future prospects of metabolic ecology.

On the other hand, this book is intended to provide a balanced, critical assessment of the current state of metabolic ecology, from both a theoretical and an empirical perspective. In this regard, it is important to distinguish between the Metabolic Theory of Ecology (MTE) as presented in Brown et al. (2004) and a more general metabolic ecology. How much of MTE as embodied in equation 1 will endure is an open question. What we do expect to endure is the perspective to look to metabolism – the uptake, processing, and allocation of energy, materials, and information – for the fundamental mechanisms of organism–environment interactions, and for the processes that link the levels of ecological organization from individuals to ecosystems. This book will have served its function if it encourages readers to approach ecology by asking three questions: (1) What currency is exchanged between organism and environment? (2) What physical, chemical, and biological principles govern the exchange? (3) What are the implications of these flows for other levels of organization, both lower and higher, and for ecological phenomena operating on different spatial and temporal scales?

Part I

Foundations

Chapter 1

METHODOLOGICAL TOOLS

Ethan P. White, Xiao Xiao, Nick J. B. Isaac, and Richard M. Sibly

SUMMARY

1 In this chapter we discuss the best methodological tools for visually and statistically comparing predictions of the metabolic theory of ecology to data.

2 Visualizing empirical data to determine whether it is of roughly the correct general form is accomplished by log-transforming both axes for size-related patterns, and log-transforming the y-axis and plotting it against the inverse of temperature for temperature-based patterns. Visualizing these relationships while controlling for the influence of other variables can be accomplished by plotting the partial residuals of multiple regressions.

3 Fitting relationships of the same general form as the theory is generally best accomplished using ordinary least-squares-based regression on log-transformed data while accounting for phylogenetic non-independence of species using phylogenetic general linear models. When multiple factors are included this should be done using multiple regres-sion, not by fitting relationships to residuals. Maximum likelihood methods should be used for fitting frequency distributions.

4 Fitted parameters can be compared to theoretical predictions using confidence intervals or likelihood-based comparisons.

5 Whether or not empirical data are consistent with the general functional form of the model can be assessed using goodness-of-fit tests and comparisons to the fit of alternative models with different functional forms.

6 Care should be taken when interpreting statistical analyses of general theories to remember that the goal of science is to develop models of reality that can both capture the general underlying patterns or processes and also incorporate the important biological details. Excessive emphasis on rejecting existing models without providing alternatives is of limited use.

1.1 INTRODUCTION

Two major functional relationships characterize the current form of the metabolic theory of ecology (MTE). Power-law relationships, of the form $y = cM^\alpha$ (Fig. 1.1A,B), describe the relationship between body size and morphological, physiological, and ecological traits of individuals and species (West et al. 1997; Brown et al. 2004). The Arrhenius equation, of the general form, $y = ce^{-E/kT}$ (Fig. 1.1C,D), characterizes the

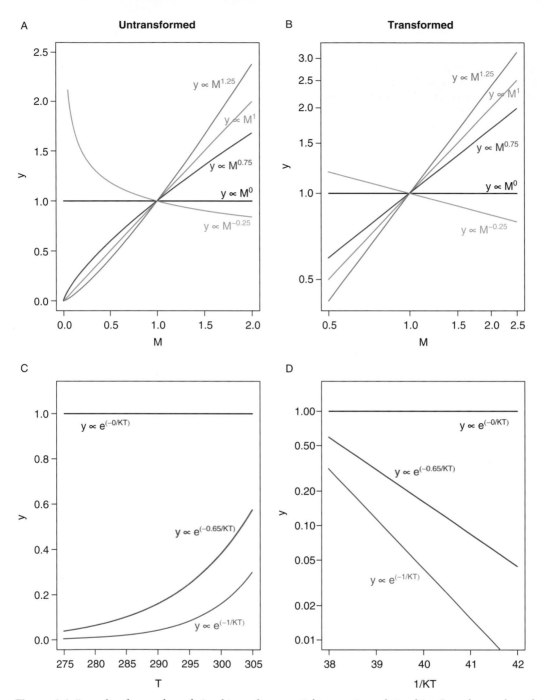

Figure 1.1 Examples of power-law relationships and exponential temperature relationships. Several power-law relationships are shown on untransformed (A) and logarithmically scaled (B) axes. Power-law relationships with exponents equal to one characterize direct proportionalities, which are linear relationships with intercepts of zero. Several temperature relationships are shown on untransformed (C) and Arrhenius plot axes ($1/T$ vs. logarithmically scaled y) (D). Power laws with exponents greater than 1 are described as superlinear because their slope is increasing in linear space and power laws with exponents less than 1 are described as sublinear because their slope is decreasing. Relationships that have exponents equal to zero do not change with the variable of interest and are therefore described as invariant with respect to mass or temperature. Note that in the Arrhenius plots different coefficients are used to allow for clear presentation.

relationships between temperature and physiological and ecological rates (Gillooly et al. 2001; Brown et al. 2004). In addition to being central to metabolic theory, these empirical relationships are utilized broadly to characterize patterns and understand processes in areas of study as diverse as animal movement (Viswanathan et al. 1996), plant function (Wright et al. 2004), and biogeography (Arrhenius 1920; Martin and Goldenfeld 2006).

Methodological approaches for comparing metabolic theory predictions to empirical data fall into two general categories: (1) determining whether the general functional form of a relationship predicted by the theory is valid; and (2) determining whether the observed values of the parameters match the specific quantitative predictions made by the theory. Both of these categories of analysis rely on being able to accurately determine the best fitting form of a model with the same general functional form as that of MTE, so we will begin by discussing how this has typically been done using ordinary least-squares (OLS) regression on appropriately transformed data. Potential improvements to these approaches that account for statistical complexities of the data will then be considered. We will discuss methods for comparing the fitted parameters to theoretical values and how to determine whether the general functional form predicted by the theory is supported by data. This will require some discussion of the philosophy of how to test theoretical models. So we will end with a general discussion of the technical and philosophical challenges of testing and developing general ecological theories.

1.2 VISUALIZING MTE RELATIONSHIPS

Before conducting any formal statistical analysis it is always best to visualize the data to determine whether the model is reasonable for the data and to identify any potential problems or complexities with the data.

1.2.1 Visualizing functional relationships

The primary model of metabolic theory describes the relationship between size, temperature, and metabolic rate; combining a power function scaling of mass and metabolic rate with the Arrhenius relationship describing the exponential influence of temperature on biochemical kinetics.

$$I = i_0 M^\alpha e^{-E/kT} \tag{1.1}$$

See Brown and Sibly (Chapter 2) or Brown et al. (2004) for details.

Most analyses of this central equation focus on either size or temperature in isolation, or attempt to remove the influence of the other variable before proceeding. As such, the most common analyses focus on either power-law relationships, $y = cM^b$, or exponential relationships, $y = ce^{-E/kT}$, both of which can be log-transformed to yield linear relationships (Fig. 1.1).

$$y = cM^\alpha \Rightarrow \log(y) = \log(c) + \alpha \log(M) \tag{1.2a}$$

$$y = ce^{-E/kT} \Rightarrow \log(y) = \log(c) - (E/kT) \tag{1.2b}$$

The linear forms of these relationships form the basis for the most common approaches to plotting these data and graphically assessing the validity of the general form of the equations. Plots of these linearized forms are obtained either by log-transforming the appropriate variables or by logarithmically scaling the axes so that the linear values remain on the axes, but the distance between values is adjusted to be equivalent to log-transformed data. In this book all linearized plots will used log-scaled, rather than log-transformed, axes. Relationships between size and morphological, physiological, and ecological factors are typically plotted on log-log axes and relationships between temperature and these factors are displayed using Arrhenius plots with the log-scaled y variable plotted against the inverse of temperature (Fig. 1.2A,B). If the relationships displayed on plots of these forms are approximately linear then they are at least roughly consistent with the general form predicted by metabolic theory.

When information on both size and temperature are included in an analysis to understand their combined impacts on a biological factor, this has been displayed graphically by removing the effect of one factor and then plotting the relationship for the other factor (Fig. 1.2C,D). The basic idea is to rewrite the combined size–temperature equation so that only one of the two variables of interest appears on the right-hand side.

$$\frac{y}{ce^{-\frac{E}{kT}}} = M^\alpha \Rightarrow \log(y) + \frac{E}{kT} - \log(c) = \alpha \log(M) \tag{1.3a}$$

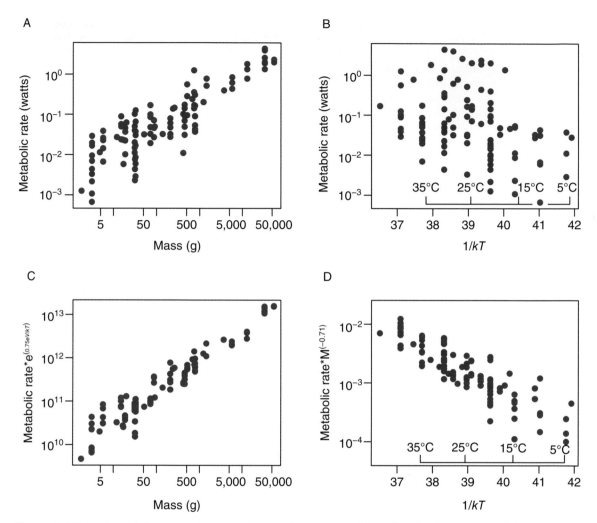

Figure 1.2 Plots of metabolic rate as a function of mass and temperature. (A) Log-log plot of mass vs. metabolic rate not accounting for temperature. (B) Arrhenius plot of temperature vs. metabolic rate not accounting for mass. (C) Log-log plot of mass vs. metabolic rate accounting for temperature. (D) Arrhenius plot of temperature vs. metabolic rate accounting for mass. Data is for reptiles from Gillooly et al. (2001).

$$\frac{y}{cM^\alpha} = e^{-E/kT} \Rightarrow \log(y) - \alpha\log(M) - \log(c) = -\frac{E}{kT}$$

(1.3b)

The value for the dependent variable (i.e., the value plotted for each point on the vertical axis) is then determined by dividing the observed value of y by the appropriate transformation of temperature or mass for the observation and log-transforming the resulting value. This is equivalent to the standard approach of plotting the partial residuals to visualize the relationship with a single predictor variable in multiple regression. Often in the MTE literature the theoretical forms of the relationships ($\alpha = 0.75$, $E = -0.65$) have been used rather than the fitted forms based on multiple regression. For reasons discussed below we recommend using the fitted values of the parameters, or simply using the partial residuals functions in most statistical packages, to provide the best visualization of the relationship with the variable of interest.

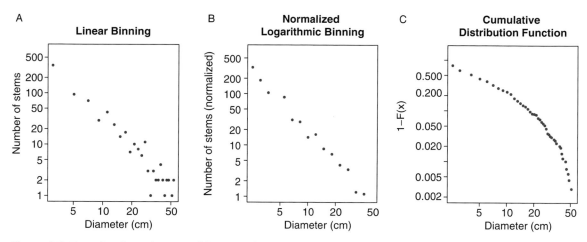

Figure 1.3 Examples of visualizations of frequency distributions. Methods include (A) linear binning, (B) normalized-logarithmic binning, and (C) linearizing the cumulative distribution function. Data are from the Nosy Mangabe, Madagascar, site of Alwyn Gentry's tree transect data (site 201; Phillips and Miller 2002).

1.2.2 Frequency distributions

In addition to making predictions for the relationships between pairs of variables – e.g., size, temperature, and metabolic rate – metabolic ecology models have been used to make predictions for the form of frequency distributions (i.e., histograms) of biological properties such as the number of trees of different sizes in a stand (Fig. 1.3; West et al. 2009). The predicted forms of these distributions are typically power laws and have often been plotted by making histograms of the variable of interest, log-transforming both the counts and the bin centers and then plotting the counts on the *y*-axis and the bin centers on the *x*-axis (Fig. 1.3A; e.g., Enquist and Niklas 2001; Enquist et al. 2009). This is a reasonable way to visualize these data, but it suffers from the fact that bins with zero individuals must be excluded from the analysis due to the log-transformation. These bins will occur commonly in low probability regions of the distribution (e.g., at large diameters), thus impacting the visual perception of the form of the distribution. To address this problem we recommend using normalized logarithmic binning (*sensu* White et al. 2008), the method typically used for visualizing this type of distribution in the aquatic literature (e.g., Kerr and Dickie 2001). This approach involves binning the data into equal logarithmic width bins (either by log-transforming the data prior to constructing the histogram or by choosing the bin edges to be equal loga-

rithmic distances apart) and then dividing the counts in each bin by the linear width of the bin prior to graphing (Fig. 1.3B). The logarithmic scaling of the bin sizes decreases the number of bins with zero counts (often to zero) and the division by the linear width of the bin preserves the underlying shape of the relationship. Another, equally valid approach is to visualize the relationship using appropriate transformations of the cumulative distribution function (Fig. 1.3C; see White et al. 2008 for details), but we have found that it is often more difficult to intuit the underlying form of the distribution from this type of visualization and therefore recommend normalized logarithmic binning in most cases.

1.3 FITTING MTE MODELS TO DATA

1.3.1 Basic fitting

Since the two basic functional relationships of metabolic theory can be readily written as linear relationships by log-transforming one or both axes, most analyses use linear regression of these transformed variables to estimate exponents, compare the fitted values to those predicted by the theory, and characterize the overall quality of fit of the metabolic models to the data. Given the most basic set of statistical assumptions, this is the correct approach.

Specifically, if the data points are independent, the error about the relationship is normally distributed when the relationship is properly transformed (i.e., it is multiplicative log-normal error on the untransformed data):

$$\log(y) = \log(c) + b\log(M) + \varepsilon, \varepsilon \sim N(0, \sigma^2) \quad (1.4a)$$

$$y = cM^b e^\varepsilon, \varepsilon \sim N(0, \sigma^2) \quad (1.4b)$$

and there is error (i.e., stochasticity) only in the y-variable, then the correct approach to analyzing the component relationships is ordinary least-squares regression.

Given the same basic statistical assumptions, analyzing the full relationship including both size and temperature should be conducted using multiple regression with the logarithm of mass and the inverse of temperature as the predictor variables. This approach is superior to the common practice of using simple regression after correcting for the influence of the other variable (see, e.g., Gillooly et al. 2001; Brown et al. 2004) because it appropriately allows for correlation between the predictor variables, thus yielding the best simultaneous estimates of the parameters for each variable and the appropriate estimates of the confidence intervals for those parameters (Freckleton 2002; Downs et al. 2008).

In many cases the assumptions underlying these basic statistical analyses may be reasonable, and these methods are often robust to some violations of the assumptions. However, there are also a number of instances in common MTE analyses where substantial violations of assumptions related to the independence of data points, and even the basic form of the error about the relationship, may necessitate the use of more complex methods to obtain the most rigorous results.

1.3.2 Log-transformation vs. nonlinear regression

While most analyses utilize the fact that log-transforming one or both sides of the equation yields a linear relationship, allowing appropriately transformed data to be modeled using linear regression (log-linear regression), it has recently been suggested that analysis on logarithmic scales is flawed and that, instead, analysis should be carried out on the original scale of meas-

urement using nonlinear regression (e.g., Packard and Birchard 2008; Packard and Boardman 2008, 2009a, 2009b; Packard 2009; Packard et al. 2009, 2010).

One fundamental difference between log-linear regression and nonlinear regression on untransformed data lies in the assumptions that the two approaches make about the nature of unexplained variation. In nonlinear regression the error term (i.e., residuals) is assumed to be normally distributed and additive, $y = \alpha x^b + \varepsilon, \varepsilon \sim N(0, \sigma^2)$, while log-linear regression assumes the error term is log-normally distributed and multiplicative (equation 1.1). The form of the error distribution in the empirical data determines which method performs better, with the method that assumes the appropriate error form (i.e., nonlinear regression with additive error, and log-linear regression with multiplicative error) yielding the best results (Xiao et al. 2011).

Throughout this chapter we recommend that the form of the error distribution be explicitly considered, when possible, in deciding which methods to use (Cawley and Janacek 2010; Xiao et al. 2011). However, log-normal error is substantially more common than normal error in physiological and morphological data (Fig. 1.4: Xiao et al. 2011; see also Gingerich

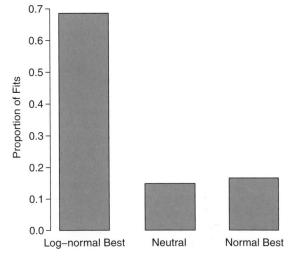

Figure 1.4 Likelihood analysis comparing the fits of normal vs. log-normal error to 471 biological power laws shows that most morphological and physiological relationships are better characterized by multiplicative log-normal error and therefore that traditional log-transformed regression is better in most cases than nonlinear regression (Xiao et al. 2011).

2000; Kerkhoff and Enquist 2009; Cawley and Janacek 2010), which implies that for most metabolic theory analyses log-linear regression is appropriate. This is good news because log-linearity allows the implementation of some approaches discussed below which cannot readily be implemented in a nonlinear context.

1.3.3 Alternatives to ordinary least-squares regression

The ordinary least-squares (OLS) approach is just one of several available choices for fitting a linear relationship between X and Y variables, with each method making different assumptions about the variation around the regression line. Understanding which of these methods to use can seem complicated because these choices depend on information about the sources and magnitude of variability around the regression line, the nature of the relationship between X and Y, and the goal of the analysis. In addition, there is conflicting advice in the literature regarding when to use which method, and uncertainty about best practice has led to many studies reporting regression slopes determined using more than one approach (e.g., Coomes et al. 2011).

The main alternative to OLS regression is commonly known as reduced major axis (RMA) regression. Whereas OLS assumes that residual variation occurs only in the vertical direction, RMA allows for variation also in the horizontal direction by minimizing the sum of the products of deviations in the vertical and horizontal directions. For most datasets, slopes estimated by RMA are steeper than those estimated by regression (Smith 2009). Other alternatives include major axis (MA), which generates estimates of the slope that are intermediate between RMA and OLS regression, and the OLS bisector, which determines the average of the slope of X on Y and the slope of Y on X (Isobe et al. 1990). OLS, RMA, and MA are all special cases of a general model in which the ratio of the error variances in X and Y can take on any value (Harvey and Pagel 1991; M.P. O'Connor et al. 2007a).

A common argument for the use of alternatives to OLS in allometric studies is that it is inappropriate to assume that X is measured without error, as implied in OLS regression (e.g., Legendre and Legendre 1998). However, this argument relies on the assumption that all of the variation about the regression line is due to measurement error, which is unlikely to be the case in biological systems (e.g., Sokal and Rohlf 1995; Smith 2009) and even then this argument is not valid in most situations (Warton et al. 2006; Smith 2009). Recent advice regarding when it is appropriate to use RMA (or a related alternative) vs. OLS is based on a combination of the goal of the analysis and the causal relationship between the variables (Warton et al. 2006; O'Connor et al. 2007a; Smith 2009). For an excellent treatment of the logic behind RMA vs. OLS see Smith (2009). All line-fitting techniques discussed can be implemented using the SMATR package in R (http://www.bio.mq.edu.au/ecology/SMATR/).

1.3.4 Which method(s) should I use?

Our interpretation of the recent discussion on which method to use is that, for the majority of cases in metabolic theory, OLS regression on log-transformed data is the correct approach. Most analyses in metabolic theory are causal in nature – the hypothesis is that the size and temperature of an organism determine a broad suite of dependent variables. In the case of hypothesized causal relationships we are logically assigning all equation error (i.e., variability about the line not explained by measurement error; Fuller 1987; McArdle 2003) to the Y variable and therefore should be estimating the form of the relationship using OLS (Warton et al. 2006; Smith 2009). In addition to causal relationships, OLS regression is also most appropriate in cases where one wants to predict unknown values of Y based on X (Sokal and Rohlf 1995; Warton et al. 2006; Smith 2009). Metabolic theory is often used in this context to estimate the metabolic rate of individuals based on body size (e.g., Ernest and Brown 2001; White et al. 2004; Ernest et al. 2009). The fact that OLS is appropriate for many metabolic theory predictions is convenient because variants on simple bivariate relationships (e.g., phylogenetic correction, mixed effects models) are typically based on OLS.

There are some cases where directional causality between the two variables being analyzed is not implied by metabolic models. For example, predictions for the relationships between different measures of size (e.g., height and basal stem diameter in trees) do not imply a direct causal relationship between the variables but an "emergent" outcome of a process affected by two interdependent variables. In this case, the choice of which variable to place on the *x*-axis is arbitrary. In this case (and in many similar cases in other areas of allometry; e.g., the leaf economics spectrum) RMA or a related approach is more appropriate for analysis because we want to partition equation error between X and Y, rather than assigning it all to Y.

1.3.5 Phylogenetic methods

A common goal of analysis in metabolic ecology is to understand the relationship between two morphological, physiological, or ecological properties, across species. The data points in these analyses are typically average values of the two properties for each species, which leads to a potential complication. Because there are limits to how quickly traits can evolve, closely related species may not be statistically independent due to their shared evolutionary history. This lack of independence among data points violates a key assumption of ordinary least-squares regression (and general linear models more broadly).

The problem of phylogenetic non-independence is well known in evolutionary biology, and a method known as independent contrasts (Felsenstein 1985) remains popular for correcting for the phylogenetic signal in comparative data. Independent contrasts have been recently superseded by phylogenetic general linear models (PGLMs), which allow a wide range of evolutionary scenarios to be modeled (Garland and Ives 2000).

The current implementation of PGLMs was devised by Mark Pagel (Pagel 1997, 1999). There are three parameters, λ, κ, and δ, each of which can be specified a priori or estimated from the data. The most important of these is λ, which is a measure of the strength of the phylogenetic signal in the data. Suppose some trait(s) have been measured in five species for which an evolutionary tree, i.e., phylogeny, is available, as shown in Figure 1.5A. If the pattern of trait variation among these species is consistent with random evolutionary change along the branches of the phylogeny, then λ is said to be 1. At the other extreme it is possible that close relatives are no more similar to each other than distantly related species. It is then as if all species were completely independent, equally distant phylogenetically from their common ancestor, as shown in Figure 1.5B. In this case λ is said to be 0. Most analyzed cases fall in between these two extremes and find that some proportion λ of the variation is accounted for by the phylogeny, the rest being attributable to recent independent evolution, as in Figure 1.5C. Parameters κ and δ provide a way of scaling the rates of evolutionary change along the branches of the phylogeny. For example, κ = 1 corresponds to gradual evolution, and κ = 0 is a model in which evolution is concentrated at speciation events. Parameter δ, which is rarely used, measures whether the rates of evolution have increased, decreased, or stayed constant over time. The best mathematical account of the method is provided by Garland and Ives (2000, p. 349) where it is referred to as the generalized least-squares approach. A recent guide to the use and misuse of PGLMs is given in Freckleton (2009).

The traits of interest in metabolic scaling analyses tend to show strong phylogenetic signals. For example, in mammals, λ = 0.984, 1.0, and 0.84 for basal metabolic rate, mass, and body temperature, respectively

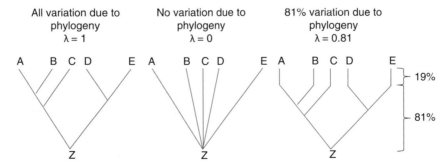

Figure 1.5 The Pagel λ approach to modeling the evolution process. (A) shows the phylogeny of five species A–E, which are descended from a common ancestor Z ; (B) shows how evolution is modeled if the species appear to be independent; (C) shows the type of intermediate model currently used (the Pagel λ model).

(Clarke et al. 2010). However, estimates of scaling parameters from PGLMs and conventional GLMs tend to be similar, converging on the same answer when the explanatory power (R^2) approaches 1.

Despite their promise, PGLMs are currently difficult to use. They require that a phylogeny, ideally with branch lengths, be available or assembled for the species of interest. They also assume that the form of the phylogeny and the assumed models of evolution are accurate. However, little analysis has been done to determine the impacts of error in either of these inputs on the outcome of the analysis. In addition, while software is available for conducting PGLM analyses, including BayesTraits (http://www.evolution.rdg.ac.uk/BayesTraits.html) and several packages in R including ape (http://cran.r-project.org/web/packages/ape/) and caper (http://r-forge.r-project.org/projects/caper/; Orme et al. 2011), the documentation is fragmentary and utilizing these packages can be difficult for new users.

In general we recommend that PGLMs be used when quality phylogenies are available. However, in cases where the relationship between two variables is strong this is unlikely to have a demonstrable influence on the results. If no phylogeny is available, an alternative is to use taxonomy as a proxy for phylogeny in a hierarchical (mixed effects) model (e.g., McGill 2008; Isaac and Carbone 2010). We also caution that factors other than phylogenetic relationship, such as similar body size or environment, can potentially be additional causes of non-independence of data in species-level analyses.

1.3.6 Methods for fitting frequency distributions

The predicted form of MTE frequency distributions is typically power law, $f(x) = cx^{\theta}$ (also known as the Pareto distribution in the probability and statistics literature), and the fit of these predictions to empirical data has typically been evaluated by fitting a regression through the data generated using histograms for visualization (i.e., binning the values of the variable of interest, counting how many values occur in each bin, log-transforming the counts and the position of the bin, and then fitting a relationship to those data points using linear regression). An example of this would be fitting a regression through the points in Figure 1.3A or 1.3B. While this seems like a reasonable approach

to this problem, it fails to properly account for the structure of the data, which can result in inaccurate parameter estimates (Clark et al. 1999; Edwards 2008; White et al. 2008) and incorrect estimates of the quality of fit of the model to the data (Newman 2005; Edwards et al. 2007; Clauset et al. 2009).

The correct approach for fitting frequency distributions in metabolic theory to data is based on likelihood (Edwards et al. 2007; White et al. 2008). Maximum likelihood estimation determines the values of the parameters that maximize the likelihood of the model, given the data. In the case of the metabolic theory this is typically finding the best-fitting exponent of a power-law frequency distribution. Determining the best parameters using maximum likelihood estimation for power laws is straightforward in most cases, requiring only a simple calculation. In the most common case where there is a meaningful lower bound (e.g., trees <1 cm are not measured) and the upper bound is assumed to be infinite, the exponent is determined simply by

$$\theta = -1 - \left[\frac{1}{n} \sum \left(\log \left(\frac{x}{x_{min}} \right) \right) \right]^{-1} \qquad (1.5)$$

where the summation is over all values of x. In other cases the calculations may be different, so care is required to confirm that assumptions being used to determine the MLE of the parameters are consistent with the data to which the calculation is being applied. In the case of the power-law frequency distributions predicted by metabolic theory, MLEs for all possible detailed forms are available in White et al. (2008; see Johnson et al. 1994, 2005, for more technical treatments).

1.4 ARE THE FITTED PARAMETERS CONSISTENT WITH THEORETICAL PREDICTIONS?

Having fit a relationship of the same general form as the MTE predictions using the methods above, the next step in evaluating the MTE is to determine whether the fitted parameters are consistent with the specific quantitative predictions of the theory.

In regression-based analyses this is typically done by determining whether or not the 95% confidence interval (CI) about the best-fitting parameter includes the theoretical prediction. This is a well-established practice and easy to apply (most statistical software that will generate parameter estimates will also generate confidence intervals for those estimates). However, hypothesis testing of this kind is not intended to determine whether two values of a parameter are similar. The appropriate interpretation of a CI containing the theoretical value is that we cannot reject the model, but this is not the same as supporting it. Alternatives that focus on determining whether or not two values are meaningfully similar are available (i.e., equivalence testing; Dixon and Pechmann 2005) but have never been applied to metabolic theory and are only rarely used in ecology in general.

Comparing the parameters of frequency distributions to those predicted by theory can also be done using confidence intervals, which can be determined accurately for all forms of power-law distribution when the number of data points is large (see appendix in White et al. 2008) and for small sample sizes for the most common form of the distribution (the Pareto; Johnson et al. 1994; Newman 2005; Clauset et al. 2009). Confidence intervals can also be calculated using bootstrap or jackknife techniques if necessary (Newman 2005). An alternative approach is to explicitly test whether a distribution with a fitted value provides a meaningfully better fit to the data than one with the theoretical value. This can be done using likelihood ratio tests (Vuong 1989; Clauset et al. 2009).

1.5 IS THE SHAPE OF THE RELATIONSHIP CONSISTENT WITH THEORETICAL PREDICTIONS?

1.5.1 Goodness-of-fit tests

For frequency distributions it is possible to directly ask whether or not the observed form of the distribution is consistent with (i.e., not significantly different from) the form predicted by the theory. This is done using goodness-of-fit tests, where the null hypothesis is that observed data are drawn from the theoretical distribution. A number of goodness-of-fit tests are available that entail different detailed assumptions including the chi-square test, the Kolmogorov–Smirnov test, and the G-test. If the sample size is sufficiently large and data are continuously distributed, all of these tests should give similar answers.

Determining the goodness of fit of regressions is more complicated and therefore simple tests are not available. Instead, it is standard to evaluate several assumptions of regression to determine whether the regression should be used or to compare the fits of linear regressions to more complex models (see below). Evaluating the assumptions of regression is good general practice, and failure to satisfy these assumptions can indicate that the model is not sufficient for characterizing the pattern in the data. Specifically the two most relevant tests are to determine: (1) whether or not the residuals about the regression are normally distributed (which can be done using standard goodness-of-fit tests); and (2) whether the variance of the residuals does not change as a function of the value of the predictor variable (i.e., the residuals are homoskedastic).

1.5.2 Comparison to alternative models

The other approach to determining whether or not the observed relationship has the same shape as the predictions of MTE is to compare the fit of the relationship or distribution to alternative models. The most common example of this is the use of polynomial regression to determine whether or not a simple linear relationship (among log-transformed variables) is an appropriate fit to the data. The standard approach is to fit polynomial regressions that include one or more higher-order terms (x^2, x^3, etc.) and determine whether or not those terms are significant in the regression. If they are, this is typically considered to be an indication that a different, or more complex, model than the simple linear relationship (on log-transformed data) is necessary. This polynomial approach has only rarely been used in MTE analyses, perhaps for reasons discussed below (see section 1.6), but it has been successfully utilized to indicate that the current metabolic theory predictions for the relationship between temperature and species richness are not sufficient to fully characterize the observed patterns (Algar et al. 2007; Hawkins et al. 2007; but see Gillooly and Allen 2007).

A more general approach is to use likelihood and information criteria-based methods. These methods determine which of a set of models is most consistent with the empirical data and whether that model provides a meaningfully better fit than alternative models

(Hilborn and Mangel 1997; Burnham and Anderson 2002). A full introduction to this area is beyond the scope of this chapter, but the basic approach is to calculate the likelihoods of all the candidate models and then compare those likelihoods to one another, taking into account that some models have more parameters than others and are therefore more likely to provide good fits to empirical data (for ecological examples see Muller-Landau et al. 2006a and Coomes and Allen 2007). We strongly recommend Hilborn and Mangel (1997) to those looking for an accessible introduction to this area of statistics. Equivalent Bayesian methods are also available, but have rarely been applied in the context of metabolic ecology. Good examples are available in Dietze et al. (2008) and Price et al. (2009) for those interested in this approach.

In addition to testing the basic shape of the predicted relationship and the specific parameter values, these methods can be used to assess the form of the error distribution to allow for decisions to be made about whether to use log-linear or nonlinear regression (Xiao et al. 2011; see above) and to determine the degree of phylogenetic non-independence among data points that needs to be accounted for (Freckleton 2009).

1.6 THOUGHTS ON TESTING ECOLOGICAL THEORIES

It is useful and informative to compare the fits of metabolic theory models (and ecological models in general) to alternative models to see if a better characterization of the empirical data is possible. If an alternative model provides a better fit to the data there are two different conclusions that can be drawn: (1) the model is not useful and should be abandoned; or (2) the model is incomplete and requires further development. In ecology we have tended to prefer the language of rejection – any model for which data deviates from the prediction using a goodness-of-fit test, or for which an alternative model is found to provide a superior fit, is rejected. This attitude likely has its origins in an emphasis on Plattian inference (Platt 1964) and an, arguably improper (Hurlbert and Lombardi 2009), emphasis on the arbitrary definition of $p < 0.05$ as being "significant." Further discussion of how a rejected model may be improved is rarely undertaken. However, in cases where a model is based on reasona-

ble starting assumptions and makes reasonable predictions, it may be better to modify and improve that model than to abandon it. This iterative process of hypothesis refinement is considered essential for the development of ecology (Mentis 1988), and several recent attempts to refine models from metabolic ecology make valuable contributions to this process (Banavar et al. 2010; Savage et al. 2010).

The goal of theory is to provide simplified characterizations of reality; so rejecting models is only useful if it leads to better models. Testing models and identifying their flaws is a necessary, but not sufficient, part of the process. This raises questions about the merits of comparing process-based models to purely phenomenological models that lack a biological mechanism. Consistent, directional, deviations from a general theory indicate that the theory is either incomplete or simply wrong. However, studies that only demonstrate the superior performance of phenomenological over mechanistic models often yield little direct progress towards acceptable theories. In contrast, comparing theoretical predictions to mechanistic models that include either additional or alternative processes has the potential to yield improved characterizations of biological systems. An illustrative example is Fisher's sex ratio theory, which predicts a canonical ratio of $1:1$. When sample sizes are large, significant deviations are almost always observed. This does not mean the theory is wrong. Indeed, considering the direction and magnitude of the deviations (large in eusocial hymenoptera, small in humans) leads to progress in understanding the additional processes that affect sex ratios in real populations.

It is important to consider the goal of a model when determining whether it should be replaced or modified (Martinez del Rio 2008). For example, in many cases related to MTE the goal is to understand the fundamental processes that produce the first-order relationship between body size and metabolic rate. MTE is successful at characterizing the relevant empirical pattern, because a 3/4-power allometric relationship is the best-supported pattern, both when analyzing large numbers of species and when the average form of the model across taxonomic groups is determined (Savage et al. 2004b; Isaac and Carbone 2010). As such, MTE may provide information about the underlying process. However, if the goal is to accurately predict the metabolic rate of species for which data is not available then it is necessary to consider the empirical evidence of variation among taxonomic groups (e.g., Nagy et al.

1999; Isaac and Carbone 2010). In this case models that incorporate taxonomic variation are an important improvement over the more general MTE (Isaac and Carbone 2010).

Evaluating models is further complicated by the fact that general ecological theories (including MTE) typically make predictions for multiple empirical patterns (see Brown et al. 2004). This generality is desirable because it makes metabolic theory applicable in a broad range of situations, but it also makes MTE easier to reject since rejection of any prediction implies rejection of the entire theory. However, it is unreasonable to compare a model that makes a large number of predictions to a model that makes one or a few specific predictions without penalizing the more specific model for its lack of generality and resultantly larger number of parameters per prediction (Price et al. 2009). Unfortunately there are no general approaches for dealing with this type of difference among models, and the one example that we are aware of (Price et al. 2009) represents a first attempt rather than a general solution to the challenge of evaluating models that make multiple predictions.

In conclusion, the goal of science is to develop models of reality that both capture general underlying patterns and processes, and incorporate important biological details. Developing general ecological theories allows us to understand how ecological systems operate and make predictions for how they will respond to global change and other major perturbations. Rigorous statistical approaches and proper testing of theories are necessary to accomplish this result. Efforts to improve methodological approaches and to use these approaches to test existing theories should always be undertaken with the goal of improving our understanding of ecological systems.

ACKNOWLEDGMENTS

We thank Jamie Gillooly for providing the data for the metabolic rate graphs, and Alwyn Gentry and the Missouri Botanical Garden for making the data for the frequency distribution graphs publicly available. James Brown, Astrid-Kodric Brown, and Alison Boyer provided valuable comments that improved the chapter.

Chapter 2

THE METABOLIC THEORY OF ECOLOGY AND ITS CENTRAL EQUATION

James H. Brown and Richard M. Sibly

SUMMARY

1 It has long been known that metabolic rate varies with body size and temperature.

2 The effect of body mass is approximately power law, often with an exponent of close to 3/4 for whole-organism metabolic rate, −1/4 for mass-specific metabolic rate and most ecological rates, and 1/4 for most biological times.

3 The first-order effect of temperature is approximately exponential, and can be described by the Arrhenius expression, with the activation energy for respiration typically in the range $E \approx 0.6$–$0.7\,eV$.

4 The metabolic theory of ecology (MTE) expresses the effects of body size and temperature in a single, simple equation that can be easily manipulated and modified to analyze a variety of data, of which some closely match the idealized models and others deviate substantially.

5 West et al. (1997) developed a mathematical model, based on the fractal-like branching of mammal and plant vascular systems, to account for quarter-power scaling with body mass. This WBE (West, Brown, Enquist) model successfully predicts

many patterns, but has also been criticized on theoretical and empirical grounds. Several research groups are currently working to further test, modify, and generalize the model.

6 The Arrhenius formulation accounts for much of the variation in temperature dependence of biological processes. Current research is aimed at elucidating natural variation in activation energies, extending the model to account for optimal temperatures and decreases in rates at higher temperatures, and clarifying the mechanistic underpinnings.

7 Availability of materials, notably water and nutrients, also affects ecological processes. Much work remains to combine MTE and ecological stoichiometry into a synthetic metabolic theory.

8 The central equation of MTE is a simple model, based on well-established principles of physics, chemistry, and biology, that makes many quantitative predictions and provides a baseline for addressing variation. But it does not include many details of behaviors, niche relationships, and species interactions.

Metabolic Ecology: A Scaling Approach, First Edition. Edited by Richard M. Sibly, James H. Brown, Astrid Kodric-Brown.
© 2012 John Wiley & Sons, Ltd. Published 2012 by John Wiley & Sons, Ltd.

2.1 INTRODUCTION

Metabolic rate is central to ecology because all interactions between organisms and their environments involve rates of exchanges of energy and materials. Biological metabolism includes the uptake of resources from the environment, transformation of these substances within the body, allocation of the products to maintenance, growth, and reproduction, and excretion of wastes into the environment. So, to a first approximation, the metabolic rate sets the pace of life, and the rates of all biologically mediated ecological processes.

2.2 THE CENTRAL EQUATION OF METABOLIC ECOLOGY

It has been known for at least a century that metabolic rate varies with body size and temperature. Although details are still debated, there is a large and long-standing literature showing that, across the diversity of living things and ecological settings, metabolic rate scales with body size as a power law and with temperature as an exponential (summarized in Peters 1983; see also Robinson et al. 1983). Beginning in the late 1990s, researchers at the University of New Mexico and the Santa Fe Institute revived interest in the scaling of metabolic rate and its ecological consequences by deriving a simple mathematical model, based on fundamental principles of physics, chemistry, and biology, that incorporates quarter-power scaling with body mass and an Arrhenius expression for temperature dependence. This model allows 15 orders of magnitude variation in metabolic rate across diverse species spanning about 20 orders of magnitude in body mass and nearly 40 °C in body temperature to be reduced to just one order of magnitude. This impressive explanation of variation, by accounting only for mass and temperature, provides a baseline to investigate the additional factors that explain the remaining variation.

The central equation for this scaling across diverse taxa and life stages is

$$B = B_0 M^{\alpha} e^{-E/kT} \tag{2.1}$$

where B is mass-specific metabolic rate, B_0 is a normalization constant that is independent of body size and temperature, M^{α} is how body mass, M, scales with α, an allometric scaling exponent, and $e^{-E/kT}$ is the expo-

nential Arrhenius function, where E is an "activation energy," k is Boltzmann's constant ($8.62 \times 10^{-5}\,\mathrm{eV\,K^{-1}}$), and T is body temperature in kelvin (Gillooly et al. 2001; Brown et al. 2004; West and Brown 2005).

This equation also predicts the scaling of most other biological and ecological rates. A similar equation, with a different exponential term for temperature dependence, was derived by Robinson et al. (1983; see also Peters 1983) based on statistical goodness-of-fit to empirical data. Robinson et al. (1983) emphasized the differences among taxa. The New Mexico scaling group showed that combining mass and temperature dependence and normalizing to a common temperature reveals a commonality among organisms and that a narrow range of activation energies reflects a shared biochemistry. The New Mexico scaling group derived the above equation based on first principles and biological mechanisms, as the central equation of metabolic scaling. The metabolic theory of ecology (MTE: Brown et al., 2004) assumes a more specific form in which $\alpha = -1/4$ for mass-specific metabolic rate and most ecological rates, and $E \approx 0.65\,\mathrm{eV}$ for processes governed by respiration, giving

$$B = B_0 M^{-1/4} e^{-0.65/kT} \tag{2.2}$$

Whole-organism rates, I, such as metabolic rate and production rate, scale positively with body size. Because $B = I/M$, this gives

$$I = I_0 M^{\gamma} e^{-E/kT} \tag{2.3}$$

and the allometric exponent, $\gamma = 1 + \alpha$, is positive and usually between 2/3 and 1. Since the pioneering work of Kleiber (1932) and Brody and Proctor (1932), many empirical studies have obtained a value for γ close to 3/4, and $M^{3/4}$ scaling of metabolic rate has often been referred to as Kleiber's Law.

Since times are the reciprocals of rates, equation 2.1 can be modified to characterize biological times, T, such as development time and lifespan, which characteristically increase with increasing body size as

$$T = T_0 M^{\beta} e^{-E/kT} \tag{2.4}$$

where $\beta = -\alpha = 1 - \gamma$ and its value is predicted to be close to 1/4.

The more general form of equation 2.1 can be used for organisms or processes that have different values of α or E – for example, for bacteria and protists where

$\gamma \geq 1$ (De Long et al. 2010), and for metazoans where γ typically ranges from 2/3 to 1, and is often but not always very close to 3/4 (see below and see also Okie, Chapter 12, and Karasov, Chapter 17). Compared to respiration where on average $E \approx 0.65\,\text{eV}$ (Gillooly et al. 2001), photosynthesis appears to have a lower temperature dependence with $E \approx 0.30\,\text{eV}$ (Farquhar et al. 1980; Chapin et al. 2002; Bernacchi et al. 2002; Allen et al. 2005; Sage and Kubien 2007; see also Anderson-Teixeira and Vitousek, Chapter 9).

2.3 ANALYZING, PLOTTING, AND EVALUATING DATA

Although MTE is deliberately idealized and oversimplified, one powerful advantage is that the above equations lend themselves to plotting and analyzing data (see White, Xiao, Isaac, and Sibly, Chapter 1). Taking logarithms of both sides of equation 2.1 and rearranging terms gives

$$\log_{10}(Be^{E/kT}) = \alpha \log_{10}(M) + \log_{10}(B_0) \tag{2.5}$$

This allows metabolic rate to be "temperature-corrected" by incorporating the $e^{E/kT}$ term on the left-hand side. Similarly, a "mass-corrected" metabolic rate can be expressed as

$$\ln(BM^{-\alpha}) = -E(1/kT) + \ln(B_0) \tag{2.6}$$

Here we use natural logarithms, ln, so that the regression coefficient in a regression of $\ln(BM^{-\alpha})$ against $1/kT$ gives $-E$.

This method facilitates quantitative empirical evaluation of the mass and temperature dependence predicted by MTE, by incorporating the predicted scalings into the analysis and into the y-axes of bivariate plots. Equation 2.5 predicts that the logarithm of temperature-corrected mass-specific metabolic rate should be a linear function of the logarithm of body mass, with a slope of α. Equation 2.6 predicts that the natural logarithm of mass-corrected metabolic rate is a linear function of inverse absolute temperature ($1/kT$). The slope of this so-called Arrhenius plot, with sign reversed, gives the activation energy, E. Values of α and E can either be taken from empirical regressions or they can be chosen a priori based on theoretical considerations, such as $\alpha = -1/4$ for mass-specific metabolic rate or $E = 0.65$ for processes controlled by respiration.

This method of graphing data is shown in Figure 2.1. This makes it easy to evaluate theoretical predictions, because the data can be subjected to linear regression analysis, the resulting slopes can be compared to predicted values of E and α, and the value of the normalization constant, B_0, can be obtained.

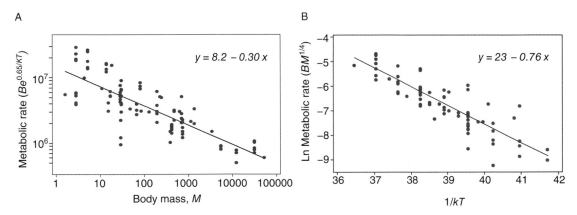

Figure 2.1 Effects of body size and temperature on mass-specific metabolic rate of reptiles. Metabolic rate was measured in watts, mass in grams, and temperature T in kelvin. (A) Temperature-corrected mass-specific metabolic rate as a function of body mass depicted using equation 2.5 to make a plot with the log-transformed rate on the y-axis and log-transformed body mass on the x-axis. The slope of the regression quantifies the mass dependence as the exponent of a power law, $\alpha = -0.30$. (B) Mass-corrected rate as a function of temperature, depicted using equation 2.6 to make an Arrhenius plot with the ln-transformed rate on the y-axis and $1/kT$ on the x-axis. The slope of the regression quantifies the temperature dependence as an "activation energy" $E = 0.76\,\text{eV}$ (with sign reversed). Replotted from data in Gillooly et al. (2001).

2.4 THE EFFECT OF BODY SIZE AND THE RIDDLE OF QUARTER-POWER SCALING

It has been known for over a century that metabolic rate increases nonlinearly with body size. The excellent book by John Whitfield (2006) provides an historical review. One of the first quantitative studies was on dogs, by Rubner (1883), who concluded that whole-organism metabolic rate scales as $M^{2/3}$, reflecting the dissipation of heat from the body surface and the 2/3-power geometric scaling of body surface area with volume or mass. This simple interpretation stood until Kleiber (1932) and Brody and Proctor (1932) measured metabolic rates of a variety of birds and animals and independently concluded that the scaling exponent was significantly greater than 2/3. Kleiber obtained a value of 0.74 and rounded it off to 3/4, whereas Brody and Proctor obtained 0.73. Subsequent studies of different taxonomic and functional groups of animals repeatedly obtained exponents very close to these values (e.g., Peters 1983; Feldman and McMahon 1983; Savage et al. 2004a; Farrell-Gray and Gotelli 2005). There were dissenters, however, most notably Heusner (1982, 1983), who continued to argue for "geometric" or $M^{2/3}$ scaling on both theoretical and empirical grounds.

Decades of active research on allometry culminated in the 1980s with the publication of four major synthetic books (McMahon and Bonner 1983; Peters 1983; Calder 1984; Schmidt-Nielsen 1984). Each of these evaluated the evidence and accepted the generality of $M^{3/4}$ scaling of metabolic rate, which had come to be referred to as Kleiber's Law. Each of these reviews also concluded that there was no satisfying general explanation for the unusual, perhaps uniquely biological, "quarter-power" scaling of metabolic rate and numerous other biological rates and times. Three hypotheses were suggested: (1) "elastic similarity" of structures such as animal bones and tree trunks (McMahon 1973, 1975; McMahon and Bonner 1983); (2) diffusion limitation in aquatic organisms (Patterson 1992); and (3) the fourth dimension is time (Blum 1977). All were unsatisfying: the first two, because they were too narrow to account for the ubiquity of the phenomenon, and the last, because the mechanism was only vaguely formulated (but see Ginzburg and Damuth 2008). In 1986 Kooijman introduced what subsequently became known as "dynamic energy budgets." In these models 3/4 is not a "canonical" value requiring a mechanistic theory, but instead a consequence of multiple processes, some scaling with surfaces as $M^{2/3}$ and others with volumes as M, resulting in an exponent with an intermediate value often close to 3/4. This approach had limited impact because Kooijman's models are very complex, with too many parameters and functions for most applications (but see Kooijman 2000; van der Meer 2006). By the 1990s, with the excitement of the molecular revolution and no theoretical, mechanistic explanation for the quarter powers, research on allometry had diminished, and much of it focused on scaling of biomechanics and hydrodynamics rather than metabolism.

2.5 THE MODELS OF WEST, BROWN, AND ENQUIST

In the mid-1990s Geoffrey West, James Brown, and Brian Enquist (WBE) began an interdisciplinary physics-biology collaboration. In 1997 they published "A general model for the origin of allometric scaling laws in biology." This first paper presented a complete, quantitative, whole-system model (the WBE model) of the important structural and dynamical features of the mammalian arterial system from the heart to the capillaries. The premise was that the properties of resource delivery networks determine the scaling of whole-organism metabolic rate (Fig. 2.2). The WBE model made several simplifying assumptions (see also Savage et al. 2008): (1) metabolic rate is proportional to the rate at which blood is pumped from the heart, because the oxygen and nutrients that fuel metabolism are delivered to the cells by the blood; (2) the arteries branch to form a symmetrical hierarchical network; (3) the network supplies all parts of the three-dimensional body; (4) the terminal units, capillaries, are invariant in structure, function, and dynamics; and (5) the system has evolved under natural selection to optimize performance by minimizing the rate of energy expenditure to pump blood from the heart to the capillaries. The model quantified the branching of the fractal network (number, radius, and length of each daughter vessel at each generation of the hierarchy can be expressed as a geometric series) and the relevant parameters of blood flow (volume flow rate, velocity, pressure, and resistance). The model used Lagrange multipliers and solved the Navier–Stokes equation for pulsatile flow through elastic vessels to derive an optimal design that minimizes the energy to pump

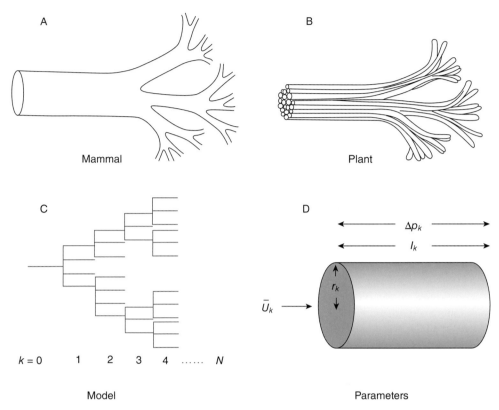

A   Mammal

B   Plant

C

$k = 0$ 1   2   3   4 ......   N

Model

D

Δp_k

l_k

r_k

$\bar{U}_k$

Parameters

Figure 2.2 Framework for the West et al. (1997) model showing diagrammatic examples of segments of biological distribution networks: (A) mammalian circulatory and respiratory systems composed of branching tubes; (B) plant vessel-bundle vascular system composed of diverging vessel elements; (C) topological representation of such networks, where k specifies the order of the level, beginning with the aorta ($k = 0$) and ending with the capillary ($k = N$); and (D) parameters of a typical tube at the kth level. (From West et al. 1997.)

blood through the network. WBE predicts that blood volume scales linearly with body size as M and metabolic rate scales as $M^{3/4}$. Because all critical anatomical and physiological features must be scaled in concert in order to perform optimally (the concept of symmorphosis; see Weibel et al. 1991, 1998; see also Karasov, Chapter 17), the model predicts the scaling of many other properties of the arterial system. Most of these predictions are well supported.

West et al. (1999b) addressed scaling of metabolic rate with body size in plants by modeling the flow of fluid through a xylem network (see Enquist and Bentley, Chapter 14). They used an approach similar to WBE, but all of the details of the branching architecture, vessels, and pump in plants are very different from those in animals. Again, however, the optimization

model predicted $M^{3/4}$ scaling of metabolic rate and many structural and functional parameters of the network. Again, most of the predictions seemed to be well supported by empirical studies in the botanical and forestry literature.

A series of subsequent papers by the New Mexico group extended their theoretical framework for quarter-power scaling to model the scaling of many phenomena, from molecules to ecosystems and in both animals and plants. This work was influential in reviving interest in allometry and stimulating new theoretical and empirical research on scaling, body size, metabolism, and related phenomena. The models have received both enthusiastic support and scathing criticism. It is still too soon for a definitive evaluation of this body of work. This book discusses applications of WBE and MTE to ecology. More generally, it highlights the spectrum of topics and the current state of knowledge that have arisen as a consequence of an explicitly metabolic conceptual framework.

2.5.1 Theoretical criticisms of WBE

Here we briefly consider some of the controversial issues with the models themselves. Most of these issues arise due to concerns about the model's assumptions and internal consistency. A model, by definition, is a deliberate oversimplification that tries to capture the essence of a more complicated reality. Most of the above assumptions of WBE are violated to some extent. The critical question is whether these differences between idealized theory and biological realism are: (1) sufficiently small so that they can continue to be ignored (e.g., Turcotte et al. 1998; Savage et al. 2004a; Farrell-Gray and Gotelli 2005); (2) sufficiently large so as to require changes in assumptions, internal logic, and predictions, while preserving the overall structure and approach (e.g., Etienne et al. 2006; Apol et al. 2008; Savage et al. 2008; Banavar et al. 2010; Kolokotrones et al. 2010); or (3) so large as to invalidate the entire conceptual framework erected by WBE (e.g., Darveau et al. 2002; Dawson 2003; Kozlowski and Konarzewski 2004; Glazier 2005, 2010; Dodds 2010). Here, we take the position that both critics and some members of the New Mexico group have identified significant problems with certain assumptions and predictions of WBE. However, many of these concerns can be corrected by changing the simplifying assumptions while preserving most of the overall framework that predicts the distinctive quarter-power scaling.

We address three problems:
1 *Fitting a network whose characteristic lengths scale as $M^{1/4}$ into a geometric organism whose lengths scale as $M^{1/3}$.* WBE assumes that lengths of the aorta and major arteries scale as $M^{1/4}$, but empirical measurements give $M^{1/3}$ as expected from the fundamentally geometric scaling of body shape (Gunther and Leon de la Guerra 1966; Peters 1983; Banavar et al. 2010). Without addressing this discrepancy, WBE cannot explain how the arteries supply the body, including distal appendages (Kozlowski and Konarzewski 2004; Etienne et al. 2006; Apol et al. 2008; Banavar et al. 2010).
2 *Fitting capillaries with invariant properties into service volumes that scale as $M^{1/4}$.* WBE assumes that capillaries are invariant with size. This is consistent in giving the number of capillaries scaling as $M^{3/4}$ and the service volume of tissue supplied by each capillary scaling as $M^{1/4}$. However, it raises questions about how capillaries of length M^0 fit into a service volume with linear dimensions scaling as $M^{1/12}$, and about the details of delivery of oxygen from the capillaries to the mitochondria within cells (Dawson 2003; Banavar et al., 2010).
3 *Changes in blood velocity from heart to capillary.* WBE recognizes that blood velocity must slow by a factor of about 1000 between leaving the heart and entering the capillaries, but predicts that the average velocity through large vessels, roughly corresponding to the arteries, is invariant with size. This constant velocity, together with linear distances scaling as $M^{1/4}$, is consistent in predicting blood circulation time scaling as $M^{1/4}$, as observed empirically. But it is inconsistent with the geometric $M^{1/3}$ scaling of lengths as mentioned above (Kozlowski and Konarzewski 2004; Etienne et al. 2006; Apol et al. 2008; Banavar et al. 2010).

It appears that these problems can be resolved by modifying WBE so that characteristic vessel lengths scale geometrically as $M^{1/3}$, and allowing the average velocity from the heart to the capillaries to scale as $M^{1/12}$ (Banavar et al. 2010; for an alternative hypothesis see Dawson 2003). This could be accommodated realistically if blood velocity is high and essentially invariant in the large arteries, only slowing in the last few generations of branching (Kolokotrones et al. 2010). This would allow blood velocity to slow when it enters the arterioles, where the cross-sectional area increases and velocity decreases by a factor of about 1000 prior to entering the capillaries. It would give higher average velocity through the network in larger mammals because the blood spends more time traveling at high speed in large vessels. This revised scenario still needs to be subjected to modeling the details of anatomy and hydrodynamics in an optimization framework similar to WBE. But the critical changes would involve the slowing of blood in the arteries and delivery of oxygen in the capillaries, which were among the least rigorously developed parts of WBE.

More generally, quarter-power scaling with velocity scaling as $M^{1/12}$ appears to be an optimal design for many resource supply networks, including animal and plant vascular systems, and human-engineered transportation and communication networks (Banavar et al. 2010). The additional "fourth dimension" arises as a consequence of the $M^{1/4}$ scaling of the service volume and the $M^{1/12}$ scaling of its linear dimensions, which allow the maximal possible metabolic rate without having total blood volume scaling superlinearly with mass. Banavar et al. (2010) show that networks with optimal quarter-power scaling do not need to have hierarchically branching structures. This raises the question of why fractal-like hierarchically branched designs have evolved so repeatedly and convergently in animals and plants. We conjecture that this is because they not only offer near-optimal solutions to resource delivery, but also allow relatively simple genetic programs to accommodate changes in size over both ontogeny and phylogeny (Metzger et al. 2008). This is consistent with recent empirical findings that quarter-power scaling may be confined to multicellular animals and plants that are large enough to require vascular systems to distribute resources within the body (DeLong et al. 2010; Mori et al. 2010).

2.5.2 Empirical issues with WBE

The renewed interest in allometric scaling of metabolic rate stimulated many empirical studies that revived old questions about the validity of Kleiber's Law. We address six issues.

1 *Comparisons across different functional or taxonomic groups of organisms reveal considerable variation in exponents.* General acceptance of Kleiber's Law dates back to the comprehensive study of Hemmingsen (1960) and the synthetic allometry books of the 1980s (McMahon and Bonner 1983; Peters 1983; Calder 1984; Schmidt-Nielsen 1984), which pointed to the predominance of studies reporting not only $M^{3/4}$ scaling of metabolic rate, but also $M^{-1/4}$ and $M^{1/4}$ scaling of many other biological rates and times, respectively. For example, Hemmingsen (1960; see also Fenchel 1974) suggested that whole-organism metabolic rates scale as $M^{3/4}$ across all organisms, spanning about 18 orders of magnitude from single-celled bacteria and protists to giant mammals (although with different normalization constants for different functional groups). More recent studies, however, have found different exponents between different taxonomic, phylogenetic, or functional groups (e.g., White and Seymour 2002, 2005; White et al. 2006; Clarke et al. 2010) and between "intraspecific" (e.g., ontogenetic or between-population variation) and interspecific relationships (Heusner 1982, 1983, 1991; Glazier 2005). Perhaps most tellingly, a compilation of new data on unicellular heterotrophic organisms reveals much higher exponents, with γ nearly 2 in bacteria and close to 1 in

protists (DeLong et al. 2010). Clearly, it is an overgeneralization to claim that Kleiber's Law, with a single "universal' or "canonical" $M^{3/4}$ scaling exponent, applies to all organisms.

2 *The relationship is not a power law with a single scaling exponent.* Several studies have shown a consistently curvilinear relationship when the extensive data on basal or resting respiration rate as a function of body mass in mammals, aquatic invertebrates, and plants are plotted on logarithmically scaled axes. In mammals, the slope increases significantly and systematically from the smallest to the largest species (Fig. 2.3A: McNab 1988; Savage et al. 2004a; White et al. 2009; Clarke et al. 2010; Kolokotrones et al. 2010). In plants, by contrast, the slope decreases consistently from the smallest to the largest individuals (Fig. 2.3B: Reich et al. 2006; Enquist et al. 2007b; Mori et al. 2010) and similarly in aquatic invertebrates (DeLong et al. 2010).

3 *Statistical issues related to using regression analysis and comparative phylogenetic methods to estimate the values of exponents and normalization constants.* Many of these are addressed by White, Xiao, Isaac, and Sibly (Chapter 1), who point out that statistical methods are constantly being refined, and many issues are far from settled. Moreover, the many empirical studies that used different methods have rarely been re-evaluated using standardized methodology. Suffice it to say that reported empirical scaling relationships should be interpreted cautiously, with the warning that biological reality may not be as straightforward as some authors have implied.

4 *Different metabolic rates with different scaling relations* (Fig. 2.4). Ever since Kleiber (1932) and Brody and Proctor (1932), the vast majority of allometric studies have focused on basal or resting metabolic rate (BMR). This is understandable, because the rate of energy use required to sustain life is fundamental to biology. Consequently, animal physiologists have developed rigorously standardized protocols for defining and measuring BMR, the rate of metabolism of an inactive, starving animal measured over a relatively short period of time, typically minutes (McNab 1997). Physiologists are also interested in the metabolic costs of various activities, so they measure the maximal rate of oxygen consumption (MMR or VO_{2max}), during strenuous exercise or thermoregulation, again usually on a timescale of minutes. Physiological ecologists interested in the average rate of energy turnover in free-living animals have developed methods for measuring field metabolic rate (FMR), usually on a timescale of days to weeks.

A

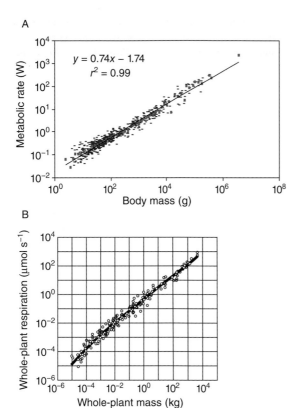

B

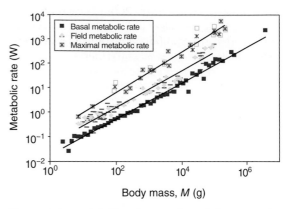

Figure 2.4 Basal, field, and maximal metabolic rates scale with body mass in mammals. Basal metabolic rate: filled dark-blue squares depict averaged data as in Fig. 2.2A. Field metabolic rate: yellow squares show data for individual species and purple bars depict averages binned by body size intervals. Maximal metabolic rate: blue squares show data for individual species and black crosses depict averages binned by body size intervals. Note the difference in the levels of metabolism as indexed by normalization constants: lowest for basal, about 3 times higher for field, and highest for maximal (about 10 times higher than basal). The slopes of the fitted regression lines also indicate somewhat different scaling exponents: basal $\gamma = 0.737$; field $\gamma = 0.749$; and maximal $\gamma = 0.828$. (From Savage et al. 2004b, by permission of John Wiley & Sons, Ltd.)

Figure 2.3 Relationships between whole-organism metabolic rate and body mass for mammals and plants are curvilinear when plotted on on logarithmic axes, so they are not precisely described by a single power law. (A) Basal metabolic rate as a function of mass for mammals: blue points depict the data for individual species and purple points are the averages for all species within a size-bin. The data show consistent deviations from a perfect power law, implying that small mammals have a lower exponent and large mammals a higher exponent than implied by the slope, $\gamma = 0.737$, of the regression line. (From Savage et al. 2004b, by permission of John Wiley & Sons, Ltd.) (B) Respiration rate as a function of mass for vascular plants: the data are fit with a smooth function, which depicts the opposite pattern from mammals: a higher exponent, $\gamma \approx 1$, in the smallest plants transitioning to a lower value, $\gamma \approx 3/4$ in larger individuals. (From Mori et al. 2010.)

These levels of metabolism are not functionally independent of each other, because they are subject to trade-offs that affect storage, supply, and utilization of oxygen and energy to support different activities on different timescales (see Karasov, Chapter 17). Not surprisingly, however, they have distinctive scaling rela-

tions: somewhat different exponents and very different normalizations. FMR is typically about threefold higher and MMR as much as 10-fold higher than BMR. It is likely, however, that none of these standardized rates measured by physiologists has been the primary target of natural selection, which should operate to optimize energy uptake, transformation, and allocation over the lifespan so as to maximize fitness in the natural environment. So the metabolic rate that has evolved by natural selection must necessarily include the energetic cost of reproduction, an essential component that is left out of the usual measures of metabolism, including FMR.

5 *Variation in metabolic rate with other factors.* There is an enormous literature documenting variation in metabolic rate with factors in addition to body size. Just in mammals, for example, statistical analyses have documented significant effects of taxon or phylogenetic lineage, geographic distribution, climate, habitat, diet, body temperature, ontogenetic stage, and metabolic level (e.g., Heusner 1982, 1983; Hayssen and Lacy

1985; McNab 1988, 2002, 2008; Lovegrove 2000, 2003; Symonds and Elgar 2002; White and Seymour 2003, 2004; Glazier 2005, 2006; Duncan et al. 2007; Sieg et al. 2009; White et al. 2009; Clarke et al. 2010). These studies all find pervasive nonlinear scaling with body size, but the residual variation and its correlates raise still unanswered questions about the "universality" of Kleiber's Law, the WBE model, and MTE.

6 *Allometric scaling of other biological rates and times.* As indicated above, because the uptake, processing, and allocation of energy and materials are so fundamental, nearly all other rates and biological times scale closely with metabolic rate. So, the long-standing literature suggesting that these rates, respectively, scale closely as the $-1/4$ and $1/4$ powers of body mass, was taken as being consistent with the overall prevalence of $3/4$-power scaling of whole-organism metabolic rate. Consequently, the dogma has been that heart and respiratory rates and ontogenetic and population growth rates scale as approximately $M^{-1/4}$ and gestation and lactation times, age at maturity, and lifespan scale as approximately $M^{1/4}$. These traits have the feature that they often implicitly integrate over longer timescales and incorporate a greater diversity of processes than BMR. Unfortunately, however, the scaling of

these other rates and times has not been subjected to the same degree of recent study and rigorous re-evaluation as the scaling of BMR. Doing so would contribute importantly to MTE by enhancing understanding of the mechanistic linkages between metabolic rate and other features of physiology, life history, and ecology.

2.6 THE EFFECT OF TEMPERATURE

It has also been known for more than a century that temperature affects the rate of metabolism and nearly all biological activities. The vast majority of organisms are active in a limited range of temperature, between approximately 0 and $40\,°C$. Physiological and behavioral performance typically exhibit a characteristic pattern: near-exponential increase over most of the range, with a peak and rapid decline as the upper lethal limit is approached (Fig. 2.5). Biologists have typically used Q_{10} to characterize the exponential phase

$$Q_{10} = \left(\frac{B_2}{B_1}\right)^{\left(\frac{10}{T_2 - T_1}\right)} \tag{2.7}$$

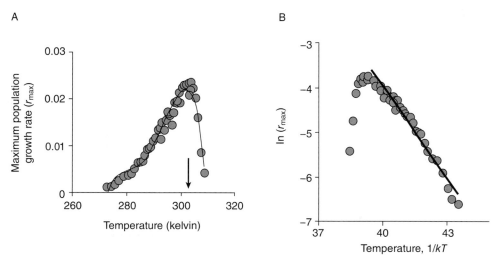

Figure 2.5 Temperature dependence of maximum population growth rate, r_{max}, of the bacterium *Pseudomonas fluorescens* plotted two ways. (A) Population growth rate as a function of temperature on linear axes, showing the typical hump-shaped pattern, increasing nearly exponentially to a peak at the optimum (arrow) and then decreasing rapidly. (B) Logarithm of rate as a function of $1/kT$, with the fitted line highlighting the near-exponential relationship at temperatures below the optimum; this is an Arrhenius plot, and the slope of the linear regression gives an estimate of the activation energy $E = 0.66\,eV$. The growth rates of microbial populations are often used as proxies for metabolic rates. (From Knies and Kingsolver 2010.)

where B_1 and B_2 are the rates of metabolism or some other process at two temperatures, T_1 and T_2, where temperature is measured in °C. The Q_{10} value gives the factor by which the rate increases with a 10 °C increase in temperature. So, for example, most biological rates exhibit a Q_{10} of approximately 2.5, meaning that they increase about 2.5 times for every 10 °C increase in temperature. A serious problem with Q_{10}, however, is that parameterizing in 10 °C increments introduces a serious distortion, even over the "biologically realistic range" of 0–40 °C.

This problem can be avoided by using the Arrhenius formulation, the $e^{-E/kT}$ term in equation 2.1. This expression, theoretically grounded in statistical mechanics and chemical kinetics (Boltzmann 1872; van't Hoff 1884; Arrhenius 1915), is the one used in physical chemistry. It also simplifies the mathematics by allowing temperature to be combined with mass in the simple expressions of equations 2.1–2.6, and it facilitates data analysis by predicting linear relationships when plotted on logarithmic axes as shown above. Temperature dependence is given by the slope of an Arrhenius plot, which with sign reversed gives the value of E.

Used in this way, E is an "activation energy," but one that indexes the overall kinetics of metabolism rather than of a single reaction as is the traditional interpretation in physical chemistry and biochemistry. Consequently MTE does not predict a single discrete value for E. Instead, based on the kinetics of the relevant rate-limiting biochemical reactions and physiological measurements of these processes, E is expected to be ≈ 0.65 eV for respiration rate and ≈ 0.30 eV for photosynthesis rate (equivalent to $Q_{10} \approx 2.5$ and ≈ 1.5 over 0–40 °C), respectively.

The above equations for temperature dependence of metabolic rate and other processes have not received the same intense theoretical and empirical scrutiny as for the mass dependence. A recent compilation and analysis of temperature-dependent processes in animals finds that most of the data are well fit by the Arrhenius expression, with the activation energy for respiration typically in the range E ≈ 0.6–0.7 eV (Dell et al. 2011). But this is unlikely to satisfy criticisms that the Arrhenius formulation is at best an oversimplification of a considerably more complicated biological reality. There are at least three issues.

First, when rate is plotted as a function of temperature, the above equations apply only to the approximately exponential increasing phase (i.e., the region where there is an approximately constant slope when the logarithm of rate is plotted as a function of $1/kT$ in an Arrhenius plot). So, the above formulation of MTE does not address the overall temperature response function, which is always hump-shaped (Fig. 2.5; see also Okie, Chapter 12). This may not be a serious problem for most organisms, because metabolic rate and other processes typically level off and then abruptly decrease only as temperatures become stressfully high, approaching the upper lethal limit. It may, however, be more of a problem for organisms that live in extreme environments and have unusual temperature dependence, such as thermophilic microbes (see Okie, Chapter 12).

Second, the response to temperature may deviate substantially from an exponential relationship, even within the biologically realistic temperature range (e.g., Clarke and Fraser 2006; Knies and Kingsolver 2010). Indeed, metabolism is a complicated process, the net outcome of a network of enzyme-mediated reactions, which differ in activation energies and are affected by factors such as enzyme properties, rates of substrate supply, and thermal history. It is well-known to animal and plant physiologists that temperature dependence is not fixed, but varies over time as organisms respond to changes in their thermal environment. These adjustments include acclimation, acclimatization, and adaptation, and occur on timescales of seconds to generations (e.g., Hochachka and Somero 2002; Clarke and Fraser 2006; Gillooly et al. 2006a; Karasov, Chapter 17). They occur in both ectotherms and endotherms. They are often expressed as temporal variation in performance of individuals living in seasonal environments and as geographic variation among populations living in different climatic regimes.

Third, there can be wide variation in metabolic rate or level within the same individual. In addition to the different metabolic rates typically measured by animal physiologists and mentioned above, many organisms are able to drastically lower their metabolism, curtail most biological activities, and enter a resting state. Examples include the cysts or resting stages of microbes, the seeds of plants, and diapause, hibernation, estivation, and facultative torpor in animals. Metabolic rates can then be orders of magnitude lower than in active states, and sometimes so low as to make it difficult to detect signs of life. The temperature dependence of metabolism can play an important role. During hibernation, estivation, and torpor, normally endothermic mammals and birds abandon thermoregulation, allow body temperature to fall, and accrue large energy savings from the reduction in metabolic rate (see Karasov, Chapter 17). On the other hand, many ectothermic insects and vertebrates use behavioral thermoregulation in which they seek out cold microenvironments to reduce metabolic rates and save energy, or warm microclimates to increase metabolic rates and be more active (see Waters and Harrison, Chapter 16).

2.7 INCORPORATING STOICHIOMETRY

MTE and the above presentation have focused on energy as the sole or primary currency of metabolism. To some extent this is justified, because metabolic rate has traditionally been measured as the rate of energy flux, directly by calorimetry or indirectly by using exchange of oxygen or carbon dioxide to estimate rates of respiration and photosynthesis. But a broader definition of metabolism includes the uptake, transformation, and allocation of materials. Especially relevant are water, nitrogen, phosphorus, and other compounds and elements that are essential components of biological structures and functions (see Kaspari, Chapter 3). The processing of these materials within organisms is an important subject of physiology, and the exchanges of these materials between organisms and the environment is the subject of ecological stoichiometry (Sterner and Elser 2002).

Stoichiometry should be an integral part of a broad synthetic metabolic ecology, but the integration has been slow to occur. This is largely because energy and materials are different, yet complementary, currencies, governed by different scientific principles, and studied by scientists with different training and interests. Specifically, materials can be converted into energy, but are also used for skeletal structures and as stores of nutrients. Historically, ecological stoichiometry has focused on the role of water and nutrients (especially nitrogen and phosphorus) as limiting resources in different environments, where they affect performance of individual organisms and ecosystem-level processes. So, for example, there have been many studies of water use efficiency in land plants, water limitation of net primary productivity (NPP) in terrestrial ecosystems, and nutrient limitation of NPP in aquatic ecosystems. Especially relevant are studies showing that phosphorus limits growth rates of animals in freshwater eco-

systems (e.g., Elser et al. 1996; Sterner and Elser 2002; Gillooly et al. 2002, 2005a). The effect of nutrient limitation – in this case, phosphorus – is readily seen as accounting for residual variation around the temperature and mass dependence of development rates of eggs of diverse phytoplankton (Fig. 2.6C; Gillooly et al. 2002). This figure also illustrates how the equations of MTE, in this case equations 2.5 and 2.6, can be used not only to make baseline predictions, but also to test hypotheses about the causes of deviations from the predictions and the variation around fitted regression lines.

Brown et al. (2004) provisionally suggested that, over the range of concentrations where availability is actually rate limiting, the concentration of a limiting substance could be incorporated as an additional linear term in equation 2.1. But this is probably too simplistic for many applications, and it does not incorporate the kinetic mechanisms explicit in the widely used Michaelis–Menten and Droop models (see Kaspari, Chapter 3, Okie, Chapter 12, and Litchman, Chapter 13).

2.8 CONCLUSION

G. E. P. Box is famous for saying "All models are wrong, but some are useful." Some comparative physiologists and physiological ecologists reject or criticize MTE, because the central equation (equation 2.1) does not take account of the above variations and complications. But is MTE useful? If so, what version is most appropriate for the particular application: the general form of equation 2.1, the more specific version of equation 2.2, or some modified version of MTE? The answer will depend largely on the question being asked, the conditions in the field or laboratory, and the precision of model or data required (see White, Xiao, Isaac, and Sibly, Chapter 1). For example, a scaling perspective which can take advantage of variation spanning several orders of magnitude may not be useful when the organisms are closely related, of similar body size, and operating at similar temperatures. As Tilman et al. (2004) and Isaac, Carbone, and McGill (Chapter 7) point out, MTE does not address many areas of traditional population and community ecology that focus on niches, the abiotic tolerances and biotic interactions that affect abundance and distribution of single or multiple species in local habitats. The topics of interest may include: (1) effects of abiotic stress, resource

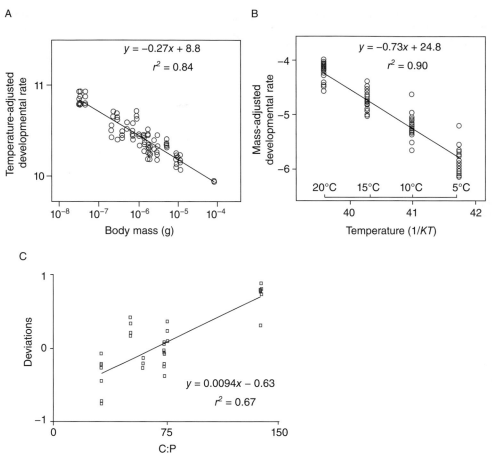

Figure 2.6 Effects of body size, temperature, and body phosphorus content on embryonic development rate of zooplankton eggs, measured here as 1/(time to hatching in days). Temperature was measured in kelvin. A. Temperature-corrected rate as a function of body mass depicted using equation 2.5 to make a plot with the log-transformed rate on the y-axis and log-transformed body mass on the x-axis. So the slope of the regression quantifies the mass dependence as the exponent of a power law, $\alpha = -0.27$. (B) Mass-corrected rate as a function of temperature, depicted using equation 2.6 to make an Arrhenius plot with the ln-transformed rate on the y-axis and $1/kT$ on the x-axis. So the slope of the regression with sign reversed quantifies the temperature dependence as an "activation energy" $E = 0.73\,\text{eV}$. (C) Effect of body phosphorus content on development rate, shown by plotting the residuals around the regression line for the temperature dependence (B) as a function of body composition (ratio of phosphorus to carbon). Note that temperature accounts for 90% of the variation in mass-corrected rate and phosphorus content accounts for 67% of the remaining variation. Redrawn from Gillooly et al. (2002).

supply, and enemies on variation in abundance and distribution over space and time; (2) foraging behavior and diet selection by consumers; (3) mechanisms of coexistence and tactics to avoid competition, predation, parasitism, herbivory, and disease; and (4) intricate mutualistic interactions and symbioses. Similarly, Hayward, Gillooly, and Kodric-Brown (Chapter 6) point

out that many studies in behavioral ecology have focused on single populations and particular interactions where fitness, rather than energy, is the explicit currency of interest.

The utility of WBE and its extensions to MTE rest on three premises. First, the theory is intended to capture only the most fundamental parameters and processes, and to explain major patterns of variation across organisms of different sizes, operating at different temperatures, and living in diverse environments. Second, the theory incorporates first principles of physics and chemistry along with fundamental processes of biology to provide a quantitative, mechanistic baseline for how body size and temperature affect metabolic rate and relevant ecological relationships. The details and variations that are of such interest to many physiologists and ecologists can be understood as refinements (but with major ecological and evolutionary implications) of the more general themes, and expressed as deviations from the theoretical predictions (e.g., Fig. 2.6). Third, the theory provides a robust framework for interpreting new data and for making and testing a priori predictions.

A telling example is the challenge to predict responses of organisms, populations, communities, and ecosystems to global climate change. There is simply not enough time nor resources to take all the details into account so as to make specially tailored predictions for every species and habitat of concern. As the global environmental temperature increases, MTE and the above equations provide first-order quantitative predictions for how metabolic rates of the microbes, plants, and animals should increase as individuals acclimatize, populations adapt, and species colonize and go extinct, and how these changes in the metabolic rate at the level of individual organisms should translate into increasing rates of photosynthesis, respiration, and water and nutrient cycling at the level of ecosystems (see Anderson-Teixeira, Smith, and Ernest, Chapter 23).

Chapter 3

STOICHIOMETRY

Michael Kaspari

SUMMARY

1 Metabolism is the use of energy and materials by life to perform work. Most of that work is applied chemistry, and so metabolic rates should covary in interesting ways with the availability and utility of chemical elements.

2 I explore the questions underlying ecological stoichiometry (ES). The 25 or so chemical elements essential to life play a variety of roles – as energy sources, structure, electrochemicals, and catalysts – and are used in rough proportion to their environmental availability. Life both tracks and creates biogeochemical gradients.

3 There are two theoretical underpinnings to ES. The first is Droop's Law, describing a positive saturating relationship between metabolic performance and availability of an element. The second is Leibig's Law of the Minimum, which posits that the element with the highest ratio of utility to availability is likely to limit metabolic performance. However, multiple nutrients are likely to limit any large-scale or complex ecological process.

4 The metabolic theory of ecology (MTE) has components related to the supply and function of essential elements. As body mass increases, so does fasting endurance; and some elements are more easily stored than others. Droop's Law predicts normalization constants that are positive decelerating functions of element supply rate. Activation energy may be a function of the availability of enzymes, and, in turn, the metal cofactors they require.

5 I conclude by addressing some particularly ripe topics for the ES component of metabolic ecology, highlighting the need for field experiments that address issues of aggregation and scale.

3.1 INTRODUCTION

Metabolism is the use of energy by organisms for the transformation of "stuff." At its most basic, metabolism is a measure of work: the conversion of glucose to ATP, CO_2, and H_2O by a cell; the transformation of materials and tissues for survival and reproduction; the breakdown of cellulose to $CO_2 + H_2O$ in an ecosystem. Stuff may be thought of as "energy" (Stephens and Krebs 1986), or a mix of carbohydrates, protein, and lipids (Simpson and Raubenheimer 2001). Ultimately, the stuff of life can be cleanly decomposed into 25 or so chemical elements. These elements are continually rearranged to build the fuel, structure, and molecular machines of life. Ecological stoichiometry (ES) explores how the availability and balance of chemical elements guides and constrains metabolism from organisms to whole ecosystems (Lotka, 1925).

Metabolic Ecology: A Scaling Approach, First Edition. Edited by Richard M. Sibly, James H. Brown, Astrid Kodric-Brown.
© 2012 John Wiley & Sons, Ltd. Published 2012 by John Wiley & Sons, Ltd.

Like MTE, ES is reductionist and integrative, testing just how far one can predict the behavior of organisms and ecosystems by knowing the inner workings of cells (Brown and Sibly, Chapter 2). To do this, ES relies on Lavoisier's conservation of mass, which says that in the generic chemical reaction

$$\text{Substrates} \xrightarrow[\substack{\text{at a given temperature, pH, osmoregularity,} \\ \text{and in the presence of catalysts}}]{} \text{Products}$$

(3.1)

elements are neither destroyed nor created (e.g., all the carbon in CO_2 at the outset of photosynthesis winds up as carbohydrate). The premise of ES is that, with a little imagination, all work performed by life can be cast as equation 3.1. The implication is that the rate (represented by the arrow) is a function of the availability of substrates, and the conditions in which those substrates are combined. The substrates are the chemical compounds, each with their elemental formula, made

available in given ratios in an aqueous medium. As we shall see, the conditions for these reactions are key, and include the temperature and the presence of catalysts.

3.2 BUILDING ORGANISMS USING ENERGY AND STUFF

There is a broad similarity in the chemical composition of life. Life has a recipe. For example, in human tissue, for every million or so P atoms, there are 38 600 Cu atoms and 1 Co atom (Fig. 3.1, Table 3.1). When we take into account mass, we also note that about 99% of life consists of just four elements: H (62.8%), O (25.4%), C (9.4%), and N (1.4%); the next seven (Na, K, Ca, Mg, P, S, and Cl) make up another 0.9%; and another 10 (Mn, Fe, Co, Ni, Cu, Zn, Mo, Bo, Si, and Se) comprise 0.1% of human mass. All are essential – shortfalls in any cause pathology.

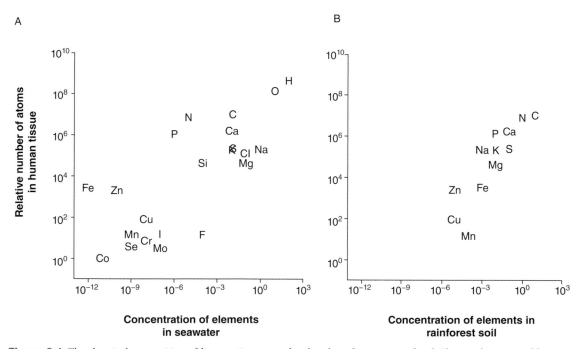

Figure 3.1 The chemical composition of human tissue correlated to that of seawater and soil. The stoichiometry of human tissue is described as relative numbers of atoms (e.g., there are 375 000 000 H atoms for every Co atom) (Sterner and Elser 2002). (A) The concentration of elements in seawater is related as mass percentage (Frausto da Silva and Williams 1991). (B) The concentration of elements in sandy soil from a Peruvian rainforest is related as mass percent dry weight, and lacks quantities of some trace metals (Kaspari and Yanoviak 2009).

Table 3.1 The 22 elements required to build a human body, their proportionate number (e.g., there are in the order of 100 000 000 H atoms for every Co atom), and some notes on their function. Data from Sterner and Elser (2002).

	Name	Number	Functions
H	Hydrogen	10^8	Solvent, energy
O	Oxygen	10^8	Solvent, energy
N	Nitrogen	10^6	Structure
C	Carbon	10^6	Energy, structure
Ca	Calcium	10^6	Structure, electrochemistry, signals
P	Phosphorus	10^6	Energy, structure
S	Sulfur	10^5	Structure
Na	Sodium	10^5	Electrochemistry
K	Potassium	10^5	Electrochemistry
Cl	Chlorine	10^5	Electrochemistry
Mg	Magnesium	10^4	Catalyst
Si	Silicon	10^4	?
Fe	Iron	10^3	Catalyst, transport O_2
Zn	Zinc	10^3	Catalyst
Cu	Copper	10^1	Catalyst, transport O_2
I	Iodine	10^1	Catalyst, signals
Mn	Manganese	10^1	Catalyst
F	Fluorine	10^1	Catalyst
Cr	Chromium	10^0	Catalyst
Se	Selenium	10^0	Catalyst
Mo	Molybdenum	10^0	Catalyst
Co	Cobalt	10^0	Catalyst

Since life evolved by natural selection, we might surmise that life is assembled from common elements that are easy and cheap to find. Indeed, life's chemical formula mimics in broad strokes the chemical proportions found in seawater and soil (Frausto da Silva and Williams 2001; Fig. 3.1). At the same time, the pumps and channels that stud cell membranes selectively acquire some elements (e.g., the C, H, and O of carbohydrates) and eliminate others (e.g., arsenic). Organisms spend considerable energy maintaining this chemistry within precise levels of tolerance (NRC 2005; Salt et al. 2008). The resulting homeostasis of intra- and intercellular environments – the concentra-

tions of substrates, structures, and the pH and osmoregularity of abiotic environments – fosters metabolic work.

3.3 WHAT DO ALL THESE ELEMENTS DO?

Metabolic ecologists include diverse processes under the descriptor "metabolic rate." These include the synthesis of protein within a cell, the wholesale breakdown of glucose in an individual, and the generation of glucose by a hectare of forest. All share a dependence on the co-availability of essential elements – as catalysts, electrochemical ions, structures, and energy sources. Table 3.1 comprises one classification of the functional roles of elements.

3.3.1 Solvent

Most of living mass consists of liquid water, H_2O. The metabolism of most living things correspondingly occurs within the range of temperatures in which water is aqueous with a pH of 7, allowing elements to be available in a soluble form.

3.3.2 Energy

Covalent chemical bonds, when broken, release useful energy. Carbohydrates (or CHOs, for carbon, hydrogen, and oxygen) are the primary energy source of life; ATP with its high-energy phosphate bond is the universal molecule for biological energy transformation. While MTE has focused thus far on photosynthesis and cellular respiration (the breakdown of long-chain organic carbon and running the resulting sugars through the Krebs cycle), bacteria and archaea offer a variety of alternatives (Okie, Chapter 12).

3.3.3 Structure

Elements can be stored in, and make up large quantities of, hard and soft tissue. N is a big component of proteins and nucleic acids; P is common in nucleic acids, ribosomes, and phospholipid membranes. C and Si reinforce plant tissue and are found in the skeletons of many animals. The S in the amino acids methionine and cysteine helps proteins fold into their ultimate, functional shape and provides thermal stability.

3.3.4 Electrochemicals

The charged ions of Na, Cl, K, and Ca are continually fluxed across life's membranes. These fluxes involve osmoregulation of intra- and extracellular fluids; they also maintain the electron potentials across the membranes. All serve to maintain the chemical gradients that allow critical substrates to accumulate and toxins to be eliminated. Some electrochemicals, such as Na, are not easily stored, are constantly excreted, and require continual uptake.

3.3.5 Catalysts

A variety of relatively rare metal atoms help enzymes combine with, orient, and bring together substrates. These atoms (often called cofactors) are functional parts of up to one-third of approximately 4000 known enzymes (Bairoch 2000; Waldron and Robinson 2009) where they rest in the pocket of the protein's active site (Fig. 3.2). There they aggressively catalyze reactions

Figure 3.2 A metallomic enzyme, carbonic anhydrase, from the chloroplast, requires a single atom of Zn (center) to function.

involving the ubiquitous small molecules key to metabolism (e.g., O_2, H_2S, and CH_4). In doing so they enhance the rate of reactions beyond those resulting from random encounters of substrates in solvent. Enzymes make life thermodynamically possible.[1]

Two other functional roles of elements bear mentioning. The Fe in heme metalloproteins serves to *transport* oxygen in vertebrates; the hemocyanin in insects serves the same function but instead uses two Cu atoms. Ca binds to and activates enzymes, acting as a *signal* away from the binding site (Clapham 2007). Another metabolic signal, thyroid hormone, contains iodine (Cavalieri 1997).

3.4 HOW ARE CHEMICAL ELEMENTS DISTRIBUTED?

Elements are arrayed along a number of gradients in space and time (Table 3.2). It would seem a safe bet that life's metabolic processes must ultimately track these gradients. But things are not that simple. Metabolism both follows and generates biogeochemistry. The most telling example starts 2.5 billion years ago, when Earth's atmosphere consisted of CH_4, SO_2, NH_3, and CO_2, much of it belched up by volcanoes. This atmosphere prevented the formation of O_2. However, as CO_2 and H_2O accrued, the stage was set for photosynthesis, the metabolic pathway that used CO_2 as a carbon source and photons as an energy source. As O_2 – a byproduct of photosynthesis – accumulated, the oxidizing atmosphere opened the door for the now ubiquitous, and remarkably efficient, aerobic respiration that runs much of life. At the same time, Fe solubility plummeted and the Fe^{3+} not bound up by the organic ligands manufactured by marine bacteria (Weaver et al. 2003; Hassler et al. 2011) sank to the dark ocean floor, away from the photosynthesizers. More on that soon (section 3.6.4).

Redfield's ratio (Redfield 1958) is another example of life shaping biogeochemistry. Redfield found that marine plankton and seawater share the formula $C_{106}N_{16}P_1$. His insight was that plankton weren't merely tracking the availability of these three essential

---

[1] Consider triose phosphate isomerase, one of the 10 enzymes in the glycolysis pathway. It is classified as "catalytically perfect" – allowing its reaction to proceed, regardless of temperature , at the rate substrates diffuse to and from the reaction site (Albery and Knowles 1976).

Table 3.2 Some gradients of elemental availability in time and space.

Driver	Elemental gradient
Sunlight and H$_2$O (Rosenzweig 1968)	Net primary productivity (NPP, gC/m^2/y) reflects the rate of CHO production by plants. NPP tends to be highest where it is sunlit and wet (in tropical rainforests and surface waters)
Ecosystem age (Walker and Syers 1976; Wardle et al. 2004)	Glaciers and volcanoes, expose new, mineral-rich bedrock, which weathers over time (see next) even as N-fixing plants accumulate and convert inert N$_2$ into soil nitrates
Wind and gravity (Chadwick et al. 1999; Smil 2000)	Soils erode, depositing their nutrients at the bottom of the ocean; the churning of the ocean lifts Na, S, and other elements into the air and overland, while dust blown from continents transports P and Fe to the oceans (Fig. 3.4)
Food webs (Luoma and Rainbow 2005)	Herbivores convert tissue with a C:N ratio >0:1 to tissue with a C:N ratio >10:1, concentrating other nutrients along the way. Predators are less likely to be nutrient stressed than herbivores

elements. There instead was an equilibrium – caused by plankton's uptake and growth and then death and release of these elements – that reflected a "global circulatory system" (Sterner and Elser 2002). P likely set the stage as most limiting, but then restricted the demand for N and C. As ecologists learned of Earth's evolving atmosphere and Redfield's ratio, it transformed the view of life from a passive, to an active, player in global biogeochemistry.

3.5 MODELS AND APPLICATIONS OF STOICHIOMETRY TO METABOLISM

Let's now explore how ES models the fluxes and stores of an element and selects *which* elements on which to focus.

3.5.1 The Droop equation

The Droop equation (Droop 1974) allows us to rigorously define *nutrient limitation*. A limiting nutrient is

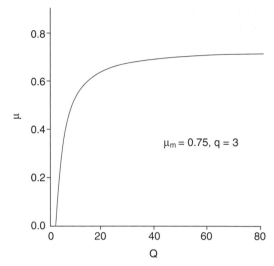

Figure 3.3 The Droop equation describes the growth rate of a cell, μ, as a function of the quota, Q, of a nutrient within the cell. Key parameters are the minimum quota to just support life, q, and the maximum growth rate, μ_m.

one which, when added to a system, increases the rate of metabolism, of growth or respiration, at the individual or ecosystem level. The Droop equation relates the elemental concentration (or quota) within the cell, to growth rate

$$\mu = \mu_m(1 - q/Q) \qquad (3.2)$$

where μ is the biomass growth rate (day^{-1}); μ_m is the maximum growth rate at theoretically infinite quota; q is the minimum quota necessary to just support life (moles per cell at zero growth); and Q is the observed quota (moles per cell, Fig. 3.3). The Droop equation is a simple model that captures a basic reality – over a range of availability, the effect of a nutrient on growth saturates. In other words, as quota increases, the part of the equation in parentheses converges on 1, giving you the maximum growth rate. According to the Droop equation, there is not a *single* magnitude of nutrient limitation; it accelerates (and growth rate decelerates) as nutrient supplies decline.[2] Droop's equation should hold for any element or biochemical essential for cell

[2]A similar equation, Monod's Law, or the Michaelis–Menten equation, describes a similar process, but explores how a cell grows as a function of resource concentration *outside* the cell.

growth (at least across the range of Q before high concentrations become harmful).

3.5.2 Liebig's Law of the Minimum

The concept of 25 Droop equations to predict a metabolic rate (one for each essential element and daisy-chained together) is almost too much to bear. The eponymous Liebig's Law of the Minimum (LLM) was the first to deal with this chemical logjam (Liebig 1855). For a set of bioavailable nutrients, he reasoned, *only* the one in shortest supply relative to demand (that is, with the largest ratio of utility to availability) will limit yield. In Liebig's Law there can be only one limiting nutrient at a time: its supply rate becomes the metabolic rate-limiting step.

Placed in the context of Droop's Law, LLM has two parts. First, elements are essential for specific functions – any concentration less than μ_∞ moves an organism leftward and downward on the Droop curve. Second, cells and organisms expend energy and resources to harvest elements, and these costs diminish as availability increases. Putting LLM into practice, ES practitioners often start with two likely candidates for the limitation of some metabolic rate. For each element, they estimate (1) the nutrient demand when growing at Droop's μ_∞, and (2) the supply rate of those nutrients. Most often, the pair of elements are C and another resource "R," where R is either N or P (White 1978, 1993; Sterner 2008; Kay 2002). For example, consider a population of planktonic crustacean, *Daphnia*, growing at their maximum rate. The baseline ratio C : P in their tissue can be used to predict how an organism will perform on diets, or in environments, of varying C : P. In lakes with low C : P, carbon-rich foods maximize growth; in lakes with relatively little P, the lack of P-rich foods limits growth (Sterner 1997).

LLM is a simple rule for finding likely candidates for nutrient limitation; it is one of ecology's most successful concepts.[3] It requires that metabolic processes are limited in a strictly hierarchical way (Sterner and Elser 2002).

---

[3] Liebig's aim was modest: to advise farmers on the best choice of fertilizer to maximize crop yield. He likened the problem to an oaken bucket with vertical staves of varying lengths. To increase the capacity of the bucket it wasn't necessary to increase the length of each stave; it was only necessary to increase the length of the shortest stave. This metaphor sold lots of fertilizer (Brock 2002).

3.6 A FEW APPLICATIONS OF LIEBIG LOGIC

3.6.1 Energy

Models of optimal diet selection in animals were some of the first applications of ES. Most focused on C as the limiting element, assuming that individuals maximizing the rate of energy harvested left more offspring (Stephens and Krebs 1986). There were a number of good reasons why energy maximization models were useful. First, in a classic case of Leibig logic, energy must be expended to maintain homeostasis and search for food; in low NPP environments, organisms that don't harvest enough carbohydrates (CHOs) do not have the opportunity to be N, P, or S limited (Sterner 1997).

Optimal foraging theory was initially tested with taxa like bumblebees and hummingbirds, as their small size, thermogenesis, expensive locomotion, and high metabolic rates all suggested energy limitation (Pyke 1978; Hodges and Wolf 1981; Stephens and Krebs 1986). More recently, aggression and spite have been recognized as energetically expensive (Kaspari and Stevenson 2008). For example, colonies of the ant *Linepithema*, when fed a high C : P diet, were more aggressive to competitors and grew more quickly, suggesting that energy limits growth and defense of colonies (Kay et al. 2010).

3.6.2 Structural

N and P predominate in studies of structural elements. If one were to search for an ecosystem in an N-limited environment, you couldn't find a better site than the Cedar Creek research station in the north temperate USA. Its grasslands combine a recent glaciation with sandy, impoverished soils. Fertilization experiments at Cedar Creek consistently show increases in biomass production with added NH_4NO_3 (Tilman 1988). Moreover, an early experiment in which plots were fertilized with a combination of other candidate elements (P, K, Ca, Mg, S, and trace metals) yielded no more plant biomass than controls (Tilman 1987), reinforcing LLM's prediction that N is the one element at Cedar Creek limiting plant production.

The *growth rate hypothesis* (Elser et al. 1996) is an elegant example of Leibig logic integrating from cellular metabolism to ecosystem respiration. It starts with the observation, across a variety of invertebrates, than an organism's growth rate was closely correlated with RNA content (Sutcliffe 1970). Now add the fact that P is a component of RNA, which in turn is a major constituent of ribosomes, the cellular machines that build proteins and thus underlie growth. The growth rate hypothesis predicts that P scarcity should decrease growth rates via decreases in ribosomal RNA, a pattern subsequently shown across a variety of invertebrates (Elser et al. 2003). It has also been applied to plants. Temperate ecosystems tend to have more available P than those in the tropics (Table 3.2). Trees from the temperate zone also tend to have higher tissue P, and grow faster, than their tropical counterparts (Kerkhoff et al. 2005; Lovelock et al. 2007).

3.6.3   Electrochemical

Many ionic elements are both common and tightly regulated. For example, plants typically don't use Na, but the herbivores and detritivores that eat plants do. This poses a dietary challenge to basal consumers, particularly in inland ecosystems where deposition and recharge of Na from oceanic aerosols is low (Fig. 3.4). There is growing evidence that Na shortage limits activity, population growth, and ecosystem respiration in the large swaths of the terrestrial biosphere >100 km from an ocean (Botkin et al. 1973). When my colleagues and I assayed ant use of salt, NaCl, and sucrose from coastal Florida to interior Amazon rainforests, we found that ants increasingly craved NaCl further inland

A

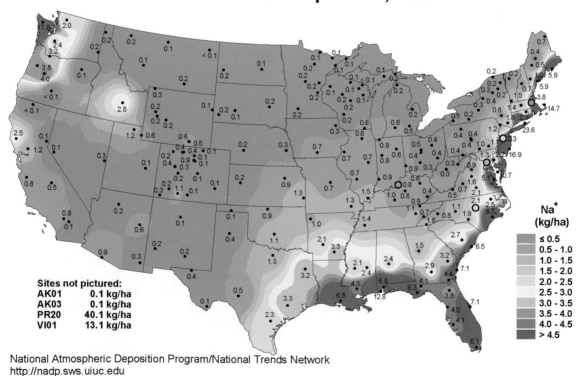

Figure 3.4 Mass transfer of elements at a geographic scale. (A) A map of Na deposition from oceanic aerosols, drawn by the US National Atmospheric Deposition Program. (B) The transfer of dust from the African Sahara to the mid-Atlantic ocean on February 26, 2000. (SeaWiFS satellite image courtesy of NASA.)

B

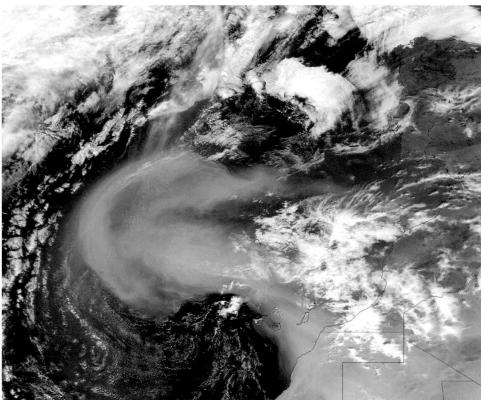

Figure 3.4 *Continued*

(Kaspari et al. 2008b). This Na limitation has ecosystem consequences: in Peruvian rainforests, termites and other decomposers accumulated on plots fertilized with NaCl, and decomposition rates were 50% higher (Kaspari et al. 2009).

3.6.4 Catalytic

Earth's evolution of an O_2-rich atmosphere had a built-in negative feedback. Fe is a cofactor for chlorophyll, and Fe supply can limit rates of photosynthesis in oceans (Sunda and Huntsman 1997). Now recall that as the O_2 accumulated, Fe oxidized and sank to the ocean floor. Ever since, ocean algae (and ocean food webs) have depended on dust blowing in from the continental interior to resupply that Fe (Fig. 3.4). This has global, in fact epochal, implications (Martin 1990). Imagine a warm CO_2-rich Earth: as droughts increase, inland winds fertilize the oceans with Fe-rich dust;

rates of photosynthesis increase and scrub the atmosphere of CO_2. The Earth cools; droughts abate, as do Fe supply rates, and CO_2 accumulates again. In a spectacular "field" test of this hypothesis (Boyd et al. 2000) $50 km^2$ of open ocean was fertilized with $3813 kg$ $FeSO_4.H_2O$. The resulting 100 parts *per trillion* increase in Fe temporarily generated a 6-fold increase in chlorophyll and doubling of algae productivity. Unfortunately for the rest of the food web, those algae produced domoic acid, a potent neurotoxin (Trick et al. 2010).

3.7 BEYOND LIEBIG: WHEN METABOLISM IS LIMITED BY MULTIPLE ELEMENTS

The above tests show the utility of LLM in identifying potential limiting nutrients.

Let's consider briefly some ways (often complementary; Saito et al. 2008) in which more than one element may simultaneously limit metabolic rates.

3.7.1   Organisms deplete multiple elements, equalizing utility : availability

LLM relies on one element unambiguously maximizing the ratio of utility to availability. But natural selection, in favoring the use of common elements to build ubiquitous structures, tends to *equalize* these ratios across the range of essential elements. Consider the challenge to an algal cell growing in the open ocean. Essential ions are roughly available at the following molar concentrations (Frausto da Silva and Williams 2001):

K^+, Na^+	10^{-1} M
Mg^{2+}, Ca^{2+}	10^{-3} M
Zn^{2+}	10^{-9} M
Cu^{2+}	10^{-12} M
Fe^{3+}	10^{-17} M

Fe^{3+} may be limiting because every ion of Fe^{3+} used to run photosynthesis requires an investment in ligands (CN-based proteins and sugars that snare the Fe^{3+} ions before they oxidize and disappear to the ocean's depths or are harvested by a competitor). But consider Zn^{2+}. Even though Zn^{2+} ions are 1 000 000 times more common in the environment, they are also metabolically ubiquitous, acting as cofactors of every functional group of enzyme (Frausto da Silva and Williams 2001).

If the ratios of utility : availability governing a given metabolic rate are converging toward a common value, there will never be the clear, enduring maximum assumed by LLM. Instead, a handful of elements may succeed one another in rapid succession, depending on local vagaries of supply and demand. In an elegant thought experiment, Saito et al. (2008) compiled data for the stoichiometry of marine phytoplankton and the biogeochemistry of three oceans. In each they found plausible cases where some combination of six elements (C, N, P, Fe, Co, and Zn), although available in widely different concentrations, could be hypothesized to limit productivity.

3.7.2   Scale and aggregation effects

A second way of reaching multi-nutrient limitation – that complements the first – happens when ecologists measure metabolic rates over larger areas, longer timespans, and more populations (Levin 1992). This is easiest to imagine at the ecosystem scale.

Consider a process like decomposition. In a square meter of forest floor, thousands of species and billions of individual bacteria, fungi, and invertebrates break down leaf litter (Swift et al. 1979). Scattered within the plot may be three kinds of substrate. Two represent the same starting point but differing periods of time: a freshly fallen leaf and a decayed leaf that is little more than lignin and cellulose. A third substrate may be a butterfly chrysalis fallen from the forest canopy. Bacteria and fungal decomposers secrete different enzymes to catalyze the breakdown of different substrates: the protein-rich fresh leaf, the lignin-rich old leaf, and the chitin-rich chrysalis. Perhaps one-third of those enzymes (or the enzymes needed to synthesize them) require a metal cofactor. With this in mind, a group of colleagues compared decomposition rates in $0.25 \, m^2$ plots embedded in a 40×40 m plot in a Panama rainforest (Kaspari et al. 2008a). These plots were fertilized with N, P, K, and a mix of micronutrients (including B, Ca, Cu, Fe, Mg, Mn, Mo, S, Zn). Decomposition increased on all but the +N plots, suggesting that at least three elements limited decomposition. Given the diversity of the leaf litter, and the duration of the study (48 days), multiple suites of metallomic enzymes were likely upregulated by the microbial community toward the conversion of tropical detritus to CO_2, H_2O, and minerals.

3.7.3   Limitation cascades

Metabolic pathways tend to occur in networks. What happens in one part of the network ramifies "downstream." If one part of the network governs the uptake of a limiting element, the lack of a metal cofactor can set up a limitation cascade: the lack of one enzyme can hinder the ability to harvest another.

One of the best examples of limitation cascades involves N fixation. Shortages of N have repeatedly been shown to limit population growth (White 1993) and ecosystem respiration (Horner et al. 1988). This puts a premium on processes that import N to ecosystems in a usable form. N_2 represents 72% of Earth's atmosphere but it is chemically inert, with both N atoms held in place by a powerful triple bond. Biological N fixation into the biologically usable form of NH_4 or NO_3 accounts for the importation of up to 120Tg $(1\,Tg = 10^{12}\,g)$, 92% of ecosystem N (lightning accounts for the rest; Vitousek 1994). But N fixation is costly. It requires ample carbohydrate as fuel. It also requires supplies of Fe and Mo to build nitrogenase enzymes. A shortage of C (via deficits of sunlight or moisture), Fe, or Mo can thus generate a limitation cascade hindering the input of N into the ecosystem. In the dark (low-C) understory of a Panama rainforest, fertilization with Mo generated 2- to 3-fold increases in N fixation by heterotrophic microbes compared to control plots (Barron et al. 2008).[4] And in an era of increasing CO_2, the resulting surplus of C should increase plant demand for N, N-fixation, and the relatively small stocks of soil Mo and Fe (Hungate et al. 2004; Barron et al. 2008).

3.8 LINKING ES AND MTE MODELS OF METABOLIC ECOLOGY

> "Organic synthesis and metabolic rate are limited by the supply rate of essential elements." (Reiners 1986)

Now that the case has been made for the relevance of ES to metabolic ecology, how do we more thoroughly integrate it into MTE? One way is to look for shared metabolic subunits. A second is to look for stoichiometric underpinnings of the MTE itself.

3.8.1 When metabolic components have a distinct stoichiometry

MTE models assume mitochondria, ribosomes, and chloroplasts are "invariant subunits": they process

energy and materials the same way regardless of the organisms that contain them (Brown et al. 2004; Allen and Gillooly 2007, 2009; Okie, Chapter 12). If so, the metabolic rates of an organism should correlate with the tissue density of these subunits. Since these subunits can have distinctive stoichiometries (Niklas and Enquist 2001; Gillooly et al. 2005a; Allen and Gillooly 2009; Elser et al. 2010), an organism's metabolic rate should be associated with its stoichiometry.

Gillooly and colleagues (2005a) used this logic to explore the 100-fold variation in [P] (phosphorus concentration) observed across organisms. They started by linking rates of protein synthesis to the chemical mechanics of RNA and ATP. As predicted, the [RNA] of a variety of taxa tended to decrease as $M^{-0.25}$ (Fig. 3.5A). Moreover, whole-body [P] decreased with M, but in a decelerating fashion, due to the increase in non-metabolic pools of P (like vertebrate skeletons) in larger organisms (Fig. 3.5B). The "invariant subunit" approach also allowed whole-body [P] to account for much of the variation in temperature-corrected growth rates among eight species of zooplankton (Allen and Gillooly 2009).

3.8.2 Stoichiometric underpinnings of the MTE equation

A second way to build a more synthetic metabolic ecology is to show how the parameters of MTE are themselves a function of ES (Brown and Sibly, Chapter 2). Recall that in the MTE

$$B = B_0 M^{\alpha} e^{-E/kT} \tag{3.3}$$

B_0 is the normalization constant, M is mass, α is an allometric scaling exponent, E is the activation energy, k is Boltzmann's constant, and T is temperature (K).

3.8.2.1 Mass

Not all mass is metabolically active. A large store of an organism's mass may be metabolically inert: surplus N can be stored intracellularly as amino acids and proteins; surplus P can be stored as polyphosphate and bone (Rhee 1978; Taiz and Zeiger 1998); surplus Fe is bound up in proteins called ferritins (Harrison and Arosio 1996). Moreover, an organism's storage capacity scales with its volume. If metabolic rates scale as

---

[4]Barron and colleagues also found increases in N-fixation on plots fertilized with P. Subsequent experiments showed this was due to Mo contamination of the phosphate (which is mined from rock). Such is a cautionary tale for one exploring the ecology of trace metals; their concentrations (at 10^{-6} the molar quantities of N and P) are difficult to measure accurately, and often difficult to manipulate without contamination.

A

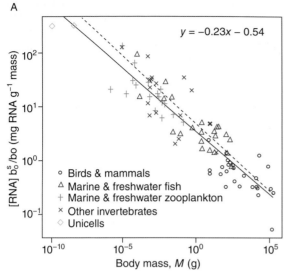

$y = -0.23x - 0.54$

B

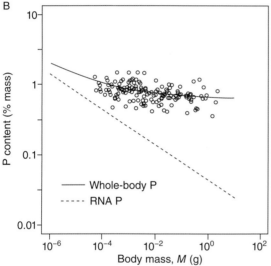

Figure 3.5 Syntheses combining WBE and ES components of metabolic ecology predicting whole-body RNA content (A) and P content (B) as a function of dry mass (Allen and Gillooly 2009).

$M^{3/4}$ and storage scales as M^1, then fasting time should increase as $M^1/M^{3/4} = M^{1-3/4} = M^{1/4}$ (Peters 1983).

This is of particular interest in environments in which resource supply (and hence the need for fasting) varies over time. Imagine a series of environments all with the same annual NPP. However, some are asea-

sonal (with that NPP evenly portioned out from January to December). Others are seasonal, with rich productive summers and barren winters. In the seasonal environments, smaller organisms with fewer fat reserves should be less able to withstand the resulting periods of resource scarcity. The starvation resistance hypothesis predicts that variable environments (in supplies of any storable essential element, not just C) will support communities of larger organisms, with consequently lower rates of ecosystem productivity and respiration (Lindstedt and Boyce 1985; Kaspari and Vargo 1995). However, not all elements are easily stored. Many of the electrochemicals (e.g., Na, Cl, and K) and elements sequestered by gut flora (e.g., Co) are more or less constantly excreted (Milewski and Diamond 2000) with excretion rates scaling as $M^{3/4}$ (Peters 1983). Periodic shortfalls of Na or K may harm small and large organisms alike (Jones and Hanson 1985).

3.8.2.2   Normalization constant

In MTE (equation 3.3) the normalization constant accounts for variability in a metabolic rate independent of the effects generated by mass scaling and the Arrhenius equation. For reasons of simple parsimony, the normalization constant is one of the first places to look for ES effects; they should show up as residuals to plots of metabolic rate vs. mass-corrected T, and temperature-corrected M (for example, in Figure 3.5A, for a given body size, there remains 100-fold variability in whole-body [RNA]).

A theory of normalization constants, based on Droop's Law, seems a good place to begin fusing ES and MTE models of metabolic ecology. For example, imagine growing size-matched plants at a fixed temperature in a common garden. Fertilize the subplots with a gradient of increasing R (where R is a nutrient) while you monitor growth rate (Tilman 1987). Normalization constants should increase in a positive decelerating fashion as one moves up [R].

3.8.2.3   Mass exponent

There remains a spirited debate as to the value(s) of the mass exponent. In at least one experiment, however, diet has been shown to influence the mass exponent (Jeyasingh 2007). Four species of *Daphnia* varying 10-fold in size were fed two diets – one diet matching

Daphnia stoichiometry (C:P c. 150) and an imbalanced, carbon-rich diet (C:P c. 800). On the balanced diet, metabolism scaled as $M^{0.83}$; on the C-saturated diet, it scaled as $M^{0.67}$. The decreased exponent resulted from a larger increase in metabolic rate in the small *Daphnia* species under P-stress. Jeyasingh suggested that the higher mass-specific requirement for P in the small species placed a premium on eliminating excess C. These small *Daphnia* effectively "burned off carbon" by upregulating C-disposal genes (Jeyasingh et al. 2011).

Jeyasingh's work is a good example of how metabolic challenges vary with the way you eat (Sterner and Elser 2002). Bacteria and fungi eat by secreting enzymes into the environment and using transport proteins to selectively absorb individual nutrients from the resulting soup. Metazoa, on the other hand, consume prey as packages, and frequently eat foods that differ stoichiometrically from themselves (e.g., herbivores have lower C:N than their host plants). Metazoa thus often need to dispose of or store nutrients when prey stoichiometry deviates from the optimum. Experiments like the one above using synthetic diets (Cohen 2004) have much potential for exploring the metabolic consequence of this postprandial processing.

3.8.2.4 Activation energy

The metabolic rates most often measured by ecologists (respiration, photosynthesis, growth, and production) integrate over hundreds of chemical reactions in organisms and ecosystems. The activation energy (and its cousin, Q_{10}) describes the temperature sensitivity of those metabolic rates. Activation energies for a given metabolic process show a central tendency (e.g., aerobic respiration averages about 0.65 eV; Brown and Sibly, Chapter 2); one study has also revealed substantial variation around the mean (Dell et al., unpublished). One step toward a theory of activation energy focuses on the availability of enzymes that catalyze chemical reactions.

Recall that perhaps 30% of enzymes require at least one metal atom to function properly (Frausto da Silva and Williams 2001). If the quotas of these metal cofactors drop below μ_∞, those shortfalls should decrease metabolic performance, especially at low temperatures. If so, metabolic rates under sufficient metal scarcity should be lower with higher temperature sensitivity (i.e., higher activation energies).

In nature, there are at least two avenues toward metal scarcity (Table 3.2). The weathered soils of older ecosystems are nutrient poor (Wardle et al. 2004). Metals also bioaccumulate: detritivores and herbivores should be more likely than their predators to consume food poor in essential metals (Jones and Hanson 1985; Milewski and Diamond 2000). As a consequence, activation energies of detritivores and herbivores should be higher than those of predators in the same environment, a disparity exaggerated in nutrient-poor ecosystems.

3.9 OPPORTUNITIES

Sterner and Elser's (2002) volume laid the groundwork for a productive decade developing the stoichiometric basis for metabolic ecology. Here are some suggestions for the next decade.

3.9.1 Describing the chemical recipe of life and its consequences

There are still plenty of holes in our understanding of the biological function of essential elements (Saito et al. 2008; Sterner 2008; Kaspari and Yanoviak 2009). This essentially descriptive enterprise is a key first step toward understanding how biogeochemical gradients shape, and are shaped by, life. Proteomics and metabolomics (Silver 1998; Bragg et al. 2006; Lupez-Barea and Gumez-Ariza 2006) and the extensive applied literature of nutrition, agronomy, and wildlife management (Jones and Hanson 1985; NRC 2000, 2005) are excellent venues for exploitation by enterprising ecologists. One continuous source of ideas is Frausto da Silva and Williams's opus *The Biological Chemistry of the Elements* (2001).

As ES databases grow, investigators are discovering that about half of the globally observed variability in stoichiometry exists among species in a given community (Kraft et al. 2008; Hillebrand et al. 2009; Elser et al. 2010). Some of these differences in ES reflect niche differences linked to metabolic subunits – high densities of ribosomes (and high titers of P) are found in bacterial taxa characterized by rapid growth (Stevenson and Schmidt 2004). To what extent do gradients of biogeochemistry undergird coexistence in communities (Tilman 1982) and what are the functional traits that provide a mechanism for these population interactions (McGill et al. 2006)?

3.9.2 Mapping availability at multiple scales

For ES to be predictive we need better maps of biogeochemistry at scales relevant to our study organisms. While ecologists are increasingly adept at measuring C availability (e.g., through remote sensing; Turner et al. 2006), direct measures of nutrient availability at the landscape scale (John et al. 2007) are still rare. Maps of mineral "licks" – deposits rich in elements like Na, Mg, and Ca (Jones and Hanson 1985) – allow for the study of mineral deficiency using the comparative approach.

Measuring absolute quantities of P, Zn, or Fe may not be enough (Weaver et al. 2003; Saito et al. 2003, 2008). Ligands – organic molecules produced by microbes to bind to and sequester metals – distinguish between an element's mere presence and its "bioavailability." Much remains to be discovered about their abundance, diversity, and function in oceans and soils.

3.9.3 Experiments: measuring performance curves for a variety of elements

Experiments remain the definitive test of nutrient limitation. With 25 or more candidate elements, the prospect of a complete screening is a little daunting. Moreover, to quantify the Droop curve, you want to measure metabolic performance across at least three levels of availability. Now imagine you wish to explore multi-nutrient space. One can begin to see one's lab bench space decrease geometrically even as one's budget increases exponentially.

At this point the reader is likely considering returning to her first love, the theater. But all is not lost. First and foremost, nothing thins out the list of possible experiments like a good set of hypotheses based on a combination of realism (e.g., testing for the effects of doubling the extant environmental concentration of R) and rules of thumb (e.g., Leibig's Law of the Minimum). Figure 3.1, comparing the composition of a vertebrate endotherm to seawater, suggests that N, P, Fe, and Zn would be worth a look as limiting elements. Comparing that same stoichiometry to the forest soils

of Peru, we see that P, Na, and Zn show LLM's requisite high ratio of use to availability.

Four elements seem especially promising for future studies. Na and I are candidates for limitation of herbivores and detritivores in inland and weathered ecosystems (Milewski and Diamond 2000; Kaspari et al. 2008b, 2009), which support a considerable proportion of the world's biota. Two macronutrients – Mg, a cofactor in numerous enzymes in the glycolytic pathway, and S, a component of two amino acids critical for protein folding – are both essential elements that are patchily distributed across landscapes and continents (Jones and Hanson 1985; Frausto da Silva and Williams 2001).

A second reason for optimism is the proliferation of useful molecular methods. You can now count and compare the number of gene copies for transport proteins for a given ion (Silver 1998). Microarrays like the Geochip (Zhou et al. 2008) quantify the expression of thousands of substrate-specific enzymes. Stress experiments detect the upregulation of stress-response genes (Webb et al. 2001) or whole metabolic cascades (Fauchon et al. 2002; Salt et al. 2008). All of these can provide valuable insights to the field ecologist provided you learn the right techniques or find the right collaborator.

3.9.4 Scaling up to communities and ecosystems

Metabolic ecology is transforming ecosystem-level models from the statistical and descriptive (e.g., Gholz et al. 2000; Mahecha et al. 2010) to those based on first principles of energetics and stoichiometry (Andersen-Teixeira and Vitousek, Chapter 9). Challenges remain. A recent review of herbivory patterns suggests one reason why (Hillebrand et al. 2009). The authors compared the ability of mass, temperature, and stoichiometric deficits to predict patterns of herbivory at the individual and population level. Consumption rates of herbivores scaled well to mass across 11 orders of magnitude of body size, but mean body size was ineffective at predicting the population consumption rate (since body size and population size were inversely related). In contrast, the stoichiometric mismatch of herbivore NP to plant NP was the best predictor of herbivory rates at the population level and within feeding guilds. We need more such studies contrasting the utility of MTE, ES, and synthetic models toward predicting metabolism at differing levels of aggregation.

Accurate ecosystem and global models are ever more important. To what extent, however, do they predict the maximum possible rates of ecosystem respiration? How do these maxima compare to a world with nutrient shortfalls across broad, predictable sections of the globe (Martin 1990; Chadwick et al. 1999; Kaspari et al. 2009)? Are organisms from inland communities, denied Na and I, slower and (given iodine's importance in brain development) dumber (Milewski and Diamond 2000)?

A key goal of this volume is to weave together models of biochemistry and energetics and by doing so inspire the next generation of metabolic ecologists. If we succeed, and I suspect we will, I predict that, as Frank Sinatra once suggested, "The best is yet to come."

Chapter 4

MODELING METAZOAN GROWTH AND ONTOGENY

Andrew J. Kerkhoff

SUMMARY

1 Ontogeny typically involves increases in size (i.e., growth) that impact the energetics of organisms. Here we review models of ontogenetic growth of multicellular animals (metazoans) based on the balance of their energetic and material inputs and outputs, with particular attention to how rates of metabolism and growth scale with body mass and temperature.

2 Studies of ontogenetic growth have a rich theoretical history relating the scaling of metabolic rate to organismal development. Multiple alternative models share similar formal structure, but are based on different underlying assumptions.

3 While patterns of ontogenetic growth are largely consistent with the predictions of the metabolic theory of ecology (MTE), growth trajectories, them-selves, are not sufficient to distinguish between alternative models. Moreover, highly variable ontogenetic scaling of metabolic rate and the "size–temperature rule" challenge the "universality" of the MTE assumptions for species-specific applications to ontogenetic growth. At the same time, the variability of ontogenetic patterns provides a wealth of opportunities for understanding the impact of different physiological, ecological, and evolutionary factors on metabolic scaling.

4 Careful examination of the assumptions of multiple, competing models may help understand not just the process of growth and ontogeny, but also the processes that underlie patterns of metabolic scaling more generally.

4.1 INTRODUCTION TO ONTOGENETIC SCALING

The study of biological scaling is rooted in problems of growth. Indeed, early work on the relative growth both brought the term "allometry" into the biological lexicon and introduced the canonical form of the power law for quantifying allometric relationships (Huxley 1932; Huxley and Tessier 1936). These early studies of allometry were concerned primarily with patterns of relative growth, such as the relationship between brain size and body size among mammals.

Metabolic Ecology: A Scaling Approach, First Edition. Edited by Richard M. Sibly, James H. Brown, Astrid Kodric-Brown.
© 2012 John Wiley & Sons, Ltd. Published 2012 by John Wiley & Sons, Ltd.

However, from the beginning, biologists applied similar ideas to patterns in the physiology, ecology, and even behavior of organisms.

Individual organisms span an amazing size range, and during ontogeny, multicellular metazoans and plants face the challenge of maintaining physiological integration over large changes in size. For example, the ratio of the mass of a large tuna to a single tuna zygote is approximately 10^{12}; that is, the adult is one trillion times heavier than when it began life. Even on a more modest scale, a fifth instar tobacco hornworm, ready for pupation, is almost 10 000-fold heavier than the egg from which it hatched less than 3 weeks before. Because all animal growth is fueled by the metabolism of food, the primary factors affecting organism energetics (i.e., body size and temperature, according to the MTE) will also affect rates of ontogenetic growth, possibly in ways that are quite regular and predictable. Thus, as organisms grow, the resulting changes in size affect their capacity for further growth, and scaling principles provide a useful basis for modeling size-dependent changes in the energetics of growth.

When applying scaling principles to problems of growth and ontogeny, it is important to remember that three fundamentally different approaches to allometry can be distinguished, based principally on the units of data used to develop the allometric relationship. *Comparative* (also called *evolutionary* or *interspecific*) allometry studies how form or function changes across species that vary in size. Species are the unit of observation, and data generally consists of species mean values of adult size and the biological variable of interest. The scaling of basal metabolic rate to body mass in mammals (Kleiber's Law), which initially inspired the MTE, is the classic example. In contrast, *intraspecific* allometries apply to the variation across individuals (or populations) within a species, generally at the same developmental stage. Finally, *ontogenetic* allometry examines variation across or within individuals of a single species that vary in size due to growth and development (Fig. 4.1).

The patterns that arise in different sorts of allometric studies do not always agree, and generally, intraspecific and ontogenetic allometries exhibit more varied scaling exponents than taxonomically broad comparative allometries (Glazier 2005, 2006). Even across different developmental stages, a single organism can exhibit variation in metabolic scaling parameters (Yagi et al. 2010). In part, this reflects the fact that the observed variation in both size and metabolism

reflect different combinations of evolutionary, physiological, ecological, and developmental processes. Moreover, statistical methods for describing scaling relationships are often sensitive to the range of variation present in the data (Moses et al. 2008a; White, Xiao, Isaac, and Sibly, Chapter 1), leading to increased uncertainty and greater potential for bias when size ranges are small.

4.2 ALLOMETRIC MODELS OF ONTOGENETIC GROWTH

Most models of organismal growth begin with the balanced growth assumption, which applies the first law of thermodynamics (conservation of mass and energy) to biological systems. That is, any change in the size of the animal must result from the balance of material and energetic inputs and outputs, and

$$\text{Growth} = \text{Ingestion} - \text{Egestion} - \text{Excretion}$$
$$- \text{Respiration} - \text{Reproduction}$$

Here, ingestion is the gross intake of material through feeding, and egestion is material that passes through the animal unassimilated; thus the net assimilation rate is ingestion minus egestion. Excretion and respiration are the degraded material byproducts of metabolism and physiological maintenance, and reproduction is the total material allocated to gametes, embryos, and other reproductive investments. Generally, growth is modeled in terms of total biomass, and we will follow that convention here. However, it is important to remember that biological stoichiometry can play an important role in organismal growth (Elser et al. 2003) and can mediate the relationships between size, temperature, and energetics that are central to the MTE (Kaspari, Chapter 3). Moreover, note that this is simply a material budget, and that while it is expedient to classify material inputs and outputs in this way, in practice things are more complicated. For example, while we group the material byproducts of respiration (by mass mostly CO_2 and water vapor) as one term in the budget, the *energy* (i.e., ATP) produced by respiration and that fuels biochemical processes is related to all of the other terms in the model. For our purposes, what is most important is that this basic balanced growth model is cast entirely in terms of rates. Since most biological rates exhibit allometric scaling, we can think of growth trajectories as representing the balance of allometric

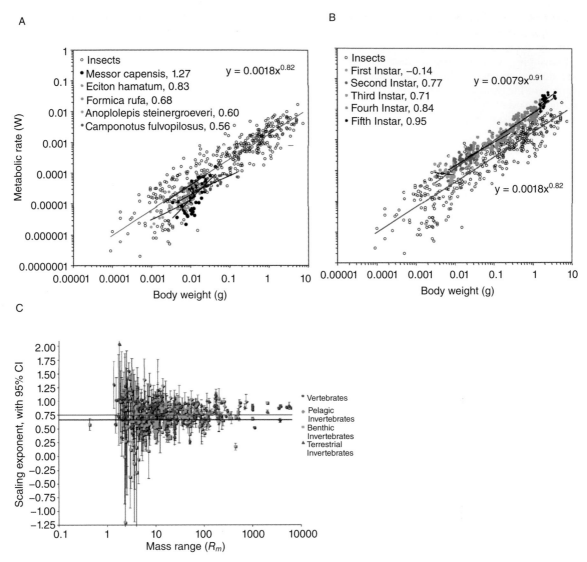

Figure 4.1 (A, B) The evolutionary (*interspecific*) allometric scaling of metabolic rate across 392 species of adult insects (open circles with blue line, compiled from the literature by Chown et al. 2007; by permission of John Wiley & Sons, Ltd). For comparison, (A) shows the *intraspecific* allometry for five species of ants with polymorphic castes, while (B) shows the *ontogenetic* allometry measured daily during development for 16 individual larvae of the tobacco hornworm, *Manduca sexta* (unpublished data from A. Boylan, H. Itagaki, and A. Kerkhoff). Separate scaling relationships were fit for each ant species and for each developmental stage (instar), with values for the scaling exponents recorded in the legend. Note that the differences in the height of the relationships (the scaling coefficient) in (B) are due to differences in temperature. The adult insect data were all corrected to 20 °C, while the *Manduca* larvae were reared at 27 °C. (C) (redrawn from Moses et al. 2008a) shows how variation in the estimate of interspecific scaling exponents is related to the range of sizes examined (R_m, the ratio $m_{min}:m_{max}$).

G = IN − EG − EX − RE − RP

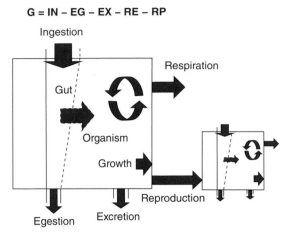

Figure 4.2 Conceptual model of balanced ontogenetic growth of a metazoan animal. Growth rate (G) is constrained by conservation of energy and mass to equal ingestion rate (IN) minus the rates of egestion (EG), excretion (EX), respiration (RE), and reproduction (RP).

changes in the various rates of input and output (Fig. 4.2).

So what do growth trajectories actually look like? Typically, they present a roughly sigmoid pattern, with accelerating growth early in development (i.e., near their initial mass, m_0) followed by a progressive slowing of growth as the organism approaches its asymptotic mass, M (Fig. 4.3). In some organisms (e.g., insect larvae), the curve appears nearly exponential, then abruptly ceases without any gradual slowing. Obviously, the sigmoidal pattern is similar to the pattern of logistic growth observed in natural populations growing under some sort of resource limitation. A generalized mathematical function that captures the pattern of sigmoid growth is

$$\frac{dm}{dt} = am^\alpha - bm^\beta \qquad (4.1)$$

where m is mass during ontogeny (and $\frac{dm}{dt}$ is thus the growth rate).

To conform to a sigmoid pattern of growth, the two allometric relationships (with coefficients a and b, and exponents α and β, respectively) must have the additional constraints that $a > bm_0^{(\alpha-\beta)}$ (which assures positive initial growth at $m = m_0$) and $\alpha < \beta$ (which leads to a cessation of growth at $m = M$). The asymptotic adult mass of the animal, M, is the point at which $0 = \frac{dm}{dt} = am^\alpha - bm^\beta$; thus $M = \left(\frac{a}{b}\right)^{\frac{1}{\beta-\alpha}}$. Likewise, the inflection point, i.e., the size at which the growth rate is maximized, is a constant fraction of the asymptotic mass, $\left(\frac{a}{b}\right)^{\frac{1}{\beta-\alpha}} M$ (Fig. 4.3).

Depending on the parameter values assigned (or derived), this general equation, which is sometimes called the Pütter equation, corresponds to several prominent models of ontogenetic growth (Ricklefs 2003). For example, if $\alpha = 1$ and $\beta = 2$, we have the classical logistic growth model. However, while the logistic model sometimes provides a reasonable fit to growth trajectory data, it is difficult to assign biological meaning to the model parameters based on the balanced growth assumption, since we know that most of the biological rates and times involved tend to exhibit scaling exponents less than 1.

Note that the overall form of the Pütter equation matches the principle derived from the balanced growth assumption, that growth rate represents the balance of allometric changes in the rates of input and output. However, the complex relationships between the components of the material budget and the pathways of energy allocation used to fuel the component processes (Hou et al. 2008) make it difficult to unambiguously map this equation onto the balanced growth assumption.

We next review several alternative models that have been derived which share this common form, but differ in their biological interpretation of the component terms.

4.2.1 The Bertalanffy and Reiss models

Ludwig von Bertalanffy (Von Bertalanffy 1951, 1957) developed a model of ontogenetic growth based on the balance of anabolic processes (represented by am^α) and catabolic processes (bm^β). Initially, he assumed that anabolic processes were limited by the surface area over which organisms assimilate resources, and that the costs of catabolism were directly proportional to the mass of the animal, i.e., $\beta = 1$. Because mass is proportional to volume (length cubed), whereas areas are length squared, the surface area assumption leads to $\alpha = 2/3$. The resulting model

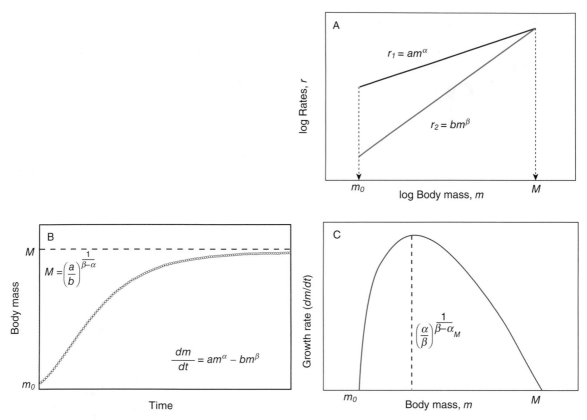

Figure 4.3 Behavior of ontogenetic growth models taking the form of the Pütter equation (equation 4.1). (A) The difference in scaling between the two components of the growth equation. Growth trajectories (B) are sigmoidal under this model, with asymptote M, while growth rate (C) peaks at a constant fraction of the asymptotic mass.

$$\frac{dm}{dt} = am^{2/3} - bm$$

has been applied widely, especially to the growth of marine fishes.

Subsequently, Bertalanffy recognized different "growth types" distinguished by whether their metabolic rates scaled with surface area ($\alpha = 2/3$), mass ($\alpha = 1$), or in some "intermediate" way (e.g., $\alpha = 3/4$). This generalization of the growth model provides a flexible tool for modeling different relationships between growth and metabolism, but it begs the question of what makes the "growth types" different. In all cases, however, Bertalanffy assumed that the exponent α reflected the scaling of metabolic rate.

In 1989, Michael Reiss developed an alternative interpretation, still based on the structure of the Pütter equation (Reiss 1989). He assumed that, instead of the

balance of anabolism and catabolism, the first term represents the scaling of assimilation (which is ingestion minus egestion in the balanced growth model) while the second term represents the metabolic cost of living, which couples respiration and excretion. While Reiss recommends parameterizing the resulting model based on empirical allometric relationships for particular taxa, he utilizes $\alpha = 2/3$ and $\beta = 3/4$ as a general solution, based on a broad survey of empirical allometries for ingestion rates and metabolic rates, respectively.

Note that although the forms of their models are quite similar, Bertalanffy assumes that metabolic allometry is reflected in the first term, while Reiss assumes it is reflected in the second term. Thus, even if the two models provide reliable fits to empirical growth data, their biological interpretations are very different.

4.2.2 The West et al. model, extending MTE

More recently, in 2001, West et al. presented a model that, while conforming to the same overall form, had yet another derivation (West et al. 2001). They began with the assumption that the total metabolic rate of an organism (B) is simply the sum of the energy devoted to growth (i.e., the synthesis of new biomass) plus the energy devoted to maintaining existing biomass

$$B = E_m \frac{dm}{dt} + B_m m \qquad (4.2)$$

where E_m is the energy required to synthesize a unit of biomass (e.g., in Jg^{-1}), and B_m is the metabolic rate required to maintain a unit of biomass (e.g., in Wg^{-1}). Based on an assumed equilibrium between the energetic (metabolic) demand and energy assimilation (ingestion minus egestion), West et al. also refer to B as the "incoming rate of energy flow" (West et al. 2001). Rearranging to solve for the growth rate, and taking into account the allometry of metabolic rate, $B = b_0 m^\alpha$, they arrive at a Pütter-style model $\frac{dm}{dt} = am^\alpha - bm$ where $a = \frac{b_0}{E_m}$ and $b = \frac{B_m}{E_m}$. Thus, a is the ratio of the metabolic scaling coefficient (b_0) to the cost of biomass synthesis, while b provides a measure of tissue turnover rate. Explicitly, West et al. also assumed that costs of synthesis (E_m) and maintenance (B_m) are independent of size. In the original derivation, they made the additional assumption that $\alpha = 3/4$, but later versions were generalized to any metabolic scaling exponent (Moses et al. 2008a).

4.2.3 Model predictions, critiques, and extensions

While the West et al. model is almost identical in form to Bertalanffy's and Reiss's, the important advance is that it provides unambiguous (and thus testable) biological interpretations not just of the exponent (α) of the growth model (i.e., it should match the observed exponent for metabolism), but also of the two coefficients (a and b), which are related to the mass-specific metabolic parameters (b_0 and B_m), and the energetic costs of biomass synthesis (E_m).

Initially, West et al. estimated E_m from the energy content of vertebrate tissue, but this fails to take into account the energetic costs of constructing tissue as well as the energetic content of the organic building blocks used to construct it, i.e., yolk or food (Makarieva et al. 2004). However, subsequent estimates of E_m from parameterized growth models for a wide variety of embryonic and juvenile vertebrates (Moses et al. 2008a) fall into a range of $800-13\,000\,Jg^{-1}$, which is quite similar to the range of empirical estimates ($1000-9200\,Jg^{-1}$) drawn from studies of embryo energetics. The authors attribute the high degree of variation in the costs of biosynthesis to differences in tissue composition of different species, as well as species-specific patterns of metabolic scaling (i.e., variation in α) and ontogenetic changes in tissue water content (Moses et al. 2008a).

Based on their application of scaling principles to patterns of ontogenetic growth, West et al. (2001) also demonstrate that individual growth trajectories are simply rescaled examples of a more general process. Specifically, when the renormalized, dimensionless mass ratio $r \equiv (m/M)^{1/4}$ is plotted as a function of the dimensionless time variable $\tau = at/4M^{1/4} - \ln[1 - (m_0/M)^{1/4}]$, all taxa should follow a single, universal growth curve. For animals as different as cod, shrimp, chicken, and dog, this appears to be the case (Fig. 4.4). Thus, in addition to providing powerful predictive tools, the model suggests an elegant unity underlying the fascinating diversity of life.

Based on the formal similarities to previous efforts to model growth trajectories, several critics were quick to point out that, while the fits of the model to data from a wide variety of species were quite impressive, support for the details of the model, and especially for the universality of the 3/4 metabolic exponent, could not be drawn from fits of growth trajectories themselves (Banavar et al. 2002; Ricklefs 2003). Furthermore, the ability to rescale growth trajectories to a universal curve is not unique to the West et al. model (Banavar et al. 2002) and, in fact, Bertalanffy had pursued a similar exercise himself (Von Bertalanffy 1951, 1957). This is an important point, because the fit of growth trajectories to a particular model cannot be taken as support for a particular explanatory model of metabolic scaling. More generally, distinguishing between alternative models of ontogenetic growth cannot be accomplished simply by comparing their fit to growth trajectories alone. Instead, their underlying assumptions must be addressed.

In the case of the West et al. ontogenetic growth model, the most contentious assumption has been that

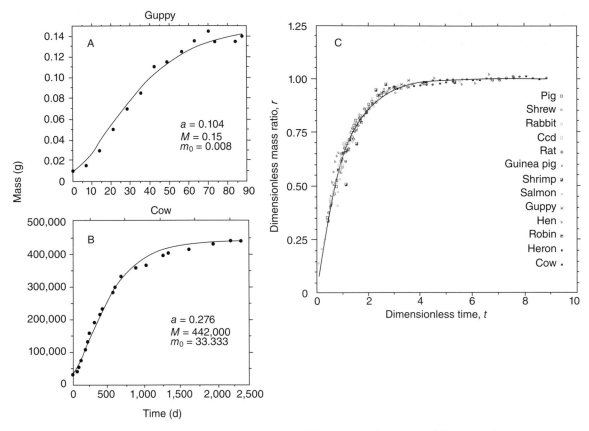

Figure 4.4 (A, B) Ontogenetic growth trajectories and fitted model parameters for two very different vertebrates, the guppy and the cow. (C) The "universal growth curve" of dimensionless mass and time variables calculated for a wide variety of taxa exhibiting both determinate and indeterminate growth. All fall quite near the same curve, $r = 1 - e^{-t}$ (all redrawn from West et al. 2001 by permission of Nature Publishing Group).

the 3/4-power scaling applies universally within, as well as between, species (Ricklefs 2003; Glazier 2005). Subsequent reassessments of the data support an overall intraspecific average of c. 3/4, especially for organisms growing over a substantial size range (Moses et al. 2008a). However, the considerable variation present across intraspecific metabolic scaling relationships has both phylogenetic and ecological components (Von Bertalanffy 1957; Glazier 2005). This suggests that generalizations of the model to incorporate variation in the metabolic scaling exponent (Moses et al. 2008a) and a fuller partitioning of energy assimilation and allocation (Hou et al. 2008) may provide a more flexible theory for understanding variation in patterns of growth across diverse taxa.

Interestingly, two other fundamental assumptions of the West et al. model have received almost no attention. First, like Bertalanffy, they assume that the scaling exponent in am^{α} reflects the intraspecific scaling of metabolic rate, even if $\alpha \neq 3/4$, but this assumption remains untested. Second, it may be questionable to assume that maintenance costs (B_m) are independent of size over ontogeny (Makarieva et al. 2004). Testing these assumptions against, for example, Reiss's assumptions that bm^{β} reflects metabolic scaling and that maintenance costs vary with size, requires careful measurements of metabolic rate, growth rate, and maintenance costs (e.g., protein turnover rates or protein:rRNA ratios) on the same animals, and thus must be pursued only on a species-by-species basis.

However, doing so provides a strong test of one of the fundamental implicit assumptions of the MTE more generally, i.e., that the scaling of most biological rates and times reflects the scaling of metabolism.

4.3 INFLUENCES OF TEMPERATURE AND STOICHIOMETRY

Incorporation of the universal temperature dependence component of MTE has focused on growth rates at sizes well below the asymptotic, adult size (M), under the assumption that the maintenance term (bm) is negligible under these conditions. In this case, the model reduces to a form that is directly analogous to the MTE central equation (Gillooly et al. 2002)

$$\frac{dm}{dt} = am^{3/4}e^{\frac{-E}{kT}} \tag{4.3}$$

Predictions from this model have been tested (Gillooly et al. 2002) using data on the embryonic development of a wide variety of terrestrial invertebrate and zooplankton taxa. Many of the concerns put forward about the temperature dependence of metabolism (Brown and Sibly, Chapter 2) also apply to ontogenetic growth (Clarke 2004). Moreover, although developmental rates and times respond to temperature roughly as predicted in many laboratory studies, the temperature–growth rate relationship in natural populations is mediated by issues of diet quality and the stoichiometric composition of the growing animals (Gillooly et al. 2002; Woods et al. 2003; Makarieva et al. 2004; Doi et al. 2010).

Another challenge in applying MTE to growth and ontogeny is the fact that many ectotherms exhibit the "temperature–size rule," growing more quickly, but to a smaller asymptotic size, at higher temperatures (Atkinson 1994; Kingsolver and Huey 2008). This pattern is inconsistent with the MTE assumption, because in the model's current form, M is implicitly independent of temperature, since the two terms in the model share the same temperature dependence. Instead, the temperature–size rule requires either systematic variation in the temperature dependence of different ontogenetic processes, for example, growth vs. differentiation (van der Have and de Jong 1996; Walters and Hassall 2006) or evolutionary trade-offs affecting selection on adult size (Atkinson and Sibly

1997; Kozlowski et al. 2004). Both of these factors are unaccounted for in the current form of the MTE, and incorporating them requires loosening the temperature dependence assumption.

4.4 CAVEATS AND CONCLUSIONS

4.4.1 A caveat on reproduction

In the presentation above, we have ignored one component of the balanced growth model that is critical both for the persistence of organisms and for understanding ontogenetic growth models in an evolutionary context: reproduction. From birth to the age (and size) at first reproduction, all of the models outlined above apply directly. However, once the organism begins reproducing, the parameters of the equations need to be adjusted to reflect the allocation to reproduction vs. growth, because, due to the balanced growth assumption, any materials and energy allocated to reproduction are by definition unavailable for growth. Including this fundamental biological process in the model requires mathematical modifications, the details of which depend on whether the organism in question exhibits determinate or indeterminate growth.

West et al. (2001) make a number of additional assumptions about reproduction in the context of their model. In the case of so-called "determinate" growth (as observed in most mammals and birds), the size at maturity corresponds to the asymptotic size, M, so all growth takes place before reproduction begins and reproductive investments are either assumed to be supported by an upregulation of assimilatory metabolism (represented by temporal variation in a) or to trade-off against the costs of maintenance (e.g., tissue turnover, b). In contrast, so-called "indeterminate" growers, including many fish and some invertebrates, continue to grow after first reproduction. For these creatures, the issue is more complicated, because allocation to reproduction can trade off against both growth and maintenance (Sibly, Chapter 5). Thus, indeterminate growers may slow growth well below their asymptotic size (West et al. 2001). Fuller theoretical and empirical synthesis between studies of metabolism, growth, and life history, as pursued by the MTE, are key to a more integrated understanding of biology and ecology.

4.4.2 Ontogeny entails both growth and development

Strictly speaking, all of the models described above, cast purely in terms of mass, only describe one aspect of ontogeny: growth. However, ontogeny also entails development and the differentiation of cells into different functional tissues, organs, and systems. Growth is directly fueled by metabolism and should thus be related to metabolic rate as in the models described above, but differentiation is at best indirectly related to metabolism. Moreover, because tissues differ in composition, relative growth rates, and energetic demand, developmental changes have the potential to feed back on metabolic scaling and the overall dynamics of ontogeny. For example, larval tiger puffer fish (*Tagifuku rubripes*) spanning more than three orders of magnitude in body size exhibit regular shifts in their metabolic scaling coefficients (intercepts) associated with discrete developmental changes in morphology and behavior (Yagi et al. 2010). Thus, whereas growth can be described in terms of simple power-law scaling relationships, modeling differentiation may be more complicated.

On one level, this sort of pattern challenges the generality of the MTE approach: such simple models as equations 4.1 to 4.3 fail to capture potentially interesting aspects of ontogenetic growth which, in the case of the tiger puffers, have measurable consequences for mortality due to cannibalism (Yagi et al. 2010). Clearly, this sort of detail is not part of what such generalized models are intended to describe, so it is tempting to dismiss results like these as "special cases." However, on another level, results that challenge the generality of the theory provide an opportunity to develop a better understanding of the physiological, ecological, and evolutionary drivers of, for example, changes in the value of scaling exponents and coefficients. Thus, they should be embraced, rather than dismissed.

The simplest first step towards bringing a fuller consideration of development and differentiation into models of ontogenetic growth is to fundamentally distinguish between two types of growth: cell expansion vs. cell proliferation. The two processes can have different effects on the scaling of metabolism and growth through their differential effects on the surface area to volume ratio of cells (Kozlowski et al. 2003). Most generally, assuming a constant cellular metabolic rate *in vivo*, size increase purely by cell proliferation provides an expectation of $B \propto m$. The additional assumption that the cellular metabolic rate becomes surface-limited at large sizes leads to an expectation of $B \propto m^{2/3}$ for growth by pure cell expansion. Previous work has considered the ramifications of these opposing methods of size increase for producing intermediate exponent values (e.g., 0.75) in studies of metabolic scaling (Kozlowski et al. 2003; Chown et al. 2007), but the two processes have not been explicitly incorporated into an ontogenetic growth model.

4.4.3 Conclusions

Scaling approaches in biology have their origin in the study of growth processes, and extensions of MTE into the realm of ontogenetic growth join a field with a rich theoretical history linking metabolic scaling to larger-scale biological processes. The relationships among size, temperature, metabolic rate, and growth are largely consistent with models derived from the MTE, but observed patterns of growth, by themselves, cannot distinguish between multiple alternative models. The ontogenetic scaling of metabolism is substantially more variable than interspecific scaling patterns observed over broad domains of taxa and size, which challenges the "canonical" 3/4-power assumed by the MTE. Likewise, the particular form of temperature dependence assumed by MTE is inconsistent with the widely observed "temperature–size rule," which seems to be as ubiquitous as many scaling phenomena. At the same time, variability in patterns of ontogenetic scaling and growth provide unique opportunities for understanding how specific physiological, ecological, and evolutionary factors impact the relationships between size, temperature, stoichiometry, and energetics that are at the core of the MTE. Thus, careful studies of the underlying assumptions of multiple, competing models may provide a means of understanding not just the process of growth and ontogeny, but also the processes that underlie patterns of metabolic scaling more generally.

Chapter 5

LIFE HISTORY

Richard M. Sibly

SUMMARY

1 The life history of an organism is a record of the ages at which it reproduces and the numbers and sizes of offspring then produced, together with age-specific mortality rates. I review mechanistic explanations of the covariation of life-history traits with body size and temperature.

2 The metabolic theory of ecology (MTE) holds that metabolic rate sets the rate of resource allocation to the processes of survival, growth, and reproduction. As a consequence, biological rates, such as birth rates, the rate of biomass production, developmental rates, and population growth rate, should scale allometrically to the $-1/4$ power of body mass and the -0.65 power of inverse absolute temperature. Biological times should scale to the $+1/4$ power of body mass and the $+0.65$ power of inverse temperature.

3 A broad mass of data on plants, invertebrates, fish, mammals, and birds generally supports MTE predictions, and there is at present no viable alternative framework within which to interpret these results.

4 MTE provides a basic mechanistic explanation for why larger organisms and those with lower body temperatures grow more slowly, reproduce later, and are less productive when they do reproduce. The explanation is that production rate is selected to increase but is held back by constraints arising from the laws of physics, chemistry, and biology. Specifically, productivity is limited by metabolic rate because of logistical constraints in supplying oxygen and other resources around the bodies of individual organisms.

5 Despite some successes, further work is required to achieve a secure mechanistic explanation of the factors constraining life-history evolution.

5.1 INTRODUCTION

Compared to mice, elephants take a long time to reach adult size, and then they breed less frequently, but they live longer. These are particular cases of general scaling relationships between life-history traits and body size. The scaling relationships are, in general, quite tight, as will be seen. This suggests they must have mechanistic explanations. Strangely, prior to MTE there were few ideas as to what these mechanisms might be.

I begin by defining what is meant by the term life history, and then give an outline of the predictions made by the metabolic theory of ecology and other theories as to how life-history variables should scale with body mass and temperature. Then I assemble data that shows to what extent these predictions are borne out in practice.

Metabolic Ecology: A Scaling Approach, First Edition. Edited by Richard M. Sibly, James H. Brown, Astrid Kodric-Brown.
© 2012 John Wiley & Sons, Ltd. Published 2012 by John Wiley & Sons, Ltd.

5.2   WHAT IS A LIFE HISTORY?

The life history of an organism is a record of the ages at which it reproduces and the numbers of offspring then produced. To this are added the chances of survival to each age. A schematic representation of a simple life history is shown in Figure 5.1.

Some of the factors that shape life histories will now be briefly described. In the classical view (Sibly and Calow 1986; Roff 1992; Stearns 1992) the resources acquired by an organism are allocated separately to maintenance, growth, and reproduction, as shown in Figure 5.2. The total available for allocation is limited by the amount the animal eats, so if more is allocated to one function, less is available for others. I shall refer to this idea as the Principle of Allocation. Allocation of resources to maintenance is needed to allow the organism to survive, and survival chances may be further improved by additional investment in personal defense in the form of spines, shells, toxins, anti-predator behavior, and so on. According to the Principle of Allocation, allocated resources have separate destinations. However, in reality some overlap must occur and it would be desirable to refine the model based on measurements that could now be made using proteomic technologies.

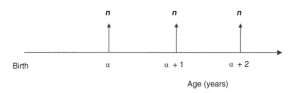

Figure 5.1 Simple life history, followed by many birds and female mammals. The animal breeds for the first time at age α and subsequently annually or at regular intervals, producing n offspring each time it breeds. In life tables, a record of juvenile and adult mortality rates is also kept. All parameters in a life table are specific to the environment in which they were collected.

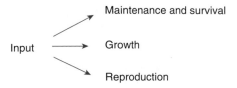

Figure 5.2 The principle of allocation. A variant of this diagram appears as Figure 4.2.

5.3   PREDICTING HOW LIFE HISTORIES CHANGE WITH BODY SIZE

Surprisingly little attention has been given to the question of how life histories might be expected to vary with body size. That they do vary systematically with body size has become increasingly evident over the last 40 years, as is described in section 5.4. We start by considering the predictions of MTE.

5.3.1   Metabolic theory of ecology

The metabolic theory of ecology (MTE) holds that (1) the laws of physics, chemistry, and biology cause mass-specific metabolic rate to scale with body mass as the negative 1/4 power, and with temperature according to the Arrhenius equation, so that the mass-specific rate of metabolism, B, is proportional to $M^{-1/4}e^{-E/kT}$, where M is body mass and T is absolute temperature (see Brown and Sibly, Chapter 2, for derivation, and Glossary for definitions of parameters E and k); (2) metabolic rate sets the rate of resource allocation to the processes of survival, growth, and reproduction; (3) so biological rates, such as birth rates, the rate of biomass production, developmental rates, and population growth rate, should scale as $M^{-1/4}e^{-E/kT}$, and biological times, such as duration of pregnancy and lactation and lifespan, being the reciprocals of rates, should scale as $M^{1/4}e^{E/kT}$ (Brown et al. 2004). So MTE makes predictions as to how life-history rates and times should scale with body mass and temperature.

In testing the predictions of MTE, the logarithm of a biological rate has generally been plotted against the logarithm of body mass and against $1/kT$. The MTE predictions are that the fitted regression coefficients will be $-1/4$ and $-E$, respectively, where the value of E will be ~0.65 eV (see Brown and Sibly, Chapter 2).

Application of MTE to life histories is based on the assumption that resources are allocated in fixed proportions to survival, growth, and reproduction. This is a special case of the Principle of Allocation (Fig. 5.2). Thus, in particular, if all species allocated the same proportion of input to reproduction, then the rate of production of reproductive biomass would scale as $M^{-3/4}e^{-E/kT}$. And population growth rate would scale similarly if it is determined by the rate of production of biomass and/or the age of first reproduction (Savage et al. 2004a; Duncan et al. 2007). Prediction of the scaling of mortality rate is more complicated. However, since most populations are neither continuously increasing nor decreasing, mortality rates must equal birth rates in the long term (Peters 1983; Sutherland et al. 1986; Sibly and Calow 1987) and so should overall be proportional to $M^{-1/4}e^{-E/kT}$ (Brown et al. 2004; Brown and Sibly 2006).

5.3.2 Predictions from other theories

Science advances fastest when rival plausible theories produce competing testable predictions, but so far we only have precise predictions for the scaling of life-history characters from MTE. Two previous studies have discussed what the mechanistic basis for quarter-power scalings of life-history traits might be. Charnov (1993) suggested that for each species the fundamental relationship is that between body size and adult mortality rate. Evolution then optimizes the life history, which is subject to life-history trade-offs deriving, for example, from the Principle of Allocation, and this leads to quarter-power scaling. However, other scaling relationships would also be possible using this approach. Kozlowski and Weiner (1997) also viewed life-history evolution as a process operating at the species level to optimize life-history characters and allocation patterns in the face of physiological and ecological constraints and trade-offs. Kozlowski and Weiner considered the implications for understanding interspecific allometries and showed that certain probability distributions of assumed constraint parameters yield results compatible with observations. While Charnov (1993) and Kozlowski and Weiner (1997) are undoubtedly correct that the evolutionary process operates at the species level, neither approach has so far yielded precise a priori predictions for the expected form of life-history allometries.

5.4 DATA

5.4.1 Historical background

It has been known for over 40 years that life-history characters vary systematically with body size. In an important early paper on life-history allometry, Fenchel (1974) showed that the maximum rate at which populations can grow is closely related to body size: populations of small-bodied animals grow faster than those with larger bodies. Once Fenchel (1974) had shown

that maximum population growth rate scaled with body size, it was only a matter of time before others started looking at the scaling of the life-history traits that together determine population growth rate. These life-history traits are age-specific birth rates, growth rates, and death rates. Birth rates are partly constrained by gestation times, growth rates determine age at first reproduction, and death rates determine lifespans. Blueweiss et al. (1978) provided an important early review of the allometries of the life-history traits of ectotherms, mammals, and birds and Western (1979) looked in more detail at the life-history allometries of the mammals. Subsequently, Peters (1983) supplied a superb compilation of early work on the allometries of life histories in book form, and Calder (1984) provided many thoughtful insights into why life-history allometries scale as they do. These early studies strongly influenced development of the MTE.

The data in these studies comes from different species, i.e., they are comparative analyses. But some species are closely related and so are not as statistically independent as distant relatives. In comparative analyses it is now accepted that species should ideally be weighted according to their phylogenetic independence (see White, Xiao, Isaac, and Sibly, Chapter 1). As phylogenies become available and analysis techniques are developed, phylogenetic information is becoming more widely used in comparative life-history analysis (see, e.g., (Gittleman 1986; Harvey and Pagel 1991; Purvis and Harvey 1995).

These early studies were not concerned with testing theoretical predictions about the scaling of life histories; indeed, there were no theoretical predictions to test. In the next sections we turn to recent work explicitly testing MTE predictions.

5.4.2 Development rate in zooplankton and fish

MTE predicts how life-history characters vary in relation to both body mass and temperature as described in section 5.3. Ideally each relationship would be plotted as a three-dimensional graph showing how each life-history character varies in relation to both body mass and temperature. Three-dimensional graphs are often hard to read, however, so here I show separate plots first in relation to body mass, where developmental rates have been adjusted for body temperature by multiplying by $e^{E/kT}$, and second in relation to body

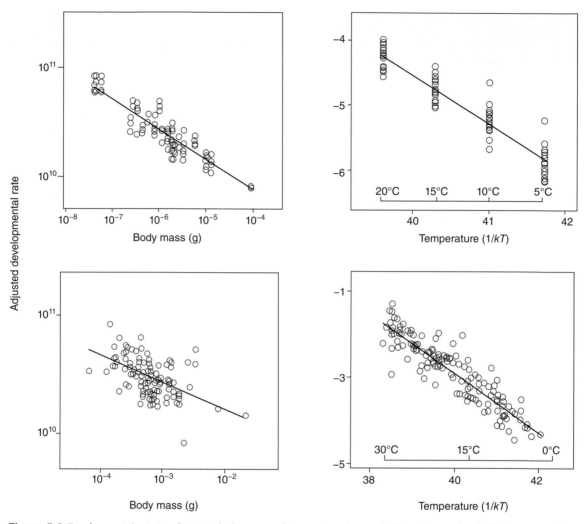

Figure 5.3 Developmental rate in relation to body mass and temperature for zooplankton (top row) and fish (bottom row). Developmental rate was adjusted for body mass and temperature by multiplying by $M^{1/4}$ and $e^{E/kT}$ respectively. (From Brown et al. 2004.)

temperature, where development rates were adjusted for body mass by multiplying by $M^{1/4}$. I begin by considering development rates of zooplankton and fish (Fig. 5.3). MTE predicts that the regression coefficient of adjusted developmental rate vs. body mass should be $-1/4$. The observed regression coefficients of -0.27 for zooplankton and -0.22 for fish were both quite close to the prediction. For temperature measured as $1/kT$, MTE predicts a regression coefficient of -0.65 and the observed regression coefficient is -0.73 for zooplankton and -0.68 for fish, again, quite close to

the prediction. To obtain an overview without stumbling on detail, statistical evaluations of goodness-of-fit are here omitted: for these the original papers must be consulted.

Similar relationships were found in the egg development times of zooplanktonic and free-swimming ectotherms from marine waters (Hirst and Lopez-Urrutia 2006). Furthermore, once mass and temperature effects were accounted for, additional lifestyle differences became apparent. Eggs that are not protected hatch three times sooner than those that are protected, by their mothers or attached to a substrate or floating in clumped masses. The authors suggest that evolution has shortened the vulnerable development period of unprotected eggs, which typically have high mortality rates.

5.4.3 Timing of the life history in mammals

The scaling of life histories of placental, marsupial, and monotreme mammals has been described by Hamilton et al. (2011). These mammal lineages diverged more than 100 million years ago and diversified rapidly on different landmasses as the supercontinent Pangaea fragmented and drifted apart. As a consequence of their independent evolution, the extant members of these lineages differ conspicuously in physiology, life history, and reproductive ecology and the differences in life history are dramatic. Monotremes lay eggs, and the energy reserves in the eggs support embryonic development until the eggs hatch. Hatching of the eggs is the equivalent of birth, and thereafter the mothers supply milk to fuel growth from hatching to independence. Marsupials have a very short period of embryonic development in the uterus, so they are born small and undeveloped. Then they are protected and lactated in a pouch for a long period before they become independent. By contrast, placentals have a more prolonged period of embryonic development in the uterus and so are born at larger size. Then, like the other lineages, they too are nourished by lactation until they become independent.

The timings of major life-history events in marsupials, placentals, and three species of monotremes are shown in Figure 5.4. The prediction of MTE is that the allometric slopes are 1/4. Slopes generally are quite close to the predicted 1/4, except for gestation time and age of first reproduction in marsupials, which are shallower. The slope for age of first reproduction in placentals is close to 1/4, as was also reported in a phylogenetic general linear model by Lovegrove (2009). Duncan et al. (2007) found it to be significantly less than 1/4, but they did not distinguish marsupials from placentals and so did not pick up the difference shown in Figure 5.4D.

Why are the scalings of gestation time and age of first reproduction so shallow in marsupials? They are shallow because neonates are born at more or less the same size, prior to transfer to the pouch, whatever the size of the mother. The reason for this is not known. It is tempting to speculate that it might be advantageous in the perhaps unpredictable environments of Australia, allowing offspring to be aborted during hard times more readily by a marsupial than a placental mother with her much longer gestation time. This idea can be discounted however, because there are now nearly as many placental as marsupial species in Australia.

Contrary to MTE predictions, Duncan et al. (2007) have shown variation between the mammal orders in the allometry of age of first reproduction. The Chiroptera (bats), Cetacea (whales, dolphins, and porpoises), and Eulipotyphla (shrews, moles, and hedgehogs) have low scaling exponents. The authors speculate this may result from adaptations to life in the air or the sea, or from the need to maintain a high metabolic rate, but it is not clear how this comes about mechanistically. It is hoped that further studies will provide answers to some of these questions.

5.4.4 Production rate

The rate at which a mother produces reproductive biomass is here termed production rate.

Production rates in relation to body mass are shown in Figure 5.5. Figure 5.5A shows temperature-adjusted production rate for a variety of endotherms and ectotherms. The overall exponent is 0.76, close to the MTE prediction of 3/4. Figure 5.5B shows mass-specific production rates of placental, marsupial, and monotreme mammals. Here production was calculated in terms of weaning mass, and mass-specific means that the whole organism production rate has been divided by the body mass of the mother. As predicted by MTE, the exponents are the same for the different lineages. This is a striking result, bearing in mind the discrepancies in life histories seen in Figure 5.2. However, the slope is not −1/4 as predicted by MTE. Instead, it is −0.37. Discussing this discrepancy, Hamilton et al. (2011) note that weaning mass is a component of production rate, but is not a linear function of adult mass, scaling instead with an exponent of 0.87 for marsupials and 0.90 for placentals. They suggest this might account for the discrepancy. The authors offer three hypotheses that might explain why relative weaning mass scales negatively: (1) Larger mammals differentially allocate resources to adult survivorship at the expense of weaning mass. This could be particularly advantageous in seasonal environments, where it helps adults

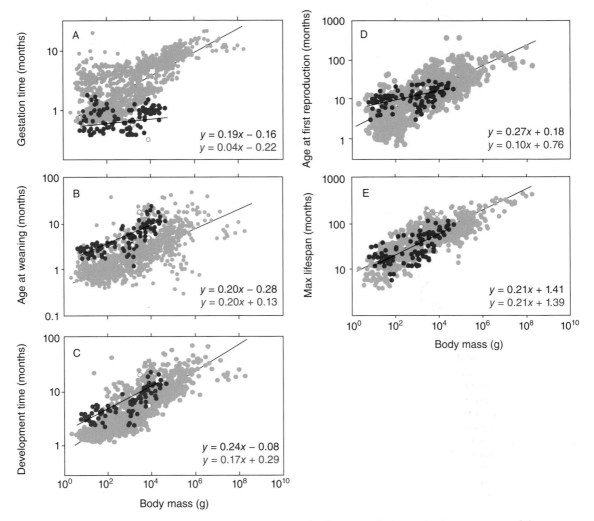

Figure 5.4 The timings of major life-history events in marsupials, shown in red, placentals, shown in gray, and three monotremes (open circles). Times are measured in months. (A) Gestation time, (B) age at weaning; (C) development time; (D) age at first reproduction; (E) maximum lifespan. (From Hamilton et al. 2011.)

to survive the winter or dry season. (2) Seasonality may place additional demands on small mammals that breed more than once a year. By producing relatively large offspring, mothers increase the chance that their offspring will survive stressful periods. (3) Fueling reproduction from stored reserves is increasingly common in larger mammals but is costly and this may favor reduced allocation to offspring. In sum, allocation of resources between mother and offspring is subject to the Principle of Allocation shown in Figure 5.2, and the resultant trade-offs may favor reduced size

at weaning in larger mammals. This may be part of the explanation why the allometric slope of mass-specific production in mammals is less than the MTE prediction of −0.25.

5.4.5 Mortality rates

Direct measurement of mortality rates in the field is difficult and has only rarely been accomplished. In mammals, for example, Promislow and Harvey (1990)

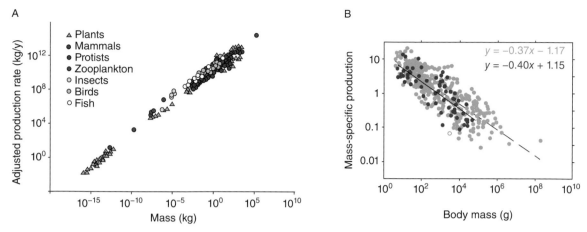

Figure 5.5 Production rate in relation to body mass. (A) Production rate, adjusted for temperature, of a variety of endotherms and ectotherms. Production rate is adjusted for temperature by multiplying by $e^{E/kT}$. Exponents for the different lineages are statistically indisinguishable, the overall exponent was 0.76. (From Ernest et al. 2003.) (B) Mass-specific production of mammals plotted using the conventions of Fig. 5.3. Thus marsupials are shown in red and placentals in gray, and a single monotreme is represented by an open circle. Slopes did not differ statistically between lineages, the overall slope was −0.37. (From Hamilton et al. 2011.)

found relevant field data for only 48 species. One problem is that mortality rates depend conspicuously on the environment in which the organism lives – some places are more dangerous than others. Maximum lifespans in captivity are, however, more easily recorded, and these can be used to estimate minimum mortality rates, as the reciprocal of maximum lifespan. An "average" mortality rate can be calculated if one accepts a suggestion in the literature that average mortality rates in endotherms are some two and a half times greater than minimum mortality rates. McCoy and Gillooly (2008) used this method to test the MTE prediction that mortality rates are proportional to $M^{-1/4}e^{-E/kT}$.

Mortality rates, temperature-adjusted where necessary, are shown in relation to body mass for invertebrates, fish, multicellular plants, phytoplankton, mammals, and birds in Figure 5.6. In general the slopes are quite close to −1/4. There is some variation about the fitted lines, but this is expected, given the difficulties in estimating mortality rates, and in any case is hardly more than in the other traits examined in Figures 5.3 to 5.5.

Mortality rates are shown in relation to temperature for invertebrates and fish in Figure 5.7. There is again some variation, but the fitted slope is remarkably close to the MTE prediction of −0.65. However Gislason et al. (2010), analysing natural mortality rates from

168 stocks of fish from marine and brackish waters, found a slope of −0.39. They attributed the difference to a correction factor for cold adaptation applied to the data analysed by McCoy and Gillooly (2008, 2009).

Mortality rates are easier to measure in plants than animals, because plants do not move around so much. Individual sessile algae and vascular plants can be tracked with great accuracy. Marba et al. (2007) examined plant mortality rates in 728 plant species ranging from tiny phytoplankton to large trees and found that mortality rates varied with plant size with a slope very close to the predicted −1/4 power. Birth rates, estimated independently for 293 species, scaled the same way as expected. However, contrary to MTE predictions, no temperature dependence was found in either mortality or birth rates. Marba et al. (2007) attributed their inability to detect temperature dependence to the limited variation in temperature that occurred between their study species in field conditions. They suggest that experimental studies are needed to identify temperature dependences and note how important this is for prediction of responses to global warming.

5.4.6 Maximum population growth rate

The patterns of reproduction and survival, i.e., the life histories of organisms, determine the rate at which

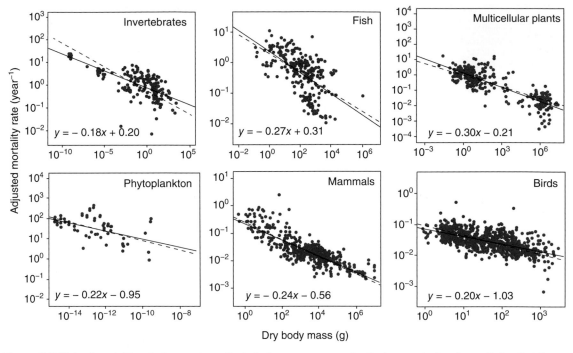

Figure 5.6 Plots of mortality rate, temperature adjusted where necessary, against body mass, testing the prediction that the allometric slope is −1/4 (dashed lines). Fitted regression equations are shown in each panel. Temperature adjustment was carried out by multiplying mortality rate by by $e^{E/kT}$. (From McCoy and Gillooly 2008, 2009 by permission of John Wiley & Sons, Ltd.)

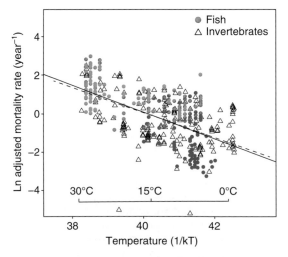

Figure 5.7 Plot of mortality rate, body mass adjusted, against temperature, for fish and invertebrates, testing the MTE prediction that the slope is −0.65 eV. The dashed line shows the MTE prediction, the solid line is fitted to the data by ANCOVA. Body mass adjustment was carried out by multiplying mortality rate by $M^{1/4}$. (From McCoy and Gillooly 2008, 2009 by permission of John Wiley & Sons, Ltd.)

their populations can increase. The maximum population growth rate occurs in ideal environmental conditions and is written r_{max}. Since smaller species generally have high reproductive rates, one would expect them to be able to increase in numbers faster than larger species. Fenchel (1974) showed that r_{max} is closely related to body size and subsequent studies amply confirmed his findings. Some examples are shown in Figure 5.8. The slope of r_{max} in relation to body mass, temperature-adjusted where necessary, is very close to the MTE prediction of −1/4 (Figs 5.8A, C; see also Karasov, Chapter 17; Isaac, Carbone, and McGill, Chapter 7), and the slope of r_{max} in relation to temperature, body-mass adjusted where necessary, is very close to the MTE prediction of −0.65 (Fig. 5.8B).

5.5 EVOLUTIONARY CONSIDERATIONS

Allocation of resources to growth and reproduction is powered by metabolism. MTE postulates that resources are allocated in fixed proportions to survival, growth,

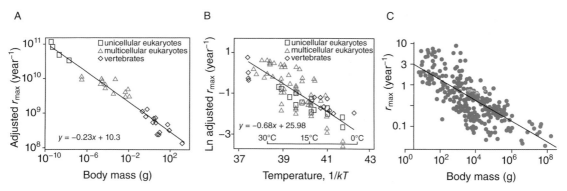

Figure 5.8 Maximum population growth rate, r_{max}, in relation to body mass and temperature. (A) and (B) show relationships for a wide variety of organisms. r_{max} was adjusted for temperature in (A) and mass in (B) by multiplying r_{max} by $e^{E/kT}$ and $M^{1/4}$ respectively. Fitted regression equations are shown in each panel. (From Savage et al. 2004a Effects of body size and temperature on population growth. *American Naturalist*, 163, 429–441. © University of Chicago.) (C) shows the relationship for mammals: the line, fitted by phylogenetic least squares, has slope −0.25. (From Duncan et al. 2007.)

and reproduction, the same in all species. MTE therefore predicts that productivity will follow the same scaling laws as metabolic rate, i.e., be governed by the central equation of Chapter 2. Organisms following these scaling laws are, however, also subject to evolutionary pressures, because organisms varying in productivity are subject to natural selection. The life histories favored by natural selection allocate resources to reproduction as fast as possible, provided this does not impair survivorship. Individuals are then selected to breed as early as possible, and maximize reproductive output at each breeding attempt.

This selectionist analysis has two interesting corollaries. In the first place, maximizing reproductive rate implies maximizing metabolic rate, because reproduction is powered by metabolism. Seen in this light, the fact that larger animals have lower metabolic rates per unit body mass, as in Kleiber's Law, seems very surprising. It requires mechanistic explanation. MTE provides such an explanation. The explanation is that production rate is selected to increase but is held back by constraints arising from the laws of physics, chemistry, and biology. Specifically, productivity is limited by metabolic rate because of logistical constraints in supplying oxygen and other resources around the bodies of individual organisms.

The second corollary discards the assumption that resources are allocated to survival, growth, and reproduction in fixed proportions. Trade-offs between life-history traits may be advantageous. Reproduction may be reduced so that the resources saved can be used to enhance adult survival between years. Offspring

number may be traded for offspring size to make bigger offspring better able to escape predators. Or offspring size may be increased to allow offspring to reach reproductive size earlier. Analysis and documentation of such trade-offs has provided a rich seam of research in evolutionary life-history studies (Rolf 2002; Sibly 2002).

Where trade-offs occur, they depart from the MTE postulate that resources are allocated in fixed proportions to survival, growth, and reproduction, the same in all species. Trade-offs are subject to the Principle of Allocation (Fig. 5.2), but allocations to different functions may vary between species. Adaptive variation in allocations may account for some of the variation in the scaling plots displayed in this chapter (see, e.g., Sibly and Brown 2007, 2009). New insights may come from the synthesis of MTE with evolutionary life-history theory and its applications, for example to explain the scatter seen in the graphs in this chapter.

5.6 CONCLUSIONS

The components of life histories, when plotted in relation to body mass or temperature, do follow in outline the predictions of MTE. Thus rates generally scale as the −1/4 power of body mass and the −0.65 power of body temperature, and life-history timings scale as the +1/4 power of body mass and the +0.65 power of body temperature. However, there is considerable scatter in many of the regressions illustrated in the figures in this chapter, and some departures from MTE predictions, for example: production rate in mammals scales as −0.37, not −0.25 power of body mass; there is variation among mammal orders in the scaling of age at first reproduction; and there is lack of variation with temperature in plant birth and death rates. Some of the scatter may be caused by adaptation of life histories to different environments under constraints stemming from the Principle of Allocation. Discrepancies between theory and data provide opportunities for a more subtle understanding of the mechanisms that underlie these scaling phenomena. Overall, further work is required to achieve a secure mechanistic explanation of the factors constraining life-history evolution.

Promising directions for future work include further analysis of parasite life cycles. These are of special interest because to some extent parasites escape limita-tions on resources, which are provided essentially free by the host. So parasites operate in a parallel universe where metabolic considerations still apply, but the constraints are different (see Chapter 19 by Hechinger, Lafferty, and Kuris). Responses to harsh and varying environments deserve study, including estivation, hibernation, and diapause (see Chapter 17 by Karasov and Chapter 6 by Hayward, Gillooly, and Kodric-Brown), and storage such as the accumulation of caches and fat reserves in animals, and water and energy in leaves, stems, roots, tubers, and bulbs in plants. It would be rewarding to study the life histories of insect colonies. Their metabolic rates can be measured (see Chapter 16 by Waters and Harrison and Chapter 24 by Moses and Forrest) and their colony sizes can be experimentally manipulated, allowing experimental tests of MTE predictions.

Chapter 6

BEHAVIOR

April Hayward, James F. Gillooly, and Astrid Kodric-Brown

SUMMARY

1 An explicitly metabolic approach provides a potentially useful conceptual framework for understanding the rates, times, and allocation decisions underlying many different animal behaviors, including those associated with resource acquisition, communication, reproduction, and cooperation.

2 Some behaviors (e.g., foraging and signaling) are directly linked to metabolism and have been shown to exhibit similar body size and temperature dependencies. Other behaviors (e.g., courtship and mating) may also be linked to metabolism through the allocation of energy and materials for the production and maintenance of specialized structures, such as ornaments and weapons.

3 Some behaviors may exhibit little or no relationship with metabolic rate. In some cases, behaviors may be primarily constrained by factors other than metabolism, such as biomechanical constraints. In other cases, behavioral plasticity may obscure or negate energetic constraints.

4 Relationships between behaviors and metabolism remain largely unexplored at this time. Still, recent research suggests that a metabolic approach holds promise both for enabling broad, quantitative, cross-species comparisons and for providing a mechanistic, theoretical framework with which to explain emergent patterns.

6.1 INTRODUCTION

Animals engage in a wide variety of behaviors, both as part of their daily routines and in association with important life-history events. In all such activities – from foraging and resting to communicating, fighting, and mating – behaviors vary tremendously across the diversity of life. Traditionally, behaviors have been studied in the context of ultimate (i.e., evolutionary) causes and explained in terms of their effect on individual fitness. While this approach has yielded many important insights into the behaviors of particular species or closely related groups of species, it has also made it difficult to identify broader patterns that might lead to a synthetic conceptual framework for behavioral ecology.

There are several reasons why such a framework is conspicuously absent in behavioral ecology. Perhaps the greatest of these is that behavior is difficult to measure and quantify in a standard way across species.

Indeed, behaviors that are associated with broadly similar life-history events can take extremely diverse forms across species. For example, reproductive behaviors can involve a variety of visual, acoustic, or chemical signals and may include the presentation of nuptial gifts, nest building, elaborate courtship displays, or the production and maintenance of exaggerated morphological traits to fend off competing suitors or to attract mates. Quantifying such behaviors using some common currency that would allow cross-species comparisons presents a difficult challenge, but is a critical step in the development of predictive models that can address the physiological, ecological, and evolutionary mechanisms that govern behavior.

Fitness has proven to be a useful metric for understanding the evolution of behaviors, but it is notoriously difficult to measure. Yet, it has long been recognized that fitness might be quantified in energetic terms. As Boltzmann (1905) noted, "The struggle for existence is a struggle for free energy available for work." Similarly, Lotka (1922) reasoned that "In the struggle for existence, the advantage must go to those organisms whose energy-capturing devices are most efficient in directing available energy into channels favorable to the preservation of the species." While a fully developed energetic definition of fitness remains elusive, all behaviors require energy. Since only some finite amount of energy is available in a day, breeding season, or lifetime, the rate at which an organism can acquire, transform, and expend energy, i.e. an organism's metabolic rate, can fundamentally constrain the rates, times, and quantities of energy associated with particular behaviors. Thus, systematic differences in metabolic rate among species could lead to predictable differences in behaviors across species. Here we explore the utility of using a metabolic framework to study animal behavior.

6.2   THE ROLE OF METABOLISM

To the extent that particular behaviors are constrained by energy availability, the metabolic theory of ecology (MTE; Brown et al. 2004) may be useful both in revealing broad-scale patterns in animal behavior and in developing predictive, mechanistic models to explain these patterns. This is because MTE models make first-order, quantitative predictions about biological rates, times, and energy allocation strategies related to

metabolism. These models rely heavily on the well-established body size and temperature dependence of metabolic rate, which has the general form

$$B = B_0 M^\alpha e^{-E/kT} \qquad (6.1)$$

(Brown and Sibly, Chapter 2). Here, B is mass-specific metabolic rate, B_0 is a normalization constant that typically varies among taxa and environments, M is body mass and α its scaling exponent, T is absolute temperature, and E and k are constants reflecting the temperature dependence of biochemical reactions. More specifically, k is Boltzmann's constant (8.62×10^{-5} eV K^{-1} or 1.38×10^{-23} J K^{-1}), and E is the average activation energy of metabolic rate. For many applications, it may be justified to use the more specific form of equation 6.1, where $\alpha = -1/4$ and $E = 0.65$ eV. Under this form, mass-specific metabolic rate scales as the $-1/4$ power of body mass and increases exponentially with temperature, such that a 10 °C increase in body temperature results in an approximately 2.5-fold increase in rate (Kleiber 1932; Hemmingsen 1960; Gillooly et al. 2001; Allen and Gillooly 2007).

Metabolic models like equation 6.1 can be extended to make predictions about other biological rates and times that are governed by metabolism by assuming that rates should be approximately proportional to mass-specific metabolic rate and that times should be inversely related to metabolic rate. Consequently, biological rates and times, including those characterized as behaviors, are predicted to vary systematically with body size and temperature. Recent studies have shown that at least some behaviors follow the predicted relationships with metabolism, body size, and temperature. In mammals, for example, cycles of rest and activity, including sleep cycles, show the same size dependence as metabolic rate (Savage and West 2007). This is hypothesized to reflect the fact that periods of rest are used for bodily repair, because the pace of repair proceeds at the rate of metabolism. Similarly, the duration of fasting, hibernation, torpor, and diapause also adhere to predictions based on metabolic models (Karasov, Chapter 17).

Metabolism in general, and MTE in particular, may prove useful in the study of behavior for several reasons. First, extending MTE models to address animal behavior may lead to general, quantitative predictions that are grounded in fundamental energetic principles.

Second, the two primary variables in MTE models, body size and temperature, are easily measured and vary substantially among organisms and across gradients of space and time. Third, relating behaviors to metabolic rate may facilitate the estimation of energetic costs. This approach has proven useful for quantifying life-history trade-offs from an energetic perspective (see Sibly, Chapter 5). Fourth, considering how energy is allocated to a particular behavior may provide insights into its fitness consequences, thus providing synthetic linkages between ultimate, evolutionary explanations and proximate, mechanistic explanations. Finally, a metabolic approach may allow animal behavior to be more fully integrated into other areas of ecology, including community and ecosystem ecology, by providing a framework for quantifying interspecific interactions.

Below we explore how the combined effects of body size and temperature on metabolic rate, as described by MTE, may offer a synthetic, quantitative approach for comparative studies in animal behavior. Specifically, we explore how body size and temperature may affect the rates, times, and allocation decisions of animals with respect to four broad classes of animal behavior: (1) foraging and resource acquisition; (2) communication; (3) reproduction; and (4) cooperation and group living. This is not intended to be a comprehensive treatment. Rather, we select a few specific examples to highlight the potential of the approach.

6.3 FORAGING AND RESOURCE ACQUISITION

Behaviors associated with resource acquisition have typically been studied in terms of their proximate and ultimate costs and benefits, and evaluated using some metric of fitness (Perry and Pianka 1997; Alerstam et al. 2003). For example, foraging behaviors are often evaluated in terms of how they maximize the rate of resource acquisition while minimizing some combination of the time and energy expended and the risk of predation (Emlen 1966; MacArthur and Pianka 1966; Pyke et al. 1977; Pykc 1984; McNamara and Houston 1992; Perry and Pianka 1997). The constraints imposed by the size- and temperature dependence of metabolic rate are less frequently considered explicitly. There is, however, a very direct linkage between feeding behaviors and metabolism because the quantity of

energy acquired through foraging must meet an organism's energetic requirements for survival, growth, and reproduction.

Metabolic rate also constrains foraging behavior in more subtle and interesting ways. For example, the length of time that air-breathing, diving vertebrates can spend foraging underwater is constrained by the size dependence of metabolic rate. This is because dive duration is determined largely by the volume of oxygen stored in the body (V_S), which scales linearly with body mass, and the rate at which oxygen is consumed, which is determined by the organism's metabolic rate (B) and scales as approximately $M^{3/4}$. Consequently, like other biological times, dive duration (t_D) is predicted to scale as the 1/4-power of body mass ($t_D \propto V_S/B \propto M^1/M^{3/4} \propto M^{1/4}$). This prediction is supported by data for a diverse range of birds and mammals (Schreer et al. 1997; Halsey et al. 2006a, 2006b; Brischoux et al. 2008; Fig. 6.1A). This same model has been extended to predict other aspects of diving behavior, including inter-dive intervals and dive depth (Fig. 6.1B; Halsey et al. 2006a). Such secondary effects of metabolic rate on travel distance may, in turn, affect other aspects of behavior, including home range or territory size (Hendriks 2007), migratory patterns, and dispersal distance in plankton (O'Connor and Bruno, Chapter 15).

Diving is just one of many behaviors related to resource acquisition that vary predictably with individual metabolic rate. In zooplankton, foraging rates vary systematically with body size and temperature (Peters and Downing 1984). In fish, the duration of suction during feeding, suction power, and the area over which suction is effective all scale predictably with body mass, reflecting a combination of metabolic and biomechanical constraints (Van Wassenbergh et al. 2006). Similarly, the closing forces of jaws of mammals, reptiles, birds, and fish and of crustacean chelae scale positively with body mass, reflecting the constraints of body and muscle size on energetics and biomechanics (Claussen et al. 2008). Even chewing and lapping rates scale allometrically in mammals, again reflecting a combination of metabolic and biomechanical constraints (Gerstner and Gerstein 2008; Reis et al. 2010). In theory, virtually all aspects of foraging, from searching for and capturing food, to consuming and digesting these resources, to excreting waste, should be constrained by individual energetics. Thus, a re-examination of foraging behavior in the

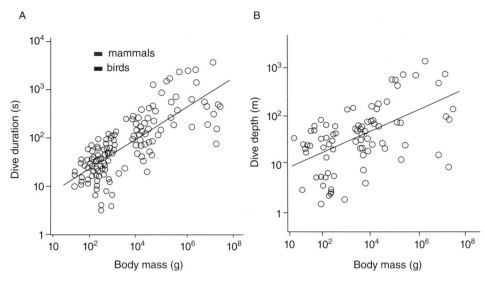

Figure 6.1 Scaling of dive duration and depth with body size for diving birds and mammals. Dive duration (A) and depth (B) plotted as a function of body mass on logarithmic axes, and fit with regressions giving slopes (allometric exponents) of 0.31 (95% CI: 0.27–0.35) and 0.25 (95% CI: 0.16–0.32), respectively. Data and analysis from Halsey et al. (2006a) by permission of John Wiley & Sons, Ltd.

light of metabolic constraints may provide new insights into old problems in foraging theory.

6.4 COMMUNICATION

Animals use a diverse array of visual, acoustic, chemical, tactile, and electric signals to transmit information to conspecifics, competitors, predators, and prey. Historically, the costs and benefits of behaviors associated with communication have been weighed in terms of their effects on individual fitness or the fitness of closely related individuals (i.e., in terms of kin selection; Hamilton 1964; Smith 1964; Dawkins 1993; Hasson 1997). Thus, classic studies of animal communication have focused on understanding how behaviors that modify predation risk, mate choice, territorial defense, or social structure affect survival, growth, and reproduction (e.g., Ryan 1988; Magnhagen 1991). Some studies have also analyzed the energetic costs of sending and receiving signals and the underlying physiological mechanisms that govern these behaviors (Ryan 1986; Prestwich et al. 1989; Prestwich 1994; Stoddard and Salazar 2011). Such studies have often concluded that communication systems can be energetically costly to use and maintain.

Animals use a wide variety of motor and sensory systems to send and receive signals, from the muscles that generate movements to the sensory systems that gather, process, and respond to signals. A common feature of these systems is that they require energy to produce or receive signals or to produce and maintain organs that are used in the production, reception, and processing of signals. Consequently, many rates, times, and allocation decisions involved in communication may scale predictably with body size and temperature. For example, many of the basic features of acoustic signals reflect the size- and temperature dependence of metabolic rate (Gillooly and Ophir 2010; Ophir et al. 2010). This includes both call rate and call power, which scale with body size and temperature as predicted by equation 6.1 across a diverse array of species including invertebrates, fishes, amphibians, reptiles, birds, and mammals (Fig. 6.2). Although there is considerable residual variation in the scaling of call features that is not explained by size and temperature, these scaling relations provide a baseline that can be used as a point of departure for exploring the relative importance of other factors. For example, residual variation in call characteristics may hint at the importance of specialized morphological features that are used for calling, such as air sacs in amphibians. These can be assessed quantitatively by first accounting for

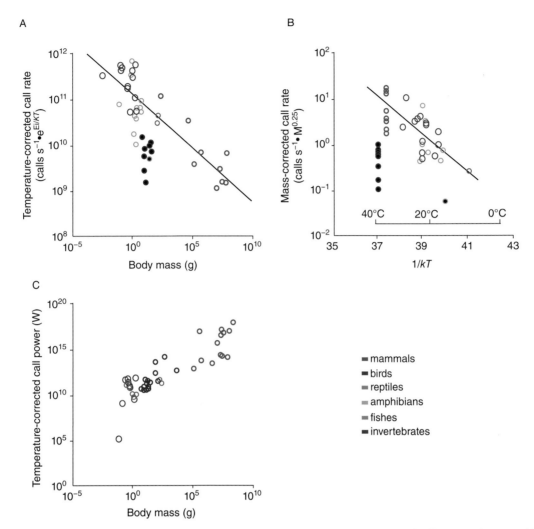

Figure 6.2 Scaling of animal calls with body size and temperature. (A) Temperature-corrected call rate as a function of body mass, plotted on logarithmic axes, and fit with a regression with a slope (allometric exponent) −0.23 (95% CI: −0.18 to 0.28). (B) Mass-corrected call rate as a function of temperature, plotted as an Arrhenius function and fit with a regression giving an "activation energy" of −0.82 (95% CI: −0.53 to 1.12). (C) Temperature-corrected call power as a function of body mass, plotted on logarithmic axes, and fit with a linear regression with a slope (allometric exponent) of 0.72 (95% CI: 0.06–0.84). For all plots, closed symbols represent species with calls described as "trills," which have significantly longer calls and lower calling rates than non-trilling species. Such species reuse air during calling and were excluded from the analyses. Data and analysis from Gillooly and Ophir (2010).

the pervasive effects of size and temperature shown in Figure 6.2 and then assessing correlates and causes of the residual variation. Future research might extend such work to examine metabolic constraints on features of acoustic communication that affect the interactions governed by these signals. Specific areas of study might include rates of courtship across gradients in environmental temperature or across species of different sizes, or the distance over which signals are typically transmitted between individuals of a given species. A metabolic approach might also be profitably applied to communication systems that use other sensory modalities, such as visual, chemical (olfactory and gustatory), tactile, and electric signals.

The example of acoustic communication may point to more profound mechanisms linking individual metabolism to the ability of species to perceive and respond to their environments. Implicit in metabolic models of acoustic communication is the assumption that both the rates of muscle contractions that generate sound production and the rates of motor neuron firing that control muscle activity are tightly linked to metabolic rate. Indeed, it has long been known that muscle dynamics vary in relation to metabolic rate, but only recently have studies emerged that link central nervous system function to metabolism. For example, neuron firing rates in birds scale with body size similarly to individual metabolic rate (Fig. 6.3; Hempleman et al. 2005). Similarly, patterns of neural activity in mammalian brains also occur at a rate that is roughly proportional to metabolic rate and therefore scale similarly with body size (Erecinska et al. 2004). Metabolic scaling of neuromuscular systems, and especially of linkages between the motor, sensory, and integrative components, may have implications for many different types of animal behavior.

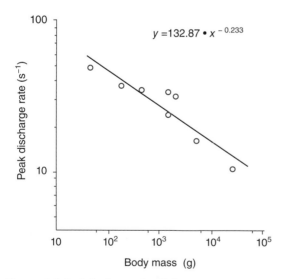

Figure 6.3 Peak discharge rate of chemoreceptor neurons as a function of body size, plotted on logarithmic axes, in several species of birds. Note that the exponent, −0.23, is very similar to the −1/4 scaling of mass-specific metabolic rate as predicted by MTE. Data and analysis from Hempleman et al. (2005).

6.5 REPRODUCTION

Animal reproduction involves a suite of diverse behaviors, from pre-copulatory displays and mate choice to post-copulatory defense and parental care. As with other behaviors, reproductive behaviors have often been addressed from an ultimate perspective and, to a lesser extent, from a more proximate perspective that incorporates the role of hormones and other aspects of physiology. For the most part, studies on the nature, timing, and duration of reproductive activities (e.g., courtship and copulation) have focused on the fitness consequences of competition within sexes, synchrony between sexes, or exposure to predation risk (Magnhagen 1991). Relatively few studies have attempted to generalize about pervasive patterns of variation across species in the rates and times of reproductive behaviors. However, many of the activities and allocation decisions associated with reproduction may be energetically constrained. Examples include the rate and duration of courtship and copulation, the metabolic power required to perform copulatory activities, and the energetic cost of alternative behaviors that result in allocation trade-offs. As such, it may be possible to link variation in reproductive behavior among species to differences in energetic constraints across species.

Recent work provides some encouraging examples. Gamete biomass production rates scale with body mass as predicted by MTE in both males and females of diverse species (Fig. 6.4; Hayward and Gilooly 2011). Similarly, the duration of lactation in mammals scales positively with body mass as predicted by MTE (Hamilton et al. 2011; see Hamilton, Burger, and Walker, Chapter 20; Fig 5.4B). Other aspects of parental care that tend to maximize reproductive success in different environments, including nest building and guarding, the incubation of eggs and young, and parental feeding are also be expected to scale in accordance with MTE models, but these behaviors have yet to be evaluated in a metabolic context.

Metabolic perspectives also provide a framework for comparing the energetic and fitness trade-offs of alternative reproductive tactics (ARTs) within populations and across species. The variation in ARTs is extensive, ranging from deer and antelopes where some males exhaust themselves fighting for and defending access to females, to some spiders that sacrifice their bodies to feed females while copulating, to some fish where tiny males devote a large fraction of their body mass to

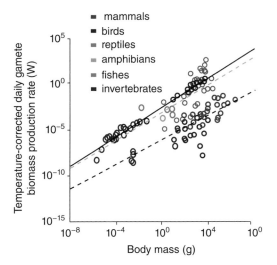

Figure 6.4 Rate of gamete biomass production as a function of body mass, plotted on logarithmic axes, in males (diamonds) and females (circles) for a variety of organisms. In both sexes, gamete production rate scales in accordance with the predictions of MTE (as 0.66 (95% CI: 0.55–0.77) for males and 0.88 (95% CI: 0.73–0.87) for females), though production rates are about 3.5 orders of magnitude lower in males (a = −6.09) than in females (a = −2.66). Data and analysis from Hayward and Gillooly (2011).

testes and a comparably large fraction of their metabolism to sperm production so as to compete as "sneakers" with large dominant males. It may be possible to extend MTE models to obtain additional insights into such patterns of reproductive behavior. At the very least, there appears to be a pronounced size dependence. For example, small males typically do not invest in metabolically costly ornaments, weapons, and courtship behaviors. Instead, they avoid direct competition with larger males by acting as sneakers, satellites, or female mimics – low-cost behaviors that allow them to gain access to mates. They also allocate energy and biomass to testis size and sperm production differently, suggesting interesting trade-offs between growth and reproduction during ontogeny. Still, very little is known about the metabolic costs of such alternative reproductive behaviors (but see Cummings and Gelineau-Kattner 2009).

Recent work extending MTE models and applying a more general metabolic framework has begun to define the role of energetic constraints in sexual and natural selection in an explicit, quantitative, and mechanistic manner. For example, it has long been known that both ornaments (e.g., dewlaps, gular pouches) that are used to attract females and weapons that are used to compete with other males (e.g., antlers, horns) typically exhibit positive allometry over ontogeny, both within species and across related species (e.g., Huxley 1932; Gould 1974; Kodric-Brown et al. 2006). Exponents typically range between 1.5 and 2.5 (Fig. 6.5), but are sometimes much higher (e.g., approximately 7.0 for the head crest of the dinosaur *Pteranodon* (Tomkins et al. 2010). These positive allometries indicate that males allocate an increasing fraction of biomass to ornaments or weapons as they age and grow. While older males with larger ornaments or weapons gain more matings and have higher reproductive success (Alonso et al. 2010), there is a trade-off between allocation to the growth and the maintenance of body mass and to the growth and maintenance of the sexually selected trait (Nijhout and Emlen 1998). Kodric-Brown et al. (2006) have recently shown that such allometric scaling of ornaments and weapons can be explained by a metabolic model of energy allocation to growth and reproduction.

Conspicuous ornaments also tend to increase risk of predation and mortality. Consequently, the scaling of exponents varies in response to local resources and predation risk, with populations in unproductive environments or those experiencing high predation rates having smaller ornaments and weapons in relation to their body size. For example, males of green swordtails (*Xiphophorus helleri*) have longer swords in environments with fewer predators (Basolo and Wagner 2004). The cost of maintaining these ornaments is elegantly demonstrated by decreased metabolic rates of males when their swords are experimentally removed (Basolo and Alcaraz 2003).

The positive allometries of sexually selected ornaments and weapons provide an interesting example of how structural traits evolve in response to trade-offs in metabolic allocation to serve a behavioral function. They also show the linkage between the energetic currency of metabolic ecology and the fitness currency of natural and sexual selection. Likewise, the relationship between gamete biomass production rate and metabolic rate suggests that fitness itself is metabolically constrained, though different strategies for enhancing fitness may be expressed through trade-offs in the size versus the number of gametes produced. These few examples hint at the extent to which a metabolic approach may be fruitfully applied to understanding

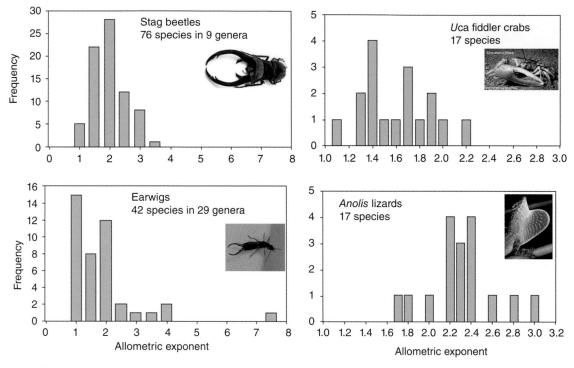

Figure 6.5 Frequency histograms of allometric exponents obtained by fitting power functions to log-transformed data for intraspecific variation in ornament size as a function of body size in different kinds of animals: horns of stag beetles (Lucanidae), claws of *Uca* fiddler crabs, forceps of earwigs (Dermaptera), and dewlaps of *Anolis* lizards. Note that in contrast to most of the allometries considered in this book, the scaling of ornament size is consistently superlinear, with nearly all exponents >1 and some >3. Data and analysis from Kodric-Brown et al. (2006).

similarities and differences in reproductive behavior, and ultimately in fitness, across species with seemingly disparate life-history strategies.

6.6 COOPERATION AND GROUP LIVING

Cooperation, group living, and sociality offer some of the most intriguing, albeit unexplored, possibilities for extending MTE into the realm of animal behavior. Virtually all taxonomic groups have representatives that live together in groups, from amoebae to vertebrates; from the temporary feeding, breeding, and migratory aggregations in some birds and mammals to the long-lived, complex societies of ants, termites, and primates (Krebs and Davies 1993; Rubenstein and Kealey 2010). Presumably, cooperating and living as a

group confers some net benefit to individual fitness (Alexander 1974, but see Sibly 1983) and a wealth of specific hypotheses have been proposed for both the benefits (e.g., resource acquisition, predator defense, parental care) and costs (e.g., disease transmission, competition) of group living (Krebs and Davies 1993; Rubenstein and Kealey 2010). Studies in this area have typically adopted an evolutionary, fitness-based approach and focused on advantages and disadvantages of cooperation as reflected in survival, growth, or reproduction at the individual or group level (Hamilton 1964; Trivers 1971; Nowak 2006; Clutton-Brock et al. 2009).

For the most part, measures employed to assess the costs and benefits of group living are species- or group-specific and the currency is often not explicitly energetic. This makes it difficult to identify more general patterns and to develop and test mechanistic hypotheses for patterns that hold across species. Exploring long-standing questions about the ecology and evolution of sociality from a more explicitly energetic perspective may offer new and valuable insights. In particular, MTE may provide a useful framework for assessing the energetic costs and benefits of group living, and thereby offer additional insights into the energy allocation and life-history trade-offs that have accompanied the evolution of varying degrees of sociality. For example, in ants and bees, the lifespan of queens varies with the breeding system. Queens in highly social species can live many times longer than those of solitary species, and queens in colonies with a single foundress tend to live longer than those with multiple foundresses (Keller 1998; Schrempf and Heinze 2007). More generally, in social insects and other animal taxa, per-capita rates of survival, growth, and reproduction vary systematically as a function of group size (Michener 1974). These life-history characteristics are generally considered to be individual-level traits, but empirical patterns suggest that they are modified by social context and perhaps represent adaptive trade-offs in energy allocation that have evolved, at least in part, by group- or colony-level selection (Heinze and Schrempf 2008; Hölldobler and Wilson 2008).

As for metabolic rate itself, one might naively expect that the metabolic rate of a group, which is by definition the sum of the metabolic rates of all members, would scale linearly with the number of individuals, and hence with total biomass. This may be the case for loosely organized aggregations, such as flocks of birds or herds of grazing mammals. Yet, two recent studies of social insects show convincingly that whole-colony metabolic rate scales sublinearly with colony mass. The first example comes from Hou et al. (2010), who compiled data on metabolism, growth, and reproduction in social insects (ants, bees, and termites). They found that whole-colony metabolic rate scaled as a function of colony mass in a manner that was quite similar to the body-mass scaling of individual metabolic rate in solitary insects, with exponents of 0.81 and 0.83, respectively (Fig. 6.6). Rates of colony growth (measured as the increase in biomass/time) and reproduction also paralleled that of solitary insects when colony mass was defined as the total mass of all workers. Similarly, Waters et al. (2010) measured the metabolic rates of colonies and solitary individuals of the harvester ant (*Poganomyrmex californicus*) under controlled conditions in the laboratory. They found

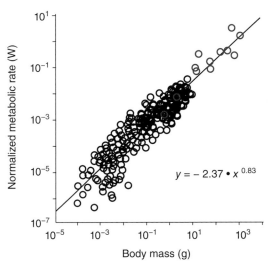

Figure 6.6 Metabolic rate as a function of mass, plotted on logarithmic axes, for whole colonies of social insects as a function of total colony mass (red circles), and for individual insects as a function of body mass (black circles). Note that both relationships are sublinear, with similar slopes (allometric exponents of 0.81 (95% CI: 0.57–1.08) and 0.83 (95% CI: 0.78–0.85), respectively). Data and analysis from Hou et al. (2010).

that whole-colony metabolic rate scaled with mass with an exponent of 0.75, significantly lower than the linear scaling that would have been predicted from the sum of the metabolic rates of equal numbers of solitary ants (Waters and Harrison, Chapter 16, Fig. 16.7). This was explained, in part, by a decrease in the average movement velocities, and hence in the metabolic rates, of individuals with increasing colony size. These startling and unexpected results imply that, at least in terms of energetics and life history, insect colonies really are behaving in some respects like "superorganisms" (Hölldobler and Wilson 2008; see also Moses and Forrest, Chapter 24).

The studies of social insects highlight the utility of using a macro-scale, energetically-based approach for investigating how differences in individual-level physiological and behavioral traits affect groups of organisms, and how individual-level physiology and behavior is affected by group living, cooperation, and social organization. Even in loosely integrated groups individuals may derive some metabolic savings from certain forms of group behavior. For example, the characteristic V-shaped flocks of geese and other flying birds are adaptive behaviors that reduce the cost of flight for individual members (Badgerow and Hainsworth 1981). Likewise, increases in group size reduce individual metabolic costs for African hunting dogs (Creel 1997). Examining cooperation, including mutualistic interactions and group behaviors, from a metabolic perspective may provide insight into the nature, costs, and benefits of such behaviors.

6.7   CONCLUSIONS

Recent research in areas such as foraging, communication, reproduction, and social cooperation has shown that variation in behaviors across species is often related to basic traits such as body size, temperature, and shared morphological features (Moller and Birkhead 1991; Kodric-Brown et al. 2006; Hendriks 2007; Martin et al. 2007; Brischoux et al. 2008; Dial et al. 2008; Hou et al. 2010; Ophir et al. 2010). This suggests close linkages to metabolism and highlights the potential for extending a comparative energetic perspective in general, and the MTE framework in particular, to animal behavior. Applications and extensions of MTE can be used to formulate hypotheses and test predictions across diverse species. Formal mathematical models, such as those developed for life-history traits and ornaments and weapons, should yield quantitative predictions for a range of behaviors that span diverse species and environments. In some cases, such models would serve as useful alternative or complementary hypotheses to known mechanisms, such as sperm competition or hormonal effects. More generally, including a metabolic perspective in studies of behavior may allow for the synthesis of the energy- and materials-based currencies of metabolic ecology and the genetic- and fitness-based currencies of more traditional behavioral ecology.

While a metabolic approach may expand our understanding of animal behavior, there are also many good reasons to expect that some behaviors will exhibit little or no relationship with metabolic rate. For example, some behaviors may be too complex or heterogeneous to be effectively described by simple mathematical

models derived from MTE. Behaviors are also easily influenced by physical and social environments such that wide-scale patterns might be obscured by context-dependent behavioral plasticity. Alternatively, the energetic cost of some behaviors may be such a small fraction of the individual energy budget that it is insignificant from an ecological or evolutionary perspective. In other cases, rates and times may be more tightly constrained by factors other than whole-organism energetics. For example, while MTE predicts that biological times generally scale as $M^{1/4}$, Stallmann and Harcourt (2006) found that copulation duration scales as $M^{-0.16}$. They attributed this finding to the fact that individuals of smaller size have relatively greater biomechanical power for grasping their mates, which would enable them to maintain copulation for longer periods than larger organisms.

Of the areas where the theoretical framework of MTE might be successfully extended and applied to animal behavior, we will mention just a few. First, the original MTE paper pointed out that rates of species interactions appear to exhibit predictable size and temperature dependence (Brown et al. 2004). For example, the rate of competitive exclusion in Park's classic flour beetle experiments showed virtually the same temperature dependence that MTE would predict on the basis of metabolic rate. Yet, to date, the numerous studies of competitive interactions, rates of predation, feeding, grooming, and courtship have not been re-examined from a metabolic perspective. Second, MTE may be useful for explaining patterns of animal movement, including home range size, foraging trips, and long-distance migration. Constraints on animal movement have often been attributed to the energetic cost of locomotion (see Alerstam et al. 2003), but few general, energetically based models have been attempted (but see Garland 1983a; Altmann 1987; McWilliams et al. 2004). Lastly, recent work showing links between metabolism and central nervous system function suggest that a metabolic perspective may be useful in understanding how animals acquire, process, and use information from their environment. We look forward to seeing how these largely unexplored areas of research progress in the years to come.

Chapter 7

POPULATION AND COMMUNITY ECOLOGY

Nick J.B. Isaac, Chris Carbone, and Brian McGill

SUMMARY

1 This chapter explores how metabolic ecology relates to population dynamics and community structure. We define two alternatives for how energy can influence populations and communities. In the narrow version, energy has influence very specifically through metabolic rate. The wide version simply hypothesizes that the signal of energy can be found.

2 There are two central parameters in population dynamics, r_{max} and K. Evidence shows that r_{max} behaves as expected under the narrow version of energetic control, except at high temperatures. It is unclear whether K conforms to the narrow version of the energetic paradigm, but certainly K is mediated by energy in the broad sense. Energetic constraints may also be helpful in parameterizing models of interacting pairs of species.

3 Communities show orders of magnitude variation across species in both abundance and mass. The narrow-sense interpretation of the role of energy predicts that abundance scales as $M^{-3/4}$ power. Since this implies that energy flux is independent of species body mass, this is also known as the energy equivalence rule. Energy equivalence appears to hold at very large spatial and taxonomic scales, suggesting it is really setting an upper limit or constraint on abundances. The majority of evidence suggests that abundance scales with an exponent shallower than $-3/4$, suggesting that larger animals gain disproportionately more resources.

4 A variety of promising new approaches have provided wide-sense linkages between energy and abundance. These include: (a) studies of allometric scaling in space use parameters; (b) comparing measures of abundance in different currencies (numbers, biomass, and energy use); and (c) looking at energy flux of whole communities, which appear to show a zero-sum pattern.

5 Community ecology also helps to explain the variation that remains around Kleiber's Law. This variation is characterized as a fast–slow continuum. Diet, life history, and climate all contribute to explain the fast–slow continuum.

6 The narrow-sense predictions of metabolic theory have mixed success in population and community ecology, but the broader insight of the importance of energy seems highly applicable. We contrast the metabolic perspective with recent "unified theories of biodiversity," and speculate on how the two approaches might be combined.

Metabolic Ecology: A Scaling Approach, First Edition. Edited by Richard M. Sibly, James H. Brown, Astrid Kodric-Brown.
© 2012 John Wiley & Sons, Ltd. Published 2012 by John Wiley & Sons, Ltd.

7.1 INTRODUCTION

This book is structured by hierarchical levels of organization: tissues, individuals, populations, communities, food webs, ecosystems. The early chapters assert that one specific physiological process, individual metabolic rate or flux of energy, is dependent on body mass and temperature in a very precise quantitative form (i.e., equation 1.1). Chapter 4 demonstrates how this process extends to ontogenetic growth rates (essentially tissue production), while Chapters 5 and 6 show how metabolic ecology influences whole organism properties of life history and behavior. This chapter evaluates the degree to which the two levels above the individual, namely populations and communities, show the signal of metabolism while later chapters focus on food webs and ecosystems.

Thus each chapter moves further away from the MTE starting point of individual-level energy flux: at each step up the biological hierarchy, the mechanistic link between energy flux and the process in question becomes weaker and less direct. Exactly how far should one expect the signal of metabolic ecology to resonate? In this chapter we define and explore the energetic paradigm for populations and communities – i.e., the notion that population and community parameters, including species abundance and population dynamic parameters, are determined by energy constraints (see also Table 2 in Brown, Sibly and Kodric-Brown, Introduction). This idea is appealing because energy provides a unifying currency with which to describe interactions across all levels of the biological hierarchy, from molecules in the vascular system to organisms competing for access to resources. However, the value and usefulness of MTE above the level of individual organisms remains hotly debated. In this chapter, we distinguish between "narrow" and "broad" versions of the energetic paradigm, and evaluate the evidence for each. In the narrow version, population and community parameters have a causal relationship with individual metabolic requirements, and take values that can be quantitatively predicted from the central equation of MTE (Brown et al. 2004; Savage et al. 2004a). In the broad version, abundance (and/or related variables) show correlations with energy but do not conform to the parameters predicted by MTE.

We start with an examination of population dynamic parameters, including the rates of population growth and carrying capacity. We then discuss how the energetic paradigm informs multispecies interactions before moving on to the way energy is partitioned within communities. We explore the relationships between body size, abundance, and other measures of animal space use. We also briefly explore the inverse of the energetic paradigm, namely, that community ecology can provide insights into metabolic ecology.

7.2 PARAMETERIZING POPULATION DYNAMIC MODELS

One of the central organizing concepts in ecology has been the population. The term is difficult to define precisely, but we consider a population to be a collection of conspecifics that interact with each other often enough to allow them to be treated as a unit. Thus the spatial extent of a population is clearly limited, although the exact limits depend both on the type of organism and on the methods employed.

The first thing one notices about a population is how abundant it is (are there many or a few maple trees?). Less obvious is that populations change in abundance over time. The rates at which populations change over time are studied using population dynamic models such as the logistic equation ($dN/dt = r_{max} N[1 − N/K]$). Such population dynamic models have been extremely successful in modeling population fluctuations over time. However, they suffer from a certain circularity. The only way to estimate the parameters of these models is to watch a population over time and fit the model; there is no theory that predicts, a priori, the values of these parameters.

MTE provides a way around this dilemma. For a population to increase in abundance it must capture or consume more energy than it uses (maintenance, i.e., basal metabolic rate), with the excess energy converted into new biomass (growth), some of which is converted into new individuals (reproduction). If a population captures less energy than it uses then abundance will ultimately decrease. Given that body size and temperature show strong relationships with physiological (Kaspari, Chapter 3; Kerkhoff, Chapter 4) and life-history (Sibly, Chapter 5) processes through metabolic ecology, one might imagine that similar relationships might exist with population parameters such as r and K. In fact there have been some successes and some failures in the efforts to find these patterns.

7.2.1 Maximum population growth rate, r_{max}

A central parameter in the logistic equation is r, which stands for population growth rate. If there are no limits on population growth, the population would grow exponentially as $N(t) = e^{r_{max} \cdot t}$. A great deal is known about how body size and temperature affect r_{max}. It has been known for some time (Fenchel 1974; Blueweiss et al. 1978; Hennemann 1983; Peters 1983; Calder 1984) that $r_{max} \propto M^{-1/4}$ which, since r_{max} has units of 1/time, conforms to the prediction of MTE (see also Sibly, Chapter 5; Karasov, Chapter 17). This phenomenon appears quite general, spanning unicellular organisms, homeotherms, and ectotherms (Hennemann 1983; Savage et al. 2004a). Unfortunately, as with many other scaling phenomena, it appears to become increasingly less predictive as the body size range decreases (Gaston 1988; Tilman et al. 2004). It is therefore likely to be of limited use for predicting, for example, the differences in r among close relatives or direct competitors.

Maximum population growth rate also varies with temperature in a manner that is consistent with metabolic theory (Savage et al. 2004a), with an activation energy of $-0.63\,\text{eV}$ (Fig. 7.1). However, the analysis by Savage and colleagues intentionally omitted very low temperatures and any temperature beyond the optimum. A more complete picture for ectotherms suggests that maximum population growth rate follows a hump-shaped response to temperature (Huey and Berrigan 2001). This pattern is common among physiological traits, with a slow rise followed by a steep decline beyond the optimum temperature (Huey and Stevenson 1979; Huey and Bennett 1987; Brown and Sibly, Chapter 2). The increasing portion is attributable to enzyme kinematics as in metabolic theory (Arrhenius 1889; Gillooly et al. 2001), with the decreasing portion due to enzyme deactivation (Schoolfield et al. 1981).

7.2.2 Equilibrium population size, K

The second parameter of the logistic equation, K, gives the equilibrium population size, i.e., the population size towards which dynamics move in the absence of perturbations. The wide-sense energetic prediction is that abundance increases with available energy. This pattern has been reported for soil invertebrates (Meehan 2006), overwintering birds in North America (Meehan et al. 2004), and among Eurasian red foxes (Barton and Zalewski 2007), but other taxa show the opposite pattern (Currie and Fritz 1993).

To make a narrow-sense energetic hypothesis about populations, we assume that the resource supply rate, R (energy flux per unit area), available to a population is independent of its population size, N, and of other species (a big assumption but perhaps approximately true if each species has a distinct niche). In this case the maximum long-term sustainable population, K, is the resource supply rate divided by the resource requirements per individual, I, i.e., $K = R/I$. Substituting the MTE equation for E we have

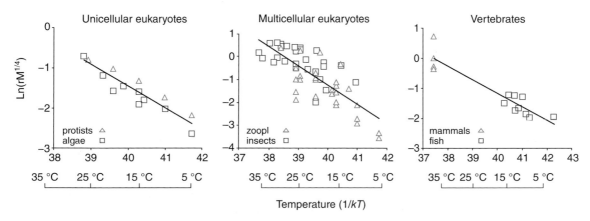

Figure 7.1 Relationship of mass-corrected population growth rate, r_{max}, to inverse temperature. As predicted, the plots are straight lines with slopes in the region of -0.63. Redrawn from Savage et al. (2004a).

$$K = RK_0 M^{-3/4} e^{0.65/kT} \qquad\qquad (7.1)$$

(where K_0 is a taxon-specific constant) as a prediction for the narrow-sense energetic hypothesis of population biology (Brown et al. 2004; Savage et al. 2004a). Note that the wide-sense energetic prediction is also produced ($K \propto R$). We explore the relationship between K and M, across species, in section 7.3.

It is possible to ask how K covaries with temperature within one population. MTE predicts that as T goes up, K should go down (roughly 9% decrease in K for a 1 °C increase in T) (Gillooly et al. 2001; Isaac et al. 2011a). This hypothesis has been relatively little studied and deserves further consideration, especially in light of climate change (Dillon et al. 2010). With the evidence available, it is not clear that K covaries with T according to equation 7.1. Gause found that a Gaussian bell curve described equilibrium abundance (K) as a function of temperature (Gause 1931, 1932). Three experimental studies (Davis et al. 1998; Fox and Morin 2001; Jiang and Morin 2004) have shown that equilibrium abundance can strongly depend on temperature, but can be an increasing, decreasing, or modal function of temperature depending on species and on the temperature range. More recently, Isaac et al. (2011a) reported that abundance of British butterfly populations is 14% lower, on average, for each degree of temperature increase: this is broadly consistent with the prediction of MTE, although non-energetic explanations are equally plausible. The mixture of increasing, decreasing, and modal responses of K to temperature appears inconsistent with both the Gaussian bell curve and the Arrhenius response. However, all three forms might be observed if each species shows a Gaussian curve but only small ranges of temperature are sampled. One mechanistic reason we may see so much variability in the relationship of K to T, unlike the relationship of r to T, is that intraspecific competition may buffer K from temperature-induced changes in r (Fox and Morin 2001).

The response of K to environmental temperature in homeotherms is necessarily different – homeotherms maintain a constant internal temperature by expending additional energy when ambient temperatures are too high or too low (Willmer et al. 2000). Meehan and colleagues (2004) showed that because conductance (rate of heat loss from the body) scales as $M^{1/2}$, one would expect that K scales as $M^{-1/2}$. This prediction is supported by data on the average abundance (a proxy for K) of North American birds in winter (i.e., the period of thermal stress; Meehan et al. 2004).

Overall it appears that K has an unknown relationship to temperature. In contrast, maximum population growth rate, r_{max}, conforms well to MTE (although not above certain temperatures). To the extent it is successful, metabolic ecology provides a useful tool for estimating population equation parameters of poorly studied species. This has the potential to be of great value to conservation. Pereira and colleagues (Pereira et al. 2004) recently showed how allometric estimation of population parameters (including r and dispersal distances) can lead to successful predictions of the extinction risk status of several hundred Costa Rican birds (see also Boyer and Jetz, Chapter 22).

7.2.3 Two species interactions

Sitting half way between populations and communities are models of two species interacting with each other. These include the classic Rosenzweig–MacArthur predator–prey model and the Lotka–Volterra competition model. Metabolic theory can also prove useful in parameterizing such multiple species equations. For example, Yodzis and Innes (1992) were able to parameterize the Rosenzweig–MacArthur predator–prey equations using allometries derived from metabolic ecology (discussed in greater depth in Petchey and Dunne, Chapter 8). Allometry can also be used to parameterize host–parasite models (Hechinger, Lafferty, and Kuris, Chapter 19).

There are also well-known relationships between predator and prey body size (Peters 1983; Vezina 1985) with predators feeding on prey that are 10% of their mass, on average (i.e., $M_{prey} \propto 0.1\, M_{predator}^1$), although there is a great deal of variability around this prediction. More generally, in all types of trophic interactions (including predation and parasitism) the body mass of the exploiter appears to scale isometrically (M^1) with the body mass of the exploited, with intercepts that depend on the exact type of tropic interaction (Lafferty and Kuris 2002). Carbone and colleagues (Carbone et al. 1999) break out one source of variability in showing a threshold of around 20 kg at which mammalian carnivores switch from small prey, compatible with the Vezina and Peters prediction, to large prey due to energetic requirements.

7.3 ENERGY AND COMMUNITY STRUCTURE

The previous section dealt with many individuals in one species (i.e., populations). This section deals with communities, which are collections of populations living in the same area. How does the energetic paradigm hold at the community level?

7.3.1 Species abundance distributions

The wide-sense prediction that total abundance (summed across all species) increases with available energy appears to hold (Hurlbert 2004; McGlynn et al. 2010). This is what Brown (1981) called a capacity rule. But an obvious second question is: How are resources and abundances distributed among species? This is what Brown called an allocation rule. It has been long known and extremely well documented that the allocation between species is highly uneven (Fig. 7.2A,B). This pattern is known as the species abundance distribution (reviewed by McGill et al. 2007). A related pattern is that the distribution of species body mass is also highly uneven (many small, few large species; Fig. 8.2C,D). Thus there is considerable variation within a community in both abundance and body size.

7.3.2 Size–density relationships and energy equivalence

We can now return to the question raised in section 7.2.2 of whether body size can explain variation in abundance across species (i.e., within a community). Although there are several ways to compare size and abundance (White et al. 2007), the most common is known as the size–density relationship (Fig. 7.3) in which estimates of population density (abundance per unit area) for each species are plotted against mean body mass on logarithmic axes.

The narrow-sense energetic hypothesis (equation 7.1), in which we assumed that the resource supply for a particular species was known (i.e., an input), is not useful to predict abundance across a community of many species. Here we are interested not in the effect of total resource supply but rather in how those resources are allocated across species. Yet the data in

Figure 7.3 (and indeed intuition) suggests that there should be some relationship. We need an additional assumption that resource supply, R, is independent of body mass, M: known as energy equivalence (Maurer and Brown 1988; Nee et al. 1991).

Early evidence supported this narrow-sense hypothesis of energy equivalence. Damuth reported a size–density relationship with a scaling exponent close to $-3/4$, using global-scale data on mammals (Damuth 1981) and later across a wide range of vertebrate taxa (Damuth 1987). This form of the size–density relationship, known as "Damuth's rule," has also been reported in aquatic (Cyr et al. 1997) and marine ecosystems (Jennings et al. 2008), and also in monocultures (Enquist et al. 1998). Data from terrestrial systems has shown that size–density relationships become tighter once resource availability (Carbone and Gittleman 2002) or trophic level (McGill 2008) are controlled for, which is consistent with the broad-sense energetic view that energy is important in determining abundance (see also Hechinger, Lafferty, and Kuris, Chapter 19).

However, a majority of recent studies have found reported size–density exponents substantially shallower (less negative) than $-3/4$, implying that bigger animals seem to get more than their "share" of resources (higher R). This pattern has been reported in both terrestrial (Russo et al. 2003; McGill 2008; Morlon et al. 2009; Reuman et al. 2009) and aquatic (Hayward et al. 2009; Reuman et al. 2009) ecosystems. Positive scaling of R might be explained by a variety of community processes, such as asymmetric competition for resources, or the possibility that large animals supplement their resource intake from outside the study community. However, the latter process cannot explain why departures from Damuth's rule have been reported at global scales (Currie and Fritz 1993), even using Damuth's own dataset (Isaac et al. 2011b). Moreover, some systems appear to depart systematically from the simple power law, with nonlinear law size–density relationships reported for more than a quarter of food webs studied (Reuman et al. 2009).

One consideration is that size–density relationships tend to be tighter (higher r^2) and more supportive of energy equivalence when data are aggregated to large spatial scales (White et al. 2007), when the range of body sizes considered is large (Tilman et al. 2004; Hayward et al. 2010) and when using maximum, rather than mean, population density (McGill 2008). This suggests energy equivalence is perhaps more of an

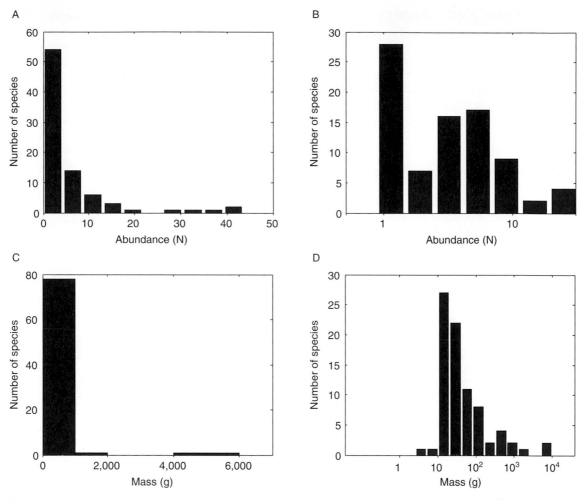

Figure 7.2 Basic patterns in the distribution of abundance and body mass in a community of 83 terrestrial bird species. Data from the North American Breeding Bird Survey (Sauer et al. 2011) from a 25-mile route in Maine, USA. (A) Histogram of abundances on arithmetic abundance scale. (B) The same as (A) on a logarithmic abundance scale. (C) Histogram of body mass (highest mass is 5800 g found in one species). (D) The same as (C) on a logarithmic mass scale.

upper constraint than a driving force in local community assembly (Cotgreave 1993; Marquet et al. 1995), which makes sense from a theoretical perspective (Blackburn and Gaston 2001).

The size–density relationship is highly dependent on the taxonomic resolution (Nee et al. 1991; McGill 2008). Thus in birds of Britain the −3/4 slope is found only for the whole class Aves and becomes less steep within taxonomic orders. It is fairly well-known that abundance varies at all taxonomic levels, while body mass varies most between higher taxonomic levels (e.g., orders) and relatively little within genera or even

families (Harvey and Pagel 1991; McGill 2008). Thus it is a mathematical necessity that size–density relationships will become increasingly noisy at lower taxonomic levels.

7.3.3 Allometric scaling of animal space use

Abundance is just one way of measuring how organisms use space. In particular, home range size and day range (daily distance traveled, or average speed) are

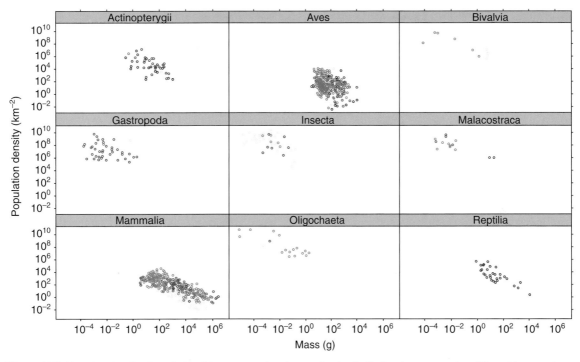

Figure 7.3 Negative size–density relationships among nine classes of animals. Each color represents a different taxonomic order. Data replotted from Damuth (1987) except birds, which are taken from Cotgreave and Harvey (1992) by permission of John Wiley & Sons, Ltd.

commonly reported measures of space use, especially in the mammalian literature. For both variables, scaling theories have been developed that use individual metabolic requirements (i.e., MTE) as a point of departure.

Jetz et al. (2004) presented a model for the scaling of home range. They distinguished between different measures of home range (area per group or per individual) and explicitly modeled overlap among adjacent ranges of conspecifics. Their model predicts isometric scaling of individual area use (home range divided by group size) and that home range exclusivity should scale as $M^{-1/4}$. Predictions were broadly validated by data from a broad range of mammal species, particularly among herbivorous species (Jetz et al. 2004).

Carbone et al. (2005) developed separate models of day range scaling in herbivores and carnivores. Day range was predicted to scale close to $M^{1/4}$ for most terrestrial species, but more steeply among predators when prey group size scales with predator body mass.

Data from 200 mammal species was broadly consistent with these predictions (Carbone et al. 2005).

7.3.4 Beyond energy equivalence

Energy equivalence has become an important concept in metabolic ecology (see also Storch, Chapter 11). Indeed, the notion that it is an ecological "rule" with predictive power has become somewhat entrenched. However, it is a poor descriptor of abundance patterns in communities (see above), and also suffers from conceptual and statistical problems as well (Marquet et al. 1995; Isaac et al., in press). Perhaps, then, energy equivalence is no longer a useful concept in ecology.

Moreover, the degree to which abundance patterns reflect the allocation of resources is highly dependent on the causal relationship between abundance and energy, and on the roles of other processes that affect abundance (e.g., predation, parasitism). Taper and Marquet (1996) showed that variance in abundance (i.e., the spread of the species abundance distribution) is a reliable indicator of the variance in energy flux only where abundance is causally dependent on species' body mass (i.e., following equation 7.1). This may be a reasonable approximation for terrestrial vertebrates (Morlon et al. 2009), but the direction of causality may be different for plants (Enquist et al. 1998) and for animals with indeterminate growth (Reuman et al. 2009).

Recently, research on allocation rules within communities (section 7.3.1) has focused on the distributions of biomass and energy flux among species, in addition to numeric abundance (Connolly et al. 2005; Ernest 2005; Morlon et al. 2009; Henderson and Magurran 2010). Tokeshi (1993) argued that since allocation between species is based on competition for resources, histograms based on energy should be preferred. These various currencies show qualitatively similar patterns (most species are numerically rare, have low biomass, and use little energy), but the differences may be informative. For example, Connolly et al. (2005) showed that distributions based on energy appeared to show more complete sampling of the rare species than those based on abundance. Henderson and Magurran (2010) showed that the plot of total species biomass against abundance is restricted to a triangular region, from which the histograms of each currency can be derived, and the shape of these histograms is determined by the distribution of organisms in space and the intensity of sampling.

Can we find a more general prediction deriving from the intuitive importance of energy to community structure? Perhaps the energy equivalence rule was applied at the wrong level. Instead of trying to explain allocation of energy between species it should be reinvented at the emergent community level. Thus, aggregate community properties such as energy flux and species richness should remain relatively constant, but the distributions of abundance and body size among species should fluctuate from year to year (i.e., any increase in total abundance must be offset by a decrease in average size; White et al. 2004). Such "zero-sum" dynamics have their origins in the idea that organisms increase in population and adapt (within constraints) so as to use all available energy (Van Valen 1973; see also Hubbell 2001; Damuth 2007). Early evidence from desert mammals and tropical trees seems to support the zero-sum prediction for

ecological communities (White et al. 2004; Ernest et al. 2008, 2009).

7.4 CAN ECOLOGY EXPLAIN THE SCATTER IN METABOLIC SCALING?

The bulk of this chapter explores how metabolic theory informs population and community ecology. However, there is at least one important topic where the information flows in reverse. While it is undeniable that body size provides the single best first-order explanation of variation in metabolic rate (i.e., Kleiber's 3/4 scaling law), there remains considerable variation around the line. Metabolic rates show about an order of magnitude variation around the Kleiber line within mammals (McNab 1986) and birds (McGill 2008).

Much residual variation in metabolic rate appears to be associated with species life history (the "fast–slow continuum," see Gittleman and Stephens, Chapter 10). Species above the Kleiber line tend to show a suite of traits associated with "fast" life histories: short gestation, high fecundity, small neonates, rapid ontogenetic growth, early maturity, and short lifespan. Species below the line tend to show the opposite set of "slow" traits. One explanation is that fast–slow continuum is a response to climatic variability, with highly unpredictable environments selecting for low metabolic rates (presumably the better to survive bad times) and vice versa (Lovegrove 2000, 2003; but see Anderson and Jetz 2005). Another explanation focuses on diet (McNab 1986, 2002, 2003) with species eating fruits, leaves, and invertebrates having slow metabolisms for their body size and nut, grass (grazers), and vertebrate eaters having fast metabolisms. McNab also notes that lifestyles have an effect, with arboreal and burrowing species having slow metabolisms and aquatic species having fast metabolisms (see also Sibly and Brown 2007).

A final approach examines life-history theory, which assumes that mortality rates drive the position on the fast–slow continuum (Read and Harvey 1989; Promislow and Harvey 1990; Charnov 1991; Kozlowski and Weiner 1997). These studies especially emphasize that it is not just metabolic rate but a whole suite of life-history traits (mentioned above) that covary on the fast–slow continuum (see Gittleman and Stephens, Chapter 10). Thus, it may not be possible (or desirable) to decisively choose one cause as primal over the other. Yet it is clear that bringing in some ecological understanding provides a great deal of insight into understanding the fast–slow continuum which explains much of the variation in metabolic rate that remains after controlling for body size.

7.5 PROSPECTS

At first glance, it is not clear what metabolic theory has in common with the ecology of communities and populations. The emphasis of metabolic ecology is on broad patterns across species spanning a wide range in body mass and temperature, while population and community ecology focus on interactions among organisms of similar size living at the same temperature. Many population and community ecologists are somewhat skeptical about the merits of broad-scale theories and generalizations, and instead value the detailed insights that can be gained from specific systems (Paine, 2010). Increasingly, this distinction is being broken down as macro-scale perspectives are successfully applied to micro-scale research questions, and MTE has played an important role in transposing between these divergent perspectives.

We have seen that the broad-sense version of the energetic paradigm is widely supported by the available data, but evidence for the narrow-sense version is patchy. We can now return to our original question, about how far up the biological hierarchy can the effects of metabolism be observed. Impressively, and perhaps surprisingly, the signal of metabolism remains clear among populations and communities. Less surprisingly, the signal is much weaker than at the organismal level. This was nicely demonstrated by Tilman et al. (2004), who showed that the variance explained (i.e., r^2) by body size is 95% or greater for individual metabolic rate, but falls to just 10–50% of the variance for population and community parameters, reflecting the growing importance of other (non-metabolic) factors. But 10–50% is not trivial, and few other theories in biology have any kind of predictive power so far from the point of origin. Thus, metabolic theory has had some great successes in explaining patterns of ecological variation, but also some notable failures and a large number of equivocal results. We expect this pattern to continue in the future. New theory will emerge that further integrates population and community ecology into metabolic theory, some of which will be more successful than others.

What, then, will this theory look like? One clue comes from a review of so-called "unified" theories of biodiversity (McGill 2010), which showed how diverse branches of mathematics have, in parallel, been applied to explore the linkages between the species abundance distribution, the species area relationship, and other macroecological patterns. One of these, Maxent (Harte et al. 2008), includes a constraint on total energy and assumes 3/4-power metabolic scaling, from which it derives both the species abundance distribution and the size–density relationship. Is it possible that other axes of variation, such as the fast–slow continuum, might be added to generate a unified theory of organismal diversity? A major challenge for such theories is that key parameters of interest cannot be predicted but are taken as inputs (McGill 2010): most unified theories of biodiversity start with a value for the total species richness, and MTE takes species' body masses as a given. The lack of prediction about these inputs (richness or body mass) suggests that ultimately we will need to adopt a broader framework in which evolutionary theory plays a central role (Isaac et al. 2005; White et al. 2010).

Empirical research in this field also faces challenges ahead. Increasingly sophisticated techniques are being applied to test scaling patterns, with relatively little critical assessment of assumptions and implications of the models and theory under investigation (see also White, Xiao, Isaac, and Sibly, Chapter 1). Too many studies test a single causal explanation for a given pattern, and we encourage pluralism in hypothesis testing. For example, non-energetic explanations for the size–density relationship have been proposed (Blackburn et al. 1993; Cotgreave 1993), but are rarely tested (see also Hechinger, Lafferty, and Kuris, Chapter 19). Collective effort is required to ensure that the strongest possible inferences are drawn from the available data (McGill 2003).

ACKNOWLEDGMENTS

We are grateful to David Storch for stimulating discussions and insights. April Hayward, Ryan Hechinger, Owen Petchey, and the editors provided constructive criticism on previous versions of the manuscript. Van Savage kindly contributed data. NI was partially supported by a NERC Fellowship (NE/D009448/2).

Chapter 8

PREDATOR–PREY RELATIONS AND FOOD WEBS

Owen L. Petchey and Jennifer A. Dunne

SUMMARY

1 This chapter addresses how knowledge of the body sizes of prey and predators can augment understanding of feeding ("trophic") interactions among individuals, and the structure and dynamics of complex food webs.

2 Foraging characteristics of individuals, such as visual acuity and movement speed, can be combined with population-level allometries to predict how frequency of feeding depends on prey and predator body sizes.

3 Body sizes of prey and predators have long been thought to structure food webs. Cutting-edge models can tell us the types of food webs and interactions that are well predicted by organismal sizes, and conversely when other traits need to be considered and modeled.

4 Results of population dynamic models show how constraints placed by body size on metabolic rates and interaction strengths can allow otherwise unstable complex food webs to persist.

5 Situations in which body size appears not to influence trophic relations and food webs may have diverse causes. Ontogenetic diet shifts and averaging of sizes can obscure effects of individual size. Some interactions seem not to depend on size.

6 Body size has fundamentally important effects on individual interactions, food web structure, and food web dynamics. Effects specific to particular taxonomic groups and ecosystem types are unclear at present, but systematic quantitative analyses are emerging. More will be required to fully understand the trophic consequences of metabolism.

8.1 INTRODUCTION

The metabolic theory of ecology (MTE) concerns how body size and temperature affect the metabolic rate of individual organisms, and the ramifications for population-, community-, and ecosystem-level processes and patterns (Brown et al. 2004). This chapter focuses on explanations of community-level patterns, in particular the structure, dynamics, and stability of networks of trophically interacting species. Trophic interactions and food webs have occupied a prominent place in ecological research, partly because consumption has been

Metabolic Ecology: A Scaling Approach, First Edition. Edited by Richard M. Sibly, James H. Brown, Astrid Kodric-Brown.
© 2012 John Wiley & Sons, Ltd. Published 2012 by John Wiley & Sons, Ltd.

one of the most studied types of interspecific interactions (e.g., Elton 1927; Holling 1959; Emlen 1966; MacArthur and Pianka 1966; Root 1967; Schoener 1971; Luckinbill 1973; Hassell et al. 1976; Cohen 1977; Holt 1977; Pimm and Lawton 1977; Paine 1988; Polis 1991). Consumption is also one of the most obvious, easiest to document, and important interactions for both ecological and evolutionary dynamics.

One of the most influential mathematical theories in ecology, the complexity-stability theory proposed by May (1972), concerns how population stability is related to three properties of food webs: the number of species, the number of interactions, and the average strength of interaction. May's theory suggests that large complex food webs should be unstable, and challenged ecologists to discover the processes that stabilize the diverse food webs that persist in many natural ecosystems (McCann 2000). As well as addressing May's rather theoretical challenge, studies of food webs have demonstrated that food web properties, such as connectance, are important determinants of responses to environmental change and species extinction (Dunne et al. 2002; Paine 2002). Thus, an understanding of the processes that govern the structure and stability of food webs has the potential to inform about how ecological communities have responded to and will respond to a rapidly changing world (e.g., Petchey et al. 1999; Montoya et al. 2006; Bukovinszky et al. 2008; Harmon et al. 2009; Woodward et al. 2010a).

The general importance of body size in determining the structure of food webs was recognized by Elton (1927), who observed that predators are usually larger than their prey. This chapter can be considered a review of recent advances that explore and elaborate on Elton's observations. The first section, on trophic relations, considers how the size of two individuals affects whether or not one eats the other. The following section, on food web structure, discusses how body size, through its effect on trophic interactions, affects food web structure. Finally, in the section on food web dynamics and stability, we explore how the effects of body size on trophic interactions influence the dynamic stability of complex food webs. Thus, the three main sections describe how effects of body size trickle (or flood) up through levels of ecological organization from individuals, to interactions, to whole networks (Fig. 8.1).

Throughout this chapter we provide pertinent examples, but due to limited space we cannot under-take a comprehensive review. For further examples and discussion we recommend, as a beginning, reviews such as Warren (2005), Woodward et al. (2005a), Dunne (2005), Bersier (2007), Raffaelli (2007), Brose (2010), Stouffer (2010), Petchey et al. (2010b), and Yvon-Durocher et al. (2011a). We use the terms predator and prey throughout, rather than the more generic terms consumer and resource, as most body-size related research thus far has focused on traditional predator–prey interactions, such as fish eating other fish, fish eating zooplankton, or aquatic insect larvae eating other smaller invertebrates. While many of the patterns and concepts we discuss may equally apply to other consumer–resource interactions, such as some types of aquatic herbivory where phytoplankton are eaten by zooplankton and small invertebrates, they will apply less to others, such as uptake of resources by plants (see Enquist and Bentley, Chapter 14) and parasitism (see Hechinger, Lafferty, and Kuris, Chapter 19).

8.2 TROPHIC RELATIONS

The fundamental building blocks of food webs are the links formed when one individual consumes another, for example an individual bat eating a moth, or a ladybird larva eating an aphid (Fig. 8.1). These trophic links are fundamentally metabolic, since they represent the channels along which energy and materials flow. Components of these interactions, such as encounter, capture, handling, and digestion rates, scale with body size. These properties of trophic links make possible a number of approaches for a metabolic and size-oriented research.

8.2.1 Metabolic requirements

Consumption must satisfy energy and mass balance; for example, the metabolic requirements of an individual multiplied by the trophic transfer efficiency gives the energy needed from its prey. Dividing this by the energy content of each resource gives the "killing rate" (number of prey eaten per time) required to sustain metabolism, which can be related to prey size and predator metabolic rate (Peters 1983). A similar approach uses individual body size to give metabolic rate which, multiplied by the trophic transfer efficiency, gives required daily ingestion rate of an individual, which gives required killing rates (Emmerson et al.

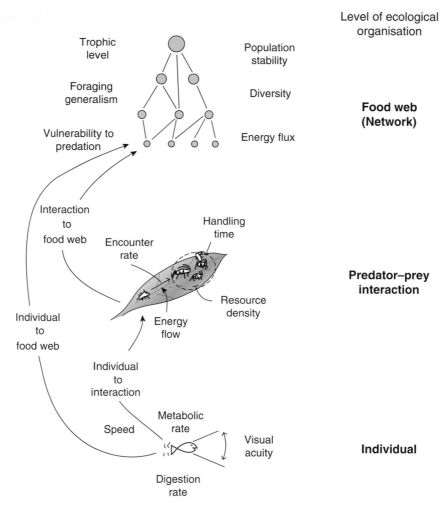

Figure 8.1 Food webs are composed of individuals and the trophic interactions among them. Their structure, dynamics, and response to perturbations depend on characteristics of the individuals within them, such as size, movement rate, visual acuity, and digestion rate, and also properties of interactions between pairs of individuals, such as handling times and attack rates. The scaling with body size of characteristics of individuals and interactions places considerable constraints on the structure, dynamics, and response to perturbations exhibited by models of ecological networks. We illustrate fish at one level and insects at another to suggest that the allometric scaling of foraging traits is not limited to aquatic organisms or ecosystems.

2005). Simple calculations such as these, based on the fundamental metabolic nature of consumption, can be very powerful. For example, these killing rates can be transformed into various measures of per-capita interaction strength and energy flow, and thus estimates of strengths that can explain the stability of complex food webs (de Ruiter et al. 1994; Neutel et al. 2002).

8.2.2 Foraging trait allometries

Killing or consumption rates can also be explored using a more bottom-up approach, by breaking consumption down into a chain of events (e.g., Gergs and Ratte 2009). First a predator must encounter its prey. The frequency with which it does so depends at least on the

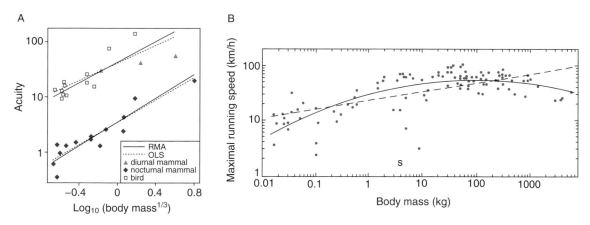

Figure 8.2 The scaling of individual foraging relevant traits with body size. (A) The distance over which organisms can detect or perceive a prey individual depends on body size. (Reproduced from Kiltie 2000 by permission of John Wiley & Sons, Ltd.) (B) Running speed of mammals increases with body size, at least up to an optimum size, and then decreases. (Reproduced from Garland 1983a.) (Also see appendix VIb in Peters 1983.)

density of individuals of the prey species and the predator's ability to detect a prey individual. Population density scales negatively with individual size, such that smaller individuals are generally more common in an area than larger individuals (White et al. 2007; Reuman et al. 2008). Smaller prey individuals may, however, be harder to detect than larger ones, as a result of access to more, smaller, hiding places. These effects of body size will to some extent compensate for each other, and reduce the effect of prey size on the rate at which they are encountered. (Note that considerable care is required when mathematically defining encounter rates; for example, some definitions include capture success, which will also depend on predator and prey size (Brose 2010).) Thus, two allometries, one about a population characteristic (density of prey) and one about an individual characteristic (visibility of prey to predators), can be combined to predict the effect of prey sizes on encounter rate, an important component of predator–prey interactions.

Characteristics of predator individuals, including their body size, will also affect encounter rates via their effect on the ability to detect prey individuals. McGill and Mittelbach (2006) provide a simple geometric model of encounter rates as a function of prey size and predator size. In their model a predator's ability to detect prey individuals is influenced by its visual acuity, which scales positively with body size (Kiltie 2000). Visual acuity determines the distance at which a prey

of a particular size can be detected or encountered (i.e., reactive distance; Brose 2010) (Fig. 8.2A). The speed at which organisms move also scales positively with size at least up to an optimum (Fig. 8.2B), hence larger predators can search an area for prey faster than smaller predators (Garland 1983b; Andersen and Beyer 2006). McGill and Mittelbach's model predicts that encounter rate scales positively with both predator and prey size. They provide three tests of the predictions of the model, all focused on data from predatory fish. First, the model predicts correctly the observed functional form of the relationship between encounter rate and body size. Second, the model correctly predicts lower importance of predator size and greater importance of prey size in more complex habitats. Finally, the model makes good numerical predictions of observed intraspecific variation in encounter rates, with $r^2 = 0.69$. Although the model is based on fish foraging in open water and has only been tested with intraspecific data, these results are encouraging. It seems possible that encounter rates can be predicted using allometric relationships and relatively simple mathematical assumptions.

Many additions could be made to increase the realism (and complexity) of this type of model. One is an effect of prey and predator size on the likelihood that a predator can successfully capture a prey individual. A small predator may have a hard time subduing a prey individual larger than itself, and a large

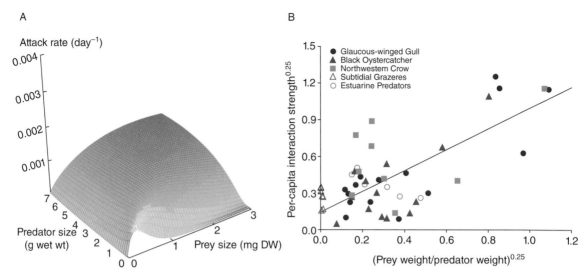

Figure 8.3 The scaling of interactions with predator and prey size. (A) Larger predators attack prey more frequently, all else being equal, though very large predators may be less likely to capture very small prey, leading to a humped relationship between successful encounter rate and predator size. (Reproduced from Aljetlawi et al. 2004 by permission of John Wiley & Sons, Ltd.) (B) Per-capita interaction strengths (the log-ratio measure of interaction strength, log(resource abundance in presence of consumer/abundance in absence of consumer)/consumer abundance) scale positively with predator–prey size ratio. (Reproduced from Wootton and Emmerson 2005, copyright 2011 Annual Reviews.)

predator may have trouble handling a very small prey (filter feedings excepted). These ideas predict a hump-shaped scaling of capture success with both predator and prey size (Weitz and Levin 2006), and Brose (2010) provides numerous examples of hump-shaped scalings (e.g., Fig. 8.3A). Brose points out that these hump-shaped scalings of capture success (and hence encounter and attack rates) are consistent with Elton's observation that there is an optimum size of food usually eaten. Therefore documenting and integrating the allometries of vision, locomotion, abundance, and capture success can provide predictive models about whether and how often predator–prey interactions occur (Gergs and Ratte 2009; Hayward, Gillooly, and Kodric-Brown, Chapter 6).

Another addition would be to tackle the scaling of handling times, defined in foraging theory as the component of foraging time spent consuming and not searching. One can easily imagine that handling time increases with prey size and decreases with predator size. There are, however, few if any datasets about interspecific variation in handling times (Petchey et al. 2008a; Rall et al. 2011). Furthermore, the concept itself may be difficult to operationalize due to the many

biological processes that can influence handling times (e.g., biomechanics, ingestion rate, egestion rate, and digestion rate).

8.2.3 Predator–prey size ratios

In addition to this first-principle, bottom-up approach, one can observe allometries of trophic interactions directly. For example, the empirical relationship between mean prey size and mean predator size across various mammal, bird, lizard, and amphibian taxa indicates that predators eat prey with 0.2 to 10% of their body size, and that prey body size scales with predator size with an exponent of about 1.17 (i.e., $M_{prey} \propto M_{predator}^{1.17}$). Many other near-linear predator–prey size scaling relationships have been documented (e.g., Vezina 1985; Warren and Lawton 1987; Cohen et al. 1993; Memmott et al. 2000; G. C. Costa 2009; Brose et al. 2006a). A recent compilation of predator–prey size ratios supports these findings, but also indicates an exponent of less than 1 in stream communities (i.e., prey of large predators are relatively small; Riede et al. 2011). A recent compilation of prey and predator (mostly fish and squid) sizes goes one step further, by reporting the observed body sizes of interacting individuals (Barnes et al. 2010), rather than mean values for interacting populations. The allometry results from these two approaches can differ (Woodward and Warren 2007), as a result of intraspecific (often ontogenetic) size variation and phenomena such as the smallest individuals of the predator species not eating the largest individuals of the prey species (also see Isaac, Carbone, and McGill, Chapter 7). A focus on individuals' body sizes indicates that, in most cases, prey become relatively smaller as predator size increases (i.e., predator–prey body-size ratios increase with predator body size). (Though the opposite seems to be true of terrestrial carnivores; Vezina 1985.) This leads to the prediction that trophic transfer efficiency (production of predators divided by production of prey) decreases with increasing predator size (Barnes et al. 2010), an example of how one foraging allometry can make predictions about another (also see Jennings, Andersen, and Blanchard, Chapter 21; Hechinger, Lafferty, and Kuris, Chapter 19).

Predator–prey size ratios might represent a shortcut for estimating interaction strengths. Wootton and Emmerson (2005) reported scalings of per-capita interaction strengths with predator–prey size ratios for avian predators, subtidal grazers, and estuarine predators (Fig. 8.3B). These findings have been somewhat corroborated in a recent study (O'Gorman et al. 2009). Very recent studies of feeding rates confirm effects of both size and taxonomic identity (Rall et al. 2011). At present the causes of this interspecific variation in the allometries of interaction strength are poorly understood.

8.2.4 Other consumer–resource interactions

The bulk of research on allometries of feeding relationships has focused on a particular type of predator, sometimes termed "true predators" or carnivores. Herbivores, parasites, parasitoids, pathogens, and other types of consumer exhibit different allometries (e.g., see Hechinger, Lafferty, and Kuris, Chapter 19). For example, the sizes of emerging parasitoids are strongly positively related to aphid host size, with an allometric scaling exponent close to 0.75 for primary, hyper- and mummy parasitoids (Cohen et al. 2005). The foraging allometries of the full range of consumer–

resource types needs to be documented and modeled with the same detail as more classic predator–prey relationships reviewed here.

8.3 FOOD WEB STRUCTURE

Food webs characterize complex networks of trophic interactions among large numbers of co-occurring species. When visualized as species (nodes) connected by trophic links (lines) representing energy and material flow due to food consumption, food webs can appear overwhelmingly complex. However, careful presentation and analysis can make sense of this apparent complexity, revealing structural regularities and organizing principles (Fig. 8.4A).

Investigating "community-level" allometries is one method for revealing such regularities. For example, one can examine how trophic level, generalism (richness or diversity of prey of a species), and vulnerability (richness or diversity of predators of a species) vary with its average body size. Trophic level of predators has been observed to increase with body size (Riede et al. 2011; Romanuk et al. 2011), and to also show no consistent pattern (Layman et al. 2005). Foraging generalism tends to increase with size of predator, and vulnerability to predation to decrease with predator size (Digel et al. 2011). These (and other) community-level allometries are reviewed in Yvon-Durocher et al. (2011a), Jennings, Andersen, and Blanchard (Chapter 21) and Isaac, Carbone, and McGill (Chapter 7). Rather than re-review these empirical patterns, we will discuss how such patterns may emerge from interspecific variation in body size.

8.3.1 The cascade, niche, and related food web models

Very simple models have been used to generate food-web-like network structures representing the feeding relationships among species. One of the earliest and simplest models of food web structure, the cascade model, ranked species along an arbitrary axis, and allowed them to feed only on species with lower rankings (Cohen et al. 1990). More recent variations include the niche model (Williams and Martinez 2000), the nested hierarchy model (Cattin et al. 2004), and generalized versions of the cascade model (Stouffer

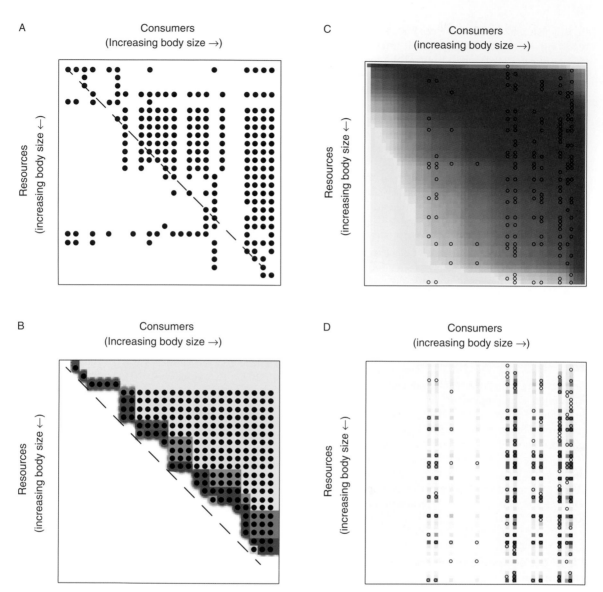

Figure 8.4 Examples of when body size more or less explains who eats whom in ecological networks. Each panel shows a network in the form of a predation matrix. Consumers are in columns and are ordered by body size from left to right. Resources are in rows and are ordered by body size from top to bottom. Feeding interactions are indicated by a black circle. Thus, for example, a consumer's diet is given in a column of the matrix. (A) The observed interactions of Coachella Valley food web (Polis 1991) and (B) the predicted interactions based on a deterministic model of allometric foraging behavior and optimal foraging theory. (Reproduced from Petchey et al. 2008a by permission of the National Academy of Sciences.) (Darker colors indicate a more energetically profitable interaction.) (C) and (D) both show observed interactions (black circles) in a laboratory-based food web – "Sheffield" in the supplementary material of Rohr et al. 2010). Darker reds indicate the predictions of models, so that coincidence of black circles and dark red indicates good prediction. (C) Interactions are poorly predicted by body size. (Reproduced from Rohr et al. 2010 by permission of University of Chicago Press.) (D) Inclusion of traits other than body size greatly increases predictive power. (Reproduced from Rohr et al. 2010 by permission of University of Chicago Press.)

et al. 2005), all of which do a remarkably good job of generating networks that have structure similar to that observed in food webs from a variety of habitats.

These somewhat more complicated models retain the cascade model's ranking of species on a one-dimensional axis, with predator species feeding mostly on species with lower ranks. Subsequent analyses of the structure of food webs have revealed that species and their diets can be closely mapped onto a single dimension (Stouffer et al. 2006). It has been repeatedly suggested that this single dimension may correspond to body size. Indeed, as section 8.2.3 reviews, it has been empirically observed that non-parasite predators tend to feed on organisms smaller than themselves, which would result in a feeding hierarchy in a food web. Also, predators tend to feed on prey within a certain body-size range (Stouffer et al. 2011; Zook et al. 2011), which corresponds to the feeding niche concept embodied in the niche model (Williams and Martinez 2000; Stouffer et al. 2006) but not the other models. Thus, body size appears related to both the hierarchical and niche structuring of empirical food webs.

8.3.2 Size-based food web models

The models mentioned above are simple and do not include body size explicitly. Since the probability of feeding can depend on the sizes of the potential predator and prey, this may be useful to include in more realistic models. This type of model can be fit to observed data about food web structure and species body sizes, and provides a measure of how well the model can explain the data. The fitting procedure is similar in principle to fitting a power-law equation to the relationship between metabolic rate and body size. Such models could be termed "allometric food web structure models." (Allometric food web dynamics models will be discussed in section 8.4.) Rohr et al.'s (2010) allometric food web structure model (of many of the same empirical food webs) displayed quite variable explanatory power; at worst it predicted 2.9% of trophic links correctly, and at best 49%. However, it only includes three parameters so is quite parsimonious.

More complex models are of course possible! For example, one can first translate predator and prey sizes into encounter rates, handling times, and ener-

getic contents. These traits can then be fed into models of diet choice, such as the contingency model of optimal foraging, to produce a prediction of which of the available prey species a predator species will feed upon (Emlen 1966; MacArthur and Pianka 1966). Such a model of the structure of 15 empirical food webs found that from about 5% to 65% of the trophic links were successfully predicted (e.g., Fig. 8.4B) (Petchey et al. 2008a; Thierry et al. 2011). More phenomenological models (e.g., Allesina 2011) have similar explanatory power and it remains to be seen whether optimal foraging is really necessary for predicting food web structure, or whether simpler phenomenological models are just as predictive and useful (Petchey et al. 2011). Furthermore, trade-offs between time spent foraging and on other activities may be important (see Hayward, Gillooly, and Kodric-Brown, Chapter 6).

8.3.3 Hybrid models

Other approaches for testing the importance of body size for food web structure are combinations of the two general approaches outlined above. For example, the "probabilistic niche model" (Williams et al. 2010) fits the three parameters of the niche model (Williams and Martinez 2000) for each species in order to maximize the match between predicted and observed food web structure. An allometric version of this probabilistic model reproduces fewer links correctly than a non-allometric version, but at the benefit of having 29 fewer parameters estimated.

8.3.4 Does size explain food web structure?

The results of these modeling exercises suggest that sometimes body size strongly explains food web structure, and other times it does not. This is not simply a consequence of the common perception that aquatic food webs are size structured, and terrestrial food webs are not. Indeed, size provides a very good explanation of the structure of a terrestrial food web of plants, insects, arachnids, and other predators, in the Coachella Valley (Polis 1991; Petchey et al. 2008a) (Fig. 8.4B). Two terrestrial food webs were, however, quite rather poorly predicted, though this is probably because they are dominated by host–parasitoid interactions, for which the contingency model of optimal foraging is inappropriate.

Another surprise was the relatively poor ability (~40% of links correctly predicted) of the model to explain the structure of the Broadstone Stream food web (a food web containing macroinvertebrate predators). Interactions in this food web are known to be extremely size-dependent (Woodward et al. 2005b). However, if the Broadstone food web is recompiled by size only, ignoring individuals' taxonomic identity, the model was then able to predict over 80% of the trophic links correctly . The model failed to predict the taxonomic version because aggregation of individuals into species ignored large ontogenetic changes in body size and diet. Hence, the use of individual- versus population-level data can fundamentally alter our understanding of body-size effects on food web structure (Werner and Gilliam 1984; Woodward and Warren 2007; Woodward et al. 2010b; Yvon-Durocher et al. 2011a). More generally, we must carefully ensure that assumptions of our models are not violated by the data we test them against.

Differences in the explanatory power of size-based foraging could have at least two other sources. First, organismal traits other than size may determine who interacts with whom. For example, herbivorous insects may be more limited by plant chemical defenses than by some measure of plant size. Rohr et al. (2010) were able to greatly increase the explanatory power of their food web model by allowing each species to be represented by its body size and two other traits (Figs 8.4C, D). These other traits, related to phylogenetic differences, modified how likely were predators to consume and prey to be consumed. It is possible that they described differences in biological characteristics such as microhabitat use, camouflage, specialized defenses, and offensive adaptations. A second explanation of poor performance of these models is that they have allowed only one allometry per food web. Perhaps size has different implications for different types of interactions and taxa. Differences in allometries of foraging traits among different types of predators, such as homeotherms and ectotherms, are not included in current models, and indeed are poorly documented (Petchey et al. 2008a). However, adding different allometries and/or foraging strategies will increase the complexity of the model, which may help to explain specific deviations from a general pattern, but draws us further from simple, general explanations.

8.4 FOOD WEB DYNAMICS AND STABILITY

Characterizing the structure of food webs as a network of the feeding interactions (links) among species (nodes) is a simple and powerful way to think about complex relationships among species. However, each species and each interaction is also changing through time in ecosystems; for example, the population of a species increases and decreases. These food web "dynamics" are influenced by body size, which in turn influences the stability of the whole system of interaction species. Understanding the dynamics of food webs is helping us to understand how many species can coexist and persist, even as their populations fluctuate (Dunne et al. 2005), and how species interactions mediate ecosystem functioning (Cardinale et al. 2008). Because it is very hard to track many co-occurring species' abundances through time in natural ecosystems, modeling has been a very important tool for understanding the relationship between dynamics and body size.

8.4.1 A bioenergetic foundation for dynamics and stability

In a seminal paper in 1992, Yodzis and Innes presented an allometric approach to modeling consumer–resource dynamics. Each population has a particular metabolic type and a characteristic body size, and is modeled as a stock of biomass that decreases due to losses to predation and metabolism, and increases due to gains from feeding or primary production. Most metabolic parameters are allometrically scaled: intrinsic growth rate of the resource, respiration rate of the consumer, and maximum consumption rate of the resource by the consumer all scale with body size to the 3/4 power (Peters 1983; Brown et al. 2004). Thus, the Yodzis and Innes "bioenergetic" approach uses abstractions based on a few interdependent, easy to measure, body-size based parameters as a way to model complex interspecific dynamics in a reasonably simple and biologically plausible way. This provides for greater biological realism than that expressed by randomly parameterized Lotka-Volterra models, and at the same time reduces the "plague of parameters" that typically afflicts multispecies population dynamic models (Yodzis and Innes 1992).

A number of studies based on this approach examined how different factors affect the stability of populations in small trophic modules, for example in short food chains (reviewed in Dunne et al. 2005). Subsequent research increased the size of the networks modeled and updated the allometric equations used (Brose et al. 2003; Williams and Martinez 2004; Martinez et al. 2006). These studies explored the impact of factors such as diversity and network structure on the persistence of species in model systems, a key measure of community stability (Martinez et al. 2006). However, these studies lacked a realistic body size distribution: they implicitly assigned a single body size to all taxa, leaving unanswered the question of how predator–prey body-size ratios affect species persistence in food webs.

8.4.2 Predator–prey size ratios and stability

Body-size ratios among certain kinds of predators and prey are well documented, for example invertebrate predators are ~10 times larger than their prey, and ectotherm vertebrate predators are ~100 times larger than their prey (Brose et al. 2006a). When a dynamic bioenergetic model includes various predator–prey body-size ratios, from inverted ratios (predators 100 times smaller than prey; sometimes referred to as micropredators) to very large (predators up to 100 000 times the size of their prey), the persistence of species tended to saturate at body-size ratios quite close to the empirically observed values (10–100; Brose et al. 2006b) (Fig. 8.5A). The effect of predator–prey body-size ratio on persistence was much stronger than effects of network structure, metabolic type, species richness, or other factors.

Thus, predator–prey body-size ratios appear fundamentally important for the persistence and coexistence of species in complex trophic networks. But why does allometric scaling, including body-size ratios, facilitate species persistence? Perhaps it is simply that predators that forage on only very small prey cannot meet their energetic requirement, and predators that forage on large prey are condemned by the paradox of enrichment (population stability decreasing with increasing productivity; Oksanen et al. 1981). High persistence in model systems appears restricted to intermediate range of possible predator–prey body-size ratios (Otto et al. 2007). Empirical data focused on predator–prey interactions support these conclusions, as 97.5% of invertebrate food chains from five food webs fall within a "stability domain" of intermediate body-size ratios (Fig. 8.5B) (Otto et al. 2007). In model food webs, allometric scaling increases intraspecific competition relative to metabolic rates for higher body-sized species and reduces biomass outflow from prey to predator when the predator is larger than the prey, resulting in increased persistence.

8.4.3 Simplicity from complexity

The "allometric food web dynamics model" approach described before (Brose et al. 2006b) has been used to explore other aspects of the dynamics and stability of complex ecological networks with realistic topologies. For example, studies have examined keystone effects (Brose et al. 2005), competitive exclusion among producer species (Brose et al. 2008), the paradox of enrichment (Rall et al. 2008), the relative roles of network structure, models of primary production, functional response, generalist behavior (Williams 2008), and interaction strength (Berlow et al. 2009). The latter study systematically conducted every possible single-species knockout in 600 networks with S (species richness) of 10–30, as a means of quantifying the interaction strength for each species pair, in terms of the change in biomass of one species in response to the removal of another species. Per-capita interaction strengths were well-predicted by a simple linear equation with only three variables: the two species' biomasses and the body size of the removed species (Fig. 8.5C; Berlow et al. 2009). When tested with empirical data from an experiment in an intertidal ecosystem, the equation successfully predicted the interaction strengths of taxa when trophic interactions, rather than spatial competition, dominated (Berlow et al. 2009). The relationship between interaction strength and body size may be connected with a recent finding that the keystone status of a species is closely linked to its body size (Berg et al. 2011).

8.4.4 Evolutionary dynamics added

Other kinds of dynamical models have also explored the importance of body size for the evolution and persistence of species in complex ecological networks (e.g., Loeuille and Loreau 2006; Guill and Drossel 2008; Rossberg et al. 2008; Stegen et al. 2009). These studies

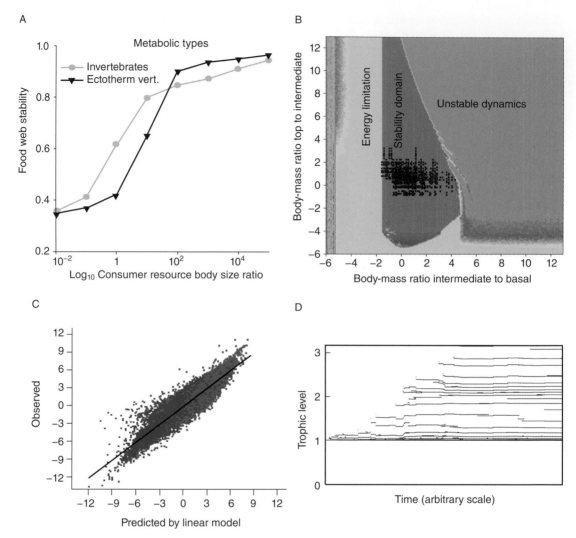

Figure 8.5 Allometries at the individual and interaction level constrain the dynamics of ecological networks. (A) Persistence of species as a function of predator–prey body-size ratios, in dynamical models of food webs parameterized for invertebrates and ectotherm vertebrates. (Reproduced from Brose et al. 2006b by permission of John Wiley & Sons, Ltd.) (B) The stability domain of species coexistence in tritrophic food chains is limited by energy availability and unstable dynamics towards low and high body-size ratios, respectively. The black points illustrate food chains of the Skipwith pond food web as an example. (Reproduced from Brose 2010 by permission of John Wiley & Sons, Ltd.) (C) Observed $\log_{10}$ | per capita interaction strength | versus that predicted by a multiple linear regression (based on population biomasses of the removed and target species and the body size of the removed species) in an "allometric food web dynamics model". Red points correspond to low population biomasses of the removed species, and blue points to high population biomasses. (Reproduced from Berlow et al. 2009 by permission of the National Academy of Sciences.) (D) Trophic levels (y-axis) of species that evolve during 10^8 time steps (x-axis) at high niche width and competition intensity, in an eco-evolutionary model in which adaptation works on body size. (Reproduced from Loeuille and Loreau 2005 by permission of the National Academy of Sciences.)

assembled food webs via stochastic speciation events, and the resulting species interacted according to population dynamic equations. In one model, adaptation worked on a single trait, body size, and resulted in persistent, complex food webs (Loeuille and Loreau 2006). The relative roles of predation, which resulted in species with distinct body sizes, and interference competition, which resulted in homogenization of body sizes, influenced food web structure by altering the degree to which species sorted into more or less distinct trophic levels (Fig. 8.5D). An extension of this approach incorporated metabolic temperature dependencies (Allen et al. 2006) as a means of extending metabolic theory to macro-evolutionary patterns (Stegen et al. 2009). Evolutionary models have also explored the emergence of body-size–abundance scaling relationship (Rossberg et al. 2008), and have linked the niche value of the niche model (Williams and Martinez 2000) to species body size and used evolutionary game theory to determine individuals' foraging efforts (Guill and Drossel 2008). All of these approaches produced complex, persistent food webs and provide some initial insights on the interplay of ecological dynamics, evolutionary dynamics, and body size.

8.4.5 Stability explained?

Thus it seems that body size and its effects on vital rates and interactions among species provides an elegant, if not completely understood, explanation for the stability (persistence) of at least some complex food webs comprising predators and their prey. Furthermore, allometric scaling appears to be related to other stabilizing mechanisms, such as weak interactions (McCann et al. 1998) and weak long loops (Neutel et al. 2002). Perhaps allometric scaling across levels of ecological organization, and feedback among levels, will provide one of the most powerful responses to May's diversity-stability challenge. Ultimately, however, we need to include in explanations of species persistence and coexistence the diversity of trophic interactions in food webs (Lafferty and Kuris 2002), and also the diversity of non-trophic interactions among organisms.

8.5 WAYS FORWARD

We hope that the previous sections have convinced readers that body size places constraints on predator foraging behavior, on interactions between prey and predators, on the structure of food webs, and on the dynamics and persistence of populations and communities. As mentioned, we have explicitly focused on particular types of interspecific interactions, primarily predation, and it remains to be seen what constraints body size places on other types of trophic interactions, or indeed other types of interspecific interaction (e.g., parasitic and mutualistic interactions) (Stang et al. 2006; Ings et al. 2009; Warren et al. 2010).

Other holes in need of filling include the following.
1 The importance of body size for the response of ecological networks to perturbations such as climate change (Rall et al. 2008; Petchey et al. 2010a), species loss and secondary extinctions (Dunne and Williams 2009), species invasions (Romanuk et al. 2009), and changes in ecosystem function (Cardinale et al. 2008) have yet to be widely studied or experimentally tested (though see O'Gorman et al. 2009; Woodward et al. 2010a). Experimental manipulative tests of interaction and food web models are lacking. While some models have faced such tests (e.g., Emmerson and Raffaelli 2004), many remain experimentally untested.
2 Effects of temperature on trophic interactions and food web structure are relatively unexplored, either theoretically or experimentally. Recent theoretical analyses suggest different activation energies (see White, Xiao, Isaac, and Sibly, Chapter 1) of metabolism and consumption will lead to stabilizing effects of temperature on population dynamics (Vucic-Pestic et al. 2011). Predictions of the effects of temperature on food web structure and metabolic balance of ecosystems are also sensitive to differences in activation energies of different processes (Petchey et al. 2010a; Yvon-Durocher et al. 2010). These findings suggest that warming might change dynamics above and beyond simply speeding them up.
3 A systematic review of foraging allometries, and continued experimental investigation of these allometries and their response to environmental change, would provide the data required to make specific predictions about effects of adaptation to environmental change (Brose 2010). At present we can make general predictions, but have insufficient knowledge about foraging allometries to make predictions that account for taxonomic composition of a specific food web.
4 Although general opinion says that aquatic communities, at least pelagic ones, are more size structured than terrestrial ones, thorough analyses comparing ecosystems are rare (Digel et al. 2011; Riede et al.

2011). Furthermore, analyses across multiple levels of ecological organization would be extremely informative (Yvon-Durocher et al. 2011a).

5 A greater appreciation of intraspecific variability, ontogenetic diet shifts, and their consequences for food web persistence is needed. It is abundantly clear that the perception of food web structure (Woodward et al. 2010b) and dynamics (Persson et al. 1998; Rudolf and Lafferty 2011) are fundamentally altered by size/stage-structuring, yet most observed food webs do not include any kind of data about intraspecific variation. Particularly interesting approaches involve elegant mathematical modeling of individual-level processes that result in community structure and dynamics (e.g., De Roos et al. 1992; Hartvig et al. 2011).

6 Dynamical models can be augmented and improved through insights from empirical and experimental work (e.g., Arim et al. 2010; Vucic-Pestic et al. 2010) and theoretical work (Weitz and Levin 2006). For example, the population dynamics of such models do not systematically incorporate hump-shaped attack rates, certain parameters are not yet allometrically scaled, and differences among predator groups are not addressed (Brose 2010).

7 The relationship of species persistence to diversity and complexity in dynamical models remains unclear and seems to be sensitive to small differences in the modeling approach (Brose et al. 2006b; Kartascheff et al. 2010). Furthermore, it is unclear what are the feedbacks and paths of causation that connect individuals, interactions, and population persistence. Do prey–predator size ratios drive dynamics, or does selective extinction drive observed size ratios?

ACKNOWLEDGMENTS

This research was partially supported by a Royal Society University Research Fellow awarded to OLP, by the University of Zurich, and NSF grants DBI-0850373 and DBI-1048302 and a National Academies/Keck Futures Initiative grant to JAD.

Chapter 9

ECOSYSTEMS

Kristina J. Anderson-Teixeira and Peter M. Vitousek

SUMMARY

1 This chapter addresses how the metabolic rates of individuals – which vary with body size, temperature, water availability, and stoichiometry – shape ecosystem-level processes.
2 The metabolic theory of ecology (MTE) yields insights into how the sizes, and respective metabolic rates, of the individuals in an ecosystem shape the cycling of energy and materials in that ecosystem by combining individual allometries with those describing the size distribution of individuals.
3 Temperature is a key driver of ecosystem dynamics. MTE provides a valuable framework for understanding (a) the scaling of the inherent temperature responses of production and respiration across spatio-temporal scales and (b) how effective temperature dependence is shaped by interacting temperature-dependent processes, both of which remain important areas for research.
4 Water strongly shapes both individual metabolism and ecosystem energetics, yet its effects remain little explored from the MTE perspective.
5 Biogeochemical cycling of key nutrients (such as nitrogen and phosphorus) shapes whole-ecosystem metabolism and is shaped by biotic metabolism.
6 Metabolic ecology holds potential to address many exciting questions about how body size distributions, temperature, water, and nutrient availability shape ecosystem energetics through their interactive effects on individual metabolism.

9.1 INTRODUCTION

As has been discussed in previous chapters, metabolic rates of organisms – that is, the pace at which they take up, transform, and release energy and materials – shape their interactions with the environment. In turn, metabolic rates are shaped by body size, temperature, and resource concentrations.

At the ecosystem level, we are concerned with how entire ecological communities – the assemblage of many individuals – absorb, transform, and store energy and materials as they interact with the abiotic environment. The idea that individual metabolism shapes ecosystem-level processes is by no means unique to metabolic ecology; there is a rich tradition of research linking individual physiology to ecosystem-level processes. What is unique in metabolic ecology's approach is that it applies certain unifying principles that quantitatively link across all levels of biological organization – and explicit mathematical descriptions of these principles – and examines their implications for ecosystem-level phenomena. This gives new

Metabolic Ecology: A Scaling Approach, First Edition. Edited by Richard M. Sibly, James H. Brown, Astrid Kodric-Brown.
© 2012 John Wiley & Sons, Ltd. Published 2012 by John Wiley & Sons, Ltd.

insights into the functioning of ecosystems, and also feeds back to our understanding of how individual metabolism is shaped by the environmental context.

In this chapter, we review how body size, temperature, water, and nutrient availability – through their influence on individual metabolism – shape the flux, storage, and turnover of energy and materials at the ecosystem level. We discuss research that explicitly links ecosystem functioning to metabolism using an MTE framework (Brown et al. 2004), and set this work in the broader context of other research addressing similar questions from other perspectives.

9.2 BODY SIZE

Ecosystems differ in the size distribution of their individuals; for example, compare the average size of primary producers of the open ocean (unicellular algae) with those of a tropical rainforest (40 m trees). Because of the pervasive influence of body size on ecological rates (Peters 1983), differing size distributions imply differences in energy and material cycling at the ecosystem level. Fluxes of energy and elements at the individual level are controlled by metabolic rate and therefore often exhibit the same $M^{3/4}$ scaling as metabolic rate. For example, the following scale as approximately $M^{3/4}$: uptake and release of CO_2 by plants and animals (Ernest et al. 2003; Nagy 2005; Niklas 2008), water flux in animals and plants (Enquist et al. 1998; Nagy and Bradshaw 2000; Meinzer et al. 2005), and nitrogen fluxes through insect and mammalian herbivores (Meehan and Lindroth 2007; Habeck and Meehan 2008). Moreover, elemental composition is linked to body size (Kaspari, Chapter 3); for example, concentrations of elements associated with key metabolic structures (e.g. chloroplast-associated N in leaves, ribosome-associated P) appear to scale as $M^{-1/4}$ (Fig. 3.4) (Allen and Gillooly, 2009; Kaspari, Chapter 3). Allometries such as these, when combined with information on the numbers and sizes of individuals in an ecological community, provide a basis for estimating ecosystem-level cycling of energy and materials. Leaving in-depth discussion of how body size shapes community structure and function to other chapters (Isaac, Carbone, and McGill, Chapter 7; Enquist and Bentley, Chapter 14), we here highlight some findings that are particularly pertinent to ecosystem functioning.

9.2.1 Body size dependence of ecosystem-level fluxes

Ecosystem-level fluxes – such as gross primary productivity (GPP, or whole-ecosystem photosynthesis), ecosystem respiration (summed respiration of all organisms in the ecosystem), or uptake and release of elements – shape both the internal dynamics of ecosystems as well as their roles in global biogeochemical cycles. While the corresponding individual-level fluxes scale predictably with body size (generally, $M^{3/4}$), whole-ecosystem flux can be more-or-less independent of body size. Within a population, size-class of individuals, or even-aged stand, changes in abundance and metabolic rate are often inversely correlated, such that net flux of energy and materials by a population is independent of body size ("energetic equivalence"; Damuth 1981; Belgrano et al. 2002; White et al. 2004, 2007; Ernest et al. 2009). For example, in mammalian herbivores, the $M^{3/4}$ scaling of nitrogen excretion (Fig. 9.1A) and the $M^{-3/4}$ scaling of population density (Fig. 9.1B) cancel such that population N flux is independent of body size (Fig. 9.1C) (Habeck and Meehan 2008). Similarly, because xylem flux (a metric of plant metabolism) and plant density scale inversely with mass, ecosystems dominated by plants of very different sizes can have similar whole-ecosystem metabolism (e.g., transpiration; Enquist et al. 1998; Belgrano et al. 2002). This somewhat counterintuitive result tells us that ecosystem-level fluxes will typically vary more strongly with temperature, water, and resource availability than with average plant size.

9.2.2 Biomass turnover

Biomass turnover – that is, how quickly any given unit of biomass (e.g., a gram of leaf, a whole plant, or a microbe) dies and is replaced – is shaped by the body sizes of individuals within an ecosystem. As discussed in other chapters, rates of individual growth, reproduction, mortality, and the intrinsic rate of population increase scale allometrically as approximately $M^{-1/4}$. As a result, the turnover rate of biomass in ecosystems exhibits quarter-power scaling with average plant mass (Fig. 9.2) (Allen et al. 2005). Simply put, aquatic ecosystems with algae as primary producers turn over carbon at high rates whereas the carbon stored in trees of forests persists in the ecosystem for a much longer time. These findings emphasize that much of the

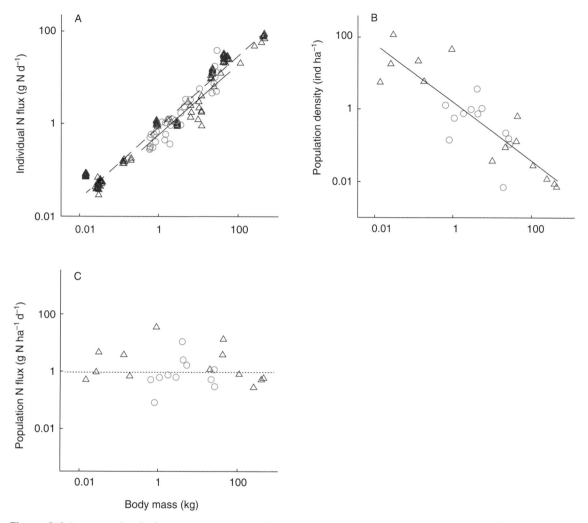

Figure 9.1 In mammalian herbivore populations, the allometric scaling of (A) individual nitrogen flux ($M^{3/4}$) combines with (B) the $M^{-3/4}$ scaling of population density to yield (C) mass-independence (M^0) of population N flux. Eutherians and metatherians are represented by triangles and circles, respectively, and by dashed and solid lines in (A). (Redrawn from Habeck and Meehan 2008; © 2008 Blackwell Publishing Ltd/CNRS.)

observed variation in rates of biomass turnover among biomes (grasslands, shrublands, forests) can be predicted based on the size dependence of metabolic rate. Size-dependent changes in turnover rates of biomass and individuals may also contribute to the observation that *species* turnover through time (changes in the set of species present) is generally faster for communities comprised of smaller-bodied organisms (Schoener 1983). This effect can be observed in ecological communities developing on a new substrate or recovering

from a disturbance; rates of species turnover tend to be highest during early stages of succession when plants are relatively small and short-lived (Anderson 2007).

9.3 TEMPERATURE

Temperature is one of the major drivers of energy flow and biogeochemical cycling in ecosystems. From a physical chemistry standpoint, the rate (V) of any

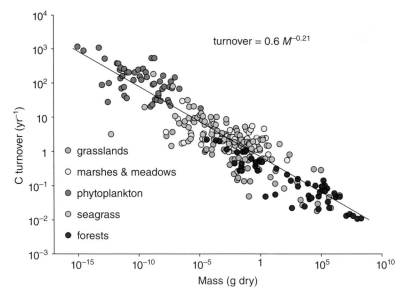

Figure 9.2 Carbon turnover in ecosystems varies as a function of average plant mass ($\propto \bar{M}^{-1/4}$). (Modified from Allen et al. 2005.)

biochemical reaction increases exponentially with temperature (T), and also depends upon reactant concentration (R):

$$V = v_0 e^{-E/kT} [R_1]^{m_1} [R_2]^{m_2} \qquad (9.1)$$

Here, v_0 is a normalization constant, and $e^{-E/kT}$ is the Arrhenius expression where k is Boltzmann's constant ($8.62 \times 10^{-5}\,\text{eV}\,\text{K}^{-1}$), and E (eV) is an activation energy characterizing the temperature dependence. Equation 9.1 is similar to the basic Arrhenius equation, only modified to include the effects of reactant concentration. The last two terms, $[R_1]^{m_1}$ and $[R_2]^{m_2}$, are the concentrations of "reactants", R_1 and R_2, raised to powers, m_1 and m_2, which characterize the effects of their concentrations on V.[1] If we apply this to ecosystems, [R]s may symbolize a range of resources and metabolic structures, including biologically available limiting nutrients (e.g., nitrate, phosphate), key enzymes, organelles (e.g., mitochondria, chloroplasts),

or structures that house the organelles (e.g., microbes, leaves) (Anderson-Teixeira et al. 2008). Importantly, different reactions are characterized by different Es, implying variation in the temperature-sensitivity of the biochemical and geochemical reactions that govern ecosystem dynamics.

Much of the challenge of understanding temperature dependence at the ecosystem level lies in the fact that multiple reactions with different Es co-occur, indeed interact, in any ecosystem. For example, as discussed below, photosynthesis and respiration differ in their fundamental sensitivity to temperature. In the long term, however, the flux of carbon through respiration is constrained by the amount of carbon fixed in photosynthesis. At the same time, heterotrophic[2] respiration is the main pathway by which nutrients are recycled in ecosystems – so that photosynthesis can be constrained by respiration through its influence on nutrient supply. More generally, the availabilities of multiple resources ([R] in equation 9.1) often co-vary with temperature, particularly when [R] is shaped by other temperature-dependent chemical reactions.

---

[1] For simplicity, we describe the effects of [R] as [R]^m. The Michaelis–Menten model provides a more realistic way to model the effects of [R] for enzyme-mediated reactions. Note that $0 < m < 1$ gives a decelerating reaction rate with increasing [R], as is commonly observed for enzyme-mediated reactions.

[2] Heterotrophs are organisms that obtain energy by feeding on organic matter produced by other organisms (e.g., bacteria, fungi, animals).

Some of these may be abiotic geochemical reactions, such as the weathering of minerals that in the long term supplies many of the nutrients that organisms require (Kaspari, Chapter 3).

These interactions can alter the observed temperature dependence of ecosystem fluxes substantially (Anderson-Teixeira et al. 2008). In this chapter, we use E to refer to the inherent activation energy of a chemical reaction in the absence of resource limitation, and ε to refer to the effective temperature dependence of processes within and across ecosystems (Anderson-Teixeira et al. 2008).

9.3.1 Inherent temperature dependence of respiration and production

On a biochemical level, aerobic respiration in all organisms, including both plants and heterotrophs, responds more strongly to temperature than does photosynthesis (GPP on an ecosystem level). MTE has often characterized the temperature dependence of respiration and production using an Arrhenius expression with Es of 0.65 and 0.3 eV, respectively (Fig. 9.3) (Allen et al. 2005). This contrast has implications for both the

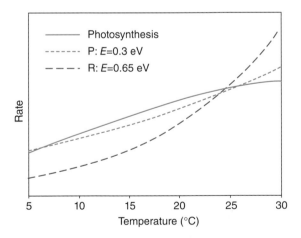

Figure 9.3 Schematic illustration of MTE's commonly used approximations of the temperature dependence of aerobic respiration (red dashed line, $R \propto e^{-0.65/kT}$) and terrestrial primary productivity (green dashed line, $P \propto e^{-0.3/kT}$; Allen et al. 2005). Also shown is the temperature dependence of Rubisco-limited terrestrial C_3 photosynthesis (solid green line; Bernacchi et al. 2001), which is the basis for the approximation of Allen et al. (2005).

energy and carbon balances of ecosystems. As temperature increases, the ratio of photosynthesis to respiration decreases, implying increasing energetic limitation at higher temperatures. In addition, the balance between carbon dioxide (CO_2) uptake through photosynthesis and CO_2 release shifts such that CO_2 release is favored at higher temperatures.

The differential temperature sensitivities of photosynthesis and respiration are significant on a global scale, particularly in the face of climate change. It is increasingly important that we understand mechanistically how temperature shapes whole-ecosystem energetics and biogeochemical cycling. The effect of temperature on carbon cycling in ecosystems is particularly significant because terrestrial ecosystems influence climate in part through the exchange of CO_2 (a greenhouse gas) with the atmosphere. When GPP exceeds ecosystem respiration, the ecosystem is acting as a net CO_2 sink, thereby providing a climate benefit. In contrast, if ecosystem respiration exceeds GPP, CO_2 is released from the ecosystem and contributes to climate change. If changing temperatures shift the balance between production and respiration – as we may expect, based on the different temperature sensitivities of photosynthesis and respiration (Fig. 9.3) – rising temperatures could trigger a positive feedback to climate change (Kirschbaum 1995). Therefore, understanding how the temperature responses of production and respiration interact and play out across a range of spatio-temporal scales is a central question in both metabolic and global change ecology. Of course, understanding the carbon balance of ecosystems is more complex than simply scaling the temperature dependence of GPP and respiration – and we will consider some of that complexity here – but a simple metabolic approach provides a useful framework against which these complexities can be tested and understood.

9.3.1.1 Ecosystem respiration

Typically, aerobic respiration displays a temperature dependence characterized by an Arrhenius equation with $E \approx 0.65$ eV (Gillooly et al. 2001). While this value has become somewhat canonical in the MTE literature, it is valuable to consider whether the dominant contributions to ecosystem respiration (soil microbial respiration and plant respiration) actually conform to this prediction.

The temperature dependence of soil microbial respiration is typically well characterized by an activation

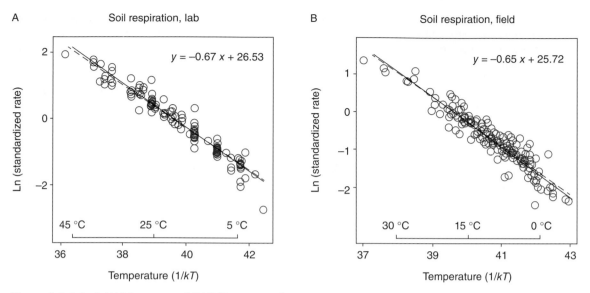

Figure 9.4 In both (A) laboratory and (B) field settings, soil respiration increases exponentially with temperature. In these Arrhenius plots, the slope of approximately −0.65 indicates that $E \approx 0.65 \, eV$. Soil respiration from multiple studies was standardized to account for differences between studies (see Allen et al. 2005 for details). (Redrawn from Allen et al. 2005; © 2005 British Ecological Society.)

energy of $E \approx 0.65$–$0.7 \, eV$ (Fig. 9.4) (e.g., Allen et al. 2005; Reichstein et al. 2005b). However, temperature sensitivity can vary with complexity of substrates, temperature, and environmental conditions. This variation is driven both by variation in the complexity of the C substrate's molecular structure and by environmental constraints such as freezing or moisture limitation (Davidson et al. 2006; Medvigy et al. 2010). Moreover, soil microbes secrete extracellular enzymes to break organic matter into compounds that they can absorb, and the weaker temperature dependence of these enzymes ($E \approx 0.3 \, eV$) affects the relationship between soil respiration and temperature (Allison et al. 2010; Wallenstein et al. 2011). All of these factors may cause the apparent activation energy (ε) to differ from the inherent one (E). Understanding more specifically how various biochemical pathways regulate respiration under different conditions may yield new insight into variations in the intrinsic temperature dependence of soil respiration, but the canonical value of $E \approx 0.65 \, eV$ is not bad for soil respiration.

In contrast, the temperature dependence of plant respiration decreases from about $E = 0.73 \, eV$ at 5 °C to $E = 0.39 \, eV$ at 35 °C (Tjoelker et al. 2001; Atkin and Tjoelker 2003). This response is probably driven by substrate limitation at high temperatures. Over longer

timescales, plants acclimate to temperature. Specifically, they adjust such that the fraction of daily photosynthesis consumed by respiration is approximately constant across the range of temperatures that they normally experience (Atkin and Tjoelker 2003; Atkin et al. 2007). Because photosynthesis responds less strongly to temperature than does respiration (Fig. 9.3), this implies that the effective temperature dependence of plant respiration is reduced over long timescales. Accounting for this acclimation is important to understanding the role of plant respiration in the carbon balance of ecosystems (e.g., Allen et al. 2005).

9.3.1.2   Photosynthesis and primary production

The temperature dependence of photosynthesis is complex, depending on photosynthetic pathway (C_3, C_4, or CAM), light conditions, water availability, and atmospheric CO_2 concentration (Campbell and Norman 1998). For the purpose of comparing the temperature dependence of photosynthesis with that of respiration, a simplification can be useful. Allen et al. (2005) estimated that the temperature response of Rubisco-limited terrestrial C_3 photosynthesis under current CO_2 concentrations – as described for a soybean model system (Bernacchi et al. 2001) – may be approximated using an Arrhenius equation with $E = 0.3$ eV across the temperature range of 0–30 °C (Fig. 9.3). This approximation, which was designed to characterize the dominant form of photosynthesis at a global scale (Allen et al. 2005), also has become somewhat canonical in MTE (e.g., Anderson et al. 2006; López-Urrutia et al. 2006; Anderson-Teixeira et al. 2008; de Castro and Gaedke 2008; Yvon-Durocher et al. 2010). However, $E = 0.3$ eV is by no means a universal constant characterizing the temperature dependence of primary production. Rather, the effective activation energy varies depending upon a host of factors: photosynthetic pathway, atmospheric CO_2 concentration, water limitation, light conditions, acclimation by the plant, and the temperature range over which the response is approximated (because temperature response of photosynthesis is not exponential; Fig. 9.3). Moreover, we should not expect this approximation to apply to aquatic systems. More research is needed to better characterize the temperature dependence of primary production in the MTE framework (Price et al. 2010). In general, however, we can say with confidence that the inherent temperature dependence of photosynthesis is less than that of respiration – and that this contrast will affect the response of ecosystems to climate change (Anderson-Teixeira, Smith, and Ernest, Chapter 23).

9.3.2 Effective temperature dependence across time or space

When we move beyond inherent, short-term temperature responses of production and respiration to con-

sider longer timeframes and broad spatial scales, there are a number of complicating factors that may cause the effective temperature dependence to differ from the inherent temperature dependence. Notably, if resource concentrations vary with temperature (equation 9.1), the ε will be the sum of E and parameters characterizing the temperature dependence of resource concentrations (Fig. 9.5) (Anderson-Teixeira et al. 2008). As an example of how this works, let us consider how whole-ecosystem leaf respiration (R_{leaf}) may vary with temperature across different forests. Let us assume that R_{leaf} increases with temperature as in Figure 9.3 ($R_{leaf} \propto e^{-0.6/kT}$), and that it also increases linearly with the "concentration" of leaf biomass (g/m^2) ([R] in equation 9.1). If leaf biomass is independent of temperature, the overall temperature response of R_{leaf} should match the inherent kinetics ($\varepsilon = E = 0.6$ eV; Fig. 9.5A). But what if leaf biomass increases with temperature (say, $\propto e^{-0.3/kT}$)? In this case, when we combine the inherent kinetics with the temperature gradient in [R] in equation 9.1, we find that the overall temperature dependence of R_{leaf} is strongly inflated ($\varepsilon = E + E_{[R]} = 0.9$; Fig. 9.5B).

9.3.2.1 Temperature dependence through time

The relationship of metabolism to temperature is dependent upon many factors, including resource availability (e.g., soil moisture, nutrients, photosynthate), microbial abundance, seasonal phenology, and physiological adjustments to changing temperatures (Reich 2010). Through time, resource concentrations can vary with temperature, thereby altering the

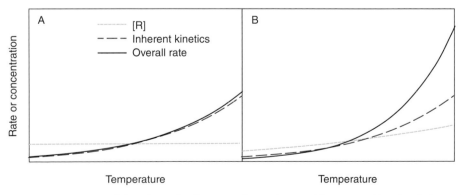

Figure 9.5 Schematic illustration of the relationship between inherent kinetics (characterized by E) and observed overall temperature dependence (characterized by ε) under two scenarios: (A) when resource availability, [R], is independent of temperature and (B) when [R] varies with temperature.

observed temperature dependence (Davidson and Janssens 2006; Davidson et al. 2006) (Fig. 9.5). For example, because plant production increases with temperature, the supply of the photosynthate that fuels ecosystem respiration (carbon allocation to roots, litterfall) tends to be highest during warm periods. Because the concentration of this resource increases with temperature, the overall temperature dependence of respiration is increased (Reichstein et al. 2005a; Sampson et al. 2007).

A number of studies have used an MTE framework to characterize temperature responses of metabolism in ecosystems through time. In some cases, temperature dependence responds as $\varepsilon \approx 0.65$ eV, matching the E of aerobic respiration. Examples include ecosystem respiration in some terrestrial temperate- and boreal-region sites (Enquist et al. 2003) and in aquatic mesocosms (Yvon-Durocher et al. 2010). In another case, however, the metabolism of lake plankton showed only a weak response to temperature (de Castro and Gaedke 2008). One possible explanation for these mixed results is that covariation of resource concentration with temperature on a seasonal basis may cause the observed and inherent temperature responses of production and respiration to deviate substantially in some cases ($\varepsilon \neq E$), but have minimal effects in others ($\varepsilon \approx E$).

A number of recent studies have quantitatively separated short-term ("inherent") temperature responses of ecosystem respiration from seasonal effects of a changing resource environment. These studies show that E, as characterized over short time periods, can differ substantially from ε, as characterized over an entire year. Depending on whether resource availability is greatest during the warm or the cool season, ε may be increased or decreased relative to E, respectively (Fig. 9.6) (Reichstein et al. 2005a). This occurs because reference respiration (respiration at a certain temperature) varies seasonally with carbon inputs (e.g., Groenendijk et al. 2009). Interestingly, ecosystem respiration appears to display a universal short-term temperature dependence characterized by an $E \approx 0.3$ eV (Reichstein et al. 2005a; Groenendijk et al. 2009; Mahecha et al. 2010), which is lower than the expected $E \approx 0.65$ eV typically observed for aerobic respiration. We might expect this value for plant respiration, which acclimates to a consistent fraction of plant production, but not for soil respiration. Why this difference is observed remains a question that is both theoretically interesting and highly significant, in that the temperature dependence of ecosystem respiration holds impor-

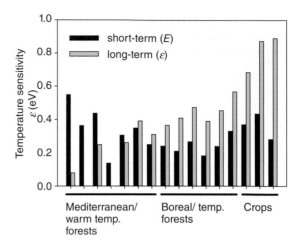

Figure 9.6 Short-term and long-term temperature sensitivity of ecosystem respiration in multiple different terrestrial ecosystems (each represented by one pair of bars). Short-term temperature sensitivity (characterized by E) generally differs from long-term temperature sensitivity, as measured over an entire year (characterized by ε). In some cases, $\varepsilon = 0$. The direction and magnitude of the difference between short- and long-term temperature responses depends upon the seasonality of resource availability. (Data from Reichstein et al. 2005a.)

tant implications for terrestrial ecosystem feedbacks to climate change (Anderson-Teixeira, Smith, and Ernest, Chapter 23).

9.3.2.2 Temperature dependence across climatic gradients

Across climatic gradients, the responses of ecosystems to temperature are confounded by several factors. Differences in soil types, water availability, and nutrient availability may cause systematic variations in resource concentrations that covary with temperature. For example, global-scale patterns in terrestrial primary productivity are strongly shaped by precipitation (e.g., Luyssaert et al. 2007), which often covaries with temperature across climatic gradients.

Moreover, seasonal, diurnal, and even stochastic variations in temperature present a challenge in that a metabolic response computed based on an average temperature differs from the average response computed based on seasonally varying temperatures (Savage 2004; Medvigy et al. 2010). Say, for example, we want to use an Arrhenius equation to compute the average rate of soil respiration at a site near Fairbanks,

Alaska (which has strong seasonal temperature variation). If we first calculate respiration based on a continuous temperature record, and then average, we will get a true approximation of average soil respiration; however, if we first average temperature and then compute respiration, our answer will be half the correct value (Savage 2004). If we do this for many sites across the globe, we will consistently underestimate average respiration in highly seasonal environments and misrepresent the global-scale relationship between respiration and temperature.

In order to characterize how temperature shapes ecosystem metabolism across climatic gradients, MTE studies have expressed metabolic rate in terms of daylight hours in the growing season and quantified how this varies with average growing season temperature (Allen et al. 2005; Anderson et al. 2006; Enquist et al. 2007a). There are two main hypotheses regarding the temperature dependence of whole-ecosystem metabolism across climatic gradients. Some studies have found that, at a global scale, the temperature dependence of production reflects that of photosynthesis ($\varepsilon \approx E \approx 0.3\,eV$) when water is not limiting (Fig. 9.7a) (Allen et al. 2005; Anderson et al. 2006; López-Urrutia et al. 2006). Presumably, we observe this pattern because the biochemical reactions of photosynthesis are phylogenetically conserved and because concentrations of key photosynthetic enzymes (e.g., Rubisco) do not vary systematically with temperature on a global scale. In contrast, other studies show that temperature dependence is far weaker or negligible across these global scales (Fig. 9.7B) (Enquist et al. 2003, 2007a; Kerkhoff et al. 2005). The explanation for this is that acclimation, species adaptation, and community assembly result in compensatory increases in efficiency in colder climates (Enquist et al. 2007a). Reality likely contains elements of both of these hypotheses. In either case, of course, water availability – alone and in its interactions with temperature – plays a dominant role in shaping global patterns in whole-ecosystem production and respiration. More research is needed to fully understand global-scale temperature dependence of whole-ecosystem metabolism and how it is shaped by other variables such as precipitation, nutrient availability, acclimation, adaptation, and species composition.

Although ε does not necessarily equal E over longer timescales and across climatic gradients, it does appear that the differential responses of respiration and production to temperature often play an important role in shaping the carbon balance of ecosystems. For example, across a range of elevations in New Mexico, the temperature dependence of ecosystem respiration exceeds that of GPP, implying that the balance shifts towards respiration at higher temperatures (Fig. 9.8) (Anderson-Teixeira et al. 2011). Similarly, the stronger temperature dependence of respiration relative to that of production (Fig. 9.3) leads us to expect relatively faster decomposition rates and less carbon storage in detritus and soil organic matter in warmer sites – the pattern that in fact we observe (Jobbágy and Jackson 2000b; Allen et al. 2005; Raich et al. 2006; Anderson-Teixeira et al. 2011).

9.3.2.3 Interactions among temperature-dependent processes

Much of the challenge of understanding ecosystem-level temperature dependence lies in the fact that ecosystems involve multiple interacting temperature-dependent processes. Both within sites and across environmental gradients, temperature-driven, biologically mediated gradients in resource availability can alter the effective temperature dependence of ecological processes (Anderson-Teixeira et al. 2008). In some situations, positive feedbacks among multiple processes may occur such that the observed temperature dependence, ε, is an aggregate of the activation energies of many separate contributing processes. This is the case for early primary succession on the lava flows of Mauna Loa, Hawai'i, where ecosystem development is driven by feedbacks among a number of temperature-dependent processes such as biotic growth and metabolism, rock weathering, soil formation, and nitrogen fixation. The observed temperature dependence of ecosystem development is substantially greater than most of the component processes (Fig. 9.9). This occurs because resources ([R]s in equation 9.1) – for example, N fixed by the lichen *Stereocaulon vulcani*, P derived from rock weathering, C fixed through photosynthesis – accumulate faster at warmer temperatures. This creates gradients in [R] across the temperature gradient; in ecosystems that are approximately a century old, N and P concentrations in leaves and soil increase with temperature across the climatic gradient (Fig. 9.10). In turn, the effective temperature dependencies of net primary productivity, litterfall, decomposition, and nutrient uptake are inflated across the gradient. This effect is inflated through time as feedback effects accrue, such that rates of forest growth and soil development in

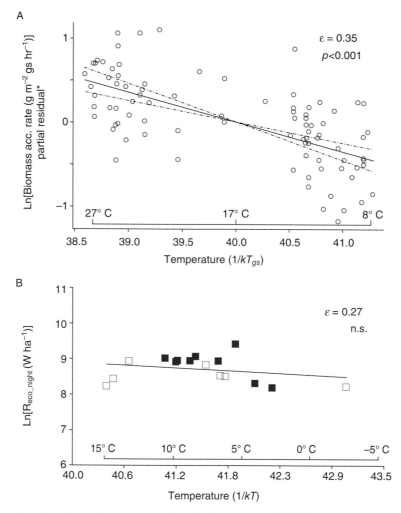

Figure 9.7 Temperature dependence of ecosystem level carbon fluxes across global climatic gradients. (A) Rate of above-ground biomass accumulation in regrowing forests (*corrected for the effects of other variables) as a function of average growing season temperature. (Redrawn from Anderson et al. 2006; © 2006 Blackwell Publishing Ltd/ CNRS.) This represents one example where whole-ecosystem CO_2 flux varies significantly with temperature across climatic gradients. (B) Annual night-time R_{eco} as a function of average annual night-time temperature for both North American (open symbols) and European (closed symbols) sites. (Redrawn from Enquist et al. 2003; © 2003 Nature Publishing Group.) This represents one example where whole-ecosystem CO_2 flux does not vary significantly with temperature across climatic gradients.

warm, low-elevation environments dwarf those of their high-elevation counterparts (Fig. 9.9) (Anderson-Teixeira et al. 2008).

9.4 WATER

Water is essential to life and metabolism and strongly shapes the structure and functioning of most

terrestrial ecosystems. In plants, photosynthesis is strongly tied to transpiration (release of water vapor from plants to the atmosphere). In order to obtain CO_2 from the atmosphere, plants must open their stomata, thereby losing water. In this way, photosynthesis and plant metabolism are limited by water availability. Of course, plants have evolved many adaptations – including alternative photosynthetic pathways (C_4 and CAM) – that increase their water use efficiency, allow-

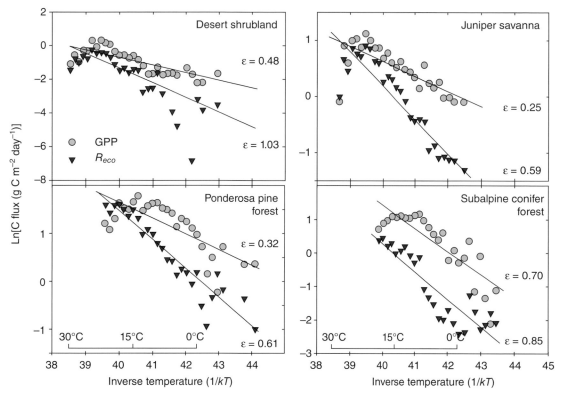

Figure 9.8 Differential response of daily GPP (green circles) and R_{eco} (red triangles) to average daily temperature at four sites in New Mexico. Data for an entire year are binned into 1 °C intervals and averaged. (Data from Anderson-Teixeira et al. 2011.)

ing more carbon fixation per unit of water lost. Nevertheless, without water, photosynthesis stops, and plants burn carbon stores (and eventually die) under protracted water stress (e.g., Adams et al. 2009; Breashears et al. 2009). Microbial respiration is likewise tied to water availability. Microbes can operate at lower water potentials than most plants and can escape harsh conditions in suitable microsites (Orchard and Cook 1983; Foster 1988) but shut down under excessively dry conditions.

Through its effects on metabolism, water shapes whole-ecosystem function. Primary productivity increases with precipitation (up to a point) on a global scale (Lieth 1973; Luyssaert et al. 2007). Ecosystem respiration also increases with precipitation, and the carbon balance of ecosystems is driven by differences in their responses. Although soil respiration responds more rapidly and to lower levels of soil moisture than does production (e.g., Jenerette et al. 2008; Inglima et al. 2009), it generally loses its advantage as moisture

increases (Austin 2002), and sustained increases in water availability tend to favor production. Therefore, in water-limited environments, carbon uptake typically increases with increasing water availability (e.g., Huxman et al. 2004a; Muldavin et al. 2008; Anderson-Teixeira et al. 2011). In contrast, because production responds with more sensitivity to drought than does respiration (Schwalm et al. 2009), water-limited ecosystems often lose carbon during a drought (e.g., Ciais et al. 2005; Arnone et al. 2008; Schwalm et al. 2009), and these effects may exhibit time lags that persist beyond the drought because of loss of leaves and damage to canopy function (Reichstein et al. 2002; Arnone et al. 2008). Over longer spatio-temporal scales where communities are adapted to climatic conditions, both individual plants and whole ecosystems struggle to maintain a favorable carbon balance under dry conditions, but are able to sequester substantial amounts of carbon when moisture is abundant. As a result, both net ecosystem productivity (annual net

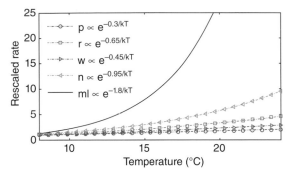

Figure 9.9 Comparison of the average temperature dependence of ecosystem development on Mauna Loa (ml; ε = 1.8 eV) with the rates typically observed for some of the constituent processes: nitrogen fixation by *S. vulcani* (n; $E \approx 0.95$ eV), basalt weathering (w; $E \approx 0.4$–0.5 eV), aerobic respiration rates (r; $E \approx 0.65$ eV), and photosynthesis (p; $E \approx 0.3$ eV). For display purposes, values are rescaled to normalize the rates to 1 at 7 °C. (Redrawn from Anderson-Teixeira et al. 2008; © 2007 the National Academy of Sciences of the USA.)

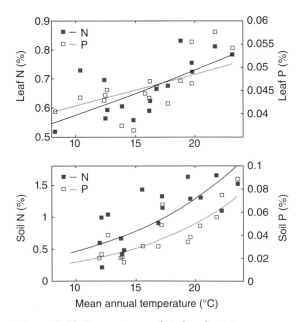

Figure 9.10 Temperature-mediated gradients in concentrations of key resources – leaf and soil nitrogen (N) and phosphorus (P) – in relatively young ecosystems (109–138 years) spanning an elevational gradient on the lava flows of Mauna Loa, Hawai'i. (Redrawn from Anderson-Teixeira et al. 2008; © 2007 the National Academy of Sciences of the USA.)

carbon uptake by the ecosystem) and biomass increase with precipitation (up to about 1500 mm/y) on a global scale (Anderson-Teixeira et al. 2011).

Although water is central to the regulation of metabolism, few studies have considered the role of water within an MTE framework (Enquist et al. 1998; Anderson-Teixeira et al. 2008; Jenerette et al. 2008). There is great potential to improve our understanding of water's role in shaping both individual metabolism and ecosystem energetics.

9.5 NUTRIENTS

Biogeochemical cycling of nutrients is closely coupled to the carbon cycle and energetics of ecosystems. The supply of N and/or P limits whole-ecosystem metabolism in most mesic ecosystems (Gallardo and Schlesinger 1994; Vitousek 2004; LeBauer and Treseder 2008), and carbon cycling cannot be understood in isolation from the movement of these elements within ecosystems. Rather, understanding ecosystem dynamics requires linking models that focus on energy or carbon with those that focus on nutrients (Reiners 1986).

A linkage between carbon and other elements may be accomplished through consideration of stoichiometry as the subcellular structures and processes that drive metabolism rely upon specific elements (Allen and Gillooly 2009; Kaspari, Chapter 3). For example, because of the key role of N-containing compounds in photosynthesis, the photosynthetic capacity of leaves increases with leaf N concentration (Field and Mooney 1986). Likewise, P-rich ribosomes are the organelles that carry out synthesis of new structures, and organismal growth rates often increase with P concentration (Sterner and Elser 2002). Thus, linking MTE and ecological stoichiometric theory could provide numerous opportunities for improved understanding of ecosystem dynamics (Allen et al. 2005; Kerkhoff and Enquist 2006; Elser et al. 2010).

In addition to these biochemical mechanisms, a full accounting of the influence of nutrient availability needs to consider the processes that bring nutrients into ecosystems, or transform them into forms that organisms can use. Mineralization – the biologically mediated breakdown of organically bound nutrients to abiotic and available forms that recycles N and P in ecosystems – is closely coupled to the respiration of soil biota. However, ecosystems are open systems that lose nutrients to their surroundings via processes such as leaching, erosion, and volatilization. If ecosystems are to be sustained over time (or, in the case of primary succession, if they are to develop from abiotic antecedents), there must be inputs of nutrients. While some of the processes that bring nutrients to ecosystems are at most indirectly related to temperature (e.g., atmospheric deposition of nutrients dissolved in rain), others are temperature dependent.

The most important temperature-dependent reactions that bring N and P in particular into ecosystems are biological N fixation and mineral weathering. Biological N fixation is interesting in that the activation energy of its fundamental enzyme system (nitrogenase) has a strongly biphasic E (Ceuterick et al. 1978), with a very high value of $2.18\,eV$ below $22\,°C$ and a value close to respiration (~$0.65\,eV$) above $22\,°C$. This striking difference could contribute to constraining N fixation, and so enhancing N limitation, in many temperate and boreal ecosystems (Houlton et al. 2008).

The weathering of minerals is the main pathway by which new P becomes available to terrestrial ecosystems. It is generally an abiotic reaction; however, it can be strongly influenced by biologically produced acidity and the mechanical action of biota (Kelly et al. 1998). Mineral weathering is also temperature dependent, with E in the range of $0.3–0.8\,eV$ in a wide range of studies of different minerals and assemblages in both field and laboratory conditions (White and Blum 1995; White et al. 1999). Thus, in many cases, the input of limiting nutrients to ecosystems is a temperature-dependent process.

Biota play an important role in introducing, recycling, and retaining elements within an ecosystem. Because biotic metabolism is temperature dependent, it shapes the relationship between resource concentration and temperature (recall the example of Mauna Loa; Fig. 9.10; Anderson-Teixeira et al. 2008). As discussed above, any covariance between the availability of limiting nutrients and temperature alters the observed temperature dependence of an ecological process (equation 9.1, Fig. 9.5). Thus, the importance of nutrients in shaping the temperature dependence of ecosystem metabolism cannot be understated. Better understanding the interactions among biotic metabolism, climate, and biogeochemical cycling remains an exciting research frontier.

9.6 CONCLUSION

As reviewed above, metabolic ecology provides a powerful framework for understanding the dynamics of ecosystems. At the same time, ecosystem ecology is an area with lots of potential for profitable expansion of MTE. Linking body size distributions to ecosystem structure and function remains an interesting and ongoing challenge (e.g., Enquist et al. 2009; West et al. 2009). There remain many unresolved and important questions regarding the effects of temperature: for example, how to best characterize the temperature dependence of production within an MTE framework, how temperature responses scale across spatiotemporal scales, and how interacting temperature-dependent processes shape nutrient supplies and alter observed temperature responses. In addition, only a few studies have considered the role of water within an MTE framework, and we have much to learn about how water shapes individual- and ecosystem-level metabolism. Moreover, there are many exciting questions regarding the interactions among biotic metabolism, climate, and biogeochemical cycling of limiting nutrients. Many of these topics are not only interesting from a theoretical perspective, but will also be crucial to understanding how global change will alter ecosystems and the services that they provide.

Chapter 10

RATES OF METABOLISM AND EVOLUTION

John L. Gittleman and Patrick R. Stephens

SUMMARY

1 The rate of evolution is intrinsically important because a multitude of characteristics, ranging from molecular variation to life histories to taxonomic diversity, are influenced by relative rate differences.

2 The metabolic theory of ecology (MTE) predicts that patterns and processes of evolutionary rates are related to resource uptake from the environment and resource allocation to survival, growth, and reproduction.

3 Although classic studies in evolutionary biology date back to the work of G. G. Simpson and J. B. S. Haldane, consistent patterns for why and how organisms differ in evolutionary rates have not developed nor provided a solid empirical understanding or theory.

4 Recently, advances have been made in more rigorously measuring rates of evolution by synthesizing dated phylogenies with explicit models of evolution.

5 MTE may shed light on some evolutionary rates problems. Organisms with higher metabolic rates appear to have faster molecular rates of evolution, rapid rates of behavioral relative to morphological change, and more rapid diversification rates.

6 Future studies of MTE and evolutionary rates will be exciting as experimental systems are developed, more comprehensive databases are created, and new questions arise as to how the adaptive nature of evolution rates across biodiversity are impacted by global changes.

10.1 INTRODUCTION

In time everything changes; the question is what are the null expectations for change, what are observed differences in rate change, and are these differences consistent across the diversity of life. The metabolic theory of ecology (MTE) provides a basis for tackling these rate issues: the rate at which organisms take up and use energy influences survival, growth, and reproduction. In turn, such allocation will impact how eve-

rything changes from ecology to evolution. Here, we look at whether metabolic rates relate to and predict various types of evolutionary rates across taxa including molecular, phenotypic traits, and patterns of diversification.

Rates of evolution are profoundly important because they may underlie "the reasons for the great diversity of organisms on the earth" (Simpson 1953). Although many studies have been devoted to understanding rates of evolution, a consistent body of work remains

Metabolic Ecology: A Scaling Approach, First Edition. Edited by Richard M. Sibly, James H. Brown, Astrid Kodric-Brown.
© 2012 John Wiley & Sons, Ltd. Published 2012 by John Wiley & Sons, Ltd.

elusive (Gingerich 2009). Reasons for this are many but primarily relate to paucity of complete databases across taxa, lack of reliable time sequence information, methodological problems in measuring rates of evolution, and the panoply of variables relevant to evolutionary rate change. Nevertheless, issues pertaining to rates of evolution have synthesized problem areas. In his seminal work on the topic, Simpson (1953) "executed a brilliant tactical maneuver" (Laporte 2000, p. 129) by using the phenomenon of evolutionary rates to pull together two disparate areas, paleontology and genetics. Today, in similar fashion, questions about evolutionary rates may synthesize new approaches for the MTE.

Differential rates of evolution are well known. For example, mammals have evolved faster than mollusks (Stanley 1973) and, within mammals, carnivores have evolved faster than primates (Mattila and Bokma 2008). In addition to taxonomic differences, rate changes are observed among traits. Quantitative behavioral or ecological characteristics such as social group size or home range size generally evolve at faster rates than morphological traits (Gittleman et al. 1996).

Such observed differential rates are not well understood. This is mainly due to uneven data availability and inconsistent trends across taxa or traits. For example, across mammals, by far the best-known taxonomic group, life-history information has been reported for only around 30% of species and even body size has not been measured in hundreds of species (Jones et al. 2009); this is especially problematic when considering metabolic hypotheses because direct measures of physiological usage or capacity are sparse. Data availability is only part of the problem. In some of the most well-studied systems, empirical trends occur in many directions. For example, morphological evolution in body size across Platyrrhini (monkeys) and Phalangeriformes (opossums) may be significantly faster in large and small species, thus not revealing consistent reasons for rate differences (Cooper and Purvis 2009).

Of course, there is no reason that rates of evolution must occur in the same form across taxa or traits. As Simpson (1953) forewarned, "Evolution involves changes of so many sorts that measurements of its rates must necessarily be complicated if they are to cover important aspects of the subject in an unambiguous and instructive way" (p. 4). Cooper and Purvis (2009) recently summarized factors known to influence rates of evolution (Fig. 10.1). Many of these are

iconic relationships in what we know about evolutionary change. For example, small species with faster generation times and shorter lifespans are known to have rapid evolutionary rates and form some of the logic behind r and K-selection theory (Gould 1983; Calder 1984). However, even when we restrict our attention to morphological evolution, we find that body size, interspecific competition, geographic range, and ecological specialization may be equally influential and, as stated above, may occur in various directions. As Cooper and Purvis (2009) discuss, there are many other important variables which influence rates of evolution and further study is needed to sort out real from artificial patterns due to uneven data.

Generally, the relationship between metabolic rate and evolutionary rate has not received much attention though there is a natural fit between the two rates: differences in behavioral and ecological change, ontogenetic differences at individual levels, and responses to environmental changes are all strongly influenced by metabolism. The focus of this chapter is to consider whether and how some core topics of evolutionary rates – molecular evolution, trait evolution, and macroevolution such as speciation and extinction – are tied to metabolic rates.

10.2 METHODS FOR MEASURING EVOLUTIONARY RATES

There is a surprising variety of methods for measuring evolutionary rates in the literature, and little consensus as to which methods are to be preferred. Here, we present a brief review of methods for quantifying evolutionary rates. We focus primarily on evolutionary rate defined as a change in the value or variance of a trait over a given time interval (e.g., a change in the nucleotide composition of a DNA sequence, or in the value of a morphological trait). We also briefly consider methods for quantifying variation in diversification rates among lineages, and discuss confusion in the literature about the relationship between evolutionary rate and phylogenetic signal.

10.2.1 Measuring rates of sequence evolution

Measuring rates of molecular evolution has become fairly straightforward with modern techniques of

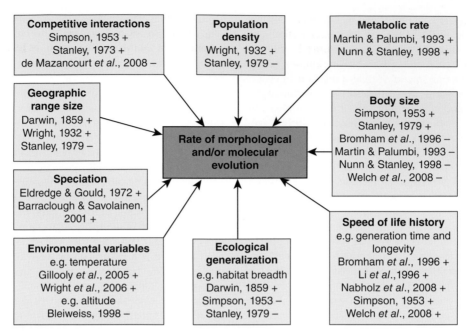

Figure 10.1 Diagram showing the variables which are hypothesized to affect rates of morphological and/or molecular evolution along with selected references for the hypotheses. (+) Hypothesized positive relationship between the variable and evolutionary rate; (−) hypothesized negative relationship between the variable and evolutionary rate. Note that these variables are themselves often interconnected, directly or indirectly, but the relationships between them have been omitted to increase clarity (Darwin 1859; Wright 1932; Simpson 1953; Eldredge and Gould 1972; Stanley 1973, 1979; Martin and Palumbi 1993; Bromham et al. 1996; Li et al. 1996; Bleiweiss 1998; Nunn and Stanley 1998; Barraclough and Savolainen 2001; Gillooly et al. 2005; Wright et al. 2006; de Mazancourt et al. 2008; Nabholz et al. 2008; Welch et al. 2008). Redrawn from Cooper and Purvis (2009) by permission of John Wiley & Sons, Ltd.

molecular systematics. By comparing a molecular phylogeny with branch lengths in units of number or proportion of nucleotide substitutions to a time-calibrated phylogeny (i.e., a chronogram with branch lengths in units of millions of years since divergence), it is possible to derive an estimate of the absolute rate of molecular evolution for a lineage. It is important that an appropriate model of sequence evolution be used when estimating a phylogeny, and there exists some debate as to which model selection procedures are most accurate (Posada and Crandall 2001; Posada and Buckley 2004). There are also a number of competing methods available for estimating a chronogram using molecular sequence data and geological calibration points (e.g., Sanderson 2003; Drummond and Rambaut 2007). However, there is little disagreement that substitution rate is an appropriate measure of rates of sequence evolution. In contrast, the best measure of rates of morphological evolution remains unclear.

10.2.2 Measuring rates of morphological evolution

The classic unit of morphological evolution in past studies of evolutionary rate is the "darwin," first described by Haldane (1949). It is defined as an evolutionary change by a factor of e (the base of natural logarithms) per million years (Gingerich 1993). The time interval separating a given change in trait values can be estimated directly from the fossil record, or based on the phylogenetic distance separating species in a time-calibrated phylogeny (Gittleman et al. 1996).

Haldane (1949) described several additional measures of morphological evolution. As an alternative to measuring rate in units of e he suggested using the phenotypic standard deviation within a population, since this is the raw variation which will be targeted by natural selection. Gingerich (1993) dubbed this measure the "simpson." Haldane further suggested that number of generations would be a more informative unit of time than years, since generation time scales so differently with absolute time in different organisms. Gingerich defined the "haldane" as a change in trait variance by a factor of one standard deviation per generation. While these measures of evolutionary rate likely more directly reflect the process of evolution than the darwin, they have rarely been used because the information needed to quantify them is often unavailable for a lineage of interest. Generation times for many species, particularly fossil taxa, are unknown. The population variance of a trait is even more rarely known. In contrast, the darwin requires only an estimate of time elapsed and the change in species average trait values.

In 2009 Ackerly introduced a new measure of evolutionary rate, the "felsen," defined as an increase of one unit per million years in the variance among sister taxa of ln-transformed trait values. It is calculated from standardized independent contrasts of a ln-transformed trait estimated on a tree with branch lengths in units of millions of years. The method has advantages over previous methods for measuring evolutionary rates. An *F*-test can be used to see if different clades have significantly different evolutionary rates, and a number of methods can be used to assign confidence limits to a rate estimate (Ackerly 2009).

Finally, Evans et al. (2011) introduced a new measure of evolutionary rate called "clade maximum rate," which is defined as the rate of change in a specified extreme value of a character (either the minimum or the maximum) for a clade within a given time interval. An advantage of this measure over the felsen is that it does not require a resolved phylogeny, only the recognition of distinct clades (e.g., monophyletic taxonomic groups). It is thus more easily applicable to fossil data. In the first empirical application of clade maximum rate, Evans et al. showed that very large (e.g., orders of magnitude) evolutionary decreases in mammalian body mass (such as extreme insular dwarfism) can happen over 1/10 fewer generations than large increases.

10.2.3 Evolutionary rate and phylogenetic signal

In addition to the myriad methods available for quantifying rates of morphological evolution, there exists some confusion about the relationship between evolutionary rate and phylogenetic signal. Phylogenetic signal can be broadly defined as a tendency for the similarity of the traits of species to be inversely correlated with their phylogenetic distance (i.e., more closely related species are more similar). A prevailing hypothesis is that traits evolving more rapidly and/or that are evolutionarily labile will tend to show weaker phylogenetic signal (e.g., Blomberg et al. 2003; Rheindt et al. 2004; Silvertown et al. 2006), and thus that low signal necessarily indicates rapid trait evolution. However, under the most commonly assumed model of character evolution, pure Brownian motion, this assumption is often not valid (Ackerly 2009).

The Brownian motion model is based on models of random particle diffusion in a liquid. It is a widely used neutral model of character evolution under genetic drift, and is assumed by most phylogenetic comparative methods (Felsenstein 1985, 1988; O'Meara et al. 2006). For continuous characters there is no expected relationship between rate and signal under pure Brownian motion (Hansen and Martins 1996; Freckleton and Harvey 2006; Revell et al. 2008; Ackerly 2009). Other models of character evolution such as bounded Brownian motion and the Ornstein–Uhlenbeck model of evolution (Butler and King 2004) have been shown in previous mathematical and simulation studies to potentially produce correlations between phylogenetic signal and evolutionary rate that are negative, absent, or positive (Hansen and Martins 1996; Freckleton and Harvey 2006; Revell et al. 2008; Ackerly 2009). Thus phylogenetic signal is not a reliable indicator of whether trait evolution is rapid or slow unless the mode of trait evolution is known. Phylogenetic signal and evolutionary rate are related but distinct patterns, and researchers wishing to test hypotheses concerning evolutionary rates would do best to quantify them directly using methods discussed above.

10.2.4 Estimating diversification rates

One of the central goals of evolutionary biology is to understand the factors that explain variation in the tempo and mode of evolution among lineages. Until recently, rates of speciation and extinction could only be estimated from the fossil record (e.g., Simpson 1944; Stanley 1979). However, with the rise of molecular phylogenetics, it is now possible to estimate diversification rates using data from extant species. Nee et al. (1992) were the first to estimate diversification rates from a time-calibrated molecular phylogeny, and numerous subsequent studies investigated diversification rates using broadly similar methods (e.g., Baldwin and Sanderson 1998; Rüber and Zardoya 2005; Harmon et al. 2008; Phillimore and Price 2008). One pattern that is commonly documented is a slowdown in diversification rates over time (i.e., a slowing of lineage accumulation towards the present, reviewed in Cusimano and Renner 2010). This pattern is generally attributed to density-dependent rates of cladogenesis and niche filling (Stanley 1979; Nee et al. 1992). However, Cusimano and Renner (2010) argued that apparent slowdowns in diversification rate may be driven by a tendency for systematists to sample phylogenetically distinct taxa, thus resolving a greater proportion of basal than recent divergences. Based on the results of simulations, they argued that diversification rates cannot be reliably estimated from phylogenies that contain less than roughly 80% of all extant species.

Additional concerns have been raised about estimating diversification rates from molecular phylogenies. Ricklefs (2007) argued that the tendency of studies to focus on large (i.e., speciose) clades has produced a distorted view of the process of diversification, and that in reality most clades exhibit much lower rates of diversification than have typically been measured. Rabosky (2009, 2010) criticized the assumption that diversification can proceed indefinitely, and argued that in many cases ecological factors will set the maximum species richness of a clade. If clades generally have a maximum potential species richness, estimates of diversification rates based on modern species richness and clade age can be quite misleading. Old clades would be more likely to have hit their diversity ceiling, causing a slowdown in diversification, regardless of their maximum potential rate of cladogenesis. We note, however, that it seems unlikely that ecological limits on clade diversity are commonplace, given the observation that clade age is generally a strong correlate of clade diversity (McPeek and Brown 2007). If such limits exist, it seems that clades frequently do not reach them.

Liow et al. (2010) criticized both purely molecular and purely fossil-based studies of diversification rates. They showed in a simulation study that estimates of diversification rate based on either will be biased in some circumstances, but that molecular and fossil studies will tend to be biased in different ways; an approach combining both sources of information may be the most appropriate general method for assessing diversification rates. The estimation of diversification rates remains an active area of research and debate.

10.3   PROBLEMS AND HYPOTHESES

Although MTE is relatively new, some interesting questions have already explicitly developed in relation to evolutionary rates. The following examples seem especially relevant because they integrate various phylogenetic and macroecological approaches, synthesize data at broad spatial and temporal scales, and have received enough attention that hypotheses are more focused on specific causal mechanisms. Until now, few studies that are promising for MTE have explicitly used the analytical methods for measuring rates (i.e., those described above), thus providing opportunities for future research.

10.3.1   MTE and rates of molecular evolution

One of the major features of molecular evolution is that rates of sequence divergence vary considerably among different lineages. Differences in mutation rate and generation time are both commonly invoked to explain this variation (reviewed in Bromham and Penny 2003; Rand 1994), and metabolism has a clear potential link to mutation rates. Mutagenic byproducts of the metabolic process, such as oxygen radicals, would be expected to reach higher concentrations in species with higher mass-specific metabolic rates, potentially producing faster mutation rates through higher rates of oxidative damage to DNA (Shigenaga et al. 1989; Rand 1994). Numerous studies have shown that smaller animals generally have higher rates of sequence evolution (e.g., Bromham et al. 1996; Nunn and Stanley 1998; Bromham 2002), which Bromham and Penny (2003) interpreted as evidence of the influence of metabolism. However, generation time also shows strong allometric scaling, so this observation alone cannot be taken as strong evidence of the effects of metabolic rate.

Further evidence comes from studies that have compared endotherms and ectotherms across a similar range of body sizes, which generally show that ectotherms have much slower rates of molecular evolution for their size (Martin and Palumbi 1993; Martin 1999; but see Seddon et al. 1998). Comparison of Foraminifera that occur at different latitudes also showed that the rate of sequence evolution was directly proportional to the temperature at which Foraminifera occur, and thus presumably their metabolic rate (Allen et al. 2006). Finally, a recent study showed that equations based on

MTE and the neutral theory of molecular evolution can accurately predict variation in the rate of sequence evolution among species with different body sizes (Gillooly et al. 2005b). As opposed to the somewhat more controversial links between metabolic rate and rates of diversification and morphological evolution, the importance of MTE for understanding variation in rates of molecular evolution among species seems well established.

10.3.2 MTE and trait evolution

Intuitively, behavioral characteristics such as group size or home range movements should evolve faster than morphological or physiological traits. Variation in traits at population and individual levels along with relative heritability measures suggest more rapid behavioral than morphological evolution. Indeed, the remarkable, long-term domestication experiment by the Russian geneticist, Dmitry K. Belyaev, demonstrated over a 40-year period of trait selection in 45 000 silver foxes (*Vulpes vulpes*) that behavioral changes such as fear or shyness occur much earlier than morphological or physiological changes (reviewed in Trut 2001). Direct selection for domestication in silver foxes shows that, in a unique population of 100 foxes bred between 30 and 35 generations, changes in

behavioral development occurred earlier than morphological change. Importantly, the developmental changes in behavior were related to physiological and hormonal effects. Over the years, others have hypothesized that behavioral and ecological traits evolve at faster rates than morphological rates (see Brooks and McLennan 1991). Nevertheless, despite considerable theoretical discussion (see West-Eberhard 1987), few empirical tests have shown actual differences in rates of evolution among traits, especially while controlling for phylogenetic differences across taxa and using accurate methods for rate measurements.

Comparing relative rates of evolution among traits suggests that MTE is important for observed differences in trait evolution. Using comparative data of six morphological (body and brain size), life-history (gestation length, birth weight), and behavioral/ecological (home range and population group size) traits, Gittleman et al. (1996) calculated darwins for all of these traits using dated molecular trees (see Fig. 10.2). Given expected trends for quantitative trends in other studies (see above discussion of darwins) three findings were consistent: (1) traits reveal an expected inverse relationship between evolutionary rate and timescale; (2) slopes are more negative with group size than body weight, though the values are shallow; and (3) "instantaneous" rate change is greater in group size than body weight. All of these findings are consistent with the

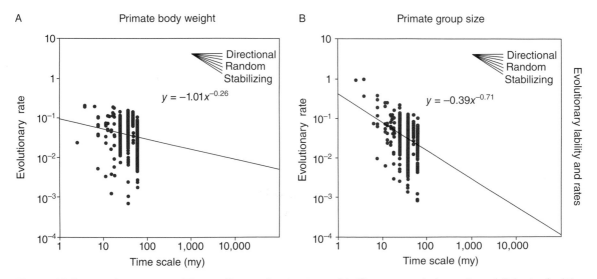

Figure 10.2 Log evolutionary rate ("darwins") versus log time interval (million years ago) observed across Primates for (A) body weight and (B) group size. Values of slope represent directedness of evolution and values of intercept represent rates over a million years of evolution. Redrawn from Gittleman et al. (1996).

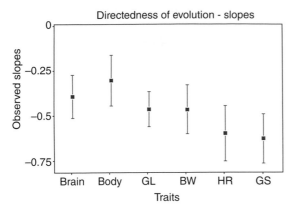

Figure 10.3 Combined values of slopes across eight mammalian taxa (Primates, Ceboidea, Carnivora, Canidae, Bovidae, Cervidae, Certacea, and Arvocolinae) observed for each quantitative trait. Traits are denoted by the following: Brain, brain weight; GL, gestation length; BW, birth weight; HR, home range; GS, group size. Means and standard errors are included for each trait. Redrawn from Gittleman et al. (1996).

intuitive prediction that behavior evolves differently from morphology. Specifically, evolutionary change in the behavioral traits of home range movements and group size is more of a random walk and relatively more directional in the morphological traits (see Gingerich 1993, for discussion of evolutionary rate models). The important observation here related to MTE is that comparing the slopes of rate change across variables the morphological traits (body size, brain size) approximate three-quarter values in contrast to the more plastic behavioral and life-history traits (Fig. 10.3). This study, however, involved small sample sizes (<40 species) and did not standardize timescale among phylogenies. Clearly, examining other behavioral and morphological traits in a greater diversity of taxa would give a more complete view of evolutionary rates.

Another promising area for future research is the relationship between metabolic ecology and rates of behavioral evolution. Organisms with a metabolic surplus may be free to evolve more complex behaviors than organisms with more constrained energy budgets. For example, shrews have such a high mass-specific metabolic rate that they are forced to spend almost all of their time foraging just to support basic life functions. By contrast, frogs, though often of similar size to shrews, as ectotherms do not need to spend as much time foraging to support basic metabolism. This may

explain why costly display behaviors to attract mates are widespread in anuran species, but are absent in shrews. We note, however, as a counter-example that hummingbirds have both extremely high mass-specific metabolic rates and costly display behaviors.

Conversely, many types of complex behavior, particularly certain types of social behavior, require relatively large brains. Brain tissue is metabolically extremely expensive to maintain (Leonard et al. 2007), and it could be that only endothermic organisms that have large energy budgets are likely to evolve large brains. This could in turn explain why complex social behavior and learned behavior are so much more common in mammals and birds than in other animal groups. The relationship between metabolism and rates of behavioral evolution is a potentially rich area of inquiry that has rarely been considered.

10.3.3 MTE and diversification rates

Allen et al. (2002) first proposed that MTE could explain global diversity gradients such as the "latitudinal diversity gradient" that occurs in many groups (see Storch, Chapter 11). MTE predicts that, due to metabolic effects on rates of evolution, diversification rates will be directly proportional to temperature. This theory is similar to Rohde's (1992) "evolutionary speed" hypothesis, but makes more explicit predictions. One of the predictions of MTE is that diversity (log species richness) should scale with temperature with a slope of approximately −0.65, and Allen et al. (2002) presented preliminary analyses demonstrating that this relationship roughly holds across a range of groups. Subsequent studies have tested for this relationship in a diversity of organisms, with mixed results. While some studies have demonstrated the scaling between environmental temperature and diversity predicted by MTE (e.g., Allen et al. 2006; Wang et al. 2009), other studies have failed to find it (e.g., Algar et al. 2007; Latimer 2007; McCain and Sanders 2010). For example, Hawkins et al. (2007) looked at temperature and diversity relationships across 46 groups and found that the predictions of MTE were only borne out in, at most, five of them. In a response to Hawkins et al., Gillooly and Allen (2007) noted that a scaling relationship between environmental temperature and diversity is only one of the predictions of MTE, that the prediction only applies to ectotherms, and that differences in water availability have the potential to diminish or even reverse expected relationships between temperature and diversity. Cassemiro and Diniz-Filho (2010) further argued that pure MTE is too simple to explain the majority of global diversity patterns, but showed that minor derivations of it can explain a much wider range of patterns.

One clear prediction of MTE, as well as Rohde's evolutionary speed hypothesis, is that diversification rates should be higher in warmer areas. While many studies have examined the relationship between diversity and temperature (reviewed in Cassemiro and Diniz-Filho 2010), surprisingly few have explicitly considered diversification rate. Negative correlations between latitude and diversification rate, or differences in diversification rate between temperate and tropical lineages, have been shown in birds (Ricklefs 2006b), amphibians (Wiens 2007), and angiosperms (Jansson and Davies 2008). Allen et al. (2006) also showed that speciation rate scales with temperature in marine Foraminifera. It remains to be seen how general the relationship between diversification rate and temperature is, and we suggest that this represents an important avenue of future research.

10.4 CONCLUDING COMMENTS

Just as Simpson integrated genetics and paleontology by developing concepts and methodologies around questions of evolutionary rate (Laporte 2000), so too may rates questions serve as an integrative approach to questions about metabolic ecology. In this chapter we have summarized some theoretical problems, emphasizing certain methods that are useful for problems of why some taxa and traits seem to have divergent rates of evolution.

A few areas seem especially interesting for future work. First, as described in this chapter, most studies to date have considered evolutionary rates and MTE from macro perspectives. It will be increasingly important to adopt experimental studies to pinpoint causal mechanisms. For example, selection experiments for metabolic rate and its influence on body size or behavioral development, similar to early experiments on inbred strains of rats to study brain–body size evolution (Atchley 1984), may be instructive. Second, more comprehensive databases that provide data on metabolic rate, morphology, and behavior with rates tests in mind will be informative. Applying a phylogenetic perspective will be particularly valuable to incorporate

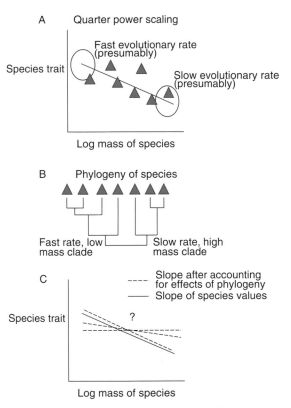

Figure 10.4 How do phylogeny and evolutionary rates relate to metabolic scaling? When species exhibit quarter-power scaling (A) can this drive differences in rates of molecular or morphological evolution (B) among clades? Conversely, when the influence of phylogeny is taken into account (C), does the scaling exponent remain similar or change?

into future studies of scaling and MTE (see Fig. 10.4). For example, Symonds and Elgar (2002) and White et al. (2009) both showed deviations from quarter-power scaling of metabolic rates in some cases when influence of phylogeny was taken into account. Last, considerable attention is appropriately being directed at problems of humanity's increasing "ecological footprint." How quickly can organisms adapt to increases of global climate change of 2 °C, which species will adapt, what genetic-phenotypic characteristics will allow for greater adaptation? These issues inherently involve rates of evolution and, as developed in this chapter, MTE is an important hypothesis for getting at these questions.

Chapter 11

BIODIVERSITY AND ITS ENERGETIC AND THERMAL CONTROLS

David Storch

SUMMARY

1 Biological diversity is affected by a multitude of evolutionary and ecological processes, but diversity patterns are quite universal across taxa; diversity generally increases towards low latitudes and towards environments characterized by high temperature and energy availability.

2 The metabolic theory of ecology (MTE) assumes that diversity is affected both by amount or supply rate of resources which positively affects total number of individuals, and by the positive effect of temperature on diversification rates. Although these assumptions are reasonable, this theory has several conceptual problems and the empirical patterns support only some of its predictions.

3 Species richness does not seem to be strongly affected by the total number of individuals.

4 Diversity patterns are certainly also affected by processes which are not accounted for by the MTE, most importantly range dynamics associated with the evolution of species climatic tolerances, which is affected by the level of the conservatism of ecological niches of species.

5 However, temperature appears as the most important driver of diversity patterns, possibly due to the temperature dependence of most biological processes.

11.1 INTRODUCTION

Biodiversity, or biological diversity, comprises all the variation of life on Earth, from genetic and molecular diversity, through diversity of species and higher taxa, to the diversity of whole ecosystems. Biological diversity is the most prominent feature of life on Earth, yet its distribution on the Earth's surface and across evolutionary lineages is unequal. Patterns in biological diversity have been affected by multiple processes acting at multiple scales, ranging from biotic interactions within local ecological communities to evolutionary radiations of evolutionary lineages within whole continents. All these processes potentially can be affected by energy availability and biological rates, and thus the considerations concerning organismal metabolism seem very appropriate when trying to understand them. However, the question is to what extent these simple considerations are useful for explaining or even predicting contemporary biodiversity patterns, given that their causes

Metabolic Ecology: A Scaling Approach, First Edition. Edited by Richard M. Sibly, James H. Brown, Astrid Kodric-Brown.

are certainly complex. From this point of view I will mostly explore spatial patterns of biodiversity, as these have been the most thoroughly studied.

Perhaps the clearest ecological generality concerning spatial diversity patterns is that diversity closely correlates with climate (Fig. 11.1). In particular, abiotic variables related to energy availability and productivity (namely, temperature and water availability) appear to drive the most prominent biodiversity trend on land, the latitudinal diversity gradient (Currie

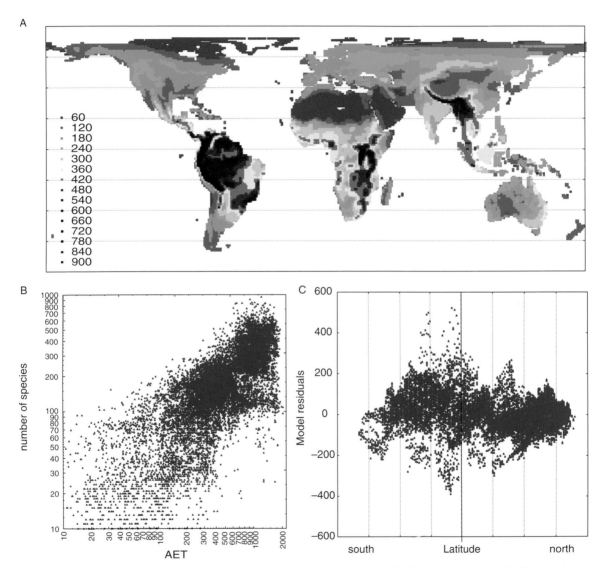

Figure 11.1 Relationship between diversity and climate. (A) Global distribution of bird species richness (defined as number of species within grid cells of approximately 1° latitude × 1° longitude) shows the highest levels in tropical areas, especially in tropical mountains. (B) Relationship between actual evapotranspiration, a measure of water and energy flow through an ecosystem, and bird species richness. (C) After controlling for AET, there is no remaining latitudinal signal of species richness, indicating that AET fully accounts for the latitudinal variation in diversity. (Data from Storch et al. 2006 by permission of John Wiley & Sons, Ltd.)

1991; Francis and Currie 2003; Hawkins et al. 2003; Buckley and Jetz 2007; Kreft and Jetz 2007; Hortal et al. 2008). Although the importance of individual factors may vary regionally (e.g., water availability may be more important at lower and temperature at higher latitudes; Hawkins et al. 2003), the role of these major climatic factors appears to be globally consistent across taxa. Three major explanatory frameworks have recently emerged to address spatial diversity patterns.

11.1.1 Hypotheses based on environmental limits of species coexistence

This class of hypotheses assumes that the number of species which can coexist at a given site is constrained by the total number of available ecological niches or by the total number of individuals which can be sustained under given energy input. The total number of available niches is hard to estimate, given the circularities inherent in niche definitions (Chase and Leibold 2003, but see Walker and Valentine 1984). However, the possibility that the total amount of resources set by environmental productivity limits the total number of individuals is quite straightforward. A higher number of individuals can be divided into more species with viable populations, and sites which support more individuals will then tend to support more species (Wright 1983). This theory, which is referred to as the species-energy theory, has been refined and restated as the more-individuals hypothesis (Gaston 2000).

11.1.2 Hypotheses based on species diversification rates

According to this class of hypotheses, diversification rates are faster in hot and humid environments, resulting in higher number of taxa in the tropics (Rohde 1992). Diversification rate may be driven by mutation rate, which is in turn dependent on temperature-dependent metabolic rate (Allen et al. 2006, 2007; Gittleman and Stephens, Chapter 10).

11.1.3 Hypotheses based on historical climate and species niche and range dynamics

Diversity patterns are affected by Earth's history and the history of individual evolutionary lineages, their evolution and spread across the Earth's surface (Ricklefs 1987; Ricklefs and Schluter 1993). Most species may live in humid warm climates simply because tropical areas historically dominated the Earth, or were more stable over geological time, so that individual taxa had enough time to adapt to these environments (Latham and Ricklefs 1993; Ricklefs 2006a). This explanation necessitates the assumption of niche conservatism (i.e., species do not adapt too quickly to new environments; Wiens and Donoghue 2004), and represents in a sense a null explanation of diversity patterns, as it assumes no particular processes generating diversity besides historical legacy.

Clearly, all of these three explanatory frameworks are essentially incomplete. Hypotheses on environmental limits on coexistence ignore the findings that the diversity of each local community is strongly affected by regional/historical effects (Ricklefs and Schluter 1993; Caley and Schluter 1997; Ricklefs 2008), so that local environmental limits only partially affect diversity of species assemblages. Hypotheses on diversification rates ignore the fact that diversity is also given by species spreading out of evolutionary sources and by extinction dynamics (Jablonski et al. 2006), and that species richness is necessarily limited by limits of species geographic ranges. Hypotheses based on historical climate and species niche and range dynamics ignore unequal diversification rates in different regions and possible environmental limits of species richness. So none of the hypotheses are mutually exclusive and they complement each other.

Although biodiversity patterns have almost certainly emerged due to all the above-mentioned processes, the striking generalities concerning the role of climate and energy are challenging. The energetic and metabolic controls on biodiversity patterns are therefore worth exploring. I will first explore in more detail the energetic controls on the number of coexisting species, and then I will consider the role of metabolic control of diversification rate and diversity patterns.

11.2 THE MORE-INDIVIDUALS HYPOTHESIS AND ITS LIMITATIONS

The assumption that energy availability constrains the total number of individuals which can coexist in an environment, consequently constraining the number of species, represents the most straightforward explanation of species richness patterns. Although this

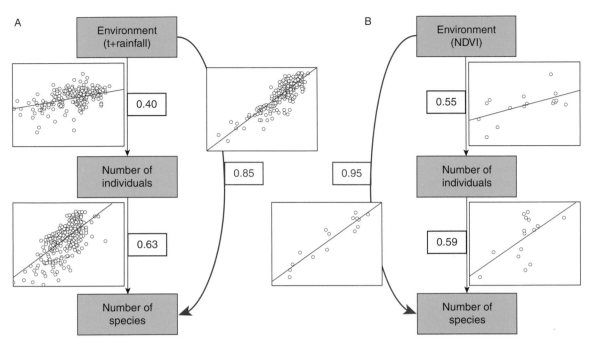

Figure 11.2 Relationships between environmental variables, total assemblage abundance, and species richness for (A) global forest plots and (B) South African birds. The numbers in boxes refer to correlation coefficients for individual relationships. The *x*-axis in the biplots always refers to the variable (or the combination of variables in A) which is assumed to be independent and causally responsible for the dependent variable (note, however, that in fact this may not be the case, e.g., abundances may be higher due to higher species richness, and not vice versa). The relationship between environmental variables and species richness is apparently stronger than both the relationships between environment and total abundance, and between total abundance and species richness, casting doubt on the more-individuals hypothesis. Data from Šímová et al. (2011) by permission of John Wiley & Sons, Ltd, and Storch et al. (unpublished).

hypothesis has never been properly formalized (but see Wright 1983), it is reasonable to believe that a higher number of individuals can be divided into more species with viable populations, and conversely that decreases in the total number of individuals due to lower energy availability would lead to generally lower abundances and consequent extinction of populations of rarer species. Since the causal chain goes, according to the hypothesis, from energy availability to the total number of individuals, and then to the number of species, it predicts that the relationship between energy availability and number of individuals should be relatively tight, and that the relationship between the number of individuals and number of species should be tight as well. Conversely, the relationship between energy availability and species richness should be weaker, as these variables should be related only indirectly, through the number of individuals (Currie et al. 2004).

However, the opposite has been typically observed: species richness is quite tightly related to energy availability, whereas the number of individuals is only loosely related to both number of species and available energy (Currie et al. 2004; Šímová et al. 2011; Fig. 11.2). Moreover, the tight relationship between energy availability and species richness is observed even if the number of individuals is controlled for (Hurlbert 2004; Sanders et al. 2007; Šímová et al. 2011).

It thus appears that species–energy relationships are generally not mediated by the number of individuals. Although the very low numbers of individuals sustainable in extremely unproductive areas may limit the number of species found there, the more-individuals hypothesis does not seem to be the universal or even primary explanation of biodiversity patterns. The observation that the number of individuals is often correlated to the number of species (e.g., Kaspari et al.

2003; Evans et al. 2005) can be explained in other ways. For instance, the number of species may be determined by some other effects, and number of individuals is simply higher where the number of species is higher (see Long et al. 2006). The causal chain between number of species and number of individuals can thus in fact be reversed (Šímová et al. 2011).

On the other hand, regardless of the exact direction of causality, number of individuals is certainly linked to the number of species, and thus the pattern of relationships between climate (or energy availability) and number of individuals in an assemblage is worth exploring. Such patterns do not appear consistent between taxa. Whereas the total abundance of birds increases with productivity in approximately the same way as species richness (so that mean population size of species is more or less independent of productivity and of assemblage species richness; see Pautasso and Gaston 2005; Evans et al. 2008), total abundance does not vary much with climatic gradients in trees (Enquist and Niklas 2001; Šímová et al. 2011) nor, perhaps, in invertebrates (Novotny et al. 2006). This results in lower population densities of individual species of ectotherms at low latitudes and/or more productive regions. This discrepancy can be addressed by the metabolic theory, for which the difference between endotherms and ectotherms is crucial.

11.3 METABOLIC THEORY OF BIODIVERSITY

The idea that temperature affects diversification rates, and consequently the major diversity gradients, is older than the metabolic theory of ecology (Rohde 1992). Interestingly, the first formal connection between the metabolic theory and diversity patterns was not explicitly based on evolutionary rates, but instead on the relationship between temperature-dependent metabolic rates and the controls of abundances (Fig. 11.3). Allen et al. (2002) derived the relationship between temperature and species richness within a given area, assuming a generalized version of the energetic equivalence rule (EER; see Isaac, Carbone, and McGill, Chapter 7). EER (Damuth 1987; Nee et al. 1991) states that population energy consumption of individual species per unit area is independent of body size, since species with larger body size (which have higher metabolic

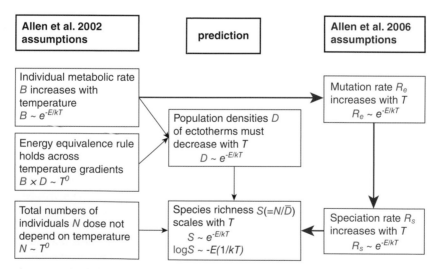

Figure 11.3 The logic of the metabolic theory of biodiversity. Originally (black arrows) the prediction of the species–temperature relationship was based on a generalized energy equivalence rule and an assumption that total assemblage abundance is constant. However, more plausible causation comprises the effect of temperature on diversification rates (red arrows). The fundamental assumption (in both cases) is that individual metabolic rate, B, scales with temperature according to the relationship $B \sim e^{-E/kT}$, where k is the Boltzmann constant ($8.62 \times 10^{-5}\,\mathrm{eV\,K^{-1}}$), T is temperature in kelvin and E is activation energy of metabolic reactions, which should vary between -0.60 and $-0.70\,\mathrm{eV}$ (Brown et al. 2004).

rate) have, on average, correspondingly lower population densities. Its generalized version according to Allen et al. (2002) assumes additionally that energy consumption of individual species' populations is independent of temperature. Ectothermic species living in warmer regions have higher metabolic rates (i.e., higher per-capita energy consumption; Gillooly et al. 2001); therefore, they need to have lower population densities in warmer regions to fulfill the generalized version of EER. Assuming additionally that the total number of individuals within an area, N, is independent of temperature, the total number of species, S, must be higher in warmer areas. The reason is that mean population density, $\bar{D}$, is given by $\bar{D} = N/S$, so that if $\bar{D}$ decreases with temperature and total assemblage abundance N is constant, number of species, S, must increase to balance the decrease of mean population density, $\bar{D}$, with temperature. The number of species should thus scale with temperature in the same way as mean population density and metabolic rate (Fig. 11.3).

Allen et al. (2002) provided some evidence that mean population size of ectotherms indeed scaled inversely with temperature, as predicted by the generalized version of EER, as well as evidence of the predicted scaling of species richness with temperature. However, all the reasoning mentioned above is quite problematic. There is no reason to assume that the *total number of individuals per unit area*, N, is independent of temperature, and simultaneously that *per-species mean density* decreases with temperature (Storch 2003). If total densities of ectotherms do not depend on temperature, total supply rate of resources must increase with temperature to support the same total community size (total number of individuals), given that every individual consumes more resources. And if the supply rate of resources increases with temperature, there is no reason why the densities of individual species' populations should be lower in warmer regions to follow the generalized energy equivalence rule. In an effort to address these problems, the theory has been reformulated in terms of evolutionary rates (Allen et al. 2006, 2007): metabolic rate affects the rate of all biological processes including mutation (Gillooly et al. 2005b) and speciation (Allen et al. 2006), and higher speciation rates in warmer environments should lead to higher number of species (Fig. 11.3, red lines).

A current formulation of the metabolic theory of biodiversity (Allen et al. 2007) assumes that energy availability affects species richness in two independent

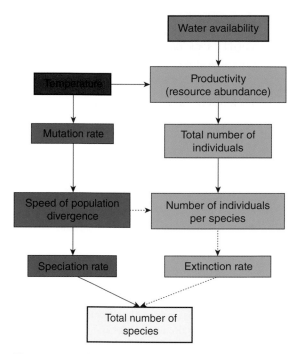

Figure 11.4 Chains of causality showing how temperature and productivity can affect diversity (negative effects are marked by dashed lines; this scheme follows Allen et al. (2007) but differs in some respects). There are multiple and sometimes counteracting effects (e.g., temperature increases total number of individual due to its positive effect on productivity, but may lead to the decrease of abundance per species due to its positive effect on population divergence), so that exact predictions are difficult to formulate and test. Note that this is still a simplified picture of possible causal links. In fact, temperature may not only increase total number of individuals via increasing productivity, but may simultaneously exert a counteracting effect on the numbers of individuals, as individuals in warmer regions have higher energy consumption due to higher metabolic rate – and thus their population-carrying capacities may be lower in warmer environments (Allen et al. 2002).

ways, combining the more-individuals effect with the effect of temperature on speciation/diversification rates (Fig. 11.4). According to the theory, there are two fundamentally different forms of energy (see also Clarke and Gaston 2006): kinetic energy of molecules expressed as temperature, and potential energy of chemical bonds, i.e., energy stored in biomolecules. Whereas temperature affects mutation, and consequently diversification rates, potential energy is equivalent to the amount or supply rate of resources (environmental productivity), and thus affects the number of individuals which can persist in an environment. However, environmental productivity is also affected by temperature, whose effect combines with the effect of nutrient and water limitation. Predictions of the theory are thus not straightforward. Population sizes of individual species are predicted to be generally positively affected by temperature through increasing productivity, but negatively by increasing diversification rates, so that the final expectation depends on the balance of these processes.

However, the theory does provide predictions for situations in which some effects are essentially constant. Most importantly, it predicts differences in species richness patterns between endotherms and ectotherms. Environmental temperature should not have a direct effect on diversification rates in endotherms, whose body temperature is constant, which may explain the observation that species' mean population sizes of birds do not change systematically with latitude or productivity (Pautasso and Gaston 2005). Conversely, population sizes of individual species of trees or insects may vary along these gradients due to the variation in diversification rate, as mentioned above. The other prediction concerns the situation in which the total number of individuals is constant, for example, due to resource limitation unrelated to temperature (e.g., forest trees limited by space). In such a case, the metabolic theory of biodiversity predicts that the causal chain which concerns potential energy (i.e., productivity effects; the right-hand column in Fig. 11.4) is not relevant, and species richness should be simply related to diversification rates, and should scale with temperature similarly to metabolic rate (Allen et al. 2006).

11.4 CONCEPTUAL PROBLEMS OF THE CURRENT FORMULATION OF THE METABOLIC THEORY OF BIODIVERSITY

Although the theory depicted above is compelling, because it deals explicitly with multiple pathways leading to observed relationships between climatic variables and biological diversity, it has several problems, both conceptual and empirical. One problem related to its original formulation using the generalized EER has been dealt with above, but this is not relevant to the current formulation of the theory. However, other problems are substantial, as follows.

11.4.1 Relationship between standing species richness and speciation rate

The metabolic theory assumes that speciation rate is proportional to metabolic rate per unit mass, and that species richness is proportional to speciation rate. Whereas the former proportionality has some empirical support (Allen et al. 2006), the relationship between speciation rate and species richness is much less straightforward. The equilibrium number of species is the net result of both speciation and extinction, similar to the way that equilibrium population size is the net result of natality and mortality. A direct proportionality between speciation rate and species richness is expected only under quite restrictive conditions (Fig. 11.5). If we assume that speciation rate is a variable which can be attributed to individual species, and that both speciation and extinction rates are dependent on mean population size (which is the case depicted in Fig. 11.5), both these rates should be dependent on species richness. The reason is that if there is a constant total number of individuals N (determined by a constant supply rate of resources), then mean population sizes which drive speciation and extinction rate must decrease with increasing number of species (since mean populations size $= N/S$). Then the species richness is proportional to per-species speciation rate only in special cases of fine-tuned dependencies of both the rates on mean population size (Fig. 11.5). Allen et al. (2006, 2007) viewed speciation rate as a variable attributed to individuals (i.e., per-capita speciation rate) instead of species, following the formalism of the neutral theory of biodiversity and biogeography (Hubbell 2001). The situation is then somehow different, but the neutral theory does not predict the proportionality between speciation rate and species richness either. Instead, Hubbell's fundamental biodiversity number, theta, should be proportional to speciation rate, and thus this characterization of biodiversity could represent a better way to build a metabolic theory of biodiversity than by using species richness as the measure of biodiversity – with a caveat that the formalism of the neutral theory may not be universally valid and acceptable (McGill et al. 2006).

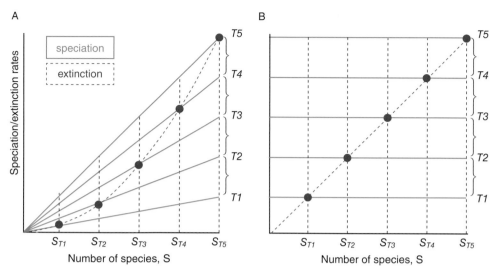

Figure 11.5 Relationships between speciation and extinction rates and the number of species. Assume that per-species speciation rate increases linearly with some variable. In such a case, total speciation rate for the species assemblage increases with the number of species, and the lines denoting the relationship between the number of species and the total speciation rate for a given variable (here, different temperatures T1–T5) will be straight lines with slopes proportional to the temperature term (A). This is represented by constant increment of the endpoints of these lines on the vertical axis. The equilibrium number of species is given by intersection of the speciation curve with an extinction curve specifying the relationship between the number of species and total species extinction rate (i.e., number of species extinct in a given time interval). Clearly, equilibrium species richness is directly proportional to speciation rate only if the extinction curve intersects individual speciation curves/lines at points equally spaced on the horizontal axis. For other shapes of the extinction curve, equilibrium species richness is not proportional to speciation rate. Therefore, standing species richness may not be directly related to speciation rate. The same reasoning applies if we assume that total speciation rate does not vary with the number of species (B) (which is assumed in Hubbell's (2001) neutral theory in which total number of individuals is constant and speciation rate is defined on a per-capita basis). In this case, total extinction rate must increase linearly with the number of species to keep the proportionality between speciation rate and species richness.

11.4.2 Scale dependence of species richness

Metabolic theory predicts a particular quantitative relationship between temperature and species richness (Fig. 11.3), namely that the logarithm of number of species should decrease linearly with $1/kT$ (where k is Boltzmann's constant and T is temperature in kelvin), with the slope equal to the activation energy of metabolism (around 0.6–0.7 eV) (Brown et al. 2004; Allen et al. 2006, 2007; Brown and Sibly, Chapter 2). However, the spatial grain at which this prediction should hold is not specified. Species richness is scale-dependent, and if some richness–environment relationship holds for areas of, say, 100 km² , a different slope would likely be observed in smaller or larger

areas. In fact, a constant (area-independent) slope of the richness–temperature relationship could be observed only if the species–area relationship had a slope which was independent of temperature. If this is not the case, i.e., if number of species increases more rapidly with area in, for example, warmer regions, then the number of species would be relatively higher in larger areas in these regions, and the overall richness–temperature relationship would be steeper for large areas. This effect was demonstrated by Wang et al. (2009) for richness patterns of woody plants (Fig. 11.6). The slope of the species–area relationship is higher in warmer areas, and consequently, the species–temperature relationship is steeper for larger areas (in statistical terms, there is a positive interaction between temperature and area in their effects on

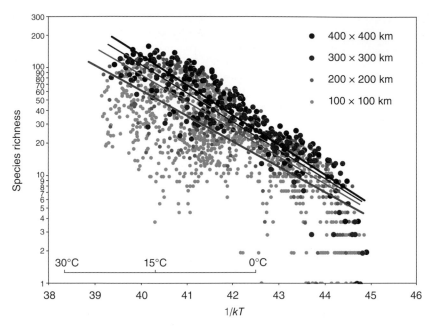

Figure 11.6 The relationship between temperature expressed as $1/kT$ and tree species richness in North America for different grain sizes (grid areas). The slope of species–temperature relationship (which is equivalent to the activation energy E) increases with grain. There is thus a positive interaction between temperature and area in their effect on species richness, with the consequence that the slope of the species–area relationship increases with temperature. Data from Wang et al. (2009).

species richness). This concurs with the findings of Harte et al. (2009) that the slope of the species–area relationship depends on the ratio between the total number of individuals and number of species within a given assemblage (N/S, which is equivalent to mean population size, see above): the higher the ratio, the lower the slope of the species–area relationship. Since the number of species increases with temperature, whereas the total number of individuals (in this case the density of trees in forests) is not too sensitive to temperature, the ratio is lower in warmer regions, leading to a steeper species–area relationship. The exact relationship between temperature and species richness thus cannot be universal across spatial scales.

11.4.3 The problem of taxonomic invariance

Any macroecological pattern can be truly universal only if it is invariant against changing taxonomic delimitation, i.e., if it can simultaneously hold for a given taxon as well as for its subtaxa (Storch and Šizling 2008). This is hardly the case for the above-mentioned exponential relationship between temperature and species richness. Imagine a taxon composed of two or more subtaxa (such as South American mammals including placentals and marsupials). The exponential relationship between temperature and species richness mentioned above could hold for the taxon and all subtaxa only if the respective equations have exactly the same parameters. Whenever one of the subtaxa deviates, either having a different form of the relationship (e.g., linear instead of exponential), or the same form with a different slope (given by activation energy E), the resulting relationship for the larger inclusive taxon cannot be exponential, simply because the summation of exponentials for different subtaxa gives an exponential for a larger taxon only if the parameters are the same. Since some taxa certainly deviate from predicted patterns (see below), it is unlikely that the exact predicted relationship can hold for species richness of any larger taxon.

11.5 EMPIRICAL PATTERNS: EVIDENCE AND COUNTEREVIDENCE

Allen et al. (2002, 2006) and Brown et al. (2004) published several graphs concerning the relationship between the logarithm of number of species and $1/kT$ in support of their theory. This has stimulated a wave of studies trying to support or refute it (e.g., Algar et al. 2007; Hawkins et al. 2007; Keil et al. 2008). Although some studies found the predicted relationship between species richness and temperature, including the slopes, it appears to be far from general (Hawkins et al. 2007). Species richness generally increases with temperature (the exceptions tend to be narrowly defined taxa with particular habitat requirements), but this increase may not always be exponential, and even in this case the slope varies much more than predicted by the theory. In fact, this is not surprising, given the conceptual problems mentioned above. Moreover, it seems that tests comprising just the exploration of the bivariate relationships between temperature and species richness are not appropriate. The theory makes specific assumptions concerning the absence of other factors affecting diversity, namely that the number of individuals does not vary due to variation in resource availability (Allen et al. 2007; Gillooly and Allen 2007; Cassemiro and Diniz-Filho 2010). These effects have rarely been controlled for (although, paradoxically, this was a problem in the original Allen et al. (2002) study as well). Šímová et al. (2011), however, have shown that the relationship between tree species richness and temperature in forest plots deviated from the predicted exponential relationship, even if water availability and number of individuals were accounted for.

The other problem with tests of the predictions of the metabolic theory of biodiversity comprises the conditions and scales under which the predictions should hold. Since the most recent formulation of the theory is based on evolutionary rates, we would expect that the relationship would have emerged during evolutionary timescales and should be observed mostly on large spatial scales. However, the cases in which the pattern did agree with the predictions often comprised much smaller scales, and apparently emerged in much shorter time intervals than would be necessary for evolutionary changes. Hunt et al. (2005) reported pronounced temporal changes of species richness of benthic Foraminifera associated with temperature changes (in the direction predicted by the theory)

during the last 130 000 years, i.e., a much shorter timescale than would be enough for evolutionary changes such as altered speciation rates. Also, the diversity patterns observed in terrestrial organisms inhabiting the temperate regions of the Northern Hemisphere must be relatively recent, as the species have immigrated to these areas after the end of last glacial, i.e., during the last 12 000 years. One could argue that these recent patterns are driven by the differential immigration of species from large species pools, whose niches are conservative and whose evolution was determined by temperature as predicted. However, the relationship between richness of the species pool (i.e., regional richness) and local community richness is not straightforward (Ricklefs and Schluter 1993), and is confounded by all the scale issues mentioned above.

These issues can be generalized. The major problem of the theory is that it is focused exclusively on the temperature dependence of diversification rates. It ignores other effects, including migration from source areas and population spatial dynamics in general, as well as other processes which may also be temperature-dependent, albeit in a different manner (Stegen et al. 2009). It is therefore reasonable to ask what the true achievements of the theory are.

11.6 MERITS OF THE METABOLIC THEORY OF BIODIVERSITY

Despite many problems mentioned above, there are several insights and contributions of the metabolic theory of biodiversity which makes it a useful and lively approach to biodiversity science. Firstly, it is the only theory which gives exact quantitative predictions of species richness patterns. Most theories concerning biodiversity patterns provide only qualitative predictions (e.g., species richness should increase with productivity/area/temperature), and the predictions of individual theories are thus not distinguishable from each other. The possibility that the increase of species richness is predictable from the knowledge of the increase of temperature is intriguing, but even more important is the fact that this prediction provides opportunity to test the theory (with the caveats mentioned above).

Second, the metabolic theory of biodiversity is the only theory which explicitly predicts an exponential increase of species richness with temperature.

Although, as has been argued above, this pattern is not always observed, this exponential relationship occurs quite often (e.g., da Silva et al. 2007; Sanders et al. 2007; Wang et al. 2009; Cassemiro and Diniz-Filho 2010), so it is reasonable to consider it to be a rule rather than an exception. Moreover, although the slopes of the relationship between log (species number) and $1/kT$ cannot be the same for all taxa and all scales – indeed, they are not – the slopes seem to converge on the predicted value (Stegen et al. 2009). This is rather paradoxical, given that the theory is so far intrinsically incomplete, without the ability to fully address all the processes producing diversity patterns.

Third, the theory explicitly stresses the role of temperature in contrast to other energy-related variables such as productivity, or actual or potential evapotranspiration. Although these variables are often closely correlated in terrestrial environments, so that their effects are difficult to disentangle, it appears that the effect of temperature is more important than productivity. This can be more easily demonstrated in marine systems, where temperature and productivity are decoupled. Fuhrman et al. (2008a) have shown that the diversity of marine planktonic bacteria increased with temperature, but was independent of productivity or bacterial biomass. Temperature is generally a better predictor of marine diversity than other environmental variables (Tittensor et al. 2010). Within terrestrial forest plots, diversity correlates with productivity only if temperature is accounted for when estimating productivity; otherwise there is no effect of productivity on tree species richness (Šímová et al. 2011). Similarly, local ant diversity is predicted by temperature but not productivity (Sanders et al. 2007). Large-scale diversity patterns in vertebrates are related to both productivity and temperature, but the relative role of temperature is stronger at larger spatial scales (Belmaker and Jetz 2011), conforming to the metabolic theory, which involves evolutionary processes rather than factors affecting local species coexistence. On the other hand, temperature appears also to predict large-scale richness of endotherms (Davies et al. 2007; Belmaker and Jetz 2011), which contradicts the metabolic theory, and indicates that the role of temperature may be different and more complex than assumed by the theory so far.

Regardless of the role of temperature on spatial diversity patterns themselves, diversification rates seem to be latitude- and thus temperature-dependent (Svenning et al. 2007; Wiens 2007; Jansson and Davies 2008; Wright et al. 2010). Although this generally conforms to MTE, similar temperature dependence was observed in endotherms (Cardillo et al. 2005; Ricklefs 2006b; but see Weir and Schluter 2007). This may indicate that diversification processes are more complex, being affected by temperature indirectly as well as directly. Moreover, Davies et al. (2004) have shown that although temperature was strongly associated with plant diversification rates as well as with plant species richness, diversification rates did not appear responsible for species richness patterns. Therefore, although the role of temperature is apparent in many aspects of ecology and evolution, the way MTE deals with these effects has so far been overly simplistic.

11.7 TEMPERATURE AGAIN: WHICH EFFECTS AND WHEN?

Temperature is a crucial abiotic factor affecting almost all aspects of organismal biology. Even the oldest considerations concerning global diversity patterns invoked temperature as an essential driver. Indeed, Alexander von Humboldt (1850) attributed higher species richness of tropical organisms to their limited cold tolerance (Hawkins 2001). Current findings support this conclusion. Minimum temperature, in contrast to mean values, appears to be the best predictor of tree species richness (Šímová et al. 2011; Wang et al. 2011). Biological diversity in terrestrial environments is obviously also limited by water availability, so that a combination of temperature and water availability predicts species richness patterns. However, minimum rainfall again appears more important than mean values (Šímová et al. 2011), indicating that, rather than simple multiplicative or additive effects of both variables, some nonlinear and threshold-like effects play a role. Additionally, species-poor regions are typically inhabited by a subset of higher taxa, which are often younger, more derived and phylogenetically clustered (Hawkins 2010; Machac et al. 2011), suggesting that only a few evolutionary lineages were able to overcome climatic constraints. All these findings can be interpreted as evidence for the role of climatic limits of individual species distributions, i.e., as support for the third of the above-mentioned hypotheses, comprising history of niche evolution and the spreading of species out of the tropics and their lower diversity in colder and drier climates limited by the difficulty of adapting to extreme abiotic conditions. Species niches are to a large extent conservative (Wiens and Donoghue 2004), and it is reasonable to assume that many species are adapted to historically prevailing warm humid conditions (but see Algar et al. 2009).

Temperature will thus undoubtedly have several independent and interacting effects, influencing diversification of lineages, range dynamics, and the spreading of taxa, and potentially also the limits of species coexistence (i.e., the major processes affecting global diversity patterns mentioned in section 11.1). The effect of temperature on local species coexistence is much less clear than in the case of the other processes. One could argue that higher temperatures should increase the rate of all interspecific interactions including competitive exclusion, and thus decrease rather than increase diversity. However, organisms living at higher temperatures could also have greater potential to avoid competition (or predation), since temperature keeps all the physiological processes active, enhancing the range of possible strategies for thriving in an environment and how and when resources can be utilized (Sanders et al. 2007). Warmer environments thus promote multidimensionality and complexity of biotic interactions, all of which supports high biological diversity (Schemske 2009). Low temperature, on the other hand, simplifies the interactions between an organism and environment, promoting simple and directional changes in community composition. More generally, higher temperature may be associated with greater complexity in possible ways of life, thus enhancing diversity (see Anderson and Jetz 2005).

Even more generally, diversity of life can be universally understood as a result of the interplay of processes which enhance it (such as evolutionary diversification, coevolution, co-adaptation, and the emergence of novelties), and processes which suppress it, namely population extinction including competitive exclusion. It is reasonable to assume that the first set of processes will be more closely – and positively – related to temperature, as all these processes arc essentially biological, and thus their rate is dependent on metabolic rate. On the other hand, the processes which reduce diversity are more closely associated with temperature-independent environmental stochasticity. Their speed will depend negatively on population sizes, the later being effectively independent of temperature, but positively dependent on energy availability, in accord with the basic framework of metabolic theory

of biodiversity (Fig. 11.4). Diversity patterns could thus be viewed as the outcome of differential rate of diversity-enhancing and diversity-suppressing processes, largely controlled by temperature.

11.8 CONCLUSIONS

A metabolic theory of biodiversity is a work in progress. It is helping to shed light on fundamental relationships between biological rates, resource supply and utilization, and numbers of individuals and species. But current versions of the theory (various MTE applications) have serious problems – logical inconsistencies and failures to account for empirical patterns. Although spatial biodiversity patterns are related to energy availability and temperature, the exact causal chains are difficult to disentangle. Energy availability apparently does not affect biological diversity simply through its effect on the number of individuals, as assumed by the more-individuals hypothesis. Metabolic theory provides a more elaborate explanation of biodiversity patterns, but so far it does not represent a logically consistent theory. It is mainly confined to temperature-dependent speciation rates, which seems reasonable, but far from complete. There is no universal slope of the relationship between temperature and species richness, and such a universal relationship cannot exist at all, given the scale dependence of species richness patterns. However, the role of temperature appears strong and essential, apparently more important than productivity (especially at large spatial scales), the form of the relationship between temperature and species richness roughly conforming to the MTE predictions. Temperature obviously is not the only factor affecting biodiversity patterns – water availability appears to be at least equally important – but it has potentially multiple effects, ranging from temperature-dependent diversification rates to limits of range expansions dependent on minimum temperature.

ACKNOWLEDGMENTS

This contribution has been supported by the Grant Agency of the Czech Republic (P505/11/2387) and by the Czech Ministry of Education No. LC06073 and MSM0021620845.

Part II

Selected Organisms and Topics

Chapter 12

MICROORGANISMS

Jordan G. Okie

SUMMARY

1 The biological activity and diversity of prokaryotes and unicellular eukaryotes is extraordinary.
2 The metabolic ecology of these microorganisms is governed by five fundamental physical and biological dimensions of life: (a) thermodynamics; (b) chemical kinetics; (c) physiological harshness and environmental stress; (d) cell size; and (e) levels of biological organization, including host–endosymbiont mutualisms, consortia, biofilms, multicellular prokaryotes, and multi-domain superorganism complexes.

3 The metabolism and chemical kinetics of the higher levels of biological organization emerge from the complex interactions of the energetics of the individuals and their biochemical reactions.
4 Identifying shifts in metabolic scaling across major transitions in ecological and evolutionary organization can elucidate some of the most fundamental features of bioenergetics that shaped the early evolution of life and shape the ecology of microorganisms today.

12.1 INTRODUCTION

Microscopic organisms are of macroscopic importance. Microorganisms are everywhere. They make up a majority of the biomass on Earth. Prokaryotes alone have an estimated abundance of $4–5 \times 10^{30}$ cells, a global carbon mass 60–100% that of plants, and global nitrogen and phosphorus masses about 10-fold more than plants (Whitman et al. 1998). The metabolic activities of microorganisms have crucial roles in local and global biogeochemical cycles. Our food industry, biotechnology, medicine, agriculture, and health rely on the biological activities of microbes.

The majority of explicit research in ecological theory has been conducted on macroorganisms. In several respects, however, microorganisms harbor the greatest biological diversity – in biochemistry, in phylogeny, in habitat, in metabolic lifestyle, in resource use, and in range of body size. Thus the greatest challenges and most promising advances for ecological theory arguably lie in its applications and extensions to understanding the ecology of bacteria, archaea, and microbial eukaryotes.

Microbes exhibit an astounding range of values along multiple dimensions of diversity, and this documented variety continues to increase as we look more carefully at the microbial world (e.g., Brock et al. 2011). The size of their cells spans 16 orders of magnitude (a factor of ten quadrillion or 10 000 000 000 000 000), from the tiniest bacteria

Metabolic Ecology: A Scaling Approach, First Edition. Edited by Richard M. Sibly, James H. Brown, Astrid Kodric-Brown.
© 2012 John Wiley & Sons, Ltd. Published 2012 by John Wiley & Sons, Ltd.

weighing $\sim 10^{-15}$ g to the largest unicellular protists weighing ~ 1 g (Table 12.2). Collectively these tiny organisms harness a huge diversity of metabolic pathways, substrates, and lifestyles that use dozens of different elements as energy sources. They maintain physiological activity across the widest range in temperatures (from -40 to $122\,°C$), pressures, salinity, and pH, and inhabit nearly every location in the Earth's crust where free energy is available, rocks up to kilometers deep underground and microscopic liquid veins kilometers deep in Antarctic glacial ice (Morita 1980; Rothschild and Mancinelli 2001; Price and Sowers 2004). Microorganisms organize themselves across multiple levels of organization – growing as single reclusive cells, multi-species social consortia, multicellular organisms, and members of multi-domain superorganisms. This biodiversity is hardly surprising given that prokaryotes and unicellular eukaryotes occupy a minimum of two-thirds of the tree of life – at least 50 different phyla. The evolutionary and phylogenetic diversity of microbes resulting from the billions of years of evolution of their lineages has allowed them to generate and conserve novel metabolic niches and occupy every corner of the Earth's crust explored by scientists.

Given the importance of microbial metabolic processes and their remarkable biological diversity, some of the most significant applications of ecological theory are in identifying the major ecological dimensions governing the metabolism of microbes and determining how metabolism scales across extremes along these dimensions. Because of their high abundances and fast biological rates, microbes offer a useful model system for metabolic ecology. Sufficient data can be generated in short periods of times from field and laboratory studies. Their high rates of mutation and horizontal gene transfer mean that evolutionary and ecological perspectives must be integrated. And big-picture ecological, biogeographic, and evolutionary experiments can be conducted that would never be possible in higher organisms.

So, in order to develop a metabolic theory of ecology that addresses the geographically heterogeneous distribution of phylogenetic and metabolic diversity on Earth we must study microbes. Their integration into metabolic theory is necessary in order to unify biological theory across levels of organization. The question is: what are the major dimensions of the metabolic ecology of prokaryotes and unicellular eukaryotes? In other words, what sets of variables must be considered

in order to understand the role of energy in the interactions between organisms and their environments? I shall employ a scaling perspective to explore the five fundamental dimensions that characterize a metabolic theory of ecology of microorganisms:

1 Thermodynamics
2 Chemical kinetics
3 Physiological harshness and environmental stress
4 Cell size
5 Levels of biological organization, including host–endosymbiont mutualisms, consortia, biofilms, multicellular prokaryotes, and multi-domain superorganism complexes.

Each dimension influences the metabolic rate of microorganisms and thus the interaction between microbes and their environments. Explicit consideration and application of these dimensions in the development of metabolic theory has great potential. It provides a basis for extending metabolic theory to explain patterns in biodiversity, such as diversity gradients and community assembly rules. After presenting an abbreviated history of the metabolic ecology of microbes, I will delve into the foundations of the energetics of individual cells that must be considered in order to develop a quantitative metabolic theory of microbial ecology. Then I will discuss the first four intrinsic dimensions as they affect individual cells. And I will end by considering the fifth dimension (levels of organization) and its interaction with the other dimensions.

12.2   BRIEF HISTORY OF METABOLIC ECOLOGY OF MICROBES

Microbial ecologists have long studied the energetics and metabolic ecology of microbes. They have investigated the temperature dependence of microbial growth and respiration employing the Arrhenius equation (e.g., Johnson and Lewin 1946; Ingraham 1958; Goldman and Carpenter 1974; Button 1985; Davey 1989; Price and Sowers 2004). They have investigated how substrate and growth conditions affect growth rate and efficiency (e.g., Droop 1973; Button 1978; Panikov 1995). Protist biologists showed early interest in the effects of body size on biological rates (e.g., Fenchel 1974; Fenchel and Finlay 1983); prokaryote biologists have showed less interest. Often performed in laboratory experiments and bioreactors, much of the research has been motivated (explicitly or implicitly) by applications to medical, industrial, food, and environmental technology. Historically, much of microbial ecology has advanced relatively independently of theoretical developments in macroorganism and ecosystem ecology – the exception being ecologists studying phytoplankton, who have extensively studied the effects of cell size and resource stoichiometry on growth rate and community structure (e.g., Sheldon et al. 1972; Droop 1973; Fenchel 1974; Litchman et al. 2007; Yoshiyama and Klausmeier 2008; Litchman, Chapter 13).

Relatively speaking, a formal metabolic theory of ecology (MTE) for microbes is in its infancy. A few MTE papers have made important initial steps. The integration of the effects of body size with kinetic effects of temperature on metabolic rate into one equation was a particularly important step in the development of metabolic ecology (Gillooly et al. 2001; Brown et al. 2004) and in the metabolic ecology of microbes (López-Urrutia et al. 2006). Subsequently, microbial ecologists have sought to integrate the core MTE equation (Brown and Sibly, Chapter 2) with the effects of resource availability and stoichiometry (López-Urrutia and Morán 2007; Sinsabaugh et al. 2009, 2010, 2011; Sinsabaugh and Shah 2010; Sinsabaugh and Follstad Shah 2011). These scaling and metabolic perspectives, together with exciting advances in prokaryote and eukaryote cell physiology, may provide the necessary stimulus to begin to develop an integrated, unified, and quantitative understanding of physiological and metabolic ecology spanning the three domains of life.

12.3 PHYSIOLOGICAL FOUNDATIONS

All organisms require energy and materials to build and maintain their complex structures far from thermodynamic equilibrium. Enzymes have evolved to harness energy from a variety of sources: sunlight, organic carbon, and energy-yielding (exergonic) geochemical substrates. Carbon is one of the essential elements used in building physiological infrastructure. It can be obtained from organic sources or carbon dioxide. So the most basic classification of trophic lifestyles is according to the energy and carbon sources utilized by an organism (Table 12.1). All of the major trophic groups of life are used by the Archaea and Bacteria, the two domains making up the prokaryotes, whereas the Eukarya cannot perform lithotrophy without the assistance of prokaryote symbionts.

Table 12.1 The major metabolic lifestyles of life.

Energy source	Carbon source	Terminology
Light	Carbon dioxide	Photoautotroph
Light	Organic compounds	Photoheterotroph
Inorganic chemicals	Carbon dioxide	Lithoautotroph
Inorganic chemicals	Organic compounds	Lithoheterotroph
Organic carbon	Organic compounds	Organoheterotroph

Organoheterotrophs and photoautotrophs are often referred to as heterotrophs and phototrophs, for short. Mixotrophs use a mix of different energy and/or carbon sources. Lithotrophs and organotrophs together are called chemotrophs; the prefix *chemo* encompasses both *litho-* and *organo*.

In order for an individual organism to maintain cellular integrity and function, the power supply (energy per unit time) available to an organism, R_{org}, must be sufficient to fuel the whole-organism minimum metabolic rate, I_{min}, required to repair macromolecular damage (Price and Sowers 2004). R_{org} must be even greater in order to supply the power used to support basic metabolic functions and activities, known as the maintenance metabolic rate, I_{maint} (more or less comparable to inactive metabolic rates called "standard metabolic rate" or "basal metabolic rate" in macroorganisms). Even more power is required for a cell to actively create new biological material, grow, and reproduce, known as the active or growth metabolic rate, I_{grow}. Thus, $I_{grow} > I_{maint} > I_{min}$[1]. In order for an environment to be habitable for life, over a reasonable period of time R_{org} must be greater than or equal to I_{grow}: $R_{org} \geq I_{grow}$ (Hoehler 2004, 2007; Hoehler et al. 2007; Shock and Holland 2007). The closer R_{org} is to I_{grow}, I_{maint}, and I_{min}, the more extreme and relatively unsuitable the environment

[1]The ratios $I_{grow}:I_{maint}:I_{min}$ have been estimated to be of the order of $10^6:10^3:1$ in bacteria communities in situ (Price and Sowers 2004); $I_{grow}:I_{maint}$ seems to more typically have maximum values of 1–2 orders of magnitude when bacteria species isolates are measured and in protist species (DeLong et al. 2010).

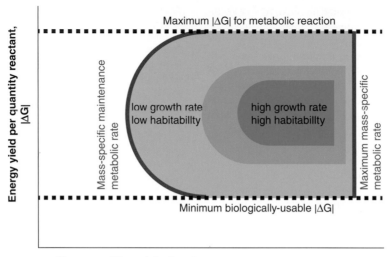

Mass-specific metabolic rate

Figure 12.1 Conceptual diagram illustrating the effects of energetics as mediated by thermodynamics, kinetics, and environmental stress on the productivity of an organism or ecosystem and on the habitability of its environment. For simplicity, the figure is presented for an organism or ecosystem with one single energy source. The living thing can only produce biomass in the green area. A living thing must be able to obtain energy from its environment at a rate greater than its biomass-specific maintenance metabolic rate (I_{maint}/M, left-hand side, red curved line) in order to produce biomass. The metabolic design of the living thing, physicochemical conditions, and resource availability impose an upper boundary on its metabolic rate (right-hand side, red line). In order for a reaction to provide biologically usable free energy, the reaction must have an energy yield $|\Delta G|$ equal to or greater than the minimum $|\Delta G|$ and less than the maximum $|\Delta G|$. Metabolic reactions close to max $|\Delta G|$ induce greater oxidative cellular damage and reactions close to the minimum $|\Delta G|$ require more complex and expensive metabolic machinery, thereby increasing maintenance energy requirements (a_{maint} and I_{maint}/M) and decreasing the amount of energy allocated to growth. Therefore, a living thing's mass-specific growth rate and biomass production tend to increase as its power and reaction's $|\Delta G|$ value approach the middle right region of the plot.

(Fig. 12.1). Thus the difference or ratio of R_{org} to I_{grow}, I_{maint}, and I_{min} determines an environment's habitability. An environment may be extreme because R_{org} is low or because I must be high in order for an organism to survive, maintain biological activity, and grow. Because of the challenges of studying microbes in the field, there is still fragmentary knowledge of the different metabolic rates and associated growth and survival rates of microorganisms in situ.

How is the metabolic rate of an individual defined and quantified? There are no intrinsically superior definitions of metabolic rate. An organism obtains power from exergonic chemical reactions or phototrophy. This supply side of an individual's metabolism could be considered its metabolic rate. This rate may be the most general and theoretically useful rate, at least in microorganisms, and so this definition is used in this chapter. Some of the supplied power may then be coupled to

ATP synthesis. So the total power used for whole-organism rate of ATP synthesis could also be considered the metabolic rate. The cell's ATP molecules are used to power endogenic reactions, so the total power produced through ATP hydrolysis could be considered the metabolic rate. The rate of the membrane electron transport chain may be used to provide a more general and encompassing measure than ATP synthesis since the electron transport chain powers both ATP synthesis and other activities, such as bacteria flagella and secondary active transport. However, some exergonic reactions may in fact power ATP synthesis without the use of an electron transport chain (substrate-level phosphorylation) or power anabolic reactions without the use of ATP hydrolysis, for example, by using the biosynthetic pathways associated with glycolysis. Therefore, the rate of ATP synthesis, ATP hydrolysis, or the electron transport chain reaction may not always provide the most useful measure of metabolic rate and may not always give an accurate measure of the total power expended by an organism.

In fact, in microorganisms the population growth rate μ (number of divisions per unit time, known in microbiology as specific growth rate) or biomass production rate P (biomass produced per unit time by a cell or population of cells) may provide a meaningful proxy for metabolic rate. These rates have been widely used by microbiologists (e.g., Dawson 1974; Fenchel 1974; Panikov 1995; Ratkowsky et al. 2005). The value of μ reflects supply-side metabolic power and the energetic efficiency H by which this energy is used to power the organism's biological and reproductive activities:

$$\mu = H \times (I/M) \tag{12.1}$$

where M is cell mass. H is in dimensions of mass per unit energy (e.g., g J^{-1}) and can be thought of as the amount of energy required for an organism to produce a unit mass of biomass. Metabolic rate is partitioned between energy use for growth, p_{grow}, and energy use for maintenance activities, I_{maint}, giving $I = p_{grow} + I_{maint}$ and $a_{maint} = I_{maint}/I$, where a_{maint} is the proportion of whole-organism metabolic rate that is allocated to maintenance. The partitioning of energy between maintenance processes and growth affects H and μ, as shown by the following commonly used mass-balance equations (e.g., Pirt 1965; Panikov 1995):

$$\mu = Y \left(\frac{I - I_{maint}}{M} \right) \text{ and} \tag{12.2}$$

$$\mu = Y (1 - a_{maint})(I/M) \tag{12.3}$$

where Y is the growth efficiency or growth yield – the efficiency by which growth-allocated energy p_{grow} is converted into new biomass (since $\frac{I - I_{maint}}{M} = \frac{p_{grow}}{M}$). The relationship between H and Y is

$$H = Y (1 - a_{maint}) \tag{12.4}$$

These mass-balance equations show that a decreased mass-specific metabolic rate or increased allocation to maintenance metabolism leads to a linear decrease in population growth rate and division rate. H and other comparable measures of metabolic efficiency, such as growth yield and ATP yield, are fundamental quantities of great interest to microbiologists and ecologists (e.g., Pirt 1965; Dawson 1974; Panikov 1995; Russell

and Cook 1995; Chapin et al. 2002; Maier et al. 2009).[2]

12.4 QUANTITATIVE OUTLINE OF THE DIMENSIONS OF METABOLISM

Whole-organism metabolic rate is a function of ΔG, the energy yielded per unit quantity of reactant molecule by each type of exergonic reaction supplying energy to the organism,[3] the rate of each reaction r (number of product molecules produced per unit time), and the number of different reactions, n:

$$I = \sum_{i}^{n} |r_i \Delta G_i| \tag{12.5}$$

where brackets denote the absolute value of $r_i \Delta G_i$ (since energy-yielding reactions have by definition negative values of ΔG). This equation can be combined with equation 12.3, giving

$$\mu = Y (1 - a_{maint}) \left(\sum_{i}^{n} |r_i \Delta G_i| \right) \Big/ M \tag{12.6}$$

These are core equations quantifying the dependence of metabolic and growth rate on an organism's biochemistry, energy partitioning, and efficiency. The major dimensions of the energetic ecology of microbes underlie the variables in these equations: (1) n and ΔG constitute the thermodynamic dimension; (2) r is the dimension of chemical kinetics; (3) the dimension of physiological harshness reflects the niches of species and has important effects on a_{maint}, and can also influence the other variables; (4) cell mass unavoidably constrains r and μ, and also may have a positive effect in

---

[2] They are also important parameters relevant to maximizing the efficiency of industrial processes that depend on microbial metabolism. This quantitative framework can be generalized to modeling any substrate use, not just energy use. The growth yield, also referred to in microbiology as the cell yield, biomass yield, or growth efficiency, is generalized to the amount of biomass produced per unit amount of substrate consumed.

[3] ΔG is the Gibbs free energy. It is the difference between the potential energy of reactants and products. A negative ΔG means that the total free energy (potential energy) of the products of the reaction is lower than the total free energy of the reactants. Reactions with more negative ΔG values yield more energy per mole of reactants.

prokaryotes on n and ΔG; and (5) the level of biological organization can influence all variables in these equations. Equation 12.5 can be rewritten as

$$I = n\,|\langle r\Delta G\rangle| \tag{12.7}$$

where $|\langle r\Delta G\rangle|$ is the average reaction metabolic power, highlighting the linear dependence of metabolic rate on the number of energy-yielding reactions and the average metabolic power of energy-supplying reactions.

12.5 DIMENSION 1: THERMODYNAMICS

The enormous metabolic diversity of microbes is striking. Organotrophic microbes can consume a multitude of organic carbon compounds too recalcitrant or toxic for animals. Phototrophic prokaryotes have several different kinds of pigments for harvesting light energy. They can harness wavelengths from 385 nm to over 800 nm, which affects the ΔG of the photoreactions. New pigments and types of photoreactive centers are still being discovered (Fuhrman et al. 2008b). Lithotrophic prokaryotes can derive energy from hundreds to thousands of geochemical reactions, using a variety of minerals and elements functioning as electron acceptors and donors (e.g., Kim and Gadd 2008; Shock et al. 2010).[4]

The first step towards understanding biological activities and their dependence on the environment is to elucidate an organism's possible available number of metabolic pathways for obtaining energy (n) and how much energy is yielded by each metabolic pathway, ΔG. This is a vibrant area of research in the fields of geomicrobiology and systems biology (e.g., Amend and Shock 2001; Price et al. 2004; Inskeep et al. 2005; Spear et al. 2005; Raymond and Segre 2006; Hall et al. 2008; Shock et al. 2010). The study is necessary in order to predict whether or not a microbe can persist in a particular environment, essential to microbial biogeography, and to predict energy and biogeochemical fluxes in particular environments, essential to ecosystem ecology.

[4]Examples of inorganic electron donors include hydrogen, ammonia, nitrite, sulfur, hydrogen sulfide, ferrous iron, oxygen, carbon monoxide, arsenite, manganese, and uranium.

ΔG varies widely between reactions and environments (Fig. 12.2). It is dependent on the energy in the bonds of the substrate molecules, the thermodynamic activity of the reactants and products (which depend on the concentrations and activity coefficients of the reactants and products, pH, and the ionic strength of the aqueous environment), temperature, and pressure according to principles of thermodynamics and theoretical geochemistry. A naive prediction would be that temperature has a linear effect on the overall Gibbs free energy change, ΔG, based on the thermodynamic equation $\Delta G = \Delta G^0 + RTln(Q)$, where ΔG^0 is the standard-state Gibbs free energy change reflecting the reaction's thermodynamic properties, R is the gas constant, T designates temperature, and Q denotes the activity product, which is calculated from the concentrations and thermodynamic activities of the reactants (Amend and Shock 2001). In reality, temperature's effect on ΔG is complex because temperature influences the concentrations and activities of dissolved reactants and products, in addition to its direct effect on the thermodynamic favorability of the reaction (Amend and Shock 2001; Hammes 2007). However, because the logarithmic term diminishes the effects of variation in Q, ΔG may often tend to vary approximately linearly or very little with temperature over biochemically relevant temperature ranges. Over such ranges, ΔG may be more strongly dependent on pH and is well approximated by a linear function of pH (Fig. 12.2; Amend and Shock 2001; Shock et al. 2010).

An exergonic reaction must have a sufficiently large enough energy yield in order for an organism to be able to exploit it (Thauer et al. 1977; Schink 1997). There are also constraints on the maximum Gibbs free energy change that can be harnessed by organisms. Reactions with high absolute values of Gibbs free energy change are more likely to cause the cell oxidative damage and there are biophysicochemical limits to the ability of enzymes to catalyze high-energy-yielding reactions (Hoehler 2007). There appear to be at least two orders of magnitude variation in the ΔG values of life's exergonic metabolic reactions (Hoehler 2007).

What are the constraints on exergonic metabolic diversity, the number n of energy-yielding metabolic reactions used by an organism? n depends on the diversity of enzymes an organism has that are involved in exergonic reactions. In prokaryotes, number of kinds of enzymes and metabolic reactions scales positively with genome size, which in turn scales with cell size;

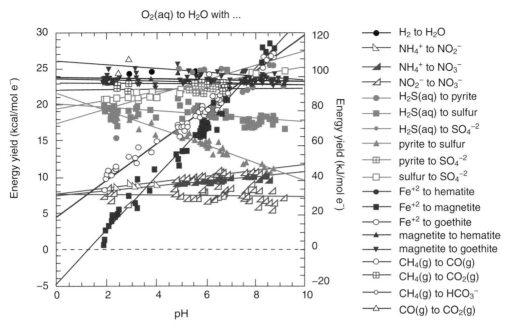

Figure 12.2 The energy yield of potential metabolic reactions varies greatly between reactions and with pH according to the principles of thermodynamics. Plotted here are geochemical reactions with O_2 as the electron acceptor and H_2, NH_4^+, NO_2^-, H_2S, S, pyrite, Fe^{2+}, magnetite, CH_4, and CO as the electron donor (as listed on the right-hand side). The energy yields are for hot springs in Yellowstone National Park, USA, were determined based on geochemical data and thermodynamic calculations, and are reported in terms of energy per mole of electrons transferred. These reactions potentially provide sources of metabolic energy to chemotrophic microorganisms in these hot springs. Many prokaryotes are known to utilize these pathways for catabolism. Modified from Shock et al. (2010) by permission of Elsevier.

in eukaryotes, number of kinds of enzymes and metabolic reactions is weakly, if at all, dependent on M (Molina and Van Nimwegen 2008; DeLong et al. 2010). n reflects the number of available metabolic reactions having negative values of ΔG and so depends on thermodynamic conditions, as discussed above. Also, the chemical diversity of an organism's environment ultimately imposes an upper boundary on n. Thus in chemotrophs n may be a positive function of the chemical diversity of its ecosystem.

12.6 DIMENSION 2: CHEMICAL KINETICS

Systems biology, an important area of cell biology, molecular biology, and microbiology, is determining how the conditions affecting reaction kinetics, the

effectiveness of the enzymes catalyzing the reactions, and the network properties of an organism's biochemical network influence the metabolic and growth rate of cells (Price et al. 2004; Westerhoff and Palsson 2004). The metabolism of an organism operates in heterogeneous and non-mixed spaces, such as along the surfaces of membranes and in the fractal-like volume of the cytosol. Metabolism comprises a complex network of reactions, and cells respond dynamically to changes in substrate availability and temperature. Therefore often the assumptions underlying the application of basic physical chemistry and biochemistry to organism metabolism are not upheld (e.g., Savageau 1995; Berry 2002). It is essential to determine which assumptions are robust and which are violated governing the biological kinetics of an organism. Despite the complexity of the cell, basic physicochemical models have been found to provide useful models for the kinetics of organism metabolism.

12.6.1 Temperature

The rate of a simple uncatalyzed reaction, r, scales with temperature according to the Arrhenius equation as $r \propto e^{-E/kT}$ (otherwise known as the Boltzmann factor), where E is the activation energy, k is Boltzmann's constant, and T is temperature in kelvin (Atkins and De Paula 2009). In theory and in practice, reaction rate has a more complicated temperature dependence (Johnson et al. 1974). However, this temperature dependence can often be approximated by the Arrhenius equation because variation in the other effects of temperature are comparatively small over biologically relevant temperature ranges (e.g., $-40\,°C$ to $130\,°C$). For a complex enzyme-catalyzed reaction, reaction rate scales over some limited temperature range approximately according to the Arrhenius function with the activation energy of the rate-limiting step (Stegelmann et al. 2009). However, as temperature increases, the reaction rate will increasingly deviate from the Arrhenius equation, because temperature will increasingly have a negative influence on enzymes and catalysis by increasing the probability that the enzymes are in their denatured states as opposed to their native and active states (Ratkowsky et al. 2005; see also Daniel and Danson 2010). At some threshold temperature, the rate of reaction begins to decrease, usually quite steeply.

Metabolic rate and growth rate exhibit comparable temperature dependences; however, in organisms temperature also has important effects on the functioning of physiological infrastructure, such as bilipid membranes supporting electron transport chain reactions or the compartmentalization of reactants. Thus, the temperature response curves for physiological rates depend on the properties of the enzymes and of the bilipid membranes. These temperature response curves vary greatly between organisms and reflect the evolutionary optimization of enzyme and membrane function for a particular temperature range (Fig. 12.3).

Biologists have long used the Arrhenius model to quantify the temperature dependence of biological rates over the increasing phases of temperature response curves, in particular of respiration, produc-

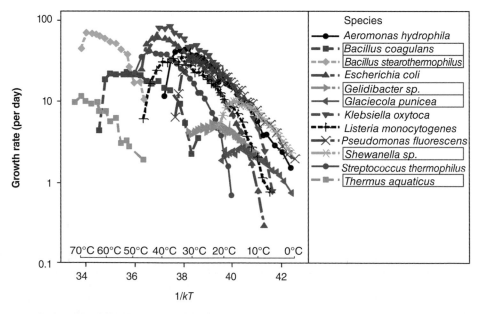

Figure 12.3 Arrhenius plot of the temperature dependence of the growth rate of different species of bacteria. The optimum temperature for growth varies from $15\,°C$ or less in cold-adapted species (psychrophiles; species names boxed in blue) to $65\,°C$ or more in hot-adapted species (thermophiles; species names boxed in red). k is Boltzmann's constant in units of eV K^{-1} and T is temperature in kelvin. Data are from Mohr and Krawiec (2005) and the work of Ratkowsky et al. (2005) and graduate students at University of Tasmania.

tion, population growth rates, and division times (Johnson and Lewin 1946; Ingraham 1958; Button 1985; Davey 1989; Price and Sowers 2004). They have also modeled the entire temperature response curve using empirically derived models (Ratkowsky et al. 1982, 1983; Rosso et al. 1995). Recently, in order to advance understanding of the temperature dependence of growth and respiration rate in microbes, Ratkowksy (2005) developed quantitative theory to incorporate the effects of temperature on the stability of enzymes, thereby modeling the entire temperature response curve. The effects of temperature on the fluidity and integrity of the cell membrane, given its importance for energy transduction and transport, must also be incorporated into biophysical models. In many microbial habitats, temperature will vary enough that consideration of the shape of the temperature response curve beyond the Arrhenius regime will be necessary in order to accurately model microbial responses to temperature.

Although the kinetic effects of temperature are important, ultimately, understanding of the temperature dependence of the power of a reaction and of an organism's metabolism is sought. The power produced by a reaction is equal to $|r\Delta G|$, so it is a more complex function of temperature that reflects both kinetic and thermodynamic dimensions.[5] Researchers have sought to rigorously combine these effects into one model of respiration (Jin and Bethke 2003, 2007; LaRowe and Helgeson 2007). Although much work remains, these models are laying the grounds for developing a foundation for a quantitative theoretical biogeochemistry and metabolic theory of microbes that integrates thermodynamics and kinetics.

12.6.2 Substrate concentration

Resource availability may account for several orders of magnitude variation in the metabolic rate and growth rate of a cell (Price and Sowers 2004; Glazier 2009a). As the availability of energy and essential materials increases, an organism can increase its use of those resources, thereby increasing its metabolic, growth, and reproductive rates. Initially, metabolic rate tends to

increase linearly with resource availability but eventually metabolic rate will saturate at a maximum possible rate.

Kinetic theory is used to model the dependence of the speed of a reaction, r, on the substrate concentration, $[S]$, found at the location of the biochemical reaction (Button 1985, 1998; Panikov 1995). Biologists often make the assumption that the metabolic and growth rate of a cell is proportional to the total rate of relevant metabolic reactions. Michaelis–Menten kinetics, which have been derived for modeling biochemical reactions involving enzymes, have been widely applied in microbial ecology to model the effects of resource availability on growth rate (often called the Monod equation in this case) and photosynthesis rate (Liu 2007; Litchman, Chapter 13):

$$r = \frac{v_{max}[S]}{K_M + [S]} \tag{12.8}$$

where v_{max} is the maximum possible reaction rate and K_M is the half-saturation constant. The assumptions underlying its derivation for a simple biochemical system may often not be strictly upheld when applied to an organism (Savageau 1995; Liu 2007) or community (Sinsabaugh and Shah 2010). However, it is often a good predictive model that provides an approximation of the kinetics of the complex metabolic network of organisms.

The substrate concentration of the bulk fluid surrounding a cell and in an ecosystem, $[S_o]$, is not necessarily equal to the substrate concentration at the site of an organism's biochemical reactions, $[S]$, that are involved in Michaelis–Menten kinetics. Ecologists are interested in the dependence of biological rates on $[S_o]$ because is easier to empirically measure than $[S]$ and reflects the general availability of a substrate to different organisms in an ecosystem. However, $[S_o]$ is not necessarily equal to $[S]$ when the enzymes of the biochemical reaction are immobilized by being attached to a solid surface, such as the enzymes located in the membranes of cells. In this case, substrate must diffuse from the bulk pool to the site of biochemical reactions at a flux rate F according to Fick's law, $F = k_s([S_0] - [S])$, where k_s is a parameter related to the physical conditions near the reaction site. Thus, F, $[S]$, and cell uptake rate I are codependent on each other: the flux rate depends on the concentration gradient, the gradient depends on $[S]$, and $[S]$ depends on the equilibrium between the cells' uptake/reaction rate and the flux

---

[5]This temperature dependence can often be approximated by the Arrhenius equation because variation in ΔG with temperature is comparatively small.

rate from the bulk fluid to the cell surface.[6] The interaction of these variables ultimately determines the form of the functional dependence of r on $[S_o]$. Numerous physically derived models have been successfully developed in microbiology and chemical engineering to model these dynamics under various conditions (e.g., Williamson and McCarty 1976a; Siegrist and Gujer 1985; Bailey and Ollis 1986; Patterson 1992; Bosma et al. 1996).

12.7   DIMENSION 3: PHYSIOLOGICAL HARSHNESS AND ENVIRONMENTAL STRESS

Physiological harshness of the environment greatly influences the metabolic rates of cells. Physiologically harsh environments are prevalent. One organism's mild environment is another organism's extreme environment. A tiny microenvironment of a cubic millimeter will be a macro-environment to thousands of microorganisms. An environment that may seem benign and homogeneous to us, such as the soil in a forest, may in fact harbor environments stressful to the physiologies of organisms. Therefore, in order to understand the distribution, abundance, and activity of microorganisms and ecosystems, the influence of physiological harshness must be considered.

Physiological stress arises because of the existence of inescapable trade-offs in an organism's biochemical and physiological attributes. For example, for biophysical reasons enzymes cannot perform well as catalysts at both extremely cold and extremely hot temperatures. Thus an organism will evolve to be best adapted to a particular temperature range. Temperature and chemical harshness are probably the most important kinds of physiological harshness affecting microbial metabolism. However, in many environments high levels of ultraviolet radiation and extremely low or

high pressures can also have important impacts on physiological harshness. In general, physiological harshness reduces an organism's growth rate by: (1) forcing the organism to allocate more energy to maintaining its physiological functions, resulting in reduced allocation of energy to growth and reduced energetic efficiency H, as previously discussed; and/or (2) negatively influencing the rate and energy yields of an organism's exergonic reactions. Here I illustrate the application of these general principles by discussing the specificities for chemical harshness. There are numerous different chemically extreme environments: salinity, desiccation, pH, and concentrations of heavy metals are some of the important chemical conditions that can stress the physiologies of microbes.

There are two different evolutionary and physiological strategies that chemical extremophiles may adopt in order to withstand such harshness (Rothschild and Mancinelli 2001). First, they can maintain homeostasis by keeping the external environment out. Such a strategy may require serious investment of energy in order to pump chemicals against concentration gradients or in materials in order to build the necessary structures to prevent chemicals from diffusing down a chemical gradient into the cell. Second, they can allow their cells to have the same chemistry as the outside but alter their biochemistry and physiology or enhance repair mechanisms in order for their cell interiors to withstand the chemical extreme. The first strategy necessitates an increase in energy allocated to maintenance processes, leading to an increase in a_{maint} and consequently a decrease in growth rate and biomass production. The second strategy may also require an increase in a_{maint}. Also, importantly, the altered chemical environment of the cell and macromolecules can influence ΔG and rates of reactions. Analogous considerations can be made for the other kinds of physiological harshness.

pH is one chemical condition that is of great importance in scaling the biological rates of unicellular organisms.[7] It varies greatly across the habitable areas

---

[6]There are two limit cases for the dependence of the reaction rate on environment substrate concentration. The reaction-limited regime occurs when the Damköhler number $Da = \dfrac{v_{max}}{k_s [S_0]} \rightarrow 0$, giving $r = \dfrac{v_{max}[S_0]}{K_M + [S_0]}$ (assuming equation 12.8). In the mass-transport or diffusion-limited regime, $Da \rightarrow \infty$ and $r = k_s[S_0]$. Systems that are both reaction and diffusion-limited exhibit intermediate functional dependences on $[S_0]$ and $[S]$ (Bailey and Ollis 1986).

[7]Environmental pH affects metabolism by: (1) influencing a cell's transmembrane pH gradient, which contributes to the proton motive force that powers ATP synthase; (2) increasing the energy expended to maintain non-extreme pH inside the cell; (3) affecting the structure and functioning of a cell's enzymes; and (4) influencing the Gibbs free energy changes of reactions.

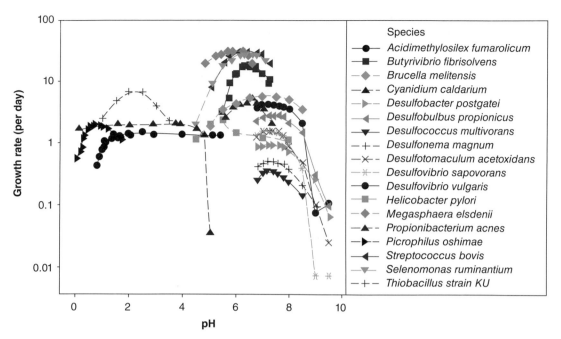

Figure 12.4 The unimodal dependence of microorganism population (specific) growth rate on pH in laboratory-grown species isolates. Most species can only grow at ±1.5 pH from their optimum pH, but species have adapted to be able to grow at pH values ranging from 0 to more than 9. Inspection of the graph suggests that species optimal pH growth rates may also be a unimodal function of pH. (Data from Doemel and Brock 1977; Hallberg and Lindstrom 1994; Kangatharalingam and Amy 1994; Rosso et al. 1995; Schleper et al. 1995; O'Flaherty et al. 1998; Pol et al. 2007).

of the planet, from 0 to 11 or more (Rothschild and Mancinelli 2001), and has significant effects on the geographic distribution of microbial diversity (Fierer and Jackson 2006). In addition to the previously mentioned thermodynamic effects of pH on the energy yield of reactions, changing the pH from the pH that a cell is adapted to is physiologically stressful, causing a decrease in growth rates (Fig. 12.4). Cells tend to favor the homeostasis strategy, keeping cytosol pH relatively independent of environmental pH; however, cytosol pH still varies from 6 in acidophiles to 9 in alkaliphiles (Ingledew 1990; Kroll 1990). Further, the cell surfaces must still deal with pH extremes, so must still be able to alter the biochemistry of some biomolecules on cell surfaces. For an individual strain, population growth rates are a unimodal function of pH (Fig. 12.4). Since microorganisms often do not inhabit their optimal pH, understanding the functional form between biological

rates and pH is an important step towards developing metabolic scaling theory for microbes.[8]

12.8 DIMENSION 4: CELL SIZE

Cell size must have an important effect on metabolic rate because it affects the total available volume and

[8]Microbiologists have used phenomenological models that successfully modeled the combined effects of pH and temperature on growth rate (e.g., Rosso et al. 1995; Tienungoon et al. 2000). An enzyme kinetic model has been developed that models the effects of pH on enzyme stability and the consequential effect on enzyme kinetics (Bailey and Ollis 1986; Antoniou et al. 1990), but more work needs to be done to develop mechanistically based quantitative models that incorporate the effects of pH on the proton-motive force and to develop an integrative model of the dependence of metabolic rate and growth rate on pH.

surface area available for biochemical reactions and the distances necessary for the transport of materials. Prokaryote cells vary from 10^{-15} to 10^{-4} grams, protist cells vary from 10^{-13} to 1 gram, and yeast vary by at least three orders of magnitude, from the typical yeast cells size of 10^{-11} grams to 10^{-8} grams[9] (Table 12.2). The study of the scaling of organismal traits with organism size is known as allometry. Given the variation in cell size in microorganisms, microbial allometry is a promising area of study.

12.8.1 Prokaryotes

In prokaryotes much of ATP synthesis occurs in the cell membrane by oxidative phosphorylation and photophosphorylation. Therefore, the naive expectation is that metabolic rate should be proportional to the surface area A of the cell membrane. If cell shape and surface roughness are not changing consistently with size, the external surface area scales with volume V as $A \propto V^{2/3}$ and so the prediction is $I \propto V^{2/3}$ and $I \propto M^{2/3}$ (assuming $V \propto M^1$). For decades, however, many biologists thought that, like other organisms, in prokaryotes metabolic rate scales as a three-quarter power function of cell mass and volume – that is, as Kleiber's Law: $I \propto M^{3/4} \propto V^{3/4}$ (Brown et al. 2004). This conclusion was based on a few bacteria species that were grouped together with protists for statistical analysis (e.g., Hemmingsen 1960; Fenchel and Finlay 1983; Peters 1983; Gillooly et al. 2001). Increases in data availability and more resolved statistical analyses have generated debate. Makarieva et al. (2005a, 2008) emphasized that the mean mass-specific metabolic rate of a group of organisms, such as heterotrophic bacteria or phototrophic unicellular protists, does not vary very much

between groups of organisms. Yet, as suggested by their analyses and demonstrated by analyses by Delong et al. (2010), whole-organism metabolic rate increases superlinearly in organoheterotrophic bacteria: $I \propto M^{\beta}$, where $\beta > 1$ (Fig. 12.5). This exceptional finding means that larger bacteria in fact have higher metabolic rates per unit body mass than smaller bacteria; and so, if the allocation of energy to growth is invariant with size, then population growth rate is expected to scale positively with cell mass as $\beta - 1$, and generation time to scale negatively as $1 - \beta$.

Delong et al. (2010) hypothesize that the superlinear scaling is made possible by a concomitant increase in an individual's number of genes, which in prokaryotes scales with cell size. In prokaryotes, cells with larger genomes have metabolic networks composed of a larger number of reactions and enzymes. This increased network size and complexity may be able to confer greater metabolic power in the following non-mutually exclusive ways: by increasing energy yields, $|\Delta G|$; by increasing reaction rates r through autocatalytic feedback pathways in reaction networks and through better-designed enzyme catalysts; or by increasing the number of substrates and reactions used as energy sources. This can explain why the metabolic scaling exponent is greater than two-thirds, but work is necessary in order to explicitly show how such network changes lead to superlinear scaling. The hypothesis proposed by DeLong et al. (2010) may apply to lithotrophic bacteria and to archaea; however, empirical scaling relations in these organisms have not been reported.

It is less obvious how this theory applies to phototrophs, since they have one source of energy. Once the effectiveness of the photosynthetic reactions and their density on the cell surface is maximized, the total photosynthetic rate will necessarily be limited by the surface area exposed to solar radiation, which scales sublinearly (Niklas 1994b). Indeed, current analyses suggest the scaling of metabolic and associated biological rates in phototrophic prokaryotes is sublinear (Nielsen 2006) or only slightly superlinear (Makarieva et al. 2008).

[9]Prokaryotic cells range in size from the tiniest mycoplasma bacteria with reduced genomes weighing about 10^{-15} g (Himmelreich et al. 1996), to the giant spherical sulfur bacterium *Thiomargarita namibiensi*, which can weigh 10^{-4} g (Schulz et al. 1999). Unicellular protists span 14 orders of magnitude in cell mass, from around 10^{-13} g in the green algae *Ostreococcus tauri* (Courties et al. 1994) to 1 g in the largest Foraminifera, Acanthophora, and Radiolaria protists. Yeasts span over several orders of magnitude variation in cell mass; typical yeast cells are 3–4 μm in diameter and the largest reported yeast cells are 40 μm in diameter in the species *Blastomyces dermatitidis* (Walker et al. 2002).

12.8.2 Unicellular eukaryotes

In unicellular eukaryotes, applying the same logic used to build an a priori expectation in prokaryotes, the expectation is that metabolic rate scales with the total

Table 12.2 The biological units created by the major ecological and evolutionary transitions of life, their ranges of reported sizes, and their number of levels of ecological and evolutionary organization.

Unit	Volume (μm^3)[a]		Number of levels of organization (cell = 1)	References
	Min	**Max**		
Prokaryote cell	10^{-3}	10^8	1	Schulz and Jørgensen 2001
Prokaryote colony	10^{1b}	10^{11}	2	Beardall et al. 2009
Multicellular prokaryote	10^1	10^3	2	Keim et al. 2007
Biofilm or microbial mat	10^{1c}	10^{5d}	2	Staley and Reysenbach 2002; Ghannoum and O'Toole 2004
Aggregate consortium	10^1	10^4	2	Orcutt and Meile 2008; Alperin and Hoehler 2009
Prokaryote endosymbionts and prokaryote host[e]	5×10^1	4×10^3	2	Von Dohlen et al. 2001
Unicellular eukaryote (without additional endosymbionts)	10^{-1}	10^9	2	Courties et al. 1994; Beardall et al. 2009
Unicellular eukaryote host and prokaryote endosymbionts	10^0	10^6	3	Curds 1975; Heckmann et al. 1983; Guillou et al. 1999
Colony of eukaryote cells	–	10^{16}	3	Beardall et al. 2009
Unicellular eukaryote host and eukaryote endosymbiont	10^2	10^{13}	3	Tamura et al. 2005; Beardall et al. 2009
Multicellular eukaryote (may include microbial symbionts)	10^5	5×10^{21}	3–4	Stemberger and Gilbert 1985; Payne et al. 2009
Eusocial colony of multicellular eukaryotes and its microbial symbionts	10^{12}	$>10^{15}$	5	Hou et al. 2010

[a] 1 μm^3 of biomass $\approx 10^{-12}$ g of biomass.
[b] As calculated for a colony of 10 small *E. coli* cells.
[c] Minimum thickness of biofilms in μm.
[d] Maximum thickness of mats in μm.
[e] Has rarely been observed.

surface area of the mitochondrial inner membranes, A_{MT}. All else being equal, $A_{MT} \propto V^1$ because cells can increase the number of mitochondria linearly with cell volume, thereby leading to $I \propto M^1$ (Okie 2011, unpublished). On the other hand, a slow rate of uptake and transport of oxygen and organic compounds from the environment and through the cell to the mitochondria could limit the total rate of activity of the mitochondria. If surface area limits the uptake of resources, then all else being equal the expectation is $I \propto M^{2/3}$ (however, see Okie 2011, unpublished, and Patterson 1992 for reasons why deviations from two-thirds may be common). If the distribution of resources within the cell is the limiting factor, network scaling theory suggests $I \propto M^{3/4}$ (West et al. 1999a; Banavar et al. 2010).

Historically, most biologists thought protists followed quarter-power biological scaling relations such as $I \propto M^{3/4}$. Few studies have investigated metabolic scaling in yeast and mold cells, despite their ecological, industrial, agricultural, gastronomical, and medical importance. Larger and higher-quality datasets have led to a re-evaluation of scaling relations in unicellular protists. In heterotrophic protists, $I \propto M^1$ (Makarieva et al 2008; DeLong et al. 2010). In phototrophic unicellular protists, biological scaling appear to follow quarter-powers, with $I \propto M^{3/4}$ (e.g., Niklas and Enquist 2001; Nielsen 2006; Johnson et al. 2009), but the

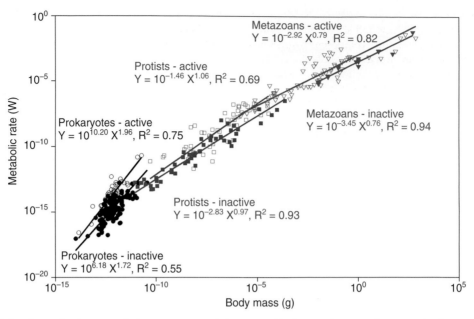

Figure 12.5 Relationship between whole-organism metabolic rate and body mass for organoheterotrophic bacteria, heterotrophic protists, and aquatic animals. Fits were determined by RMA regression on logarithmically transformed data. Unfilled symbols are for active, fed metabolic rates and filled symbols are for inactive, starving metabolic rates. Data from DeLong et al. (2010).

subject is still open to some debate. For example, Makarieva et al. (2008) found linear scaling of metabolic rate in eukaryotic microalgae. Phototrophic protists, however, likely cannot sustain linear scaling as size increases because, as in phototrophic prokaryotes, their surface areas govern their ability to harness solar energy. And packaging chloroplasts at higher densities and further within the cells leads to increased shading by surrounding chloroplasts and cytoplasm – the "package effect" (Niklas 1994b).

In sum, central to understanding allometric scaling is identifying the fundamental constraints on metabolic rate for a given size. Because these constraints may change with body size and organismal design, the metabolic scaling exponent may also shift with changes in size and major evolutionary transitions. Determining empirically and theoretically at what sizes and in what groups of organisms these scaling shifts occur is an important avenue for future research.

12.9 DIMENSION 5: LEVELS OF BIOLOGICAL ORGANIZATION

Cells in nature rarely live in isolation, and a cell's interaction with other cells profoundly alters major dimensions of its metabolism. On ecological timescales, microbes group together in tightly knit populations and communities, which I call a major ecological transition in level of organization. Increased cooperation and decreased conflict between individuals in a population or community causes levels of organization to become more permanent and integrated. Eventually this can lead to the evolution of the integration of groups of individuals into a new higher-level unit of natural selection, a process called a major evolutionary transition (Maynard Smith and Szathmary 1995; Michod 2000). Thus, ecological and evolutionary dynamics have organized life into different levels of organization (Fig. 12.6A).

The history of life is characterized by dramatic increases in the complexity and size of living things as a result of ecological and evolutionary transitions (Table 12.2; also see Payne et al. 2009). Major ecological transitions influencing microbial metabolism include the formation of microbial consortia of syntrophic species (microbial mats, biofilms, and microbial aggregates),[10] colonies, endosymbionts living within the cells of other single-celled and multicellular organisms, and unicellular microbes living in close association with multicellular organisms (Table 12.2, Fig. 12.6A). Although these complexes develop on ecological timescales, many of the species coevolve and consequently their metabolisms are the manifestation of eco-evolutionary processes. In major evolutionary transitions, prokaryotes evolved into eukaryotic cells via endosymbiosis, unicellular prokaryotes and eukaryotes into multicellular organisms via cooperation between related cells, and unitary multicellular organisms into obligatory social organisms living in eusocial colonies called "superorganisms" (Maynard-Smith and Szathmary 1995; Szathmary and Smith 1995; Michod 2000; Queller and Strassmann 2009).

These ecological and evolutionary transitions have significant effects on the metabolism of microbial cells, microbial communities, plants, animals, and ecosystems. Interactions between cells in the collection can also influence the cell's allocation of energy and materials to growth, maintenance, and infrastructure.[11] By altering the diffusion and active transport of resources and waste products, these collections influence the flux and concentration of substrates available to cells. By inducing syntrophy and the metabolic specialization of cells, these groups influence the kinds of substrates and associated Gibbs free energies available to a cell.

For example, although the species found in mature biofilms and mats may also be found as plankton in the surrounding water, the consortia cells have fundamentally different traits and behaviors from their free-living counterparts. The mats and biofilms are composed of layers of metabolically distinct species and characterized by pronounced physical and chemical heterogeneity, specialized niches, and complex spatial organization. The transport and transfer of nutrients and gases are generally rate controlling in biofilms and mats (Teske and Stahl 2002; Petroff et al. 2010), and channels in the consortia may form that function as primitive circulatory systems with water flowing through channels, augmenting the exchange of gases and resources between the consortia and environment (Davey and O'Toole 2000). Consequently, the thickness of biofilms and mats has been shown to affect the consortia's rate of metabolism and production (Williamson and McCarty 1976b). There also are many prokaryote consortia that grow in spherical aggregates in which cells in the spherical core carry out different metabolic pathways than the cells forming an exterior shell of the aggregate (e.g., Dekas et al. 2009).[12] Modeling suggests that the metabolic rates and reaction energy yields of

---

[10]Microbes form complex networks of mutualistic and competitive interactions with inter-agent flows of metabolites and toxins (Costerton et al. 1995). These communities are called microbial consortia. Their species are interdependent and many of the interactions are synergistic and syntrophic, allowing metabolic processes not possible to one individual cell. Many of the species in the consortia are so dependent on their neighbors that biologists have not yet found ways to cultivate them in isolation in the laboratory. Consortia that grow on solid surfaces such as rocks, desert soils, the bottom of lakes, on teeth, and in human lungs are biofilms and microbial mats. Biofilms are microscopic with typical thicknesses around 30–500 μm, can have geometrically complex structures, and are pervasive (Ghannoum and O'Toole 2004); microbial mats are similar but are macroscopic formations that can grow up to several centimeters in height (Staley and Reysenbach 2002). Thus the range in size of these surface-growing microbial consortia varies by at least three orders of magnitude.

[11]Prokaryote cells can communicate with each other by releasing chemical signals. Prokaryotes utilize quorum sensing, regulating the gene expression of the population in response to changes in cell population density (Miller and Bassler 2001). Bacteria use quorum sensing to regulate a variety of physiological activities, including virulence, competence, conjugation, antibiotic production, motility, sporulation, and biofilm formation, thereby leading to the coordination of the behavior of the entire community and bestowing on quorum-sensing communities qualities of higher organisms, sociality, and multicellularity (Miller and Bassler 2001; Branda and Kolter 2004; Dekas et al. 2009; Nadell et al. 2009; Queller and Strassmann 2009; Strassmann and Queller 2010).

[12]These aggregates have been observed to vary in size by at least four orders of magnitude, from $0.5\,\mu m^3$ to $8200\,\mu m^3$, and have been observed to be composed of from 60 to ~100 000 cells.

A

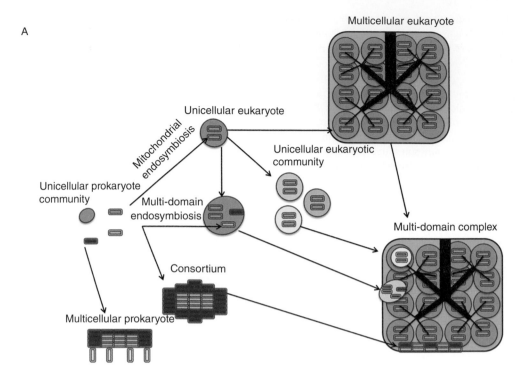

Multicellular eukaryote

Unicellular eukaryote

Mitochondrial endosymbiosis

Unicellular eukaryotic community

Unicellular prokaryote community

Multi-domain endosymbiosis

Multi-domain complex

Consortium

Multicellular prokaryote

B

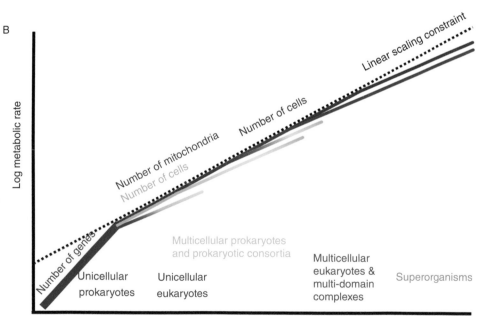

Log metabolic rate

Linear scaling constraint

Number of mitochondria

Number of cells

Number of cells

Number of genes

Multicellular prokaryotes and prokaryotic consortia

Unicellular prokaryotes

Unicellular eukaryotes

Multicellular eukaryotes & multi-domain complexes

Superorganisms

Log mass

Figure 12.6 The metabolic ecology and scaling of prokaryotes and unicellular eukaryotes inhabiting different levels of biological organization. (A) Arrows show how higher levels of organization are formed from lower levels by ecological and evolutionary processes. The larger multicellular complexes require transportation networks for exchanging resources and wastes with the environment and distributing them within the complex, as denoted by the dark green networks. (B) The theorized relationship between metabolic scaling and major evolutionary and ecological transitions. Each transition allows organisms to avoid sublinear scaling constraints at the smaller sizes. However, as size increases, surface area and distribution constraints eventually impose sublinear scaling (faded lines), at which point individuals tend to be outcompeted by individuals of the same size but at the higher level of organization. Superlinear scaling in unicellular prokaryotes (solid blue line) reflects the increase in number of genes and metabolic enzymes with cell size. This eventually gives way to a new constraint (fading blue line) imposing sublinear scaling as a result of respiratory complexes and proton pumps being localized in the cell surface. Protists overcome this constraint by incorporating respiratory complexes on internal surfaces through the endosymbiosis of mitochondria. Larger protists can accommodate more of these organelles, resulting in metabolic rate scaling linearly with cell mass (solid red line), until a new geometric constraint of surface area or transport distance limits rate of resource supply to the mitochondria, imposing sublinear scaling (fading red line). Since the smallest multicellular organisms are composed of relatively few cells and minimal vascular or skeletal systems, the scaling should initially be near-linear, as observed empirically in both animals and plants (Zeuthen 1953; Enquist et al. 2007; Mori et al. 2010). As body size increases, transport distances within organisms and exchanges of resources across surface areas increasingly come into play, leading to sublinear scaling (green lines). Similarly, small eusocial colonies, consortia, and multi-domain complexes may be able to increase their resource acquisition rate linearly with size, but the resource acquisition rate in larger complexes must be constrained by the transport of resources from the environment to and within complexes, imposing sublinear scaling in larger complexes (green and purple lines).

cells decrease significantly with increasing aggregate size, leading to sublinear scaling of aggregate metabolism (Orcutt and Meile 2008; Alperin and Hoehler 2009).

A diversity of prokaryotes, protists, yeasts, and molds also form organized colonies of varying sizes[13] and studies have found that their biological rates scale with colony size (Nielsen 2006; Beardall et al. 2009), as is also found in eusocial animal colonies (Gillooly et al. 2010; Hou et al. 2010). In addition to the well-recognized primitive multicellular organisms found in eukaryotes, there are numerous colonies in prokaryotes that are considered by many biologists to be multicellular organisms (Shapiro 1998).[14] Like the

endosymbionts that evolved into mitochondria, there are also abundant and diverse prokaryotes, protists, and yeasts that live within the cells of protists, plants, fungi, and animals (Lee et al. 1985; Douglas 2010). Many protists, bacteria, and yeast also live in the interspaces between cells of multicellular hosts and on the skin and within the guts of animals, forming multi-domain communities. Many of these interactions are mutualistic, for example, with the microbe fixing carbon and respiring oxygen (e.g., Kerney et al. 2011), providing bioluminescence, fixing nitrogen, or digesting recalcitrant organic carbons, and the host providing the microbe with habitat, protection, and resources.[15] These communities have coevolved such that the multicellular organism cannot properly function without its microbial symbionts. Thus, some of these communities could even be considered organisms or superorganisms, and the metabolic ecology of higher-level animals is in fact the metabolic ecology of a multi-domain complex involving both the plant or

---

[13] For example, cyanobacteria colonies of *Trichodesmium* spp. can attain volumes of $10^{11} \mu m^3$ and protist colonies of the Chlorophycean *Hydrodicyton* can attain volumes of $10^{16} \mu m^3$ (Beardall et al. 2009).

[14] Examples include the magnetotactic prokaryotes made up of 15–45 coordinated cells and varying over two orders of magnitude in volume, from $6 \mu m^3$ to $1020 \mu m^3$ (calculation based on diameters reported in Keim et al. 2007), heterocyst-forming cyanobacteria such as *Anabaena* spp. and *Nostoc* spp. (Beardall et al. 2009; Flores and Herrero 2009), *Proteus* spp., myxobacteria (Shapiro 1998), and *Myxococcus xanthus* (Queller and Strassmann 2009). These species have some of the hallmarks of multicellularity: cell–cell adhesion, complex intercellular communication and coordination, cell differentiation, and lack of cell autonomy.

[15] Animal gut microbes have essential roles in the metabolisms of the animal hosts. For example, understanding obesity in humans requires understanding the ecology of the microbes living within the human gut (Ley et al. 2006). Bacteria form endosymbiotic relationships with leguminous plants that are coordinated by chemical signaling between microbe and host plant (Jones et al. 2007) and such inter-domain signaling also occurs between many animal hosts and their gut microbes (Hughes and Sperandio 2008).

animal and its symbiotic microbes. In sum, the metabolism of the complex formed by an eco-evolutionary transition is more than simply the sum of its parts. An important avenue of research will be to understand how biological rates of cells and their complexes scale with the complexes' size and number of levels of organization.

In Figure 12.6B, I present a unified theory largely based on DeLong et al. (2010) to explain the relationship between metabolic scaling and the major transitions in organization. At the core of the theory is identifying the shifting constraints on metabolic scaling at different sizes and levels of organization. Scaling within each eco-evolutionary group of organisms is bounded by the linear scaling with body mass of the total membrane surface area on which membrane-bound metabolic processes are localized, leading to a potential for linear metabolic scaling at smaller sizes as observed in heterotrophic protists (DeLong et al. 2010), animals (Zeuthen 1953), and plants (Enquist et al. 2007b; Mori et al. 2010; prokaryotes being the exception to this generalization, as discussed in section 12.8). However, as size increases, geometric constraints on exchange surfaces and transport distances limit the supply of substrates to energy-yielding or ATP-synthesizing sites on the membranes, thereby imposing sublinear scaling (West et al. 1999a; Banavar et al. 2010). Each transition incorporated innovations in metabolic design that allowed newly integrated organisms or complexes to initially escape the sublinear scaling constraint by increasing the uptake and distribution of resources to the sites of catabolism, thereby allowing for greater metabolic rate, growth rate, and hence, all else being equal, greater fitness (see Brown and Sibly 2006). Also, each added level of organization requires additional allocation of materials and energy towards building and maintaining more complex metabolic infrastructures. So with each increase in level of organization, the energetic efficiency of biomass production, H, declines.

12.10 COMMUNITY METABOLISM AND THE INTERPLAY OF DIMENSIONS

An ecosystem's metabolism is a distributed network of metabolic reactions – a meta-metabolome (Raes and Bork 2008; Vallino 2010). The temperature dependence of a microbial community's metabolic rate, I_{TOT}, is the sum of the temperature dependences of the energy fluxes of all the concerned metabolic reactions or individuals in the community (e.g., see Panikov 1995), giving

$$I_{TOT} = \sum_{k}^{N} I_k = \sum_{i}^{n_{TOT}} |r_i \Delta G_i| \rightarrow I_{TOT} = N\langle I \rangle = n_{TOT}|\langle r\Delta G \rangle| \quad (12.9)$$

where N is the number of individuals, I_k is an individual's metabolic rate, n_{TOT} is the total number of reactions in the community, and brackets denote average quantities. If the community is composed of individuals all having the same temperature response curves, then the functional form of the community temperature dependence is identical to the individual temperature response curve. If the individuals have different response curves, then the community temperature response will depend on the distribution of individual response curves along the temperature axis. Microbial species can grow in an extraordinary range of temperatures, so consideration of the distribution of individual temperature response curves is important. Most ecosystems and microenvironments experience temperature fluctuations, both over short daily timescales and seasonal timescales.[16] Ecosystems with temporal temperature variation are likely to be composed of species with different temperature response curves because competition causes species to spread out along the temperature niche axis. Immigration of microbes from locations with different temperatures will also lead to variation in the thermal niches of species within the community. Therefore, most microbial communities are represented by a variety of temperature response curves. The challenge will be to determine how this variation affects community-level temperature dependence.[17]

[16] For example, the top layer of desert soils may experience diurnal temperature fluctuations in the summertime of up to 40 °C.

[17] Remarkably, despite these complexities, there are striking concordances in the responses of organism respiration and terrestrial ecosystem-level metabolic rates to temperature through the Arrhenius phases of temperature response curves (e.g., Gillooly et al. 2001; Allen et al. 2005). Such similarities suggest that confounding effects of the variation in properties of enzymes within these ecosystems are outweighed by universal thermodynamic and kinetic effects of temperature on whole-ecosystem metabolic reactions. Given the dominant contribution of aerobic respiratory and oxidative photosynthetic reactions to the energy budgets in terrestrial biomes, the observed concordance probably results from the activation energies of these reactions, which are the metabolic reactions used by most species in metabolic scaling datasets.

The higher community metabolic rates found at higher temperatures require greater amounts of resources. If these resources are unavailable, then the community-level biological rate will be lower than expected based on temperature alone. Thus resource limitation and stoichiometric imbalances at high temperatures can reduce the observed temperature dependence of the community. Examples include the limited availability of phosphorus or nitrogen in aquatic and marine planktonic communities (see Kaspari, Chapter 3; Anderson-Teixeira and Vitousek, Chapter 9; Litchman, Chapter 13). Temperature also influences the physical properties of the environment, which may in turn affect the metabolism of the cells. Thus, temperature may also have indirect effects on microbial communities that complicate the temperature dependence of the community's metabolism. Probably one of the most important effects relevant to terrestrial microbes is the increased evaporation of water at higher temperatures. Under water-limited conditions microbes must decrease their metabolic rates in order to survive, thereby leading to a reduced temperature dependence of the community's metabolic rate (Rothschild and Mancinelli 2001).

12.11 CONCLUDING REMARKS

The Earth's microorganisms harbor amazing metabolic diversity. Five major dimensions must be invoked to develop metabolic theories of the ecology of microbes. These dimensions involve the physicochemical attributes of life and its environment and the uniquely organic features of natural selection, competition, and cooperation that have organized life into hierarchical levels of organization. Exciting opportunities are available for contributing to the development of a unifying understanding of ecology that integrates across all three domains of life.

Chapter 13

PHYTOPLANKTON

Elena Litchman

SUMMARY

1 Phytoplankton are key primary producers in aquatic ecosystems and are responsible for almost half of global primary productivity.

2 Major physiological and ecological processes in phytoplankton scale with cell size and temperature. The scaling relationships and the corresponding exponents can often be derived based on first principles, such as surface area to volume relationships, and MTE considerations, such as fractal scaling of supply networks.

3 The scaling relationships in phytoplankton are sensitive to environmental conditions, resource availability in particular, and can deviate from the predicted relationships.

4 Scaling of physiological processes at the cellular level can be used to explain diverse macroecological patterns in phytoplankton communities, both at present and over geologic time.

5 Scaling relationships in phytoplankton can be sensitive to changing environmental conditions, which may be critically important in mediating the effects of global environmental change on phytoplankton communities, structure and functioning of aquatic food webs, and major biogeochemical cycles.

6 Understanding how mechanistic constraints and multiple selective pressures shape the scaling of phytoplankton physiology and ecology remains a formidable challenge.

13.1 INTRODUCTION

Phytoplankton are microscopic aquatic photoautotrophs that are ubiquitous in the world's oceans and lakes, ponds and rivers. They are extremely phylogenetically diverse, comprising representatives from the two main domains of life (Prokaryota and Eukaryota) and several major lineages of eukaryotes (Protozoa, Plantae, and Chromista). Common groups of phytoplankton include cyanobacteria (blue-green algae), diatoms, dinoflagellates, prymnesiophytes, green algae, and golden algae or chrysophytes (Table 13.1). Phytoplankton were instrumental in several major planetary changes: cyanobacteria oxygenated Earth's atmosphere about 2.5 billion years ago (Canfield et al. 2000) and green algae gave rise to terrestrial plants about 480 million years ago (Kenrick and Crane 1997). Phytoplankton play key roles in the global carbon cycle, accounting for almost half of Earth's primary productivity, and in many other biogeochemical cycles (silica, nitrogen, etc.) (Field et al. 1998; Smetacek 1999). Phytoplankton also form the base of many aquatic food webs in both marine and freshwater environments, and their community composition profoundly affects higher trophic levels, as well as material and energy transfer.

Metabolic Ecology: A Scaling Approach, First Edition. Edited by Richard M. Sibly, James H. Brown, Astrid Kodric-Brown.
© 2012 John Wiley & Sons, Ltd. Published 2012 by John Wiley & Sons, Ltd.

Table 13.1 Major groups of marine and freshwater phytoplankton, their approximate size ranges, habitats and noteworthy characteristics.

Group	Size range (approximate cell linear dimensions)	Habitat	Important characteristics
Cyanobacteria (prokaryotes)	<1–20 µm, colonial forms are often visible by eye (>1000 µm)	Marine and freshwater	Many are N-fixers, produce harmful algal blooms (HABs, mostly in freshwater and estuaries), important as picoplankton in oligotrophic ocean and lakes
Prasinophytes (primitive marine green algae)	1–20 µm, phycoma (cyst) stage 1000 µm	Predominantly marine	Gave rise to terrestrial plants, abundant as picoplankton in oligotrophic ocean; *Ostreococcus taurii* – smallest free-living eukaryote (c. 1 µm diameter)
Green algae (non prasinophytes)	2–30 µm, colonies can be much larger	Mostly in freshwater	Many unicellular forms are fast-growing, associated with high nutrients, some are colonial (e.g., *Volvox*)
Dinoflagellates	20–60 µm, range from 5 (symbiotic) to 2000 (heterotrophic) µm	Mostly marine (90%), some freshwater	Produce HABs (in marine environments), many are mixotrophs (photoautotrophy and phagotrophy) or heterotrophs, symbionts of corals, some are bioluminescent
Cryptophytes	2–100 µm	Marine and freshwater	Flagellated, many are mixotrophs
Chrysophytes (class Chrysophyceae)	2–30 µm	Mostly freshwater, some marine	Dominant in oligotrophic lakes, some colonial
Coccolithophores (class Prymnesiophyceae)	3–40 µm diameter	Marine	Contain calcium carbonate, produce dimethyl sulfide (DMS), acting as nuclei in cloud condensation; *Emiliania huxleyi* – widespread bloom-forming species
Other Prymnesiophyceae	2–20 µm, colonies can be macroscopic	Marine	Flagellated, some produce HABs, *Phaeocystis* – bloom-forming genus
Diatoms (class Bacillariophyceae)	5–2000 µm diameter, chains of cells even longer	Marine and freshwater	Responsible for c. 1/4 of Earth's primary productivity, contain silica in their frustules

Phytoplankton are mostly unicellular; they share similar organismal (cell) plans but range more than five orders of magnitude in volume (Reynolds 1984), which makes them ideally suited for scaling approaches. Indeed, phytoplankton present an excellent model system for formulating and investigating hypotheses on scaling of fundamental physiological and ecological processes such as growth, respiration, and community assembly and applying them to more complex systems. Moreover, because of their relative simplicity and well-studied physiology and ecology, phytoplankton can be used to elegantly bring together different theories of community structure and ecosystem functioning that integrate multiple levels of biological organization, such as MTE and ecological stoichiometry. Understanding how different physiological and ecological processes in phytoplankton scale with cell size and environmental temperature should also increase general understanding of the structure and function of aquatic ecosystems. For example, determining the relationship between cell size and primary productivity would allow us to infer cell size distributions from measured primary productivity and vice versa and, consequently, to better understand the energetic base of food webs in pelagic marine and freshwater ecosystems.

13.2 SCALING OF PHYTOPLANKTON PHYSIOLOGY AND ECOLOGY

Phytoplankton are relatively simple organisms, with major axes of their ecological niche being resources (nutrients and light), predators (grazers) and parasites, and gravity (sinking). Ecological and physiological processes related to these axes are dependent on cell size (Chisholm 1992). Consequently, it is of a fundamental interest to determine how all these processes scale with cell size. There have been numerous studies on these relationships. The two major approaches to deriving the scaling exponents are based on the surface area to volume relationships, leading to $M^{2/3}$ or $M^{-1/3}$ scaling and the fractal network theory predicting the $M^{3/4}$ or $M^{-1/4}$ scaling (West et al. 1997). In phytoplankton, cell size is often expressed in terms of cell volume

or carbon content, which are proportional to cell mass. Most of the allometric scaling relationships in phytoplankton are thus reported for cell volume or carbon content.

Here I outline general scaling relationships and their implications for phytoplankton ecology, physiology, and the function of aquatic ecosystems. For an indepth coverage of scaling in other unicellular organisms, including heterotrophic bacteria, see Okie, Chapter 12.

13.2.1 Growth and respiration

In the absence of resource limitation, phytoplankton populations grow at their maximum rate, μ_{max}, which was shown to scale allometrically with cell size. Scaling exponents for maximum population growth rates have been reported in many studies and appear somewhat variable and not exceptionally tight. In general, the maximum population growth rate of phytoplankton scales less strongly with cell size than in heterotrophic microorganisms (Chisholm 1992): the reported exponents range from −0.11 (volume-based) or −0.13 (mass-based) (Banse 1976) to −0.22 (Finkel 2001), while in heterotrophs it is usually close to the theoretically predicted −1/4 (Fenchel 1974) (Table 13.2; see Brown and Sibly, Chapter 2). Interestingly, even when

Table 13.2 Major scaling relationships in phytoplankton with cell size. Compilation based on references discussed in the text. S/V, predictions based on surface area to volume considerations; MTE, predictions based on the metabolic theory of ecology.

Rate/Characteristic	Predicted scaling exponent	Observed scaling exponent	Comments
Maximum population growth rate μ_{max}	−0.25 (MTE)	−0.1 to −0.22	Weak dependence, highly variable slopes
Maximum photosynthesis rate P_{max} (carbon-normalized)	−0.25 (MTE)	−0.33 to −0.25	Exponent is lower under resource limitation
Initial slope of the photosynthesis-light curve α (photosynthetic efficiency)	−0.25 (MTE)	−0.24	Some studies showed no relationship
Mass-specific respiration rate	−0.25 (MTE)	−0.28	Some studies showed no relationship or different slopes
Maximum nutrient uptake rate V_{max}	0.67 (S/V) or 0.75 (MTE)	0.67	May be consistent with either theory
Population density	−0.75 (MTE)	−0.78 or −0.79	For both lab cultures and oceanic communities

the scaling exponents are similar, the normalization constants (intercepts) for major taxonomic groups, such as diatoms and dinoflagellates, are significantly different, with diatoms growing much faster than dinoflagellates of the same size (higher normalization) (Chisholm 1992). The reasons for this relatively wide variation in phytoplankton growth as a function of cell size are not completely known (Banse 1982).

Respiration is a fundamental metabolic process that in many organisms also scales with a −1/4 exponent (mass-specific rate) (Brown et al. 2004). This was found for phytoplankton as well (Banse 1976; Finkel 2001), although other studies either did not find a significant relationship of respiration with cell size (Langdon 1988) or found the exponent to be different from the predicted −1/4 (Lewis 1989).

13.2.2 Nutrient uptake and population growth

Like other primary producers, phytoplankton require several inorganic nutrients to grow and reproduce. These include nitrogen, phosphorus, iron, and several micronutrients such as copper, manganese, and a handful of vitamins (Reynolds 1984). A standard way to describe nutrient-dependent growth and uptake in phytoplankton is the Droop equations:

$$\text{growth} = \mu(Q) = \mu_\infty\left(1 - \frac{Q_{min}}{Q}\right)$$
$$\text{uptake} = V(R) = V_{max}\frac{R}{K+R} \quad (13.1)$$

where population growth rate μ is a saturating function of internal nutrient concentration Q and the uptake rate V is a function of external nutrient concentration R, and μ_∞ is the growth rate at an infinite internal nutrient concentration (quota), Q_{min} is the minimum internal nutrient concentration at which growth equals zero, V_{max} is the maximum uptake rate, and K is the half-saturation constant for uptake (Droop 1973; Litchman and Klausmeier 2008; see also Kaspari, Chapter 3). These parameters, or physiological traits, have been measured for many phytoplankton species and their relationship with cell size explored. For many of these traits, the allometric relationships can be derived based on first principles, such as surface area to volume relationships and enzyme kinetics (Pasciak and Gavis 1974; Aksnes and Egge

1991; Litchman et al. 2007). For example, V_{max} is predicted to scale with cell volume to the power of 2/3 because of its proportionality to the number of uptake sites, that in turn is proportional to cell surface area (Aksnes and Egge 1991). Similarly, K, the half-saturation constant for nutrient uptake, is predicted to scale with cell volume to the power of 1/3 (Aksnes and Egge 1991). A compilation of these traits for marine phytoplankton revealed very close agreement between predicted and observed allometric exponents, with different taxonomic groups following the same relationship (Litchman et al. 2007) (Fig. 13.1A). Minimum cellular quotas for nutrients are also predicted to scale with cell volume to the power of 2/3 and the data for nitrogen quotas in marine phytoplankton confirm the prediction (Litchman et al. 2007). Interestingly, Allen and Gillooly (2009) analyzed the same data and suggested that the V_{max} scaling exponent is also consistent with a power of 3/4. This illustrates a common difficulty of distinguishing between exponents of competing theories, especially when using noisy data.

Different nutrient utilization traits are also correlated with each other and these correlations often manifest cellular investment trade-offs: the maximum nutrient uptake rate is positively correlated with the half-saturation constant for nutrient uptake, so that higher nutrient competitive ability can be achieved, either by increasing V_{max} at the expense of low K, or by decreasing K while ending up with low V_{max} (Litchman et al. 2007). Sommer (1984) proposed distinguishing different nutrient utilization strategies in phytoplankton, based on the relative values of the maximum growth rates, half-saturation constants, and nutrient uptake rates. Cells with low cellular maximum nutrient uptake rates, V_{max}, and low half-saturation constants for uptake, K, are "affinity" specialists that may have a competitive advantage in low-nutrient environments; cells with high maximum growth rates are "velocity" specialists; and cells with high V_{max} but low maximum growth rate are "storage" specialists (Sommer 1984). Both velocity and storage specialists may dominate in pulsed-nutrient environments (Sommer 1984; Grover 1991; Litchman et al. 2009).

The dependence of key nutrient utilization traits on cell size leads then to the association of different cell sizes with distinct ecological strategies defined by contrasting combinations of trait values (Litchman et al. 2007). For example, small cells with high maximum growth rates would be velocity strategists and small

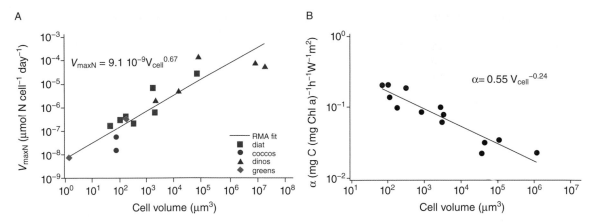

Figure 13.1 Scaling of some key phytoplankton traits with cell size (volume). (A) The maximum nutrient uptake rate V_{max} (redrawn from Litchman et al. 2007 by permission of John Wiley & Sons, Ltd). The fitted line is the RMA regression. (B) The initial slope of the photosynthesis-light curve α. The fitted line is the OLS regression. Redrawn from Geider et al. (1986).

cells with low K and V_{max} would be affinity strategists. In contrast, large cells, by having high V_{max} and low maximum growth rates, would be storage specialists. Thus, environmental conditions favoring certain nutrient utilization strategies would also impact community size distributions.

13.2.3   Light harvesting and utilization

Light is a major resource for phytoplankton and the efficiency of its utilization greatly influences the ecological success of different phytoplankton species. Phytoplankton photosynthesis is a globally important process responsible for almost 50% of carbon fixation on Earth. Therefore, understanding how photosynthetic parameters scale with cell size and temperature should help to assess the potential changes in global primary productivity under different environmental change scenarios. Photosynthetic rate is a saturating function of ambient light that can be characterized by the initial slope α and the maximum rate P_{max}. The carbon-normalized α declines with cell size (Taguchi 1976; Geider et al. 1986), suggesting that small-celled phytoplankton are more efficient at utilizing low light; the exponent appears close to $-1/4$ (Fig. 13.1B). A similar negative dependence on cell size was observed for the slope of the growth–irradiance curve (Schwaderer et al. 2011). The underlying reason for

this decline in light utilization efficiency with cell size is the photosynthetic pigment self-shading (the package effect) that increases in larger cells: light has to travel longer distance in larger cells and is, thus, attenuated more, reducing the photosynthetic efficiency per unit pigment (Raven 1984; Geider et al. 1986).

The cell-normalized photosynthetic rate P_{max} and cellular chlorophyll concentration under resource-saturated conditions are predicted to scale to the power of 3/4 with cell size, as expected from the MTE, and the experimental data support this prediction (Finkel et al. 2004). Consequently, the carbon-normalized photosynthetic rate and chlorophyll content should scale to the power of $-1/4$ (Finkel et al. 2004; Cermeño et al. 2005). However, under resource-limited conditions, the exponents are significantly different from the expected 3/4 and decline to 2/3 with increasing limitation (Finkel 2001; Finkel et al. 2004). These different scaling exponents may be explained by the packaging of light-absorbing pigments within a cell and the surface area to volume ratio influence on nutrient uptake, where small cells have higher concentrations of photosynthetic pigment and thus greater efficiency of light utilization than large cells under resource-limited conditions (Finkel et al. 2004). The deviations of the carbon-normalized P_{max} scaling from the expected exponent in natural communities can also be due to taxonomic differences in photosynthetic rates (Cermeño et al. 2005).

13.2.4 Predators and parasites

Predator (grazer) and parasite avoidance is a key feature of phytoplankton evolution that leads to tremendous morphological and life-history diversification. As with most other eco-physiological processes, susceptibility to predation depends on cell size. There is great variety of phytoplankton grazers with different preferred prey size distributions, from microzooplankton (ciliates, flagellates, mixotrophic phytoplankton) to macrozooplankton (copepods, gelatinous zooplankton, and fish larvae). Each of these groups or even species of grazers differs in its scaling relationships with phytoplankton prey (Fuchs and Franks 2010). The general notion is, however, that large cell sizes are less susceptible to grazing and that evolution of large cell size is a successful escape strategy (Smetacek 2001; Kiørboe 2008). Including size-dependent growth and grazing in models of planktonic food webs better reflects predator–prey diversity and allows a more realistic representation of the complex phytoplankton–zooplankton interactions, compared to models with no size dependence (Moloney and Field 1991; Armstrong 1994; Poulin and Franks 2010). Interestingly, size-dependent grazing by zooplankton can generate diverse size spectra in phytoplankton (Armstrong 1999; Poulin and Franks 2010).

Increasing cell size to avoid grazing comes at a cost because of the scaling of resource utilization with cell size and a consequent decrease in the efficiency of resource uptake and growth rate with increasing size. The trade-off between nutrient competitive ability and grazer resistance is thought to be important in promoting phytoplankton diversity (Leibold 1996). This trade-off can essentially be represented on the cell size axis. This cell size axis may also be important for describing higher-order (e.g., three-way) trade-offs in phytoplankton (Edwards et al., in press). A general overview of the scaling of predator–prey interactions and its role in food web stability is provided by Petchey and Dunne (Chapter 8).

Little is known about scaling of phytoplankton susceptibility to parasites or the abundance of parasites on phytoplankton hosts. Phytoplankton parasites include viruses, bacteria, fungi, and protozoa. Parasites play an important role in regulating phytoplankton population density, community structure, and energy and material fluxes in aquatic ecosystems (Park et al. 2004). The susceptibility of individual phytoplankton species to parasites critically depends on temperature and may be increased due to global warming (Ibelings et al. 2011). Hechinger, Lafferty, and Kuris (Chapter 19) discuss metabolic ecology of parasites in general.

13.2.5 Sinking

Sinking due to gravity is a significant source of mortality for phytoplankton, as most cells are slightly negatively buoyant. Sinking rate scales allometrically with cell size and can be derived based on physical constraints such as Stokes' Law (Smayda 1970; Kiørboe 1993). Sinking rate v of a spherical particle can be expressed as follows:

$$v = 0.222 g \eta^{-1} r^2 (\rho - \rho') \tag{13.2}$$

where g is the gravitational constant, r is the radius of the particle, η is the viscosity of the fluid, ρ is the density of the particle, and ρ' is the density of the fluid. Consequently, sinking rate in spherical cells scales to the power of 2/3 of cell volume (Kiørboe 1993, 2008). The consequence of this sinking dependency on cell size is that large cells sink faster than small cells and need higher turbulence to persist in the water column (Kiørboe 2008). In addition to size, sinking depends on the metabolic activity of the cell: fast-growing, metabolically active cells have significantly lower sinking rates, in part due to their lower density (Smayda 1970; Waite et al. 1997). Large cells counteract rapid sinking with the help of morphological structures and cell vacuoles that may occupy a disproportionately large fraction of cell volume compared to small cells, at least in diatoms, thus decreasing cell density (Smayda 1970; Sicko-Goad et al. 1984).

Depending on the strength of different selective pressures such as resource competition, predation, and sinking, different cell size distributions may emerge in phytoplankton communities. For example, if resource limitation is strong, there is likely a selection for small cells, as in oligotrophic (low nutrient) lakes and ocean (Chisholm 1992). With higher nutrient loading, the relative importance of grazing pressure may increase, leading to a shift in size distribution towards larger cells which are poorer nutrient competitors but less susceptible to grazing (Carpenter and Kitchell 1984, Kiørboe 1993).

13.2.6   Temperature scaling

Temperature is a universal driver of most processes in the majority of organisms, including phytoplankton. Understanding how key processes scale with temperature has become especially pressing in the context of globally increasing temperatures. A classic paper by Eppley (1972) established the relationship between phytoplankton maximum population growth rate, μ_{max}, and temperature T (Bissinger et al. 2008):

$$\mu_{max} = 0.59e^{0.063T} \tag{13.3}$$

This defines an upper bound for growth rate distributions as a function of temperature (Fig. 13.2). A new analysis of an expanded dataset using quantile regression produced a similar relationship with slightly different coefficients (Bissinger et al. 2008). The Q_{10} for growth, a scaling factor characterizing the growth rate increase with a 10° temperature increase, was found to be 1.68–1.88, corresponding to the Boltzmann activation energy E ≈ 0.38–0.48 eV, respectively), depending on the regression model used (Bissinger et al. 2008). Other fundamental processes such as respiration, photosynthesis, and nutrient uptake also depend on temperature (Raven and Geider 1988; Aksnes and Egge 1991). Interestingly, different physiological traits characterizing nutrient uptake are predicted to have a different dependence on temperature: the maximum rate of uptake V_{max} is predicted to increase exponen-

tially with temperature but the half-saturation constant for nutrient uptake K may increase more slowly with temperature than V_{max} (Aksnes and Egge 1991). This differential dependence may lead to a shift in nutrient utilization strategies in phytoplankton with increasing temperatures, where more species would utilize nutrients more efficiently by having higher V_{max} but still relatively low K. This potentially could lead to a greater nutrient depletion in aquatic ecosystems in response to global climate change.

Phytoplankton temperature optima for growth appear to depend on the latitude of origin of the strain and thus exhibit a significant latitudinal gradient (Thomas et al., in preparation). Species living at higher temperatures have higher temperature optima for growth, suggesting local adaptation on a global scale (Thomas et al., in preparation). This has also been observed for photosynthesis, with polar species having significantly lower optimal temperatures (Li 1985). O'Connor and Bruno (Chapter 15) discuss temperature scaling in other organisms (marine invertebrates).

13.2.7   General issues of scaling phytoplankton traits

As phytoplankton are unicellular photoautotophs occupying the lowest part of the organismal size spectrum, it is interesting to ask whether they have the same scaling relationships of metabolic rates as more complex multicellular photoautotrophs and multicellular organisms in general. Numerous macroecological meta-analyses include unicellular algae in analyses of relationships between body size and metabolic rate, growth rate, carbon turnover rate, and other processes (Brown et al. 2004; Marba et al. 2007), implying that the same scaling exponents derived from the metabolic theory hold over the whole spectrum of body size, from unicells to metazoans. It is a fascinating pattern, suggesting the same scaling principles at work for a wide variety of life forms, including phytoplankton. When we zoom in, however, and consider phytoplankton only, the scaling exponents appear much more variable and sensitive to environmental conditions. Phytoplankton growth rate scales differently with cell size, depending on the degree of nutrient or light limitation: under resource-limited conditions, the allometric exponents for growth deviate from the 3/4 value, ranging from 1/3 to 2/3 (Finkel et al. 2004; Mei et al. 2009). It may be that the exponents expected from the

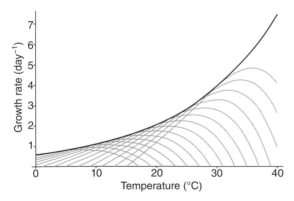

Figure 13.2 A general representation of the Eppley curve defining the maximum possible population growth rate for a given temperature (thick line) and the growth–temperature response curves of individual species or genotypes characterizing their thermal niches.

metabolic theory (West et al. 1997; Brown et al. 2004) emerge when a large range of body sizes is considered, but within the individual groups that have a much smaller size range such patterns are less readily found (Tilman et al. 2004). A consideration of major phylogenetic or functional groups separately may yield novel insights and reveal that scaling exponents do indeed differ across groups. For example, DeLong et al. (2010) found that metabolic rates of heterotrophic prokaryotes, protists, and metazoans scale with body and cell size with distinctly different slopes (Okie, Chapter 12). These differences in scaling suggest different constraints on metabolism operating at major evolutionary transitions in the organization of life (DeLong et al. 2010). Similarly, focusing on individual groups and the instances where scaling deviates from the theoretically expected relationships may be a fruitful area of research for elucidating additional constraints on metabolic processes.

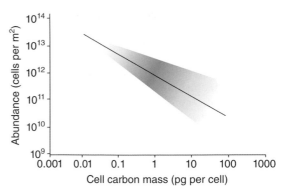

Figure 13.3 Scaling of phytoplankton abundance (number per unit area) with cell size M_C in the North Atlantic Ocean. The shaded area represents the observed data spread. The fitted line has equation $\log N = 11.91 - 0.78 \log M_C$. The slope is indistinguishable from $-3/4$. Adapted from Li (2002) by permission of Nature Publishing Group.

13.3 SCALING OF PHYTOPLANKTON COMMUNITIES

13.3.1 Scaling of phytoplankton abundance and productivity

Several studies have addressed how different characteristics of phytoplankton communities scale with cell size. Li et al. (2002) found that the total phytoplankton abundance in the North Atlantic Ocean scales with cell volume to the power of $-3/4$ (Fig. 13.3). Remarkably, similar exponents were observed for very different ecosystems in the Atlantic Ocean, both nutrient-rich coastal and nutrient-poor open ocean regions (Cermeño et al. 2006). In contrast with species abundance, scaling of primary production with cell volume in phytoplankton appears to be nearly linear, and thus not to follow the 3/4 rule (Marañón 2008). Maximum population density of laboratory phytoplankton cultures scales with cell volume with an exponent of -0.79 (Agustí et al. 1987) and the maximum biomass per unit area with a positive exponent of 0.20 under saturating light, showing that large-celled species achieve lower maximum density but higher biomass than small-celled species (Agustí and Kalff 1989). This relationship may arise from self-shading that decreases light availability to the point where growth stops, but self-shading appears to be reduced in large cells (Agustí 1991).

Species richness of phytoplankton in both natural and laboratory environments was shown to scale with surface area of the habitat sampled, similar to species–area relationships in macroorganisms (Smith et al. 2005). The scaling exponent (0.13) appears to be toward the lower end of the range reported for macroorganisms, probably indicating high immigration from regional species pools, at least in natural communities (Smith et al. 2005).

Knowledge of scaling relationships at the organismal level can be used to explain macroecological patterns at the global scale. For example, Lopez-Urrutia et al. (2006) used the scaling of phytoplankton photosynthesis and respiration with cell size and temperature and data on phytoplankton abundance to predict how the metabolic patterns in the ocean, such as the balance of primary production to respiration, depend on temperature. The differential scaling of growth rates of phytoplankton and their grazers with temperature (higher exponent for grazers) may explain phytoplankton blooms in polar regions, where zooplankton growth is too slow to control the blooms (Rose and Caron 2007; López-Urrutia 2008).

13.3.2 Geologic patterns in phytoplankton size distributions and evolution of cell size

Size distributions of phytoplankton are not static, and respond to environmental changes at different spatial and temporal scales. Changing environmental conditions may select for certain sizes, leading to cell size evolution and shifting size distributions. Analysis of phytoplankton cell size distributions at different times or locations may thus provide insights into selective pressures that have been operating recently. On a geologic timescale, significant changes in average cell sizes in different phytoplankton groups have been documented. Finkel et al. (2007) found that the mean cell size of two major phytoplankton groups, diatoms and dinoflagellates, decreased over the Cenozoic more or less in parallel, suggesting a common driver for this decline. Climate-related changes in nutrient availability could potentially drive these cell size trends (Finkel et al. 2007). A decline in the average cell size of coccolithophores, the calcifying phytoplankton, during the Oligocene may have been caused by the decrease in carbon dioxide concentration (Henderiks and Pagani 2008). Consequently, an ongoing increase in carbon dioxide concentration may lead to an increase of cell size in calcifying algae (Iglesias-Rodriguez et al. 2008).

Different scaling exponents across physiological traits may influence cell size evolution in phytoplankton communities under different environmental conditions. A faster increase of nitrogen storage capacity (Q_{max}) with cell size compared to the minimum requirements for nitrogen (Q_{min}), manifested in the larger cell size scaling exponent, leads to selection for large phytoplankton under nutrient pulses, when high storage capacity is advantageous (Litchman et al. 2009). In contrast, for phosphorus, scaling exponents between Q_{min} and Q_{max} are similar, making selection for large sizes under phosphorus limitation, even under pulsed conditions, less likely (Litchman et al. 2009).

13.4   EFFECTS OF GLOBAL CHANGE

Various scaling relationships mediate phytoplankton responses to a diverse set of anthropogenic stressors and, in turn, may be modified by these stressors. Among the most prominent global change drivers are increasing air and water temperatures. Temperature has both direct and indirect influence on phytoplankton. Increase in temperature increases the rates of most physiological processes. However, the asymmetry in most temperature dependence curves, with a sharp decline in function past the optimum (e.g., Chapter 2, Fig. 2.5, and Chapter 15, Fig. 15.2), can lead to a drastic decrease in major metabolic processes, such as

growth, photosynthesis, and others, once the temperatures are above the optimum (Eppley 1972). Therefore, rising temperatures may directly adversely affect phytoplankton, especially in the short term. If cells of different sizes have different temperature optima, changing temperatures may significantly alter the size distribution of phytoplankton communities.

In addition to directly influencing phytoplankton growth, temperature has a strong effect on the stratification and stability of the water column and, consequently, on nutrient availability for phytoplankton. High air temperatures, by warming the surface layer, generally increase the strength of stratification due to increasing density differences (warm light water at the surface and cold dense water at depth). Stronger stratification impedes wind-induced mixing and decreases nutrient supply from the depths to the surface. This leads to stronger nutrient limitation and a decline in primary productivity in both marine and freshwater ecosystems (MacKay et al. 2009). A stronger nutrient limitation usually favors small-celled phytoplankton, due to their high efficiency of nutrient uptake (high surface area to volume ratio). Consequently, this could lead to a shift in cell size distribution towards small phytoplankton, and such changes have already been reported for lakes (Winder et al. 2009) and the ocean (Li 2002; Li et al. 2009; Morán et al. 2010). Additionally, strong stratification and lack of mixing are disadvantageous for large cells because of their generally higher rates of sinking (Smayda 1970). If there is no significant mixing, large cells sink rapidly out of the euphotic zone, and die because there is not enough light for photosynthesis. Higher temperatures also decrease water viscosity, thus putting large heavy cells at an additional disadvantage by causing faster sinking (Kiørboe 1993). All these consequences of warming suggest that we may see greater dominance of small-celled phytoplankton in future lakes and oceans.

Increased nutrient loading leading to freshwater and coastal eutrophication can also change size distributions in phytoplankton by allowing large-celled species that are poorer nutrient competitors but grazer-resistant to persist in a community (Kiørboe 1996). Consequently, increased nutrient loading tends to increase overall biomass, primarily by adding larger phytoplankton size-classes (Chisholm 1992; Li 2002; Irigoien et al. 2004).

Rising temperatures, increasing nutrient inputs, changes in the physical structure of aquatic ecosystems, and other anthropogenic stressors are already altering the scaling of phytoplankton communities and may lead to drastic reorganization of aquatic food webs. Therefore, a better understanding of how scaling relationships may remain unchanged or be altered in the face of global environmental change is needed, to adequately predict ecosystem structure and function in the future.

13.5 FUTURE DIRECTIONS

Despite significant progress in using scaling approaches to understand phytoplankton physiology, ecology, and community structure, much remains to be explored. Among the challenges are the questions on scaling at the interface of ecology and evolution, reflecting the joint nature of ecological and evolutionary processes shaping phytoplankton (Litchman and Klausmeier 2008). How do multiple selective pressures on diverse traits and processes affect metabolism and allometric scaling? How do these selective pressures shape phytoplankton communities? What are new ways to address these and related questions in phytoplankton? One promising way to explore the evolution of organismal strategies under different environmental conditions, given multiple allometric and other constraints, is to use "adaptive dynamics" models. Such models combine multiple constraints and take into account interactions of multiple phenotypes (competition) to determine the best strategy with respect to trait values for a given environmental scenario (Geritz et al. 1998). Adaptive dynamics modeling has been applied to explain the

evolution of large cell size in marine diatoms, and predicted that large size arises under intermediate frequencies of nitrogen pulses due to differences in allometric scaling of key nitrogen utilization traits (Litchman et al. 2009). The adaptive dynamics framework that includes empirical or theoretical scaling relationships can also be used to explore trait evolution, not only in competitive algal communities (Litchman et al. 2009; Verdy et al. 2009), but also in food webs with multiple trophic levels.

Phytoplankton have been the model organisms for the development of the field of ecological stoichiometry, where many fundamental concepts were first applied to phytoplankton (Redfield 1958; Sterner and Elser 2002). At the same time, phytoplankton, because of their relative simplicity (unicellularity) and a small number of niche dimensions, are excellent model organisms for studying the mechanistic bases of scaling relationships. Recent efforts to integrate the two burgeoning fields, metabolic ecology and ecological stoichiometry, promise a fertile new area of research involving phytoplankton (Allen and Gillooly 2009; Beardall et al. 2009; see also Kaspari, Chapter 3). For example, including allometric scaling relationships in the models describing multi-nutrient growth dependence would elegantly combine the stoichiometric and metabolic approaches. Understanding how scaling of metabolic processes and community function in phytoplankton affects global nutrient cycling and energy transfer in aquatic ecosystems is an exciting question linking multiple levels of biological organization and is of tremendous environmental importance.

Chapter 14

LAND PLANTS: NEW THEORETICAL DIRECTIONS AND EMPIRICAL PROSPECTS

Brian J. Enquist and Lisa Patrick Bentley

SUMMARY

1 Scaling relationships are observed at multiple levels of plant biology. Metabolic scaling theory, which integrates the West, Brown, and Enquist network model (WBE), the metabolic theory of ecology (MTE), other existing network theories, and empirical knowledge, offers a unified framework to mechanistically connect scaling phenomena.

2 Over the last decade, metabolic theory has rapidly developed, matured, and evolved. Several critiques have raised important questions, some of which have been incorporated into new versions of the theory and some of which have been addressed and shown to be incorrect. Some critiques have suggested metabolic theory is incomplete or perhaps wrong; others have noted variation in scaling exponents and have questioned the ability of the theory to account for this variation. These critiques have mainly focused on quantifying scaling exponents but not the hypothesized underlying mechanisms generating scaling.

3 A review of the foundations of metabolic theory as applied to botany shows how it is not a completely new theory as presumed by several of its critics. Instead, it builds upon and unites several long-standing lines of botanical investigation.

4 We show that the theory can address several criticisms and become more predictive by relaxing many of the secondary or optimizing assumptions presented in the original WBE model (West et al. 1997).

5 An important insight is that extensions of the theory make quantitative predictions for botanical scaling exponents and scaling normalizations. The origin of their values is shown to be due to a handful of quantifiable functional traits. These traits appear to be central in regulating the scaling of whole-plant metabolism and growth and are key for connecting variation in the environment with coexisting phenotypes and for developing a more quantitative plant ecology.

6 We show that a mechanistic understanding of the allometric exponent and normalization lays the foundation by which to scale from plant cells to ecosystems. Elaboration of the original theory now encompasses three aspects of plant form and function: branching geometry/architecture, variation in functional traits, and how differential selection can act on these traits to optimize plant performance in differing environments.

Metabolic Ecology: A Scaling Approach, First Edition. Edited by Richard M. Sibly, James H. Brown, Astrid Kodric-Brown.
© 2012 John Wiley & Sons, Ltd. Published 2012 by John Wiley & Sons, Ltd.

14.1 INTRODUCTION

Since the pioneering work of Julian Huxley (Huxley 1932), questions concerning how natural selection influences specific traits within integrated phenotypes have been a prominent focus in comparative biology (Coleman et al. 1994; Murren 2002). The phenotype is a constellation of traits that often covary with each other during ontogeny. Further, organism size is a central trait that influences how most biological structures, processes, and dynamics covary with each other.

Being able to scale from specific traits of organisms to whole-organismal performance is a central question not only in plant physiological ecology but also in population biology, and community and ecosystem ecology (Suding and Goldstein 2008; Lavorel et al. 2011). For example, plants dominate the flux of carbon on the planet (Field et al. 1998; Beer et al. 2010). This flux ultimately reflects spatial and temporal variation in the traits that govern plant metabolism. Nonetheless, our ability to accurately predict spatial and temporal variation in autotrophic carbon flux remains a continuing challenge (Moorcroft 2006). In order to begin to predict and "scale up" spatial and temporal variation in the autotrophic metabolism – from individuals up to ecosystems – one must link the diversity of botanical form with variation in plant function (Moorcroft et al. 2001).

This chapter outlines how, over the last decade, metabolic scaling theory, as applied to plants, has rapidly developed and matured. The original assumptions and predictions of this theory for the scaling of plant metabolic rate with plant size were given by West, Brown, and Enquist (1999b; the WBE2 model), and for how the scaling of plant metabolism ramifies to ecology were given by the metabolic theory of ecology (Enquist et al. 1998; Brown et al. 2004). It is now becoming clear that several parts of the theory appear to be useful while other parts have needed revision. There is a need to provide a more biologically realistic theory capable of more fully incorporating a diversity of form and function. Here we argue that metabolic scaling theory provides a rich framework and a quantitative roadmap to then scale up from anatomy and physiology cells to ecosystems. Further, elaboration of WBE2 in plants has shown that it is unique in being able to explicitly identify the key traits necessary to measure in order to make explicit quantitative scaling predictions.

Here we focus on the foundations of metabolic scaling theory (WBE2 and MTE) as applied to plants. Specifically, our goal is to clarify the road forward for

developing a more predictive plant ecology but based on a trait-based metabolic theory. While we briefly touch on the role of temperature, we mainly focus on the key plant traits (which may have temperature dependencies). In doing so this chapter has five specific goals. First, we show that metabolic scaling theory as applied to plants actually has its basis in several seemingly disparate but related theoretical developments and empirical findings. Together, these lines of research form the foundation for a predictive trait-based framework for the scaling of botanical phenotypes and enable one to scale up to ecology. Second, we review development and debates of botanical scaling since the original formulations of WBE and MTE. Third, we provide a framework for a botanical scaling synthesis. This framework focuses on the origin of botanical scaling exponents and normalizations and allows one to account for ecological variation. Fourth, we show how this framework provides a framework for scaling from plant cells to populations and ecosystems. Lastly, we present several questions and challenges for moving forward.

14.2 PLANT SCALING: HISTORICAL OVERVIEW

Perhaps the most important achievement of metabolic scaling theory is the integration of disparate lines of botanical investigation under a common mechanistic framework. However, several critics of application of MTE to plants have assessed it as a completely separate theory (Coomes 2006; Petit and Anfodillo 2009). It needs to be emphasized that the WBE2 model or even its application to ecology (e.g., the metabolic theory of ecology, MTE) does not "come out of the blue." Both start with and are grounded in several foundational insights that arguably have formed a theoretical foundation for botanical scaling. This foundation can be traced to several separate lines of investigation that are the basis for much comparative botany and plant ecology (Westoby 1984; Niklas 1994a; Westoby and Wright 2006; Falster et al. 2011).

14.2.1 Allometric scaling

How organismal shape and function changes as size increases – allometry – has a substantial research history, and the similarity of many allometric scaling relationships across diverse taxa has suggested to many the promise of a unified scaling framework for biology (e.g., Huxley 1932; Thompson 1942; see also Peters 1983; Schmidt-Nielsen 1984; Niklas 1994b, 2004). In particular, organismal size appears to be a central organismal trait. Size influences nearly all aspects of structural and functional diversity by influencing how several other traits and whole-organism properties scale. Most size-related variation can be characterized by allometric scaling relationships of the form

$$Y = Y_0 M^\alpha \qquad (14.1)$$

where Y is the variable or trait of interest and Y_0 is a normalization constant that may vary across taxa and environments. An allometric approach has been a part of the botanical literature since 1927 with the foundational studies of Pearsall and Murray (Murray 1927; Pearsall 1927). As is discussed below, the central contribution of the WBE model was to offer a mechanistic theory for the origin of *both* α and Y_0.

14.2.2   Relative growth rate and trait-based literature

Plant ecologists have been notably successful in predicting variation in plant relative growth rate or RGR (Hunt 1978; Lambers et al. 1989; Poorter 1989). This theory sought to identify and link together fundamental traits, building on a long line of research going back to Blackman (Blackman 1919). Blackman detailed a "law of plant growth" based on the central assumption that it is directly proportional to leaf area, assuming that carbon assimilation per unit leaf mass is constant. Consequently, whole-plant net biomass growth rate, dM/dt, should be directly proportional to total plant photosynthetic leaf area or leaf biomass, M_L (see also Niklas and Enquist 2001; Koyama and Kikuzawa 2009), where

$$\frac{dM}{dt} = \dot{M} = \beta_A M_L \qquad (14.2)$$

where β_A is an allometric normalization and M_L is the net biomass produced per unit leaf mass. Plant relative growth rate has traditionally been shown (Hunt 1978; Lambers et al. 1989; Poorter 1989) to be influenced by three key traits: (1) leaf net carbon assimilation rate

(NAR, the carbon gain per unit area of leaf, $g\ cm^{-2}t^{-1}$); (2) specific leaf area (SLA, the leaf area per unit mass, a_L/m_L, $cm^2 g^{-1}$); and (3) leaf weight ratio (LWR, the ratio of total leaf mass to total plant mass, M/M_L). Dividing dM/dt by total mass gives the relative growth rate $RGR = (dM/dt)/M = NAR \times SLA \times LWR$ (Hunt 1978). Thus, the allometric normalization for equation 14.2 can be shown to originate in two key plant traits as $\beta_A = NAR \times SLA$. Below (section 14.5) we show that this decomposition of RGR is consistent with the WBE model but still is incomplete in terms of the critical traits that influence growth.

14.2.3   Pipe model

In two novel papers, Shinozaki et al. (1964a, 1964b) proposed a theory for the origin of several plant allometric scaling relationships. The pipe model is unique as it provided, for the first time, a hypothesis for how changes in plant size will govern scaling the total number of leaves and total plant biomass. The pipe model implicitly assumes, as point (2) above, that growth dynamics are driven by the total number of leaves. Each leaf is supplied by a given number of xylem "tubes" that extend from the leaf down to the trunk (see Brown and Sibly, Chapter 2, Fig. 2.2). As these "tubes" or "pipes" diverge at branching junctions the dimensions of the distal branches must change in proportion to the number of leaves. As the tree grows, some branches and twigs are shed so some pipes then turn into disused pipes that lose their connection to the foliage, stop growing, and become embedded in the woody structure, creating heartwood or non-conducting tissue. This basic model predicts the total number of leaves, n_L, distal to a branch of a given radius r, as $n_L \sim r^2$. Thus, the pipe model predicts that the cross-sectional area of the branching network is "area preserving" so that the total number of branches, N, distal to a given branch of radius r, scales inversely as $N \sim r^{-2}$. Further, the total above-ground biomass, M, will scale as $M \sim r^{8/3}$. The pipe model can be seen as providing perhaps the most simple model of the allometry of plant form and function that makes a number of predictions for whole-plant scaling relationships (see Table 14.1). The WBE model builds upon the pipe model but importantly adds several further assumptions for how selection has shaped the scaling of the external and internal vascular branching networks (Savage et al. 2010; Sperry et al. in preparation; von

Table 14.1 Predicted scaling exponents for physiological and anatomical variables of plant internal networks as a function of branch radius ($r_{ext,k}$) for the pipe model, the 1999 WBE model, and the Savage et al. (2010) model. Observed values for average cross-species scaling exponents (mean, 95% confidence intervals using standardized major axis (SMA) regression) are shown for literature data and measurements for oak, maple, and pine.

Internal network property	Pipe model (1964)	WBE model (1999) exponent for $r_{ext,k}$	Savage et al. (2010) model exponent for $r_{ext,k}$	Observed average interspecific exponent from the literature for $r_{ext,k}$	Observed average intraspecific exponent for all measured trees for $r_{ext,k}$
Packing (conduit frequency versus conduit radius, $r_{int,k}$, NOT branch radius)	0	0	−2	−2.04 (−2.74, −1.34)	−2.16 (−3.35, −0.97)
Conduit radius taper ($r_{int,k}$)	0	1/6 ≈ 0.17	**1/3 ≈ 0.33**	0.27 (0.20, 0.34)	0.29 (0.08, 0.50)
Number of conduits in branch segment ($N_{int,k}^{seg}$)	2	2	**4/3 ≈ 1.33**	n.d.	1.19 (0.86, 1.52)
Fluid velocity (u_k)	0	−1/3	**0**	n.s.	n.m.
Conducting-to-non-conducting ratio	0	1/3	**0**	n.d.	0.00 (−0.88, 0.88)
Network conductance (κ_k)	1/2	2	**1.84 (finite) 2 (infinite)**	1.44 (Meinzer et al. 2005)	n.m.
Branch segment conductivity (K_k)	0	8/3 ≈ 2.67	**8/3 ≈ 2.67**	2.78 (Meinzer et al. 2005)	n.m.
Leaf-specific conductivity (K_k/N_{leaves})	0	2/3 ≈ 0.67	**2/3 ≈ 0.67**	2.12 (−1.38, 5.62)	n.m.
Volume flow rate (Q_k)	1/2	2	**2**	1.77 (1.38, 2.16)	n.m.
Total number of branches	−2	−2	**−2**	−2.14 (−2.34, −1.95) (West et al. 2009)	n.m.
Pressure gradient along branch segment ($\Delta P_k/l_k$)	0	−2/3	**−2/3**	n.d.	n.m.
Total biomass, M	8/3	8/3	**8/3**	2.62 (Enquist 2002)	2.64 (Pilli et al. 2006)

n.d., no data found; n.s., non-significant; n.m., not measured.

Allmen et al. in preparation). Indeed, these recent elaborations and rephrasing of the theory go beyond resolving problems with WBE and WBE2.

The MTE model appears to accurately predict many attributes of vascular plants and provides a more realistic characterization of plant structure and function than previous models such as the pipe model. A comparison between allometric predictions with the original pipe model (Shinozaki et al. 1964a, 1964b) reveals several important differences between the more recent scaling models. The pipe model does not explicitly include biomechanical constraints, nor allow for the presence of non-conducting tissue. More critically, it does not incorporate the paramount problem of total hydrodynamic resistance increasing with increasing path length from root to leaf. Both WBE and Savage et al. build upon certain aspects of the pipe model (as reflected in similar scaling exponents for some plant traits) but importantly these models invoke additional selective drivers on whole-plant form and function not included in the pipe model. For hydraulic conductance Savage et al. calculated the predicted exponent based on a finite size network (a network with realistic range of branching generations) and infinite network (a network with an infinite number of branching levels).

14.2.4 Exchange surfaces and the classification of plants based on branching architecture

There has been a long tradition in botany of searching for general principles of plant form and function through the way selection has operated on the surface areas where resources are exchanged with the environment. For example, in 1930, Bower concluded that those who search for general principles shaping plant evolution will find that "the size-factor, and its relation to the proportion and behavior of the surfaces of transit will take a leading place" (Bower 1930, p. 225). Bower hypothesized that understanding how natural selection has shaped the scaling of resource exchange surfaces as plant size increased would lead to an understanding of size-related scaling relationships in botany. This hypothesis has been extended in the more recent work of Karl Niklas who has shown that a few branching traits can form the basis for the diverse architectures of plants (Niklas 1982, 1997). The search for general principles that have shaped plant branching architecture and vascular networks has been fundamental to understanding the integration of the plant phenotype (Horn 1971). Interestingly, only a few branching "network" or architectural designs exist in all vascular plants (Hallé et al. 1978; Niklas 1982). This surprising fact suggests that perhaps similar equations, but with different parameter values, have governed the evolution of botanical form and diversity (see Niklas 1997). As is argued in WBE and MTE, the principles of space-filling and area-preserving branching appear to characterize plant branching geometries and the way organisms fill space and compete for limiting resources. Further, a focus on a few similar branching traits also provides a central component of the development of WBE2.

14.2.5 Models on plant hydraulics

A long-standing central question in plant functional biology focuses on how plants are able to transport water and nutrients to such impressive heights (see Zimmermann 1983; Ryan and Yoder 1997). The focus has been to understand how physical and selective processes govern fluid flow within the vascular system and how they influence the evolution of xylem anatomy (e.g., Huber 1932; Huber and Schmidt 1936; Zimmermann and Brown 1971; Zimmermann 1978; Tyree et al. 1983; Tyree and Sperry 1988; Tyree and Ewers 1991; Sperry et al. 1993; Comstock and Sperry 2000). These "resistance-capacitance" models use Ohm's Law to show how anatomical and physiological attributes of plants and their environment influence the water potential gradient from root tips to leaves and the rate of fluid transport throughout the individual (Van den Honert 1948; Jones 1978; Smith et al. 1987; Tyree and Sperry 1988; see also Jones 1992; Schulte and Costa 1996). These models link how differences in the local environment (e.g., drought) can influence xylem flow resistance and vulnerability to cavitation through differences in tissue (wood) density and xylem conduit size (Hacke et al. 2001). As we show below, many of these hydraulic traits are central to the WBE model. Further, differential selection across differing environments, as reflected by variation in hydraulic traits (tissue density and xylem conduit size), will ramify to constrain the scaling of plant productivity and ecological dynamics within MTE.

14.2.6 Models of plant geometry, competition, demography, and the thinning law

In 1963 Yoda et al. showed that most plant populations and forest stands exhibit a negative relationship between size and number of plants. The self-thinning rule describes how this relationship is generated by plant mortality due to competition in crowded, even-aged (sized) stands (Yoda et al. 1963). Self-thinning is the label applied to density-dependent mortality due to competition (Harper 1977). Yoda et al. argue that three basic principles shape many dynamics in plant population and community ecology (Yoda et al. 1963): (1) there is an upper limit or "constraint" on the total leaf area or biomass that can be supported given a certain number of plants; (2) under this constraint, mortality (self-thinning) is caused by the growth rates of competing individuals; (3) the inverse relationship between size and density is general and originates from an equally general rule of how plant morphology scales with plant size. Similarly, "demographic theory" (Holsinger and Roughgarden 1985; Kohyama 1993 and references therein) has shown how the distribution of sizes in a plant population or community must ultimately reflect the outcome of size-dependent growth and mortality rates. As shown below, MTE applied to plant populations and communities builds upon these arguments. It shows how the scaling of size and number of individuals within populations and ecological communities is constrained by the geometry of fractal-like branching architectures. Further, it argues that, under resource steady state, the allometric scaling of resources use (metabolism) then constrains the scaling of turnover and mortality of individuals.

As shown below, metabolic scaling makes a modest number of additional assumptions in order to integrate each of these theories and insights into a common theoretical framework capable of making detailed quantitative predictions. In doing so it provides unique insight into how natural selection has guided the evolution and diversity of plant form and function (e.g., see Niklas 1997; Shipley 2010). As we discuss below, a mechanistic understanding of the allometric exponent α and normalization Y_0 lays the foundation by which to scale from plant cells to ecosystems. Specifically, the origin of both α and Y_0 encompass how selection has acted on these traits to optimize plant performance in differing environments (e.g., Norberg et al. 2001). Applying additional assumptions at the ecological scale then forms the basis for the metabolic theory of ecology (MTE, Brown et al. 2004) which then enables a series of additional predictions for the ramification of the scaling of metabolism and metabolic traits.

14.3 EXAMPLES OF BOTANICAL SCALING: FROM ANATOMY AND PHYSIOLOGY TO ECOSYSTEMS

Since the publication of the original WBE2 model for plants (West et al. 1999b) numerous papers have documented scaling relationships associated with metabolism. Other papers have elaborated metabolic theory. Scaling relationships are observed at multiple botanical levels. The similarities in scaling exponents (often

quarter-powers, so 3/4, −3/4, 1/4, 3/8, etc. when expressed as a function of size or −2, 2, 2/3, and 8/3 when expressed as a function of stem radius). Remarkably, scaling relationships are observed from the cellular to the ecosystem level. In the interest of space we focus on a few examples to give a sense of the breadth of patterns found.

14.3.1 Within-plant scaling relationships

Many allometric scaling relationships are observed in plants (Niklas 1994b). Robust scaling relationships within the internal vascular network of plants ultimately control the scaling of whole-plant water use and carbon flux. Figure 14.1 shows an inverse relationship between the total number and size (diameter) of the xylem conduits in a branch. This relationship holds within plants as well as across diverse taxa. It is known as the "packing rule" (Sperry et al. 2006). In general the number decreases as the inverse square of diameter. Several other scaling relationships are associated with variation in the dimensions of internal vascular (xylem) anatomy and external branching architecture. Several of these relationships are reported in Table 14.1 (see also West et al. 1999b; Savage et al. 2010).

14.3.2 Whole-plant scaling relationships and the partitioning of biomass and production

Papers by Niklas and Enquist (2002a) and Enquist et al. (2007c) have shown that interspecifically, whole-plant rates of production scale with exponents indistinguishable from 3/4 across Angiosperms and Gymnosperms (Fig. 14.2; see also Ernest et al. 2003). Similarly, total leaf area scales as the square of branch diameter (Fig. 14.3A). Until relatively recently, the general principles underlying how plant metabolic production is allocated between above- and below-ground compartments was unclear (Bazzaz and Grace 1997). As shown in Figure 14.3, biomass allocation is a size-dependent phenomenon ultimately controlled by the scaling of metabolism and growth rate. Enquist and Niklas derived the inter- and intraspecific scaling exponents for leaf, stem, and root biomass at the level of the individual plant (Enquist and Niklas 2002a; Niklas and Enquist 2002b).

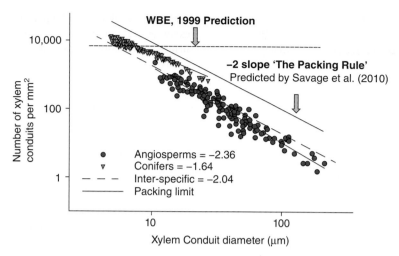

Figure 14.1 A plot showing the "packing rule" in Angiosperm and Gymnosperm plants. The relationship is measured by counting the frequency of xylem conduits and the conduit radius within and across branches. The relationship across each group varies approximately inversely with the square of conduit radius. This packing rule contradicts the WBE model's assumption (horizontal dotted line) that conduit frequency remains unchanged as conduit radii taper, decreasing in size from trunk to terminal twig. Selection for hydraulic safety and efficiency considerations have been proposed to underlie the packing rule, suggesting new theory is needed to accurately describe vascular architecture. Data from Sperry et al. (2008) by permission of John Wiley & Sons, Ltd.

14.3.3 Scaling relationships in plant ecology

Several authors have reported scaling relationships at the level of plant populations or forest stands. Several of these relationships appear to be remarkably general across diverse environments and taxa. For example, Figure 14.4 shows community size distributions for two forests – one a tropical forest with about 100 species and the other a temperate forest with 15 species. Despite the difference in physiognomy and species composition, scaling of number of stems with size is similar. Further, Figure 14.5 shows that the rate of mortality or turnover of individuals scales inversely with size. A global sample of forest plots suggests that these ecological relationships may be general (Enquist and Niklas 2001; Enquist et al. 2009).

14.3.4 Scaling relationships at the level of the ecosystem

A recent novel empirical insight is the documentation of ecosystem-level scaling relationships. Two examples are given in Figure 14.6. First, plotting whole-ecosystem biomass of plants (autotrophic biomass) versus the annual net primary production reveals a

scaling exponent significantly less than 3/4 (Kerkhoff and Enquist 2006). The shallow slope indicates that the NPP per unit biomass across ecosystems actually decreases as phytomass increases.

Below we discuss how recent developments in metabolic scaling theory not only can make sense of these relationships but also show how they are interconnected with each other and are influenced by a handful of functional traits at the organismal level.

14.4 ORIGIN OF BOTANICAL SCALING EXPONENTS: WEST, BROWN, AND ENQUIST MODEL

MTE rests upon the original network theory proposed by West, Brown, and Enquist in 1997 (but note that the temperature part of MTE does not). This model (WBE for short) was presented as a general framework for understanding the origin of allometric scaling laws in biology. Similarity in metabolic scaling relationships and allometric relationships within and across taxa are used in MTE to scale up from organisms to ecological and ecosystem attributes. In order to properly test and assess metabolic theory it is important to distinguish core predictions and assumptions from secondary predictions and assumptions. Much of the confusion in

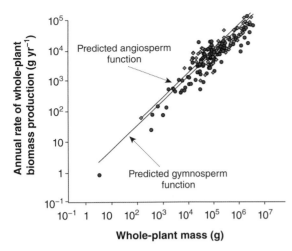

Figure 14.2 Allometric scaling of total plant biomass (roots, stems, and leaves), M, versus annual biomass production, dM/dt for both Angiosperms (red circles) and Gymnosperms (gray diamonds). Figure from Enquist et al. (2007c) by permission of Nature Publishing Group. The allometric scaling relationship for each group is indistinguishable from 0.75. The shown allometric functions (solid lines) are not the fitted function but instead the predicted allometric scaling function where the exponent (slope) and normalization (intercept) are predicted from metabolic theory. The normalization of the scaling function was calculated for each taxon based on re-sampling global values of taxon-specific mean trait values as specified by equations 14.7 and 14.8. As shown by empirical data for these same plants, we used the value of $\theta = 3/4$. The predicted allometric functions for whole-plant growth, based on trait data, provide good allometric approximations of annualized plant growth.

the literature in interpreting the various predictions of the metabolic theory stems from a lack of understanding of the differences between the core and secondary aspects of the theory. These distinctions were not clearly made in any of the original WBE papers but have been delineated subsequently (Price et al. 2007).

14.4.1 Core assumptions and hypotheses of the WBE model

Building on several of the above botanical theories and insights, there are three core assumptions of the WBE network model. First, at the heart of the model is the hypothesis that the scaling of metabolism is primarily influenced by the geometry of vascular networks that control the scaling of effective surface areas where resources are exchanged with the environment. These surface areas control the transport of resources to metabolizing tissue (West et al. 1997). This then implies that the value of *several additional* allometric scaling exponents also arises from the geometry of branching networks. The second assumption is that normalization (Y_0 in equation 14.1) is driven by traits that define the metabolic demand of "terminal metabolic units" (i.e., leaves). The third assumption is that natural selection can act to shape the scaling of metabolism and several associated allometric relationships via selection for the scaling of resource uptake and the cost of resource uptake (West et al. 1999a). In sum, the core hypothesis of the WBE model is that the scaling of many organismal, anatomical, and physiological traits (e.g., whole-plant carbon assimilation, vascular fluid flow rate, and the number and mass of leaves) is mechanistically determined by natural selection, which has shaped the geometry of the external branching network (see West et al. 1997).

14.4.2 Secondary assumptions of the WBE model

On top of the three core assumptions, WBE then also invoked several additional secondary assumptions for an "allometrically ideal plant." For simplicity, they assumed that the plant's external branching network is a hierarchical, symmetrically branching network (see Brown and Sibly, Chapter 2, Fig. 2.2). Therefore, the radii (r) and the lengths (l) of all branches within branching level k are assumed to be approximately the same. At each branching node, a parent branch (at level k) splits into n daughter branches (level k + 1). All parent branches, N_k, are assumed to give rise to the same number of daughter branches, N_{k+1}, across the tree, so that the branching ratio n_k is a constant as $n_k = N_{k+1}/N_k$. Under this framework, there are two branching traits that govern the allometric scaling within a tree. The scaling of the branch length ratio, γ, is defined by the exponent b, and the scaling of branch radii ratio, β, is defined by the exponent a,

$$\gamma = \frac{l_{k+1}}{l_k} \equiv n_k^{-b}; \beta = \frac{r_{k+1}}{r_k} \equiv n_k^{-a} \qquad (14.3)$$

As we discuss next, the above core assumptions when combined with these secondary assumptions, lead to a

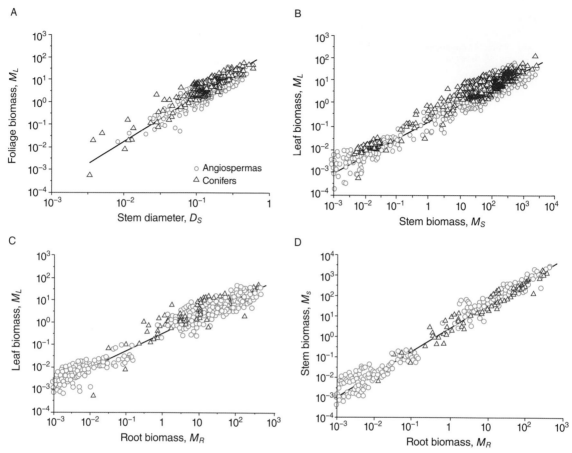

Figure 14.3 Global allometric relationships in biomass partitioning across Angiosperms and Gymnosperms. Here we plot basal stem diameter, D_S, leaf mass, M_L, stem mass, M_S, and root mass M_R. Data are from worldwide datasets as reported in Enquist and Niklas 2002a. Solid lines are reduced major axis regression curves of log-transformed data. Angiosperm and conifer species are denoted by circles and triangles, respectively. (A) M_L versus D_S (trunk diameter at breast height). (B) M_L versus M_S. (C) M_L versus M_R. (D) M_S versus M_R. See Enquist and Niklas 2002a for additional statistics. Note, the relatively larger spread in (B) and (C) is due to differences between Angiosperms and Gymnosperms. Extension of the WBE model for plants predicts the scaling exponents for each of these relationships. Predicted M_L versus M_S slope = 0.75, observed = −0.75, 95% CI = 0.73–0.76; predicted M_L versus M_R slope = 0.75, observed = 0.79, 95% CI = 0.76–0.82; predicted M_S versus M_R slope = 1.00, observed = 1.09, 95% CI = 1.05–1.13.

different set of predictions depending on how selection has shaped the branching traits a and b.

14.4.3 Core predictions of the West, Brown, and Enquist model: scaling exponents driven by branching traits

An important implication of the core assumption of the WBE model is that the values of a and b deter-

mine numerous scaling relationships within and between plants. The total metabolic rate, B, or flow through the plant network, $\dot{Q}$, scales as $\dot{Q} \propto B = b_v V^\theta$ where the allometric constant or normalization, b_V, indexes the intensity of metabolism per unit canopy or rooting volume, V_{plant}. Here, the allometric scaling exponent is

$$\theta = \frac{1}{(2a+b)} \qquad (14.4)$$

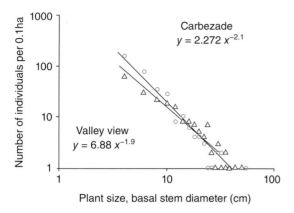

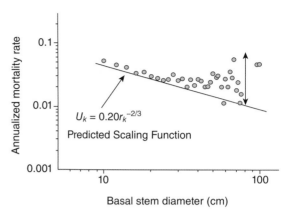

Figure 14.4 Community-level scaling relationships. Here, an inverse relationship exists between tree size and abundance within 0.1 ha forest communities. These inverse "size frequency" distributions tend to show approximate power-function scaling with a slope of −2. The 95% CI of the exponent for each distribution includes the predicted value of −2.0 based on model type I or II regression analyses. The two overlapping size/frequency distributions of a 0.1 ha sample of the South American tropical forest community located at Carbezade (−10.2° latitude; data shown as open circles) and a 0.1 ha sample of a North American community located at Valley View Glades, Missouri (38.15° latitude; data shown as open triangles).

Figure 14.5 Relationship between tree size (measured as basal stem diameter, Dk, where $Dk = 2r_k$) and annualized mortality rate, μ_k, for tagged trees within a dry topical forest, the San Emilio forest (see Enquist et al. 1999). Size-classes are binned at 1 cm resolution. The shown line is not a fitted line but instead is the predicted mortality function based on scaling of growth and the allometric relationship between stem diameter and total biomass for individual trees in this forest. Whereas the observed data are generally close to the predicted curve, there is increasing variation and deviation for the largest trees, likely because of noncompetitive sources of mortality not included in the model (see Enquist et al. 2009 for additional detail).

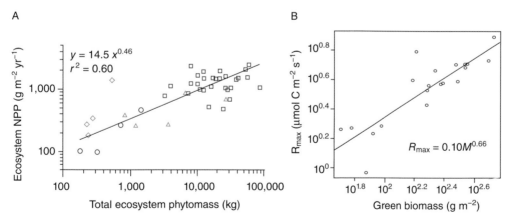

Figure 14.6 Whole-ecosystem allometries. (A) Whole-ecosystem scaling of net primary productivity across grasslands (diamonds), tundras (circles), shrublands (up triangles), and forests (squares). Figure from Kerkhoff and Enquist (2006) by permission of John Wiley & Sons, Ltd. These data show that above-ground NPP scales with total plant mass to the 0.46 power. (B) Similarly, the total above-ground plant respiration scales as a function of the total phytomass (excluding dead woody material) to the 0.66 power. Data from a study of several temperate montane meadow communities near Gothic, Colorado (Kerkhoff and Enquist, in preparation). As discussed in the main text, these ecosystem-level scaling relationships can be predicted from the metabolic theory of ecology that scales up metabolism from cells to ecosystems.

Thus, WBE predicts how various allometric relationships ultimately originate in just two branching traits, a and b, so that

$$N_N \propto V^\theta = V^{\frac{1}{2a+b}} \qquad (14.5)$$

where N_N is the total number of terminal twigs or leaves on a plant, and V is the volume of all branches of the whole tree (a proxy for mass). Similarly, the scaling of the total number of leaves is related to stem radius and length as $N_N \propto r_0^{1/a}$ and $N_N \propto l_0^{1/b}$, where r_0 is the radius at ground level, and l_0 is the tree's maximum path length from base to the most distant terminal twig (West et al. 1999b).

Importantly, the values a and b also directly determine numerous other scaling relationships, such as whole-plant respiration rate, $\dot{R}$, carbon assimilation rate $\dot{P}$, xylem flow rate $\dot{Q}_0$, and total number of leaves (n_L). Converting to plant mass we have $M = V\rho$ where ρ is the tissue density., If ρ does not vary with plant size, V, and if the water flux per unit leaf area as well as photosynthesis and respiration per unit leaf area are independent of plant size then a central prediction of the WBE model is

$$\dot{R} \propto \dot{P} \propto \dot{Q}_0 \propto n_L \propto M^\theta \qquad (14.6)$$

As we discuss below, equation 14.6 is an approximation (Savage et al. 2008) for allometric scaling for plants with a large number of branching generations (West et al. 1999b; Savage et al. 2008). So, these predictions are expected to hold for plants larger than seedlings (Reich et al. 2006; Enquist et al. 2007a, 2007b). An important question is: what sets the value of theta in any given taxon or environment?

If the branching traits, a and b, are constant within a tree and do not vary with the size of the branching network, V, then the network is self-similar. A self-similar network predicts that (1) allometric relationships will be best fit by a power function; (2) branching architecture will be fractal-like within and across trees; and (3) branching architecture will play a crucial role in shaping analogous scaling relationships for the internal vascular network (Savage et al. 2010). This leads to extensions of the WBE plant model and extensions to new models that predict similar patterns across plants, plant communities, and forest ecosystems (Enquist 2002; Enquist et al. 2009; West et al.

2009; but see Coomes 2006; Muller-Landau et al. 2006a; Russo et al. 2007; Coomes and Allen 2009).

14.4.4 Additional secondary assumptions: selection to optimize external branching network geometry and the origin of quarter-power scaling relationships

Many of the initial papers by West, Brown, and Enquist focused on the origin of the well-known relationship where theta = 3/4 (see Brown and Sibly, Chapter 2). This allometric scaling rule, between whole-organism metabolic rate and body mass, has been documented across taxa for decades (Kleiber 1932; Hemmingsen 1950; Peters and Wassenberg 1983; Schmidt-Nielsen 1984; Niklas 1994c). Within the framework of the WBE model, the 3/4 rule originates from four secondary optimizing assumptions (West et al. 1999b): (1) selection has maximized the scaling of total whole-plant leaf surface area with plant size, resulting in a branching network that is space-filling. This assumption builds upon a long-held notion in botany (discussed above) that selection for increased surface areas where resources are exchanged with the environment has likely been a central organizational principle (Bower 1930; Küppers 1989; Farnsworth and Niklas 1995); (2) selection has also acted to maximize water conductance and minimize the scaling of hydrodynamic resistance through the vascular network with plant size (see below); (3) the dimensions and physiology of leaves and petioles do not systematically vary with plant size; and (4) biomechanical constraints to elastic buckling (McMahon and Kronauer 1976; King and Louks 1978) are uniform. This assumption enforces biomechanical stability across branching levels and leads to eventual 2/3 scaling between tree height and diameter and 3/8 scaling between tree mass and diameter. Because WBE also assume that plant metabolic rate, B, is directly proportional to the number of petioles (or leaves, see equation 14.6), and using assumption #3 and assuming large number of branching levels, N, this dictates the scaling of whole-plant resource use and metabolic rate. Furthermore, assumption #1 leads to $b = 1/3$, and assumptions #2 and #4 leads to "area-preserving branching" where the sum of cross-sectional areas of daughter branches equals the cross-sectional area of the mother branch (Richter 1970; Horn 2000). Thus, $a = 1/2$. Together, the principles of space-filling and area-preserving branching set the values of *a* and *b* and then govern a suite of different allometric scaling relationships within an "allometrically ideal" plant (see Table 14.1). Using equations 14.5 and 14.6 then leads to the prediction $\theta = 3/4$.

In general the 3/4 scaling of metabolism (Niklas and Enquist 2001) appears to hold across a broad sampling of plants. Building on these arguments, Enquist and Niklas used the "allometrically ideal" predictions of the WBE model to show how the total amount of stem, root, and leaf biomass – as well as the total above- and below-ground biomass – should scale with each other (Enquist and Niklas 2002b; Niklas and Enquist 2002a, 2002b). Global data are in general agreement with predictions from the WBE model (Fig. 14.3).

14.4.5   Evolution by natural selection on the hydraulics of internal vascular network geometry

How selection has shaped the scaling of whole-plant hydraulics appears to have been central to the evolution of plant form and function. Evolutionary increases in the range of plant sizes over macroevolutionary scales (Knoll and Niklas 1985) necessitate a hydraulic cost. Selection for increase in sizes also necessitates an increase in the transport distance over which resources must be transported (Enquist 2003). As a result the hydrodynamic resistance of transport must also increase (Zimmermann 1983). This linear increase in resistance would constrain any diversification of plant size (Raven and Handley 1987; Ryan and Yoder 1997). The WBE model demands that selection has acted to minimize the scaling of resistance with increased plant size (or transport distance). The WBE model predicts that if selection has shaped internal xylem networks so that radius of xylem tubes (r_{xylem}) must taper from the base of the tree to the leaf (see Fig. 14.9), then the total xylem resistance along a given path length from leaf to trunk is minimized (see Table 14.1). If r_{xylem} scaled positively with the size of a branch, r_k, then the total resistance, which is the sum of the resistances through all branches, could be minimized and approximately independent of the number of branchings, *N*, and the total xylem transport length, l_T (West et al. 1999b). Thus, a volume-filling, area-preserving external network, that contains an internal vascular network that minimizes the scaling of resistance with path length, will result in a 3/4 scaling of

whole-plant water flux and conductance with plant mass, or a square law for the scaling with stem diameter (see Table 14.1).

14.4.6   Intra- or interspecific?

It is critical to point out that, although it was not emphasized in the original papers, WBE2 is primarily an *intra*-specific model. However, its predictions should also hold *inter*-specifically if (and only if) terminal branch, leaf, xylem, and physiological traits do not vary systematically across species. This has led to some confusion and discussion regarding how to best test the predictions and the scope of inference of WBE, as well as the specific assumptions that need to be assessed (see Mencuccini 2002).

The assumptions and optimization principles stated above straightforwardly lead to quantitative predictions of how, within an "*allometrically ideal plant*," numerous aspects of physiology and anatomy scale with plant size (see Table 14.1). Allometric exponents are predicted to be "quarter-powers" when plotted in terms of plant mass, *M* (see West et al. 1999a). These predictions can be straightforwardly converted to allometric predictions based on stem diameter, *D*, or stem radius r_k (Table 14.1).

14.5   ORIGIN OF BOTANICAL SCALING NORMALIZATIONS: MERGING OF WBE2 WITH TRAIT-BASED PLANT ECOLOGY

An important aspect of metabolic theory is that it is also capable of predicting the allometric normalization (and not just the allometric exponent). Several papers have taken the original WBE and expanded to derive the value of the allometric normalization (Enquist et al. 1999; Economo et al. 2005; Gillooly et al. 2005a; Enquist et al. 2007c). Here we show how two key plant allometric scaling normalizations originate in a handful of plant traits. A core assumption of the WBE model is that the normalization of metabolic allometry is driven by metabolic demand of "terminal metabolic units" (see also section 14.2.2 on RGR theory above). In the case of plants, the terminal "unit" is the leaf and metabolic rate depends on the properties of the leaf and how the total number of leaves scales with plant size (Eqn. 5).

14.5.1 Normalization of leaf allometry

The net carbon assimilation rate of a plant, NAR, can be rewritten as $NAR = c\dot{A}_L/\omega$, where $\dot{A}_L$ (g $C \cdot cm^{-2} \cdot t^{-1}$) is leaf-area-specific photosynthetic rate, c is the net proportion of fixed carbon converted into biomass (Gifford 2003) or the carbon assimilation use efficiency (dimensionless), and ω is the fraction of whole-plant mass that is carbon (Enquist et al. 2007c). Thus, using this expression (see also Hunt 1978; Lambers et al. 1989; Poorter 1989) for NAR, the equation for whole-plant growth (see equation 14.2) becomes

$$\dot{M} = \beta_A M_L = NAR \cdot SLA \cdot M_L = \left(\frac{c}{\omega}\dot{A}_L\right)\left(\frac{a_L}{m_L}\right)M_L \quad (14.7)$$

where β_A is an allometric normalization in equation 14.2 and is the net biomass produced per unit leaf. Its value represents several leaf level traits. An example is SLA, the specific leaf area (leaf area/leaf mass). In principle, the plant traits listed in equation 14.7 can vary. Equation 14.8 can be expanded by incorporating the importance of whole-plant size and biomass allocation into the equation for growth rate (i.e., equation 14.3).

14.5.2 Normalization of growth rate

The WBE model states that M_L scales with whole-plant mass as $M_L = \beta_L M^\theta$ (this is essentially a version of equation 14.5). Elaborations of the WBE model (Enquist et al. 1999) show that the term β_L is governed by additional functional traits and plant size. Specifically, $\beta_L = M_L M^{-\theta} = M_L(\rho V)^{-\theta}$ where again, ρ is the tissue density. The allometric constant, $\phi_L \rho^{-\theta}$, measures the mass of leaves per allometric volume of the plant body. Therefore, substituting for the M_L term in equation 14.7 yields a growth law dependent on several traits including the branching that define θ:

$$\dot{M} = \left(\frac{c}{\omega}\dot{A}_L\right)\left(\frac{a_L}{m_L}\right)\beta_L M^\theta = \left(\frac{c}{\omega}\dot{A}_L\right)\left(\frac{a_L}{m_L}\right)\left(\phi_L \rho^{-\theta}\right)M^\theta$$

$$(14.8)$$

Extensions of metabolic scaling theory shows that it is possible to predict plant growth from knowledge of a handful of traits. From above, we predict the normalization constant for the scaling relationship between plant size and growth $b_0 \approx \beta_G \approx (a_L/m_L)([c/\omega]\dot{A}_L)\beta_L$, where each of the variables corresponds to a critical trait. To predict growth rate one must first measure these traits as specified by the model. Enquist et al. (2007c) analyzed a global data compilation for the traits listed in equations 14.7 and 14.8, for a wide sampling of Angiosperm and Gymnosperm trees. Figure 14.2 shows that predictions successfully approximate, *with no free parameters*, the empirical scaling of plant growth. The lines that pass through the allometric relationship comprise the predicted scaling function based on plant traits. Interestingly, as supported by data, our model predicts that Gymnosperms have a higher value of β_A (see equation 14.3) than Angiosperms, but that both taxa have similar values of β_G due to opposing mean trait differences: (a_L/m_L) and $\dot{A}_L$.

14.5.3 Temperature

Here, due to space limits, we have not focused on the role of temperature in plant metabolism (see Anderson-Teixeira and Vitousek, Chapter 9, for more detail). However, temperature is critically important and its influence is also expressed in the scaling normalizations of equations 14.7 and 14.8 via temperature influences on traits – most notably the net carbon assimilation $\dot{A}_L$ (which includes respiration) as well as c, the carbon use efficiency (see Enquist et al. 2007c). While the Arrhenius equation appears to hold within plants (Gillooly et al. 2001; Enquist et al. 2003; Anderson et al. 2006) in our opinion an open question focuses on the relative importance of temperature acclimation, adaptation, and the replacement of taxa with different metabolic traits (as defined by equations 14.7 and 14.8) across communities (assembly) observed broad-scale temperature gradients (Kerkhoff et al. 2005; Enquist et al. 2007a; Enquist 2011). Specifically, how important are these biotic responses (what we call the three As: acclimation, adaptation, and assembly) across broad-scale temperature gradients in influencing observed variation in plant growth and physiology (Enquist et al. 2007a)? Another open question is whether the activation energy for photosynthesis is 0.6 eV or, instead, 0.3 eV (see discussion in Allen et al. 2005, Kerkhoff et al. 2005, and Marba et al. 2007). These two points may have important implications for temperature-correcting biological rates across broad gradients (Kerkhoff et al. 2005).

The theoretical framework that we have outlined here will also ultimately provide a basis for similar predictions for how variation in plant traits influences the scaling of whole-plant performance. Variation in the traits specified in equations 14.7 and 14.8 will influence the normalization (i.e., the residual scatter about the allometric function). As we discuss below (section 14.6.1.4), a trait-based elaboration of metabolic scaling theory now enables one to assess how selection for different trait values in differing environments (such as specific leaf area or allocation to leaves or roots) then must influence the scaling of whole-plant growth and resource use. This is an exciting development, as a trait-based elaboration of the WBE model then effectively integrates a long line of research in trait-based ecology (Wright et al. 2004; Westoby and Wright 2006) with MTE.

14.6 WHAT HAVE WE LEARNED SINCE 1997 AND 1999?

The publication of the WBE in 1997 and WBE2 in 1999 (West et al. 1997; West et al. 1999b) has led to numerous and multifaceted studies testing its predictions and implications in plants. On the one hand, several studies, outside of our scaling collaborators, have found general support for many of the predictions. For example, (1) xylem conduits taper as predicted, from the roots to the leaves, so as the scaling of hydraulic resistance is minimized (Anfodillo et al. 2006); (2) the scaling of branch dimensions and plant biomass generally scale as predicted (Pilli et al. 2006); (3) analysis of whole-plant physiology shows convergence in the 3/4 scaling of plant water use and growth across diverse species (Meinzer 2003; Meinzer et al. 2005); (4) plant birth and death rates also scale with quarter-powers (Marba et al. 2007); and (5) several novel methodological approaches using remote sensing have largely confirmed many of the predictions for the scaling of leaf area and partitioning of biomass within and across numerous trees and biomes (Wolf et al. 2010). On the other hand, there have been several criticisms that question its basic framework, assumptions, generality, and applicability (Harte 2004; Tilman et al. 2004; Kozlowski and Konarzewski 2005; Makarieva et al. 2005b). Further, several additional studies, as we discuss below (section 14.6.1), have highlighted seeming deviations from the model predictions and problems with the model assumptions.

Several prominent critics have argued that the WBE model cannot explicitly account for the range and origin of inter- and intraspecific variability in allometric exponents (Bokma 2004; Glazier 2005). Here we focus on the nature of these criticisms and show how revisiting the secondary assumptions of WBE and WBE2 not only provides a way to integrate these concerns but also provides additional quantitative predictions for plant scaling.

Several studies have identified a number of issues and questions with the original WBE2 1999 plant model, as follows. (1) Several authors have suggested that selection for hydraulic safety and efficiency, instead of space filling and minimization of resistance as assumed by WBE, has shaped the evolution of vascular networks (Mencuccini 2002; Sperry et al. 2008). (2) Others have questioned whether vascular safety and efficiency (McCulloh and Sperry 2005; Zaehle 2005; Petit and Anfodillo 2009) or the carbon costs associated with the scaling of plant hydraulic networks (Mencuccini et al. 2007) are adequately described by WBE. (3) Several additional studies have revealed empirical patterns that contradict some of the predictions (McCulloh et al. 2003, 2004, 2010; Mencuccini et al. 2007; Petit et al. 2009). (4) Reich et al. (2006) have questioned whether the scaling exponent for whole-plant metabolism (respiration) actually scaled with an exponent closer to 1.0 rather than 3/4 (see also Mencuccini 2002). (5) Lastly, several studies have pointed out that not all species follow the 3/4 scaling prediction. They have emphasized that there is variation in metabolic scaling relationships within and between species (Mäkelä and Valentine 2006; Russo et al. 2007) – especially in trees that grow in light- or resource-limited environments (Muller-Landau et al. 2006b) – as well as changes in leaf traits during plant ontogeny that would influence metabolic scaling (Sack et al. 2002).

Perhaps the most significant criticism of the 1999 WBE2 model is that it makes several incorrect assumptions about plant anatomy. For example, building on earlier work (Van den Oever et al. 1981), Sperry and colleagues compiled data for the xylem conduits that transport water in plants, and documented the general inverse square "packing rule" discussed above (Fig. 14.1). This "packing rule" contradicts the assumption of the WBE model that number of conduits remains unchanged as conduit radii taper, decreasing from trunk to terminal twig. Natural selection for safety and efficiency considerations have been proposed to underlie the packing rule (Sperry et al. 2008), suggesting that a new or revised theory is needed to more accurately describe the observed scaling (Weitz et al. 2006; Price and Enquist 2007; Sperry et al. 2008; Petit and Anfodillo 2009).

14.6.1 Proposed ways to advance metabolic scaling theory

Several recent papers have proposed a way forward in order to develop a more predictive metabolic scaling theory applied to plants (Price and Enquist 2006; Price et al. 2007; Savage et al. 2010). This framework can be summarized in the diagrams in Figures 14.7 and 14.8. Figure 14.7 is a schematic that shows which assumptions are necessary for which predictions of the WBE model. Specifically, these authors proposed to expand the scope of metabolic scaling theory by: (1) deriving more realistic predictions for botanical scaling exponents based on more detailed models of plant architecture and hydrodynamics; (2) deriving the nor-

malization constants for all of the scaling relationships listed in Table 14.1 by relaxing some of the secondary assumptions of the WBE model; and (3) elaborating on the network model by incorporating additional plant traits and differences in resource availability and limitation.

There appear to be four specific points that will enable the original WBE model to address several of the above criticisms. As we discuss below (sections 14.6.1.1 to 14.6.1.4), these approaches collectively will allow for a more integrative metabolic theory of ecology able to "scale up" from traits characteristic of different environments to the ecological implications of this variation. Already, several studies have begun to address these points and thus to point the way forward to a more integrative metabolic scaling theory.

14.6.1.1 Addition of further selection drivers on plant metabolism

Recently, Savage et al. 2010 proposed a unified framework to integrate divergent views on the applicability

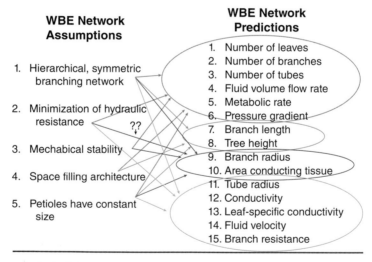

Figure 14.7 Summary of which of the secondary assumptions of the WBE network model influence which scaling predictions for several plant traits. This schematic shows which assumptions are necessary for which predictions in the WBE model. Therefore, it links how certain violations of any of the *secondary* optimizing assumptions will influence specific predicted scaling relationships. It is important to emphasize that violations of *any* of these secondary assumptions yield different values of *a* and/or *b*, and hence deviations from $\theta = 3/4$. All of these will influence plant-scaling relationships including growth and flux. As an example, for small plants, including plants early in ontogeny such as seedlings and saplings, area-preserving branching may not hold. Indeed, empirical data shows that $a \sim 1/3$ and $\theta \approx 1$. Further, in plants with unusual architectures and growth forms (such as palms, lianas, ground spreading herbs, succulents), where volume filling is absent and/or the biomechanical constraints are minimal, θ will likely deviate from the canonical 3/4 predicted by the secondary assumptions of the WBE model.

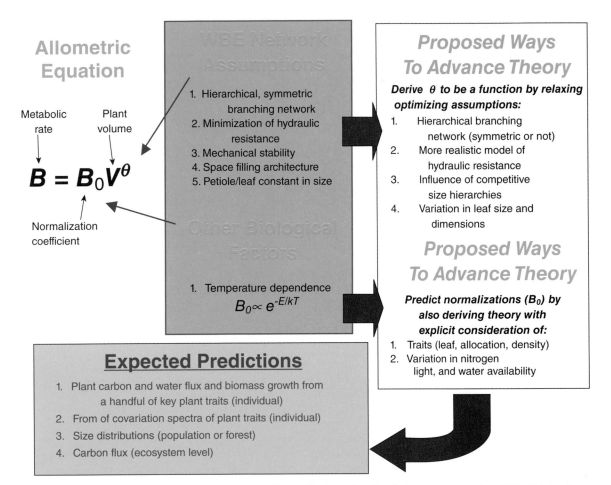

Figure 14.8 Proposed framework for incorporating additional biological and evolutionary processes into MTE. This involves relaxing the secondary assumptions of the WBE model (see Fig. 14.7). This framework will allow for a more predictive botanical scaling theory that will: (1) derive more detailed predictions for botanical scaling exponents by basing theory on more realistic models of plant architecture and hydrodynamics (Fig. 14.9); and (2) derive the scaling normalizations for all of the scaling relationships listed in Table 14.1. This is done by relaxing some assumptions of the WBE model (Fig. 14.7) as well as further elaboration of the WBE theory by incorporating key plant traits and resource acquisition. The foundations of this proposed theoretical development have recently been published (see text for details).

of the metabolic scaling theory to understanding the scaling of plant hydraulics (Fig. 14.9). In a revision to the West, Brown, and Enquist model, Savage et al. argued that the evolution of plant branching and vascular networks can be better understood to be guided by five general selection drivers (see Table 14.2). These selection drivers are hypothesized to be the central principles that have shaped the integration and scaling of botanical phenotypes: Principles #2 and #3 in Table

14.2 are more central in a more generalized botanical metabolic scaling than in the original theory (Shinozaki et al. 1964a; West et al. 1999b). Furthermore, Savage et al. elevate the importance of the principle of space filling not only to the external network but *also* to the internal vascular network, allowing one to relate conduit radius to conduit frequency (Fig. 14.1). These principles together enable metabolic scaling theory to now predict and incorporate this vascular "packing

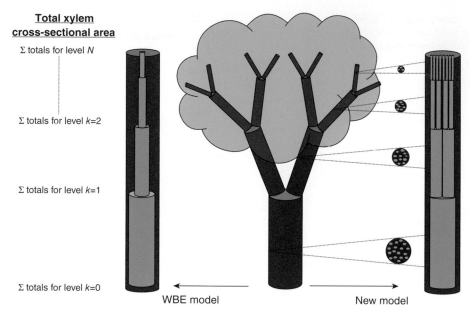

Figure 14.9 Progress in integrating criticisms of the 1999 WBE model in constructing a more flexible and realistic botanical scaling theory. Branching structures depicting the difference in internal network structure for the Savage et al. (2010) model compared with the 1999 WBE model. Trees are labeled from the base (level $k = 0$) to the terminal twigs (level $k = N$). The left and right columns represent simplified versions of the models. Both models predict conduit taper, but the Savage et al. (2010) model also allows additional section drivers on plant hydraulics so that the numbers of conduits increase and potentially fill a constant fraction of available wood area (shown to the right).

rule" (Fig. 14.1) and other vital plant properties that better match real plant networks and empirical data (Table 14.1). Table 14.1 also shows the progress in the development of plant scaling models (ranging from the pipe model, to the West, Brown, and Enquist 1999 model, and the more recent Savage et al. (2010) model) that better predicts observed plant scaling relationships (see also Fig. 14.9). One insight from the Savage et al. (2010) model is that the principle of space filling appears to apply at all scales of botanical organization – from anatomy, to canopies, to ecology – indicating that this principle is behind the patterns in Figures 14.1–14.6 and so is perhaps the most important principle shaping the range of variation in traits and sizes across scales in botany.

14.6.1.2 Incorporation of finite size effects

There appear to be two types of finite size effects that can influence plant scaling predictions. Both necessitate relaxing some of the secondary assumptions of the original theory. First, equation 14.5 is an approximation that will only hold for plants with several or

more branching generations, if not infinite, number. Savage et al. (2008) estimate that ~20 branching generations (from roots to above ground) is required for these equations to begin to approximately hold. So, the WBE and WBE2 predicted scaling relationships will change if the number of branching generations is small. So, during ontogeny, or from very small plants to large plants, the scaling exponent should actually change with N, the total number of branchings, and result in a curvilinear scaling function when viewed across all plant sizes. The curvilinearity is constant across all sizes but is subtle enough that a large range of data must be analyzed to observe it. As a result, fitting a power function to very small plants will produce deviations between the metabolic scaling predictions based on a large number ("infinite networks") or branches and data collected from networks that have just a few branches ("finite networks").

Another type of finite size effect is if the values of the branching traits a and b are not constant during ontogeny but instead change during growth. There is indication that ontogenetic changes in a and b occur within very small plants as they grow. For plants early

Table 14.2 Savage et al. (2010) proposed that the scaling of plant form and function within integrated botanical phenotypes can be better understood and predicted by a handful of general principles. Specifically, the evolution of plant branching and vascular networks has been primarily guided by five general selection drivers. Note, these build upon and detail the original WBE selective drivers but these are more specific to the way selection has shaped plant hydraulic architecture.

1	Selection for space-filling branching geometries in order to maximize carbon uptake by leaves and sap flow through xylem conduits
2	Selection to minimize the scaling of hydraulic resistance which is equivalent to maximizing the scaling of hydraulic conductance and resource supply to leaves
3	Selection to protect against embolism and associated decreases in vascular conductance
4	Selection to enforce biomechanical stability constraints uniformly across the plant branching network
5	Within plants during ontogeny and across species, terminal leaf size, physiological rates, and internal architecture are independent of increases in plant size.

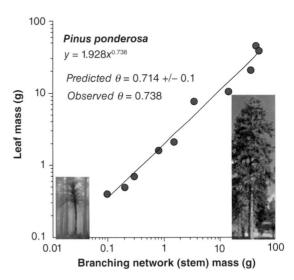

Figure 14.10 (A) Metabolic scaling theory (MST) predicts a coordinated shift in allometric exponents. Interspecific scaling for branch diameters ($2 \times r$) and lengths (l) from seedlings to trees (from Enquist et al. 2007c). As predicted, the scaling exponent changes from ≥1 for small plants and seedlings (green squares, above-ground biomass <1 g, reduced major axis (RMA) fit, $b/a = 1.8 \pm 0.12$; see text) to $b/a = 0.97 \pm 0.048$ for all the larger plants (red and brown diamonds; RMA fit not shown) to ~2/3 (RMA fit, $b/a = 0.65 \pm 0.02$) for the maximum interspecific heights achieved (brown diamonds). (B) As the scaling of branch lengths and radii changes, the scaling of total leaf biomass, M_L, as well as R and P, are then all predicted to change. Indeed, $\theta = 1.01 \pm 0.7$ ($n = 95$, $r^2 = 0.88$) for plants with a mass of <1 g and $\theta = 0.77 \pm 0.2$ for plants with a mass of >1 g ($n = 563$, $r^2 = 0.959$), which is consistent with the MST-predicted shift from $\theta = 1$ to $\theta = 3/4$ and the scaling relationships observed in Fig. 14.11A.

in ontogeny such as seedlings and saplings, gravity is relatively unimportant, so the 2/3 law is relaxed so that branch radius might then scale isometrically with branch length so $r_k \approx l_k$ and $a \approx 1/3$ rather than $a = 1/2$ (Fig. 14.10). Additionally, there are few branching levels so branches do not completely fill space and $b > 1/3$. Thus, relaxing the secondary assumptions of WBE so that a and b can vary then predicts that in the case of very small plants $\theta \approx 1.0$. However, as plants grow, gravity becomes increasingly important and volume-filling architecture develops (West et al. 1999b), so a relaxed version of the theory predicts a shift in θ from $\theta \approx 1.0$ to $\approx 3/4$ (Fig. 14.11). Indeed, empirical data (Niklas 2004) shows that $r_k \sim l_k$, leading naturally to $a \sim 1/3$ and $\theta \approx 1$, consistent with a predicted shift to isometry of R, P, and n_L. Indeed, *intra- and interspecific* scaling of total leaf mass, M_L, shows (Enquist and Niklas 2002a) a transition from $\theta \approx 1.0$ in seedlings to $\theta \approx 3/4$ in larger plants (Enquist et al. 2007b; Fig. 14.10). Further, a recent study by Mori et al. (2010) measured whole-tree respiration rate as a function of size. As expected, the scaling of metabolism was steep for small plants with an exponent close to 1.0

but then settled to a shallower exponent close to 3/4 (see Brown and Sibly, Chapter 2, Fig. 2.3; Mori et al. 2010). Further, in plants with unusual architectures and growth forms (such as palms, lianas, ground spreading herbs, succulents), where volume filling is absent and/or the biomechanical constraints are minimal, θ will likely deviate from the idealized 3/4.

14.6.1.3 Variation in branching architecture (scaling exponents)

First, as discussed above, plants exhibit a plethora of branching architectures – ranging from palm trees, to grasses and succulents, to vines and lianas, to ground spreading forbs. Each of these different growth forms clearly violates the general principles of space filling and perhaps area preserving invoked by the secondary assumptions of WBE (Price et al. 2007; Dietze et al. 2008; Koontz et al. 2009). Second, unlike the network model (see Brown and Sibly, Chapter 2, Fig. 2.2, and Fig. 14.9), many plants exhibit architectures in which the branching is asymmetric such that the daughter vessels/branches are very different sizes. Indeed, plants show a wide range of apical dominance where the main stem or branch is larger and grows preferentially over the side branches (see also Price and Enquist 2006). Strongly apically dominant trees, such as conifers, thus reflect strong branching asymmetry.

Recent work by Price et al. has shown that a relaxed version of the WBE model begins to capture a diversity of morphologies and architectures. Doing so shows that the WBE model is capable of matching observed variation in metabolic scaling exponents that range within 0.5 to 1.0. This captures the range of variation in observed scaling exponents (Price et al. 2007; see also Glazier 2010) and runs counter to the criticism that generalizations and extensions of WBE or MTE can only predict 0.75 (or quarter-power scaling). This range of values will then be reflected by shifts in branching geometry. As of yet we are unaware of any study that has started to incorporate branching asymmetries into metabolic theory (although see Turcotte et al. 1998). An exciting implication of a relaxed version of WBE (and hence MTE) is that it shows that variation in scaling exponents must be ultimately due to variation in branching geometry (see equations 14.6 and 14.7) or size-related variation in the functional traits that underlie the allometric normalization (see equations 14.8 and 14.9). Relaxing the secondary assumptions allows for a unique test of the core predictions of the WBE model (Price and Enquist 2007). Specifically, the core prediction states that if you measure the two network branching traits (a and b) then one should be able to predict exactly the scaling exponent.

As a test of the core prediction of MTE we compiled *intraspecific* data showing the allometric relationship between the mass of the above-ground branching network and leaf mass. We collected data from 10 individual *Pinus ponderosa* trees (Driscoll, Bentley, and Enquist, unpublished data). The observed reduced major axis scaling slope for total leaf mass and plant network biomass (or exponent θ) is 0.714 ± 0.2 (Fig. 14.11). We then calculated the predicted value θ based

on the branching traits, a and b, measured within the branching network of each of these trees. Our predicted scaling slope from equations 14.5 and 14.6 is 0.738, which is strikingly close to the observed exponent.

14.6.1.4   Variation in environment and traits (scaling normalizations)

Plant functional traits vary across environmental gradients (Westoby and Wright 2006). Given their importance to plant transport and construction, variation in traits must affect whole-plant growth, dM/dt (equations 14.7 and 14.8), and plant energetics. Additional studies have argued that size-dependent variation in resource availability can also influence scaling relationships. For example, light limitation should be an important component of our understanding of variation in growth rate and the scaling of whole-plant form and function (Muller-Landau et al. 2006b). Elaborations of WBE can incorporate this environmentally driven variation in traits via variation in the scaling normalizations (see equations 14.7 and 14.8; see also Kaitaniemi and Lintunen 2008). Specifically, variation in the environment will be reflected in specific trait values. Nonetheless, a still open question is: how strong is this influence and what is its form? The effect of variation in these variables will certainly shift normalization constants among taxa and environment, but it is still unclear whether leaf size and tissue stoichiometry vary systematically with plant size and thus also affect the scaling exponents.

14.7   SCALING UP TO POPULATIONS, COMMUNITIES, AND ECOSYSTEMS

The WBE model provides the basis to "scale up" from individual plants and their specific traits to populations, communities, and ecosystems. The metabolic theory of ecology or MTE extends the WBE model to ecology by invoking four additional principles or assumptions (Enquist et al. 1998). Intriguingly, these assumptions appear to have identified important organizational principles that are shared across ecological systems.

14.7.1   Resource steady state

MTE assumes that within a given plant community or population, ultimately biomass production is limited by

A

B

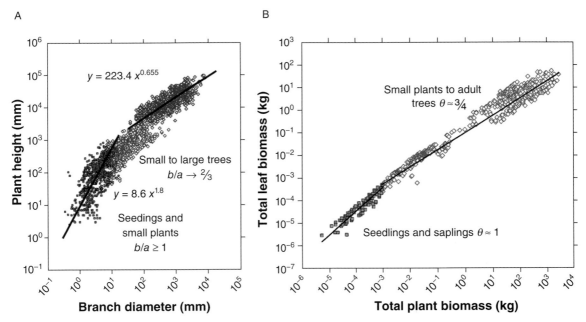

Figure 14.11 A test of the core predictions of the WBE model by measuring the branching traits (*a* and *b*) in order to predict the scaling exponent. Preliminary *intraspecific* data showing the allometric relationship between the mass of the above-ground branching network and leaf mass from 10 individual *Pinus ponderosa* trees (Driscoll and Enquist, unpublished). The scaling slope (or exponent θ) is 0.714. For each of these trees we calculated the predicted value θ based on the branching ratios measured within the branching network of each of these trees. Our predicted scaling slope from the branching ratios of the network is 0.738 which is strikingly close to the observed. Note, our theory indicates that a more detailed measure of leaf size allometry and the asymmetry of branching will provide an even more accurate prediction of θ. Further, measures of the specific traits listed in equation 14.7 will then yield the value of the *y*-intercept or normalization of this function. Thus, in principle, we can predict the entire function with no fitted parameters.

the rate of resource supply, $\dot{R}_{Tot}$ (Enquist et al. 1998). Consequently, plants tend to grow until they are limited by resources. This then leads to a trade-off between plant size and abundance. The maximum number of individuals that can be supported per unit area (N_{max}) is related to the rate of resource supply, $\dot{R}_{Tot}$, per unit area and the average rate of resource use per individual, $\dot{Q}$, so that $\dot{R}_{Tot} = N\dot{Q} \propto NM^{3/4}$, and so $N \propto (1/\dot{R}_{Tot})M^{-3/4}$. Thus, population density is predicted to scale inversely with plant size. This is the idea sometimes referred to as energy equivalence (Damuth 1981; but see also White et al. 2007).

14.7.2 Demographic steady state

Resource steady state essentially demands that the plant population or community also be in demographic

steady state. Understanding the origin and maintenance of demographic steady state requires an understanding of population transitions between size-classes as individuals grow from seedlings to the canopy. In steady state, the number of individuals within a given *k*th size-class, Δn_k, does not change with time so growth and death rates are not independent but inextricably linked (Clark 1990). Thus, as individuals grow then individuals must also die. The steady-state assumption is consistent with dynamical data and canopy gap size distribution data from several old-growth forests (Hubbell and Foster 1990; Lieberman et al. 1990; Kellner and Asner 2009).

14.7.3 Ecological space-filling

Metabolic ecology invokes the principle of ecological space-filling. Space-filling is assumed to be a necessary outcome of all plants competing for similar limiting resources and the resource steady-state assumption. Space-filling is the end result of individuals filling in space so as to use all available resource. It is assumed to imply that the total area of leaves of all individuals within any size-class k, $\Delta n_k a_k^L$, equally fills the same amount of area, across all of the size-classes. Specifically $\Delta n_k a_k^L = \Delta n_{k+1} a_{k+1}^L$ where k is a given size-class, a_k^L is the average leaf area per plant in size-class k and Δn_k is the number of plants whose age is between t and $t + \Delta t$ and whose size is between r_k and $r_k + \Delta r_k$ where Δr_k is the size of the bin used to visualize the size distribution.

14.7.4   Allometric similarity across taxa

The MTE has often made the simplifying assumption that all individuals follow the same (specifically, quarter-power) scaling relationships. For example, in applying MTE to populations and whole communities, the scaling of plant canopy and rooting dimensions as well as the scaling of growth rate and resource use is assumed to be similar across taxa. This explicitly assumes that the scaling of plant form and function (as reflected by the branching traits a and b) and the metabolic demands and traits reflected in the scaling normalization of whole-plant growth and metabolism (in equations 14.7 and 14.8) are similar within individuals and across organisms.

14.7.5   MTE predictions

As a result of the above ecological assumptions, several predictions emerge. However, as shown by the examples below, violations of any of these assumptions will lead to deviations from MTE predictions. One important prediction is the scaling of number of individuals and their size. MTE predicts that there should be an inverse relationship between size and number (West et al. 2009). Specifically, $\Delta n_k \propto 1/a_k^L \propto 1/r_k^2$. The steady-state size distribution then approximates an inverse square law with many small individuals and few large ones. This prediction appears to be largely supported by empirical data sampled across different biomes as well as following a given forest through time (Enquist and Niklas 2001; Enquist et al. 2009; see also Anderson-Teixeira and Vitousek, Chapter 9, Fig. 9.3A; see also Coomes et al. 2003; Muller-Landau et al.

2006a). It is convenient to translate the above discrete formulation into a continuum notation. The distribution function representing the number of individuals per unit trunk radius, $\Delta n_k / \Delta r_k$, becomes $f(r) \equiv dn/dr$ (see West et al. 2009 for details). When size-classes are characterized by linear binning of radii, corresponding to a constant infinitesimal Δr_k, independent of k, we now have the more general statement

$$N_k \approx \frac{\dot{R}}{(K+1)b_0} r_k^{-2} \text{ or } f(r) = \frac{dn}{dr} = \frac{\dot{R}}{r_m b_0} r^{-2} \qquad (14.9)$$

where r_m is the stem radius of the largest tree sampled. Since $r \propto m^{3/8}$, equation 14.9 is equivalent to $f(r) \propto m^{-3/4}$ reflecting the "allometrically ideal" 3/4-power scaling of metabolism and growth (see also discussion regarding transformation of size variables within discrete and continuous size distributions, in supplemental material in West et al. 2009). Note, relaxing the secondary assumptions of the WBE model (see Figs 14.7 and 14.8) would yield different scaling exponents than the canonical −2. An important implication of a metabolic theory of ecology is that the predicted scaling function of number and size explicitly predicts that the normalization of the size distribution should increase with increasing rates of limiting resource supply, $\dot{R}$ (as was recently shown by Deng et al. 2006) and decrease with increasing rates of mass-corrected metabolism, b_0. Here the value of b_0 can again be seen to be linked to fundamental plant traits detailed in equations 14.7 and 14.8 where $b_0 \approx \beta_G \approx (a_L/m_L)([c/\omega]\dot{A}_L)\beta_L$. Thus, equation 14.10 provides a basis to integrate variation in functional traits and resource supply with variation in the relationship between plant size and number. Lastly, we note that if climatic or biophysical limits constrain the maximum size of a tree so that r_m has an upper limit in equation 14.9, then this would induce another "finite-size" effect but at ecological scales. This would be expressed as a "curvy" size–frequency relationship that is "bent down" at large sizes (see truncated Pareto fit to power functions in White et al. 2008).

Another prediction from MTE applied to forests stems from the resource steady-state assumption that the mortality rate of a population, μ, or stand should scale inversely with stem radius, r, as $\mu = A r^{-2/3}$ (see Fig. 14.5). Broad support for this prediction at the population scale is reported in Marba et al. (2007). However, this prediction should also hold within diverse communities if the central assumptions of MTE hold. Further, the normalization of the mortality scaling relationship, *A*, can be shown to relate directly to the scaling of plant growth rate (see equations 14.7 and 14.8; Enquist et al. 2009).

14.7.6 Scaling up MTE to ecosystem scaling and dynamics

14.7.6.1 Disturbance and succession

A central prediction of MTE is that *even* when disturbed from steady state, the successional trajectory of a community or stand is governed by the allometry of growth and mortality but is constrained ultimately by the central assumptions of MTE. An important implication is that a forest that conspicuously violates the assumption of steady state due to a major recent disturbance should deviate substantially from theoretical predic-

tions. Over time, however, as new seedlings are recruited, grow, and fill space to reestablish a steady-state mature forest (Odum 1969), the size distribution should converge asymptotically on the canonical form, $f(r) \propto r^{-2}$ (Kerkhoff and Enquist 2007). Enquist et al. (2009) show a series of Costa Rican forests in various stages of recovery from disturbance, with steady state reached in approximately 50–100 years. Early in succession the scaling exponent is steeper than −2, while later in succession the size distribution exponent begins to approximate the resource steady-state distribution (see Anderson-Teixeira and Vitousek, Chapter 9, Fig. 9.3B). These data and previously published data on forest recovery from fire (Fig. 14.12) and other disturbances (Kerkhoff and Enquist 2007) predicted return to the inverse square law. These results support the assumption that resource steady state is a strong "attractor" that significantly guides and constrains ecological dynamics.

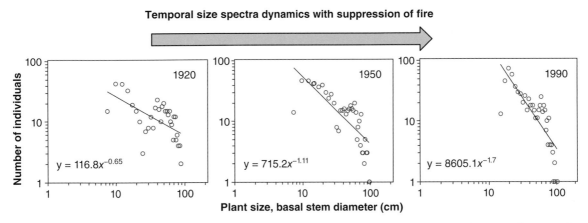

Figure 14.12 A unique prediction of MTE is that the shape of the size distribution (i.e., its exponent) can provide a general indicator of time since disturbance. Changes in the size distribution for a *Pinus ponderosa* forest in northern Arizona, from 1920 (at the dawn of active fire suppression) to 1990. With the suppression of fire, the size distribution exponent becomes steeper, approaching the energetic equivalence rule prediction of −2. Furthermore, the distribution shifts from clearly bimodal to more continuous, perhaps reflecting the relaxation of cohort dynamics that results from the typical fire return interval (data from Biondi et al. 1994). This scenario leads naturally to the hypothesis that if the processes limiting recruitment are somehow removed from the system, the forest size structure will converge toward the EER expectation. Before European settlement, this vegetation type was controlled by high-frequency, low-intensity ground fires, and stands were characterized by an open mosaic structure dominated by mature, fire-resistant trees, with little understory vegetation. With the introduction of grazing and active fire suppression in the early part of the twenty-first century, recruitment of *P. ponderosa* has been less sporadic, and the increased fuel load has led to a more catastrophic, crown-fire disturbance regime (Savage and Swetnam 1990). In accordance with our hypothesis, with the elimination of recruitment limitation by fire, the size structure of the forest appears to become steeper over time. Thus, systematic departures from scaling (both the shallow exponent and the structural size gap observed in 1920) appear to be the signature of a natural structuring process (ground fire), which in this case limits recruitment (ecological space-filling and resource steady state).

14.7.6.2   What is the size of the terrestrial carbon sink? How much organic carbon is below ground?

Recent re-evaluations of global change models used to understand how ecosystems respond to climate change as well as to calculate the amount of carbon stored in terrestrial ecosystems show that they do not sufficiently incorporate the observed plant scaling relationships (e.g., Figs 14.2 and 14.3; Wolf et al. 2011). Understanding of the role of metabolic scaling will likely greatly inform global change models. For example, in a series of papers David Robinson (2004, 2007) used allometrically informed scaling relationships between above- and below-ground biomass components (as shown in Fig. 14.3; Enquist and Niklas 2002a) to calculate how much root biomass and carbon there was in the soil. He then revised the estimates for the global root carbon pool and the results are striking. He found that, given the above-ground biomass within a site, the amount of carbon stored in plant roots could be almost 70% more than previously estimated. His work then also provides a theoretical basis to argue for a significant revision of the global root carbon pool that influences estimates of global carbon sources and sinks. Robinson predicted a global pool of at least 268 petagrams (1 Pg = 1 gigatonne), which compares with previous estimates of about 160 Pg. Based on these metabolic scaling predictions Robinson concluded that the land-based carbon sink is larger than previously thought. He estimated the sink could be 2.7 Pg per year – 0.1 Pg per year greater than current estimates, indicating a stronger role of the terrestrial environment in regulating the Earth's carbon cycle (see also Allen et al. 2005).

14.7.6.3   Ecosystem allometry

An exciting recent development in MTE is the ability to begin to predict whole-ecosystem scaling relationships. For example, given the metabolic rate of an individual, $B = b_0 r^2$, where b_0 is a normalization constant, and the size distribution function, $f(r) = c_n r^{-2}$, where c_n is another normalization constant, the total energy use of the stand, per unit area, extension of metabolic theory (see Enquist et al. 2009) shows that the total flux of energy, $\dot{B}_{Tot}$, (as well as total autotrophic net primary production) is predicted to scale non-isometrically with total stand biomass:

$$\dot{B}_{Tot} = b_0 c_n \left(5c_m / 3c_n M^{Tot}\right)^{3/5} \qquad (14.10)$$

where c_m is from the allometry of stem radius and plant biomass, m where $m = c_m r^{3/8}$. Thus, variation in whole-ecosystem autotrophic metabolism scales with phytomass, M_{tot} to the 3/5 or 0.6 power. Comparison of two datasets potting NPP and autotrophic respiration (Kerkhoff and Enquist, unpublished) bracket the predicted scaling exponent and are indeed close to the predicted value (see Fig. 14.6). Note also that the scaling normalization is given by several key plant functional traits and other scaling normalizations (see equation 14.9) and thus provides a way to link plant traits with ecosystem processes.

14.7.7   Proposed ways to advance the metabolic theory of ecology

In order for MTE to reach its potential as a quantitative foundation of plant ecology it is important to integrate several new insights and criticisms. While the above studies highlight several empirical patterns that support MTE, as we discuss below, several studies have identified some limitations and failings of MTE applied to plant ecology (Coomes 2006; Muller-Landau et al. 2006b). In order to develop a more predictive and quantitative plant ecology, it is important to cast WBE – as applied to plants – and its extension to ecology (MTE) in a proper light. We make three specific points.

First, the WBE model and the metabolic theory of ecology (MTE) is a "Fermi approach" or "zeroth-order" model (Harte 2002). It is a deliberately simplified theory in that it starts with the fewest number of assumptions and principles to make the most number of predictions. Scaling predictions from MTE builds on a rich theoretical foundation of plant ecology (Yoda et al. 1963; Shinozaki et al. 1964a) by starting from a trait-based view of the organism and the general principles of metabolism and allometry that are shared across most plants. Consequently, the theory requires no additional fitted parameter values to predict many scaling functions including: (1) the allometry of plant growth (Enquist et al. 2007c); (2) the allometry of whole-plant water flow and conductance (Savage et al. 2010; von Allmen et al., in preparation); (3) the steady-state distribution of tree sizes (Enquist et al. 2009); and (4) the scaling of mortality (Enquist et al. 2009). The above predictions run counter to arguments made by Tilman et al. (2004) who argued that MTE was sufficiently removed from more pressing questions asked by most researchers in community ecology. Predictions from MTE have strong implications for both population and community ecology because metabolism fuels growth and powerfully constrains plant form, which in turn constrains abundances, sizes, and rates of competitively driven mortality (see also Clark 1990, 1992).

Second, a zero-order framework can reveal the influence of factors in addition to metabolism and allometry, because these will appear as deviations from MTE predictions of a deliberately simplified model. So, for example, deviations from the predicted scaling function of growth or metabolism will reveal which traits are affected and hence how selection has shaped variation in growth rate in differing environments, and will also let us quantify the magnitude of effect of this variation on whole-plant performance. Similarly, deviation from the predicted mortality scaling function (Fig. 14.5) and hence the predicted −2 scaling of individuals will allow mortality to be partitioned between competitive density-dependent and non-competitive density-independent sources (see Clark (1992) and Fig. S3 in Enquist et al. (2009).

Third, metabolic theory provides a conceptual foundation, which can be fleshed out with additional idiosyncratic detail as needed to account for site- or taxon-specific variation. For example, deviations in the scaling of plant growth potentially associated with plants growing in light-limiting environments (Coomes 2006), or forests experiencing herbivory or disturbance such as fire so as to deviate from MTE predictions (Coomes et al. 2003; Muller-Landau et al. 2006a), etc., rather than providing evidence against metabolic theory, instead illustrate the value and promise of a general theory based on fundamental mechanistic features of an idealized population or community. For example, quantifying the degree of variation from metabolic theory then provides a measure of the degree to which these processes are important.

14.8 CONCLUSIONS

Metabolic scaling theory applied to plants perhaps offers some of the better examples of how the theory can integrate scaling phenomena observed at multiple biological scales. A central component of a metabolic theory is the origin of allometric relationships. The WBE model is unique in that it provides a framework for deriving the traits that not only underlie scaling exponents (most of the traditional focus of metabolic

studies) but also scaling normalizations. This work builds on the core hypothesis originated by the WBE model that branching networks are fundamentally connected to the flux of matter and energy (West et al. 1997). The geometry of branching networks then governs the scaling of energy, carbon, and water flux within branches and trees, as well as across size-classes, taxa, and whole forests (Enquist et al. 2009). These scaling relationships form a baseline for integrating how plant traits influence physiological processes and life histories. Network geometry partly controls abundance, leaf area, and ultimately ecosystem fluxes through complex feedback mechanisms. By understanding how key functional traits and network geometry relate to variation in environments, we can understand how plants interact with their environment, especially light, water, and temperature, and how networks and their traits ramify to influence ecology and how they will potentially react to future climate change.

Many recent studies are integrating metabolic theory and modifying it in order to account for much more of the rich variation in the diversity of plants and ecosystem processes that influence whole ecosystem flux. While several authors have been critical of the approach, we know of no other theoretical framework that offers the ability to predict from first principles the range of scaling functions observed as well as to mechanistically connect how variations in traits then combine to influence plant form and function as well as ecology and ecosystem dynamics. Our approach is pragmatic. It builds on the successes and failures of the original WBE and MTE approaches. By continuing to evaluate the original and secondary assumptions as well as pushing to discover the limits of the predictive ability of the theory, we are optimistic that MTE will increasingly reveal a powerful and predictive framework for plant ecology and global change biology.

ACKNOWLEDGMENTS

We thank our collaborator Van Savage for helping to influence these ideas as well as his input and encouragement on earlier drafts of this manuscript. In addition, our collaborators John Sperry and Peter Reich also influenced our thinking on several ideas expressed here. BJE and LPB were supported by a NSF Advancing Theory in Biology award. LPB was also supported by a NSF postdoctoral fellowship.

Chapter 15

MARINE INVERTEBRATES

Mary I. O'Connor and John F. Bruno

SUMMARY

1 Ocean temperature strongly controls metabolic rates of marine invertebrates. A highly conserved, nonlinear scaling of metabolic rate with temperature explains variation in metabolic rate and associated ecological and evolutionary processes over depth and latitudinal gradients.

2 Body mass and temperature explain most variation in metabolic rate with depth in accordance with the metabolic theory of ecology, although predators in sunlit surface waters appear to have evolved higher resting rates.

3 General temperature-dependence models also explain the duration of a key demographic life stage – planktonic larval development – over temperature gradients. Within a species, larval development is faster in warmer water, and the rate of acceleration is predicted by metabolic theory.

4 Comparisons of larval development rate among species reveal that effects of temperature within species are stronger than patterns among species. The among-species pattern possibly reflects adaptations to compensate for the strong effect of temperature on development.

5 The metabolic theory of ecology offers a powerful, unifying approach to explain biogeographic patterns in marine life histories and responses to climate change.

15.1 INTRODUCTION

Most animals on Earth are marine invertebrates. Across this ecologically and phylogenetically diverse group, there is clear evidence of temperature- and mass-dependence of metabolic rates. The metabolic theory of ecology (MTE; Brown and Sibly, Chapter 2) provides a conceptual framework for applying general models of mass- and temperature-dependence of metabolic rates to understanding mechanistic drivers of variation in life-history traits and metabolic rates over environmental gradients in the ocean. Though relationships between temperature, mass, and metabolic rate have been studied among species and closely related taxa for decades (Thorson 1950; Childress 1971; Emlet et al. 1987; Pearse and Lockhart 2004), the generality of the models in MTE (Brown and Sibly, Chapter 2) have inspired synthesis of scaling relationships across broader taxonomic groups and geographic regions (Hirst and Lopez-Urrutia 2006; Lopez-Urrutia et al. 2006; Seibel and Drazen 2007). These syntheses have allowed new tests of old hypotheses (Seibel and Drazen 2007), and have provided a quantitative framework for considering biological impacts of future

Metabolic Ecology: A Scaling Approach, First Edition. Edited by Richard M. Sibly, James H. Brown, Astrid Kodric-Brown.
© 2012 John Wiley & Sons, Ltd. Published 2012 by John Wiley & Sons, Ltd.

climate change (Hirst and Lopez-Urrutia 2006; Lopez-Urrutia et al. 2006; O'Connor et al. 2007; see also Anderson-Teixeira, Smith, and Ernest, Chapter 23).

This chapter provides an overview of several themes in the studies of metabolic ecology of marine invertebrates that pre-date the Metabolic Theory, and considers these themes in the context of the recent conceptual and empirical syntheses. After an outline of hypotheses for how abiotic conditions in the ocean limit metabolic rates in extreme environments, the primary focus of this chapter is how size- and temperature-scaling of metabolism in marine invertebrates has provided new insight into their ecology and evolution. Specifically, metabolic scaling theory has provided a framework that allows identification of general allometric and temperature-scaling relationships and exceptions to these patterns, and points to patterns in life-history traits and community structure that may reflect indirect effects of environmental temperature gradients. In this way, a theory of metabolic scaling in the ocean sheds new light on old problems, and points a way forward for understanding changing ocean ecosystems.

15.2 OVERVIEW OF METABOLIC THEMES IN MARINE INVERTEBRATE ECOLOGY AND EVOLUTION

Of the over one million described animal species, 95% are invertebrates, and every major animal phylum is represented in the oceans (Ruppert and Barnes 1994). The only common characteristics that all marine invertebrates share are that they live in the ocean, do not internally regulate body temperatures (are ectotherms), and their bodies lack a backbone (i.e., they do not belong to the subphylum *Chordata*). The body masses of adult marine invertebrates span 11 orders of magnitude, from the smallest metazoans – planktonic rotifers (10^{-8} g) – to the colossal squid (*Mesonychoteuthis hamiltoni*, 500 kg) (Rosa and Seibel 2010) which even outweighs the giant squid (*Architeuthis dux*). Marine invertebrates have exploited every ocean environment, including rocky shores, sunlit tropical surface waters and dark, hypoxic, cold water 5 km below sea level.

Across these environments, geographic patterns in invertebrate abundance, diversity and life-history traits have attracted the interest of scientists (Thorson 1936, 1950; Thiel 1975; Witman et al. 2004; Tittensor et al. 2010). Some of these patterns have been dubbed "rules": Thorson's rule describes the trend in numer-

ous phyla toward greater parental investment per offspring in the form of larger offspring, brooding or enhanced nutritional provisions, with increasing depth and latitude (Thorson 1950; Pearse and Lockhart 2004). At the community level, diversity declines with increasing latitude and depth, such that communities of extremely cold environments are characterized by low diversity, low abundance, slow biological rates, and large size (Witman et al. 2004; Etter et al. 2005; Rex et al. 2006). In stark contrast, warm environments are characterized by high diversity, high abundance, and a fast pace of life. Though hypotheses to explain these patterns in terms of metabolic constraints are not new, they have lacked a general framework that would unify patterns across taxa through a fundamental, quantifiable biological mechanism.

A metabolic theory of ecology that transcends system (marine, terrestrial) and taxon provides a powerful approach to understanding ecology and evolution, not only in the ocean but in all systems. Testing and developing such a theory in different cases involving unique taxa and systems allows us to explore deviations from predictions, and potentially expand the theory (Duarte 2007). Historically, marine invertebrate (and vertebrate) ecological research has emphasized the role of constraints on metabolism. Major research foci have been the roles of food limitation, oxygen limitation, and temperature limitation in driving geographic variation in life histories, size, abundance, and diversity. In the next paragraphs we outline three major hypotheses for how abiotic conditions limit metabolic rates in the oceans.

15.2.1 Resource limitation hypotheses: food limitation

The vast majority of the ocean houses consumers that feed on primary production from surface waters. Primary production in the sunlit ocean is limited by nutrient availability, and one common fate of primary production is to sink to deep, dark water where it may be consumed by deep-sea animals. In polar environments where winter months are dark and cold, food chains are sustained by primary productivity from the preceding growing season. Many adaptations toward slower rates of growth and secondary production have been interpreted as consequences of long periods without adequate supply of food (Childress 1971; Clarke 1983). Certainly food limitation is an important ecological constraint, and the biomass of higher trophic levels is ultimately constrained by primary production. The hypothesis that food limitation can act as a selective force for lower metabolic rates has received much attention, but has been rejected in several recent studies (Cowles et al. 1991; Pearse and Lockhart 2004; Seibel and Drazen 2007).

15.2.2   Resource limitation hypotheses: oxygen limitation

Though marine animals live in water, they still require oxygen for respiration. In midwater depths and in the deep sea, waters are hypoxic, meaning very low in concentration of dissolved oxygen. It has been suggested that low metabolic rate (oxygen consumption rate) has evolved to facilitate persistence in low-oxygen environments (Childress and Seibel 1998). Though some adaptations to low-oxygen habitats have been shown, including elevated gill surface areas, high ventilation volumes, and respiratory proteins with a high affinity for oxygen (Pauly 2007), it is not clear that low metabolic rate is an adaptation to low-oxygen concentration zones (Childress and Seibel 1998).

15.2.3   Metabolic constraints of cold temperatures

Declines in metabolic rate as temperature decreases ("scaling of metabolic rate with temperature") are well appreciated in marine ecological studies. Typically, this effect is approximated by a Q_{10} value of 2–3, and is equivalent to an activation energy of approximately $E = 0.5–1.0\,eV$. Early research suggested that metabolic rates in cold water were acclimated such that they occurred at rates comparable to those observed in warm water (Krogh 1916; Pearse and Lockhart 2004). This type of acclimation is termed "metabolic cold adaptation," and has been postulated to occur in other groups including insects (Addo-Bediako et al. 2002). More recently, it has been shown and argued that metabolic cold adaptation does not occur among marine invertebrates, and at cold temperatures metabolic processes are generally very slow, as would be predicted based on fundamental constraints imposed by temperature (Clarke 1983; Pearse and Lockhart 2004). Therefore, the prevailing effect of temperature on biogeographic patterns in metabolic rates in marine invertebrates is one of scaling, and tests of the Arrhenius

relationship between temperature and rates have supported the use of the Gillooly et al. (2001) MTE model for broad groups of marine invertebrates (Hirst and Lopez-Urrutia 2006; Lopez-Urrutia et al. 2006; O'Connor et al. 2007; Lopez-Urrutia 2008).

15.2.4   An opportunity for synthesis

Each of these hypotheses reflects a constraint of the environment on fundamental metabolic processes. Yet, so far, they have not been integrated into a single framework. Such a framework would relate the effects of temperature on oxygen consumption rates and resource use rates, potentially explaining patterns that appear to be exceptions to any single limitation hypothesis. The remainder of this chapter explores how metabolic theory can be used to identify unusual conditions that point to selection for higher metabolic rates when fast movement is required to catch prey, as well as selection of life-history traits that can compensate for extreme effects of temperature in cold climates. Selection for unusually high metabolic rates for a particular temperature or size effectively integrates into a gene pool the results of ecological and evolutionary processes that produce the diversity of life in the ocean. Finally, general, direct and indirect effects of temperature on metabolism may explain community-level patterns in the ocean corresponding to temperature gradients. In this way, metabolic ecology of marine invertebrates has already made unique contributions to a broader metabolic theory (Duarte 2007), and much more work remains to be done.

15.3   PATTERNS IN METABOLIC RATE WITH DEPTH, AND SELECTION FOR HIGH RATES IN SURFACE WATERS

The marine habitat extends from sunlit waters in the upper 100–200 m of the ocean to dark, cold (−1 to −4 °C) waters over 4000 m deep. Environmental conditions change dramatically over the uppermost few hundred meters, and below that pressure is high, and light and temperature are very low and decline gradually with depth. It has been suggested that environmental differences between warm, light surface water and the cold dark water below 1000 m severely constrain metabolic processes and thus limit the performance, abundance, and body size of deep-sea animals

(Thiel 1975; Rex et al. 2006). Ideas about how this gradient should impact ecology and evolution via constraints on metabolic processes preceded sufficient data to test whether and how metabolic rates actually vary with depth.

One commonly cited pattern is a reduction in metabolic rate with depth. Recently, Seibel and Drazen (2007) reviewed metabolic rates of marine animals across a range of depths to explore potential explanations for this pattern. Most of their samples were measured at 5 °C, and those that were not were standardized to 5 °C using a $Q_{10} = 2$ ($E = {\sim}0.5\,$eV in the Arrhenius equation; Gilooly et al. 2001). Controlling for temperature in this way, metabolic rate still declined with increasing depth (Fig. 15.1A). The strong effect of temperature on metabolic rate means that animals living in cold (<5 °C) polar or deep water use oxygen at rates up to 27-fold lower than some animals living in surface waters in the tropics (Gilooly et al. 2001; Seibel and Drazen 2007), and may explain how some can persist in low-oxygen environments.

For marine invertebrates, mass-specific metabolic rate varies with body size, and the slope of this relationship across all taxa was −0.22 in Seibel and Drazen's (2007) dataset, consistent with quarter-power scaling. For any given body size, there is substantial variation among taxa in metabolic rate. Despite this variability, comparison of nested, general linear models that allowed for variation in slopes, intercepts, or both among taxonomic orders shows that the same slope (−0.20) applies to all orders (Fig. 15.1B). When body size variation was accounted for in the full dataset, the trend in metabolic rate with depth was eliminated (Fig. 15.1C), possibly reflecting a trend toward smaller body sizes in deeper water (Fig. 15.1D) (Rex et al. 2006). For most groups in the study, body size and temperature explain observed declines in metabolic rate with depth among marine invertebrates.

In some groups a depth gradient in metabolic rate persists even after controlling for size and temperature (Fig. 15.1E). Some cephalopods and crustaceans (and fish) living near the ocean surface have much higher metabolic rates than expected based on their size and temperature (Seibel and Drazen 2007) (Fig. 15.1F). It is possible that an evolutionary "arms race" (or in this case, a speed race) has occurred in which selection favors speedier prey to escape predation, and speedier predators to catch prey in warm, clear water (Seibel and Drazen 2007). The visual predation hypothesis formalizes the idea that selection has favored high metabolic rates in groups of cephalopods, fish, and crustaceans that use image-forming eyes to identify and locate prey in clear, refuge-free waters (Childress 1995; Seibel and Drazen 2007). When Seibel and Drazen (2007) analyzed their data and partitioned taxa based on whether they were visual predators, the visual predators had elevated metabolic rates above what would be indicated by body-size and temperature-based expectations, and these rates did not decline with body size as steeply as in other groups (Fig. 15.1F).

In sum, across a broad thermal gradient, metabolic rate in marine invertebrates scales with temperature across species. Most deviations from this temperature-scaling expectation occur not at low temperatures, where it had been hypothesized that rates should be higher than predicted due to thermal acclimation and adaptation, but at *high* temperatures in sunlit surface waters. The work of Childress, Seibel, and their colleagues suggests that selection favors high metabolic rates in certain groups in light, warm water. Such adaptation can cause large deviations from predictions based on simple scaling theory. But for the majority of marine invertebrates and fish in the dataset, allometric and temperature-scaling relationships are sufficiently informative to explain variation in metabolic rate over environmental gradients spanning 4000 m from the surface of the ocean to the deep sea.

15.4 TEMPERATURE DEPENDENCE OF DEVELOPMENT, AND RELATED PHENOTYPIC COMPENSATION

For most marine invertebrates, the distinct life stages of egg and larval development are characterized by very small size, high risk of mortality, and passive dispersal away from adult habitats in a moving pelagic environment (Levin and Bridges 1995; Kinlan and Gaines 2003). Often, the larval period is the only opportunity for long-distance dispersal among populations and habitats. Survival and dispersal during this period contribute to spatial patterns in demography, genetic connectivity, and biogeography of marine invertebrates (Shanks et al. 2003; Siegel et al. 2003; Kinlan et al. 2005), and is critical to assessments of the potential effectiveness of conservation measures such as marine protected areas (Hart 1995; Shanks et al. 2003; Lester and Ruttenberg 2005; Laurel and Bradbury 2006).

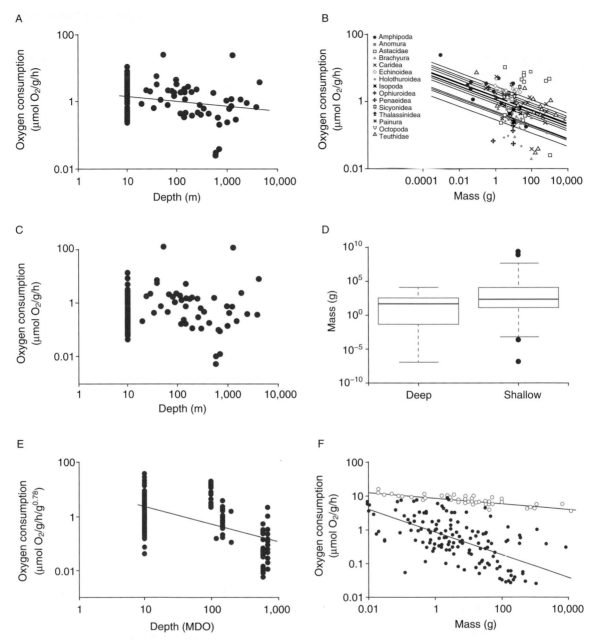

Figure 15.1 Across marine invertebrate species ($n = 145$), resting metabolic rate varies with body size and temperature (data replotted and reanalyzed from Seibel and Drazen 2007). (A) Temperature-corrected oxygen consumption rate declines with depth (minimum depth of occurrence) across a broad taxonomic group of in marine invertebrates (slope = −0.13, $p = 0.03$, $r^2 = 0.05$). (B) Temperature-corrected, mass-specific oxygen consumption rate declines with body mass within 14 taxonomic orders of marine invertebrates (slope = −0.31 within taxonomic order, and intercepts vary among orders). (C) After correcting for body size, oxygen consumption rate does not decline with depth across entire dataset. (D) The sample of deep-sea animals in this dataset suggests smaller size in deep water ($p = 0.01$). (E) Even after controlling for effects of size and temperature, metabolic rate declines with depth (minimum depth of occurrence) among cephalopods ($p < 0.001$). (F) Temperature-corrected, mass specific metabolic rate declines with increasing size more strongly for cephalopods that are not visual predators (filled symbols, slope = −0.33, $p < 0.001$) than for visual predators in surface waters (open symbols, slope = −0.08, $p < 0.001$).

15.4.1 Temperature dependence of a critical demographic life stage links metabolic ecology at the individual level to community-level processes

The egg and larval development periods are highly sensitive to temperature. For offspring collected from a single population, development occurs faster and the larval period is shorter in warmer water (O'Connor et al. 2007). The near exponential effect of temperature within a non-stressful range is well known, but a comparison of the scaling effect of temperature across a diverse set of species that spanned six phyla showed that the temperature dependence of larval duration is highly consistent with predictions from MTE (Hirst and Lopez-Urrutia 2006; O'Connor et al. 2007) (Fig. 15.2).

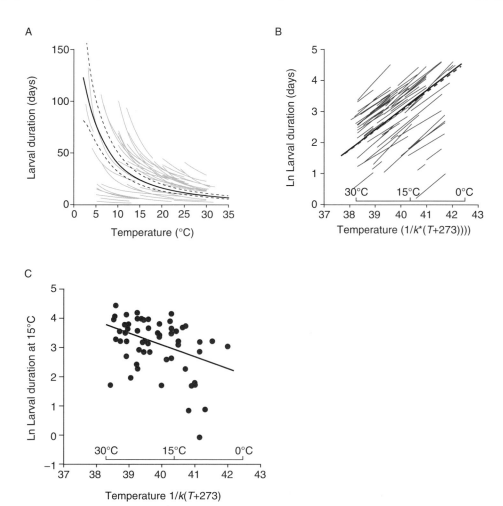

Figure 15.2 Planktonic larval duration (the inverse of development rate) is strongly and consistently affected by temperature. (A) Blue lines represent curves fit to data on individual species, and the black line is the population averaged trajectory and 95% confidence intervals (O'Connor et al. 2007). (B) Data in A replotted on Arrhenius axes (warmer temperatures to the left) and statistical model (black line) compared with MTE prediction of a slope of 0.62 eV ($p < 0.001$, red dashed line). At a single temperature, there is still variation among species (e.g., variation among blue lines at 15 °C in (A). (C) When larval duration is normalized to 15 °C for each species, there is a much weaker relationship between temperature and the normalized larval duration (slope = −0.40, $p < 0.001$). Points in (C) reflect larval duration at 15 °C for each species in (A). All data from O'Connor et al. (2007).

The generality of the effect of warming on larval duration implies potentially predictable effects of temperature on processes that scale with larval duration, including adaptation, dispersal distance, survival, and population connectivity for benthic marine invertebrates and fish (Hirst and Lopez-Urrutia 2006; Duarte 2007; O'Connor et al. 2007; Munday et al. 2009; Cheung et al. 2010).

Evolution and connectivity depend on the success of dispersing larvae, and their rates of survival and settlement. Larval settlement success depends on the probability that a larva released from a certain point (e.g., a benthic parent population) will settle at any other point. Settlement probability depends on dispersal distance and can be represented as a dispersal kernel (Fig. 15.3A). Predictable variation in dispersal

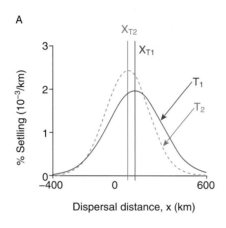

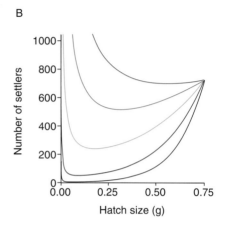

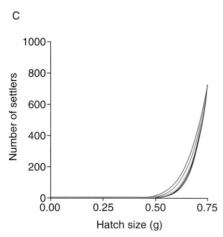

Figure 15.3 (A) A dispersal kernel representing the probability of larvae settling along a coastline (x-axis). Larvae were released from a parental population at 0 on the x-axis, and this model assumes directional alongshore currents with cross-shore transport and random diffusion (Siegel et al. 2003). Kernels are compared for two temperatures (a cool temperature T_1 in blue, and a warmer temperature $T_2 = T_1 + 2\,°C$, in red). The only difference between the kernels is the effect of temperature on planktonic larval dispersal, shown in Figure 2A. (B) Results of a temperature-dependent optimal hatch size model that assumes a trade-off between offspring size and number, and predicts optimal size in terms of the number of settlers (Kiflawi 2006). At warmer temperatures (red line), producing larvae that hatch at smaller sizes will produce the highest number of settlers. At cooler temperatures (blue and black lines), producing larger hatchlings results in the most settlers. In this model, development scales with size as $b = -0.25$, while mortality is independent of temperature and size. (C) When mortality is negatively size-dependent and positively temperature-dependent, the effect of temperature on number of settlers is eliminated in the model, and large larvae result in the most settlers at all temperatures.

with temperature might translate to quantifiable change in the probability of settlement at a particular site. In the case of a coastline with directional flow, cross-shore transport and eddies, the duration of larval development in the plankton influences the mean dispersal distance, the variance about the mean, and the survival of larvae (Fig. 15.3) (Siegel et al. 2003). By changing the development period, a small change in temperature alone could shift the dispersal kernel, therefore affecting both short- and long-distance dispersal and the associated demographic processes of population connectivity and gene flow (Kinlan et al. 2005; O'Connor et al. 2007; Munday et al. 2009). If unmodified by ecological or evolutionary processes (see section 15.4.2), temperature scaling of planktonic larval duration could lead to broader dispersal kernels in colder climates, or narrower kernels with ocean warming (Fig. 15.3A). Greater population connectivity and reduced community dispersal limitation at higher latitudes are also consistent with broader dispersal kernels in colder regions (Witman et al. 2004; Etter et al. 2005).

15.4.2 The effect of temperature on metabolic processes is generally stronger at colder temperatures

Stronger effects of a change in temperature at colder temperatures (Fig. 15.2A) suggest geographic variation in the ecological effects of temperature. At warmer temperatures (above 15–20 °C), the temperature constraint on planktonic larval duration effectively disappears for many taxa, because the nonlinear temperature-dependence relationship is nearly flat (Fig. 15.2A, temperature above approximately 25 °C).

15.4.3 The general pattern of within-species scaling of development time does not imply a pattern among species

Most tests of metabolic theory have examined patterns across species, where one datum in the analysis represents the rate and body size or temperature of an entire species (e.g., Gillooly et al. 2001; Savage et al. 2004a; Hirst and Lopez-Urrutia 2006). In contrast, the general pattern of temperature scaling of planktonic larval development is seen most clearly within species, and this effect is shared among species (Fig. 15.2A,B).

When examined among species, however, the slope of the relationship between planktonic development period and temperature is much shallower (Fig. 15.2C). Hirst and Lopez-Urrutia (2006) found even shallower slopes of $E = -0.074$ to -0.127 for planktonic egg development times.

One explanation for the absence of a strong trend among species is evolutionary compensation. The slow development rates at extremely cold temperatures would require such long larval durations that cumulative mortality could be unsustainable (Hirst and Lopez-Urrutia 2006; O'Connor et al. 2007), and could select for traits that would reduce the *planktonic* larval duration (Emlet et al. 1987). Direct reduction of development rate is constrained by extremely limited options for the fundamental biochemical components of metabolism (Gillooly et al. 2002). Instead, shorter development times can be achieved by life-history strategies that produce larvae provisioned with greater nutritional stores, or by altering traits such as size at hatch relative to size at settlement or metamorphosis, or by protecting offspring (Havenhand 1993; Charnov and Gillooly 2004; Hirst and Lopez-Urrutia 2006). Such traits are consistent with increased parental investment per offspring.

The hypothesis that selection should favor increased parental investment per offspring in very cold climates, and that this could eliminate a temperature-driven latitudinal trend in absolute development time, rests on the premise that there is a trade-off between offspring size and number if the total allocation to reproduction is not increased; larger eggs or larvae require more nutritional resources and therefore come at a predictable cost to the parent in terms of number of offspring (fecundity) (Lack 1954; Smith and Fretwell 1974; Emlet et al. 1987). In many taxa, increased offspring mortality shifts the balance of this trade-off toward increased parental investment and lower fecundity (Vance 1973; Kolding and Fenchel 1981). Therefore, producing fewer, larger offspring may be favorable at cold temperatures (Pearse et al. 1991; Yampolsky and Scheiner 1996). Indeed, for numerous marine invertebrate taxa, parental investment per offspring increases with latitude. Eggs and larvae in cold water tend to be larger, better provisioned, or partially or completely brooded (Thorson 1950; Mileikovsky 1971; Pearse et al. 1991; Clarke 1992; Laptikhovsky 2006), all minimizing the planktonic period.

The hypothesis that compensation for mortality during the planktonic stage selects for traits that minimize larval duration demands a reason for why selection would not favor such traits at all temperatures. Two factors suggest that the fitness gain of increased investment per offspring at the expense of offspring number can be beneficial at cold temperatures and not at warm temperatures. First, the nonlinear effect of temperature results in a weak effect of a change in temperature on development in warm climates, but a strong effect in cold climates (Fig. 15.2A). In warm ocean regions, temperature only weakly constrains development rate or related traits reflecting parental investment. Second, through an analysis of an optimal size model, Kiflawi (2006) showed that when planktonic larval duration scales with temperature according to MTE, the optimal larval size is large at cold temperatures and small at warm temperatures (Fig. 15.3B). Therefore, if the temperature constraint were the most important factor determining larval hatch size, in cold water each parent should release a few, large offspring that reach their settlement size quickly and spend a short time in the plankton, while at warm temperatures numerous, small hatching larvae would be more common because development times are faster (Fig. 15.3B). This model suggests that the negligible fitness benefit of increased larval hatch size to shorten the larval duration is likely not worth the cost of increased investment in warmer climates, and small offspring sizes should be most common at all temperatures except the coldest temperatures (Fig. 15.3B). Latitudinal patterns in larval and egg size within several closely related taxa are consistent with this model (Thorson 1950; Mileikovsky 1971; Pearse et al. 1991; Clarke 1992; Laptikhovsky 2006).

Kiflawi (2006) also showed that when larval mortality rates are size- and temperature-dependent, the effect of temperature on optimal hatch size is nullified (Fig. 15.3C). So, stronger predation in warmer water (Seibel and Drazen 2007) could impose a much stronger selection on larval traits than temperature constraints on development time. At warm temperatures when development is fast, larger larvae may be more likely to escape predation, and survival conferred by this size escape may outweigh the fecundity cost of increased parental investment. This prediction is consistent with comprehensive reviews of invertebrate egg size distributions (Levitan 2000; Collin 2003). Thus, a life-history model that incorporates metabolic theory predicts that the ecological factors influencing selection on larval size shift from temperature in high latitudes to predation or food availability in warmer environments (i.e., lower latitudes).

In summary, comparison of within- and among-species trends in how temperature affects development time illustrates that applying the metabolic theory of ecology to biogeographic patterns is not necessarily a straightforward endeavor. Similar findings have emerged from tests of latitudinal variation in primary production in terrestrial plant communities (Kerkhoff et al. 2005; Enquist et al. 2007a), where it appears that adaptation and acclimatization of plant physiological processes compensate for shorter growing seasons and lower temperatures to elevate annual primary production and eliminate any trend toward lower primary production in colder climates. For marine invertebrates, strong temperature-dependence of planktonic larval duration implies predictable effects of changes in temperature, on ecological timescales, on processes that scale with development. Over evolutionary timescales, adaptation of traits related to dispersal and survival during the planktonic period may compensate for the underlying metabolic scaling of development rate. Metabolic theory therefore provides a conceptual framework for uniting short- and long-term metabolic effects of temperature on population and community ecology.

15.5   MOVING FORWARD: A NEW GENERATION OF MARINE METABOLIC ECOLOGY IN A TIME OF GLOBAL CHANGE

Metabolic theory can inform population- and community-level patterns when metabolism is tightly related to a demographically critical rate, such as a larval development rate. The strategy of identifying critical rates and exploring their temperature dependence has been applied to herbivory as well (O'Connor 2009). A stronger general temperature dependence of herbivores (zooplankton and heterotrophic protists) relative to primary producers (phytoplankton) drives shifts in the structure and carbon cycling of pelagic planktonic food webs (Lopez-Urrutia et al. 2006; O'Connor et al. 2009). A similar approach in other systems could quite likely provide additional insights into how temperature affects complex ecological processes and the resulting patterns.

Scaling of metabolic rate with temperature implies predictable responses to ocean change associated with climate change. But to understand impacts of climate change, metabolic scaling theory will need to be integrated with another major theme in marine metabolic ecology – the ecology of physiological stress. Differences among interacting species in their tolerance of temperature change can strongly influence species distribution, interactions, and food webs (Harley 2003; Gilman et al. 2006; Gooding et al. 2009; Kordas et al. 2011). Through a combined approach, general effects of scaling with temperature can be related to species-dependent responses to temperature change, such as range shifts (e.g., Cheung et al. 2010).

As environmental temperature regimes change with global climate change, we must marshal our best ecological and evolutionary theory to anticipate how eco-logical systems will respond. The scaling models of the metabolic theory of ecology explain substantial variation within and among ecological communities with variation in temperature. Patterns not directly explained by metabolic scaling with temperature, such as trends in metabolic rate with pelagic cephalopods or offspring size in marine larvae, might be explained by integrating evolutionary theory with scaling theory. Such a united theory would provide a powerful approach to understanding the ecological consequences of abiotic environmental change for populations and communities.

Chapter 16

INSECT METABOLIC RATES

James S. Waters and Jon F. Harrison

SUMMARY

1 Insect metabolic rates are highly variable and are affected by environmental, behavioral, developmental, and evolutionary factors.

2 The effects of temperature on insect metabolic rates depend on their behavior, life-history stage, morphology, and size. In many cases, inactive insect metabolic rates increase with temperature in a manner consistent with the assumptions of the metabolic theory of ecology (MTE), but exceptions include insects that are flying, endothermic, or behaviorally thermoregulating. In these cases, metabolic rates may remain constant or decrease with increasing temperature.

3 Insect metabolic rates are not generally constrained by oxygen supply limitation.

4 The metabolic rates for behaviorally active insects may be elevated up to 30 times greater than their standard resting metabolic rates, an aerobic scope greater than the comparable range found among similarly sized vertebrates.

5 Nutritional state can have dramatic influences on insect metabolic rates, ranging from extreme diapause in response to starvation to nearly 10-fold increases in metabolic rate following feeding.

6 Metabolic rate correlates with insect body size both intra- and interspecifically. Individual insects as well as eusocial insect colonies share common hypometric scaling exponents, but there is extensive variation in the metabolic elevation (i.e., scaling intercept or normalization constant) of these allometric relationships. While some of this variation may be related to methodology and behavioral variation, it is likely that these patterns reflect previously unrecognized evolutionary differences in physiology and life history.

7 Extensions of MTE should include more physiological, behavioral, and evolutionary mechanisms. Future developments of MTE have great potential to identify a number of areas in which further research is highly needed including the evolution of insect endothermy, body size, eusociality, and metabolic symmorphosis.

16.1 INTRODUCTION

16.1.1 Insect physiological diversity

The insects are among the most species-rich, morphologically diverse, and physiologically complex groups of organisms on the planet. The number of documented insect species is between one and four million and some ecologists estimate that there are potentially as many as seven million species alive today (Wilson 1985; Gaston 1991). Insects have evolved adaptations that allow them to occupy terrestrial, aquatic, and

Metabolic Ecology: A Scaling Approach, First Edition. Edited by Richard M. Sibly, James H. Brown, Astrid Kodric-Brown.
© 2012 John Wiley & Sons, Ltd. Published 2012 by John Wiley & Sons, Ltd.

aerial ecosystems and environments that vary in temperature, humidity, salinity, oxygenation, and resource abundance.

Insects exhibit an impressive range of sizes over ontogenetic development, within and between species. One of the smallest adult insects is the 20 μg whitefly (Hemiptera: Aleyrodidae). Seven orders of magnitude larger, the Goliath beetle (Coleoptera: Scarabaeidae) is one of the most massive individual insects at ~50 g, larger than many birds and mammals. Among the shortest adult insects is a 0.1 mm springtail collembolan (Minelli et al. 2010). The longest may be the stick insect (Phasmatodea: Phasmatidae), which stretches over 0.5 m. Fossil records from the Paleozoic include giants such as the griffenfly (Protodonata: Meganeuridae) that had wingspans as long as 0.71 m (Grimaldi and Engle 2005). The sizes of eusocial insect superorganisms can be much larger. An average honeybee colony may weigh more than 10 kg and one single colony of ants may stretch over many square kilometers (Giraud et al. 2002). In addition to exhibiting broad variation in size, insects are among the most ecologically dominant taxa, filling crucial roles in ecosystem functioning including pollination, seed dispersal, and nutrient cycling (Fittkau and Klinge 1973; Janzen 1987). Recognizing the ecological and physiological diversity among the insects presents a great opportunity to advance the development of a comprehensive and mechanistic theory of metabolic ecology.

Insect metabolism is primarily aerobic and is fueled by catabolic substrates transported by an open circulatory system, oxidized within cells by oxygen that is directly transported from the environment in the gas phase to metabolizing tissues by a system of branching and interconnected air-filled tracheal conduits (Fig. 16.1). Although the transport capacity of the insect tracheal system was once thought to be limited by the passive mechanics of diffusive flux through stationary tubes, this is now known to be an antiquated paradigm (Chown and Nicolson 2004; Socha et al. 2010). An impressive number of active mechanisms achieve convection through tracheal systems, including convective pumping of air sacs by ventilatory muscles of the abdomen (Miller 1966; Socha et al. 2008), convection associated with thoracic volume changes during flight (Weis-Fogh 1967; Wasserthal 2001), ventilation associated with hemolymph transfer between compartments (Wasserthal 1996), and "suction ventilation" associated with the reduced tracheal pressures that occur when spiracles are closed (Miller 1981; Lighton

et al. 1993; Hetz and Bradley 2005). Furthermore, the geometry of the tracheal system is sensitive to environmental conditions and exhibits both phenotypic plasticity and evolutionary responses to compensate for changing oxygen availability (Harrison et al. 2006b; Klok and Harrison 2009).

16.1.2 Measuring insect metabolic rates

Analyses of metabolic rate patterns in physiology and ecology rely on standardized conditions for measurement. In the field of mammalian biology, basal metabolic rate is relatively well defined as the metabolic rate of resting, non-digesting animals within their thermoneutral zone, the temperature range in which metabolic rate is constant (Hulbert and Else 2004). The field of insect physiological ecology does not have a thoroughly applied or well-defined set of criteria for standardizing metabolic rate measurements. To a large extent, this is not the result of researcher negligence but rather a consequence of the broad diversity of insect behaviors and physiology.

Defining criteria for standard metabolic rate is challenging in insects due to both behavioral and physiological issues. On the behavioral side, it can be difficult to get many insects to stop moving long enough to obtain stable metabolic measurements. For example, ants or bees removed from their colonies will often search ceaselessly for a way to rejoin their colonies. While it is possible to use movement sensors and chambers with short time constants to eliminate trials or time periods with locomotion, this approach can be challenging and is impossible for some species (Vogt and Appel 1999). Decapitation eliminates most insect locomotory movements, and some insects will continue to metabolize and exhibit regular discontinuous gas exchange cycles (DGC) following decapitation (Lighton et al. 1993); however, this terminal approach is not suitable for many studies and may cause other stresses that affect metabolic rate. A variety of studies have used respiratory patterns (exhibition of discontinuous gas exchange) as a way to determine when insects are in a "resting" state (Davis et al. 2000; Klok and Chown 2005; Lachenicht et al. 2010). Lower metabolic rates do increase the likelihood of discontinuous gas exchange (Contreras and Bradley 2009), but some insects can be quite active while exhibiting DGC and some simply never show such cycles, so this cannot be used as a uniform criterion for all insects. It

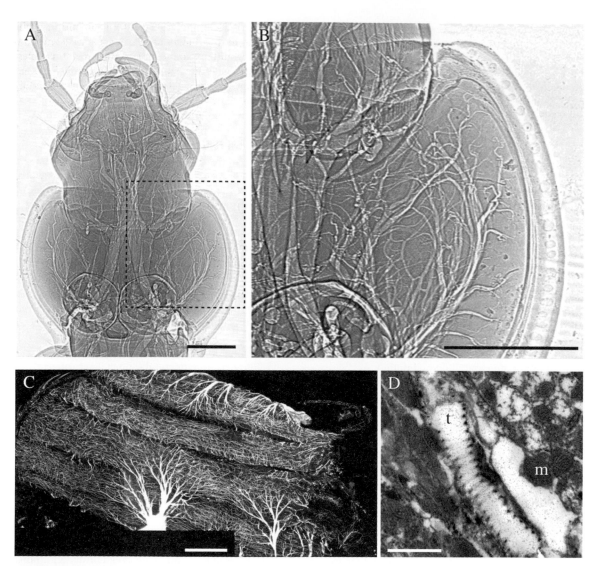

Figure 16.1 Insect tracheal systems provide the primary pathway for transporting oxygen from the environment to all of the metabolically active tissues within the body. (A) Synchrotron x-ray phase contrast image (Socha et al. 2007) of the head and thorax of the beetle, *Pterostichus stygicus*; scale bar: 1 mm. (B) Magnified view of the thorax from the region enclosed by the dotted lines in (A); scale bar: 1 mm (Socha et al. 2007). (C) Confocal microscopy image of the autofluorescent tracheae and tracheoles within the thoracic longitudinal flight muscle of a *Drosophila melanogaster* male; scale bar: 200 μm. (D) Transmission electron microscopy image of a single taenidia-reinforced tracheole (t) positioned near mitochondria (m) within the flight muscle of *Drosophila*; scale bar: 1 μm. Data for (C) and (D) were collected by the authors at the Bioimaging Facility in the School of Life Sciences at Arizona State University.

is often challenging to determine whether insects are in a post-absorptive state, as is commonly done for vertebrates, again because of the great diversity among insects. Some insects tolerate starvation very well, while in other species (e.g., honeybees), high metabolic rates lead to rapid utilization of nutrient stores and death after only a few hours of starvation at a temperature of 20 °C. The lack of a uniform definition for conditions for measurement of insect metabolic rates has two important implications. First, meta-analyses that compile data from various studies need to carefully consider such problems. Second, investigators should monitor and report behavior and time since feeding during all metabolic measurements of insects. Although variability of this sort might be expected to only add noise to analyses of metabolic rate allometry, it may also contribute bias; for example, smaller animals might be more likely to more rapidly exhaust metabolic reserves during a set period of starvation, and respiratory patterns can be size-dependent (Lighton 1991; Lighton and Berrigan 1995; Davis et al. 1999).

Another source of confusion can be terminology. Here we define isometric scaling as following the standard predictions of Euclidian geometry, with volumes scaling with mass^1, surface areas with mass$^{0.67}$, and linear dimensions (e.g., leg length) with mass$^{0.33}$. Despite these scaling exponents ranging from 0.33 to 1.0, they all represent isometric scaling. Hypermetric scaling refers to allometric patterns in which the dependent variable exhibits a significantly higher rate of change than predicted by isometry (e.g., leg length scaling with mass$^{0.5}$). Hypometric scaling refers to a significantly lower relationship than predicted by isometry (e.g., leg length scaling with mass$^{0.2}$). Since the vast majority of these patterns are nonlinear, and since the sign (positive or negative) of the scaling relationship does not by itself indicate the nature of the allometry, we have chosen to use the hypometric/hypermetric language to consistently classify the deviation of allometric relationships from the predictions of isometry.

16.2 ENVIRONMENTAL AND BEHAVIORAL EFFECTS ON INSECT METABOLIC RATES

16.2.1 Temperature

One of the primary environmental influences on insect metabolic rates is temperature. The effect of temperature, however, is highly complex and depends on behavior, life-history stage, morphology, and size. Most insects are poikilothermic ectotherms, meaning that their body temperatures vary and that the source of that variation is environmental. Nonetheless, many insects utilize behavioral thermoregulation to achieve relatively constant body temperatures over large parts of the day (Forsman 2000; Ruf and Fiedler 2002). A few insects are endothermic, often demonstrating

considerable capacity for regulation of body temperatures using heat generated by the flight muscles (Heinrich 1992). Some social bee colonies that generate their own heat and a stable core colony temperature exhibit features consistent with homeothermic endothermy (Heinrich 1981; Southwick 1985). The ability of many insects to uncouple body temperature from air temperature contributes to some of the variation in how insect metabolic rate responds to air temperature (Fig. 16.2), an important factor to consider when extrapolating from climatic models to predicted insect energetics.

The temperature dependence of metabolic rates has been analyzed with two main approaches. The MTE proposes an Arrhenius expression with a single activation energy that hypothesizes a broadly applicable, exponential effect of temperature on rate processes (Gillooly et al. 2001, 2006a) in ectothermic poikilotherms (see Brown and Sibly, Chapter 2). The classic physiological approach focuses on measuring an organism's Q_{10}, defined as the factorial change in metabolic rate for a 10-degree temperature difference (Lighton 2008). In many cases, Q_{10} is not constant, but varies depending on the specific range of temperatures being modeled (Lighton 1989; Nielsen et al. 1999; Downs et al. 2008). The intraspecific variation in Q_{10} and the interspecific variation in MTE-modeled activation energy may be due to potential behavioral, acclimatory, and evolutionary effects that cause deviations in thermal response patterns away from simple exponential models (Chown et al. 2003; Nespolo et al. 2003; Clarke 2006; O'Connor et al. 2007).

In many cases, insect metabolic rates increase with temperature in a manner approximately consistent with the assumptions of the MTE (Fig. 16.2A; "inactive insects"). Typical fitted activation energy parameters for these cases are in the range of 0.5 to 0.8, consistent with the findings of a recent meta-analysis using a much larger database of insect metabolic rates (Irlich et al. 2009) and with Q_{10} values in the range of 2–3.

These general patterns occur despite substantial variation in metabolic intensity. For example, similar thermal sensitivities of metabolic rate (i.e., slopes) are observed for scarab beetles and whiteflies, despite their very different metabolic rate at a given temperature (normalization constants, Fig. 16.2A). In the scarab study, only data from insects exhibiting DGC are included, probably explaining their relatively low metabolic rate (Davis et al. 2000), while the whitefly data are for feeding groups (Salvucci and Crafts-Brandner 2000).

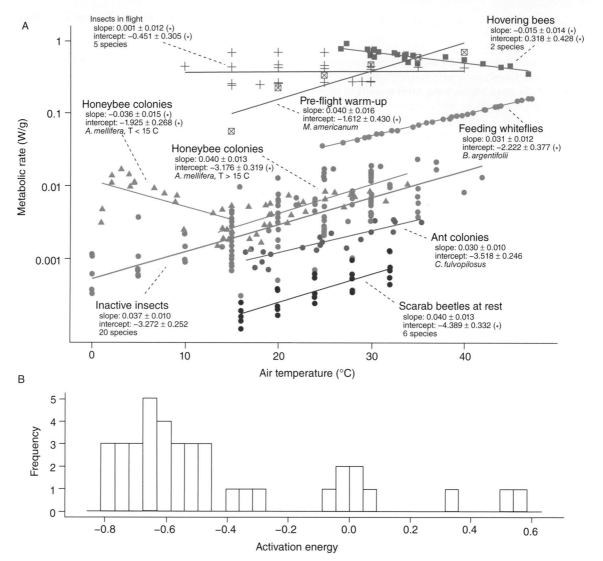

Figure 16.2 Insect metabolic rates are sensitive to temperature and dependent on behavioral and environmental factors. In (A), the mass-specific metabolic rates for 31 insect species from eight taxonomic orders and also ranging across eight orders of magnitude in body size are plotted as a function of air temperature. There is a common trend of insect metabolic rates increasing with temperature in the majority of sampled studies but there are also a number of exceptions. In particular, highly metabolically active, endothermic insects (e.g., endothermic flying insects, cold-exposed honeybee swarms) tend to show little effect of air temperature on metabolic rate, or even an inverse relationship. The coefficients of the linear regressions describing the rate–temperature relationships are provided with asterisks (*) indicating whether the slope or intercept of the fitted data are significantly ($p < 0.05$) different from the coefficients of the common regression parameters shared by the majority of inactive insects. Note that the intercept estimates and their standard error are in log-transformed units. (B) Frequency distribution of activation energies of insects. Activation energies were obtained as the slopes of OLS regressions of the natural logarithm of mass-specific metabolic rate as a function of inverse absolute temperature for the 35 analyzed datasets. References for the data analyzed in (A) and (B) include: ant colonies (Lighton 1989), flying insects (Casey and Ellington 1989), hovering bees (Roberts et al. 1998; Harrison and Fewell 2002), honeybee colonies (Heinrich 1980), inactive insects (Casey 1977; Herreid et al. 1981; Chappell 1983, 1984; Morgan et al. 1985; Casey and Knapp 1987; Schultz et al. 1992; Vogt and Appel 1999; Fielden et al. 2004; Klok and Chown 2005; Terblanche and Chown 2007), pre-flight warm-up (Casey and Hegel-Little 1987), scarab beetles (Davis et al. 2000), and whiteflies (Salvucci and Crafts-Brandner 2000).

Despite the modal trend for thermal effects on metabolic rates to be relatively well predicted by MTE (Fig. 16.2B), there are some striking exceptions that illustrate potential dangers of not considering the physiological ecology of the species in question. While metabolic rates of social insect larvae or sleeping adults indicate fairly normal responses to temperature (Schmolz et al. 2002; Petz et al. 2004), endothermic flying insects or insect colonies can exhibit constant or even decreasing metabolic rate as temperature increases (Fig. 16.2A). Because flight (foraging) costs can be a significant fraction of total metabolic rate for such insects (Harrison and Fewell 2002), and metabolic rate during overwintering can affect survival of such colonies (Harrison et al. 2006a), it is important to consider these mammal-like thermoregulatory responses of metabolic rate to temperature when considering the effect of climate on these species. Diurnal behavioral thermoregulation can result in higher than expected responses of metabolic rate to air temperature (Casey and Knapp 1987), as can testing insects outside their normal thermal ranges (Schultz et al. 1992). Exposure to naturally occurring fluctuating temperature regimes can also induce stress (e.g., oxidative damage) that increases metabolic rates even where the average temperature decreases (Lalouette et al. 2010). Furthermore, some insects exhibit seasonal and intra-seasonal variation in mass-specific and temperature-independent metabolic rate (McGaughran et al. 2009). In many of these cases, the biochemical/physiological mechanisms responsible for thermal responses that differ from MTE remain unknown.

16.2.2 Oxygen and supply limitation

Metabolism represents a balance between energy supply and demand integrated across the many tissues and systems within an organism. Energy is generated primarily by catabolism of fuels using oxygen transported by the tracheal system. One foundational concept of MTE is the proposition that allometric scaling of metabolic rate reflects a resource supply constraint (West et al. 2001). Alternatively, or additionally, the hypometric scaling of metabolic rate with body mass could relate to body-size related scaling of energy demand (Ricklefs 2003; Seibel and Drazen 2007). One way to consider the matching of oxygen supply and demand is to consider how metabolic rate is affected by ambient changes in oxygen supply. To model this effect,

it can be useful to consider the classic mass balance equation of respiratory physiology:

$$VO_2 = G \cdot \Delta P_{O_2} \qquad (16.1)$$

in which VO_2 indicates an organism's oxygen consumption rate, G the conductance of the respiratory system, and ΔPO_2 the partial pressure gradient for oxygen from atmosphere to mitochondria. G is a measure of the capacity of the respiratory system to transport oxygen, and in this simplified case represents the combination of both diffusive and convective conductance (Buck 1962). If ambient oxygen level is slowly lowered, and ΔPO_2 drops, animals will typically increase the conductance of their respiratory system (in the case of insects, by opening spiracles and increasing ventilation) to maintain a constant VO_2. Over this range, the organism is within its safety margin for oxygen transport and is not supply limited. The organism's critical pO_2 for that particular function is defined as the pO_2 when oxygen becomes limiting and below which VO_2 decreases (Fig. 16.3). We know from work

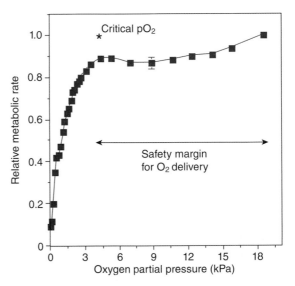

Figure 16.3 To quantify the safety margin for oxygen delivery, organism function can be measured over a range of oxygen partial pressures; the partial pressure (pO_2) at which the activity measure significantly decreases is referred to as that organism's critical oxygen partial pressure. The critical pO_2 for the metabolic rate of adult *D. melanogaster* is 3 kPa or at about 85% less oxygen than normal, with a safety margin of 18 kPa O_2 (Van Voorhies 2009).

with isolated mitochondria (Gnaiger and Kuznetsov 2002) that mitochondria themselves need very little oxygen to perform maximally (less than 1 kPa) so at the critical pO_2, the average ΔPO_2 is likely approximately equivalent to the atmospheric pO_2. Under these circumstances, the maximal capacity of the respiratory system to conduct oxygen, G_{max}, can be estimated as VO_2/critical pO_2 (Harrison 1997). Conductance varies with behavior; for example, it is much higher during insect flight than at rest due to recruitment of more active methods of ventilation (Harrison 1997). Comparison of critical pO_2 values for a given behavior across insects of different sizes can provide a direct way to test whether the ratio of oxygen supply to demand changes with body size. To our knowledge, insects are the only taxonomic group in which there have been systematic tests of the effect of body size on respiratory conductance and critical pO_2.

Most inactive insects exhibit very low critical pO_2 values (Fig. 16.4), clearly indicating that resting metabolic rate is not oxygen-limited. However, critical pO_2 values do tend to be higher when metabolic rate is elevated, as during flight (Fig. 16.4). When compari-

sons are made controlling for behavior and developmental stage, there is no evidence that critical pO_2 values are higher in larger insects, and mass-specific tracheal conductances at least match the scaling of metabolic rate. Thus there is no evidence that oxygen demand outstrips supply as insects increase in size (Greenlee and Harrison 2004a, 2005; Harrison et al. 2005; Greenlee et al. 2007, 2009). However, there is a tendency for juvenile insects tested later within the development of a single instar (when mass increases without molting) to have much higher critical pO_2 values, suggesting that size increases without molting and resizing of the tracheal system might lead to oxygen supply limitation (Greenlee and Harrison 2004b, 2005).

While oxygen supply seems to meet demand as insects increase in size, this may occur because larger insects exhibit an increased investment in respiratory structure. Larger tenebrionid beetle species have a greater fraction of their body devoted to the tracheal system, and extrapolations of these trends suggest that this pattern could explain oxygen limitations on insect size (Kaiser et al. 2007). Similar hypermetric patterns

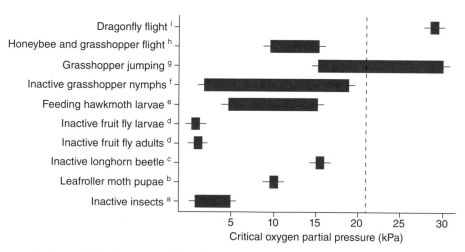

Figure 16.4 Insects often exhibit an impressively broad safety margin for maintaining measures of activity (e.g., O_2 consumption, CO_2 emission, performance) in spite of reduced partial pressures of oxygen in their environment. This figure plots the range of critical pO_2 in the literature for a diverse range of insects and their behaviors. Critical pO_2 tends to be higher in active insects. In cases where hyperoxic values are reported, this indicates that the measure of activity (i.e. dragonfly CO_2 emission and grasshopper jumping performance) increased in hyperoxia relative to normoxia. Normal pO_2 is 21 kPa, as indicated by the dotted line. The letter superscript associated with each row indicates the reference for that dataset: a (Harrison et al. 2006b), b (Zhou et al. 2000), c (Chappell and Rogowitz 2000), d (Klok et al. 2010), e (Greenlee and Harrison 2005), f (Greenlee and Harrison 2004a), g (Kirkton et al. 2005), h (Joos et al. 1997; Rascón and Harrison 2005), and i (Harrison and Lighton 1998).

of tracheal investment have been observed in grass-hoppers during ontogeny (Harrison et al. 2005; Greenlee et al. 2009). The increased investment in respiratory structure in larger insects suggests that body size influences metabolic rate via evolutionary trade-offs such as reduced proportions of active tissues per unit volume in larger insects (Harrison et al. 2010).

16.2.3 Locomotion

The metabolic rates of behaviorally active insects range from 3 to 30 times resting rates, and the maximal mass-specific metabolic rates of active insects can be more than double those of maximally active, similarly sized mammals or birds (Harrison and Roberts 2000). Flying insects exhibit the highest metabolic scopes and flight metabolic rates are approximately 10 times greater than maximal metabolic rates for running insects of a similar size (Full 1997). In some endothermic insects, transitions from rest to activity are associated with strong increases in body temperature, leading to very high metabolic scopes. For example, it has been reported that stridulating katydids (Stevens and Josephson 1977) and running beetles (Bartholomew and Casey 1977) exhibit metabolic scopes in the range of 50 to 100-fold during these behaviors as they endothermically warm their bodies by up to 20 °C.

Metabolic rates increase linearly with running speed (Weier et al. 1995; Lipp et al. 2005) and peak metabolic rates among running and flying insects scale on average with $M^{0.86}$ (Full 1997). Metabolic rates during flight have been reported to scale with $M^{0.9}$, but the degree of this allometry is likely influenced by the tendency of larger endothermic insects to have higher body temperatures and flight metabolic rates than smaller insects (Niven and Scharlemann 2005). Mechanical power output (usually estimated from the kinematics of limb or wing movements and dynamic models) scales isometrically with body size in running insects and either isometrically or hypermetrically in flying insects (Full 1997). This pattern (increasing mechanical power output relative to metabolic power input) suggests that the relative efficiency (mechanical power output/metabolic power input) of locomotion increases with insect body size. If efficiency is defined as the ratio of locomotory power output to metabolic power input, isometric scaling of power output and input would predict that efficiency is invariant with mass ($\sim M^{0.0}$). For insects in general, efficiency scales

hypermetrically (relative to the isometric prediction) with mass$^{0.12}$ and even more dramatically for honeybees, locomotory metabolic efficiency scales with $M^{0.45}$ (Harrison and Roberts 2000).

16.2.4 Nutrition and feeding

The metabolic costs of insect foraging are usually tightly linked to the energetics of locomotion. In honeybee colonies, the energetic costs of foraging are approximately 30% of the estimated whole-colony metabolic rate, but this fraction is likely much lower in terrestrial foraging species such as ants (Harrison and Fewell 2002). However, the metabolic rates of stationary ants (*Atta sexdens rubropilosa*) during leaf-cutting may be more than 30 times their inactive metabolic rate, yielding a similar aerobic scope to flight (Roces and Lighton 1995).

Metabolic rates typically increase in response to feeding and these increases may scale with both body size and meal type. The metabolic costs of post-feeding digestion can be quantified by the elevation of metabolic rate relative to baseline, the postprandial metabolic scope (~3.3 for insects) and also by the net energy expended for the duration of the specific dynamic action response (SDA), which ranges among the insect groups studied from 0.00025 to 0.102 kJ (Secor 2009). Across a broad range of invertebrate taxa, SDA scales with $M^{0.31}$, meal-mass$^{0.72}$, and meal-energy$^{0.32}$; for comparison, among mammals, SDA scales with $M^{0.32}$, meal-mass$^{0.7}$, and meal-energy$^{1.21}$ and among reptiles SDA scales with $M^{-0.08}$, meal-mass$^{1.13}$, and meal-energy$^{1.06}$ (Secor 2009). Among insects, the kissing bug (*Rhodnius prolixus*) exhibits the highest postprandial metabolic scope (10-fold increase in whole-organism metabolic rate) as well as the greatest SDA (0.102 kJ) following feeding on a blood meal (Bradley et al. 2003). The specific dynamic action for migratory locust (*Locusta migratoria*) nymphs is in the four- to fivefold range (Gouveia et al. 2000). In addition to effects of being fed or not, the characteristics of the diet can also affect metabolic rate. A high-carbohydrate diet is linked to increased metabolic rates in honeybees (Blatt and Roces 2001). In locusts, increased carbohydrate : protein intake can lead to strong elevation in metabolic rates, probably to dispose of excess energy in the diet and allow intake and assimilation of needed quantities of protein (Zanotto et al. 1997; Gouveia et al. 2000).

Restricted nutritional resource supply can have a range of effects on insect metabolic rates. Foraging honeybees given richer (higher carbohydrate) rewards exhibit higher metabolic rates during periods of foraging that include both flight and non-flight (Balderrama et al. 1992), suggesting that in this species, metabolic rate is positively influenced by nutritional supply. Similarly, starvation may decrease metabolic rates or impair flight performance (Goldsworthy and Coupland 1974; Matsura 1981; Stoks et al. 2006), but this is not always the case. In the African fruit beetle (*Pachnoda sinuata*), voluntary flight performance and duration is not inhibited by 15–30 days of starvation (Auerswald and Gäde 2000). Reduced water supply can elevate metabolic rate in growing insect larvae (Martin and Van't Hof 1988) but does not affect the overall metabolic rate of adult locusts (Loveridge and Bursell 1975). One of the reasons for the complex pattern of nutrient-supply effects on insect metabolic rates is the fact that there are often plastic physiological responses to resource deprivation including dramatic shifts in the metabolic pathways and nutrient substrates used to fuel metabolism, often without affecting overall metabolic rates (Djawdan et al. 1997; Juliano 1986; Auerswald and Gäde 2000; Renault et al. 2002; Sinclair et al. 2011). However, both comparative and artificial selection studies suggest that an evolutionary response to starvation and water stress may involve reduced mass-specific metabolic rate (Harshman et al. 1999; Marron et al. 2003).

At the extreme of environmental nutrient restriction, insects may utilize torpor and diapause to survive long dearth periods, reducing metabolic rates for extended periods of time by more than 98% (Schneiderman and Williams 1953; Hahn and Denlinger 2010). A meta-analysis of the metabolic rate scaling for insect eggs, larvae, and pupae ($62.4\,M^{0.77}$) shows a similar hypometric exponent but a significantly reduced intercept (normalization constant) relative to the allometry ($363\,M^{0.86}$) for the corresponding adult resting metabolic rates (Guppy and Withers 1999).

Social insect colonies are particularly well adapted to maintaining physiological homeostasis in response to variation in environmental resource availability. For example, workers within the colony vary their foraging activity in response to the nutritional demands of the brood (Sorensen et al. 1985; Dussutour and Simpson 2009) and many species harvest and store resources (Hölldobler and Wilson 1990). Colonies may also catabolize somatic tissue to survive resource scarcity and environmental stress (Wilson 1971; Sorensen et al. 1983; Schmickl and Crailsheim 2001). In the acorn ant (*Temnothorax rugatulus*), decreases in activity levels and increases in trophallaxis (mouth-to-mouth food transfer) are hypothesized to facilitate this species' remarkable ability for colonies to survive greater than eight months of starvation (Rueppell and Kirkman 2005). Kaspari and Vargo observed a hypermetric allometry for the duration of queen survival in the fire ant (*Solenopsis invicta*) which scaled with the size of the colony as $M^{0.21}$ (Kaspari and Vargo 1995). This capacity for resilience has been hypothesized as one of the factors involved in the evolution of eusociality, caste ratios, and variation in colony size (Michener 1964; Wilson 1968; Bouwma et al. 2006).

16.3 CORRELATIONS BETWEEN BODY SIZE AND METABOLIC RATE

16.3.1 Developmental allometries

Insect larvae represent excellent, albeit relatively unexplored, model systems for investigating the interface between physiology and ecology. As for adults, larval insects are quite diverse. Many insect larvae live underground, in leaf litter, or in decaying fruits and likely experience a range of hypoxic environments, but others (e.g., many lepidopteran larvae) forage on leaves in normoxia. Many insect larvae are solitary, but other species rear brood cooperatively, such as the bessbug (*Odontotaenius disjunctus*) which raise larvae in communal galleries carved out of decaying wood. While most insect larvae are terrestrial, some are aquatic, and of these, some have open (e.g., mosquito) and others closed (e.g., caddis fly) tracheal systems. Some aquatic insect larvae (e.g., the chironomids) have evolved the use of hemoglobin for oxygen transport (Oliver 1971). The diversity of these environments and behaviors as well as the general paucity of literature data make it difficult to draw broad conclusions about the energetics of insect larvae.

Insect larvae metabolic rate allometries have been investigated on an intraspecific basis for a number of insect species that can be easily reared. While many of these studies report hypometric scaling exponents, there is weak support for a canonical 0.75 exponent. Growing honeybee larvae increase in mass by more than 400-fold in only 4 days and exhibit metabolic rates that scale with mass$^{0.9}$ (Petz et al. 2004). Larvae of the tobacco hornworm (*Manduca sexta*) span three orders of magnitude in body mass and exhibit CO_2 emission rate scaling with a mass exponent that ranges from 0.77 (Alleyne et al. 1997) to 0.98 across the entire larval stage (Greenlee and Harrison 2005). However, individual instars show different patterns of metabolic rate scaling; as larvae grow within an instar, the mass-specific CO_2 emission decreases with age/size among early instars, but it increases with size in final larval instar (Greenlee and Harrison 2005) (see also Kerkhoff, Chapter 4, Fig. 4.1B). Similarly in grasshoppers, the pattern of CO_2 emission rate scaling varies within different instars with exponents ranging between 0.45 and 0.91 (Greenlee and Harrison 2004b), while across its entire development, metabolic rate scales with the exponent 0.73 (Greenlee and Harrison 2004a). Larvae of the flour beetle (*Tribolium castaneum*) exhibit mass-specific metabolic rates that decrease by over 90% during less than 12 days of development (Medrano and Gall 1976) and the hemimetabolous milkweed bug exhibits a 38% decrease in mass-specific metabolic rate from the first instar to adult (Niswander 1951).

Insect development from larvae to adults is associated with complex changes in body form and physiology in addition to alteration in body size. Metabolic rate scaling patterns may depend on the nature of these changes. Adult holometabolous insects are often substantially smaller than the terminal larval instar and the few studies available suggest that they have greater resting and maximal metabolic rates. Adults of vinegar flies (*D. melanogaster*) (Klok et al. 2010), fire ants (*S. invicta*) (Vogt and Appel 1999), and honeybees (*A. mellifera*) (Lighton and Lovegrove 1990; Petz et al. 2004) exhibit mass-specific metabolic rates approximately twice as high as their larvae. Are the higher mass-specific metabolic rates in adults due to their smaller size? If the adults and larvae are assumed to belong to a common mass-scaling allometry, then the ratio of their mass-specific metabolic rates can be calculated by:

$$\frac{B_2}{B_1} = \frac{B_0 M_2^{\alpha-1}}{B_0 M_1^{\alpha-1}} = \left(\frac{M_2}{M_1}\right)^{\alpha-1} \tag{16.2}$$

where symbols are defined as in equation 2.1. Equation 2.6 can be rearranged to solve for the mass ratio (ΔM) that would be necessary to generate an observed ratio in metabolic rate (ΔB):

$$\Delta M = \frac{M_2}{M_1} = (\Delta B)^{\frac{1}{\alpha-1}} \tag{16.3}$$

If the whole-animal scaling exponent (α) is 0.75, then we can predict what difference in masses would generate the observed ratio in mass-specific metabolic rates:

$$\Delta M = \Delta B^{-4} \tag{16.4}$$

In the case of a two-fold difference in mass-specific metabolic rates, the mass ratio would have to be 0.0625 for allometry to predict the observed difference in mass-specific metabolic rates. In other words, the adult stages of the ant, bee, and fly species mentioned above would have to be 94% smaller than their larval forms (or the larvae 16.67 times larger than the adults) for simple mass-scaling to explain the two-fold higher mass-specific metabolic rates in adults relative to larvae. Since adults are only approximately 10–30% less massive than larvae, the relatively high adult mass-specific metabolic rates are not simple allometric consequences of smaller body mass. An alternative ultimate explanation for the higher mass-specific metabolic rates of adults may be analogous to the higher metabolic rates of flying relative to non-flying adult insects (Reinhold 1999). The complex changes (e.g., in body tissue composition and tracheal system structure) that take place during metamorphosis in the holometabolous pupal stage apparently also enable fundamental changes in resting metabolic rate.

16.3.2 Intraspecific allometries

The relatively low range in masses among adults of a single species makes it difficult to accurately test for an intraspecific correlation between mass and adult insect metabolic rates in many species (Vogt and Appel 1999; Van Voorhies et al. 2004). Ants are somewhat exceptional in this regard, with some species exhibiting substantial variation in worker size. For example, the dry masses of *Pheidologeton diversus* workers vary by more than 500-fold (Hölldobler and Wilson 1990). In most cases, it appears that metabolic rates of such workers scale hypometrically with mass, with homogenous slopes ranging from 0.55 to 0.83 (Chown et al. 2007; see also Kerkhoff, Chapter 4, Fig. 4.1A).

In some insects, intraspecific variation is associated with morphological allometries that produce surprising

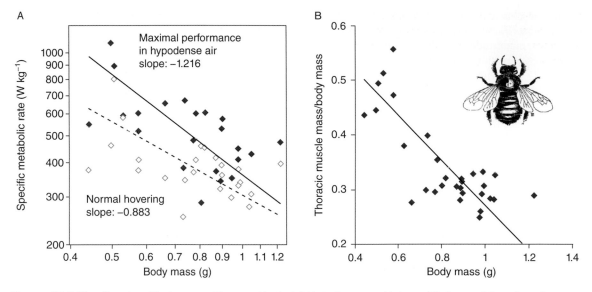

Figure 16.5 The allometry of body composition provides insight into the unusual intraspecific hypometric scaling of metabolic rate with body size in the carpenter bee, *Xylocopa varipuncta*. This figure, adapted from Roberts et al (2004), shows in (A) that body mass-specific metabolic rates decrease very strongly with body mass (so whole-organism rates scale with $M^{-0.22}$ for maximal performance in hypodense air and $M^{0.12}$ for normal hovering). This pattern occurs because the relative content of thorax muscle mass (the major site of oxygen consumption during flight) decreases with body size (B).

patterns in metabolic rate scaling and locomotory performance. Among female carpenter bees, flight metabolic rate scales hypometrically with mass $^{0.12}$ (Fig. 16.5A), a very low scaling exponent. When these bees exhibit their maximal performance, flying in the lowest density air possible (to stay aloft in thin air requires more power), the scaling exponent is −0.22, meaning that both the mass-specific and absolute metabolic rates are lower in larger individuals than smaller individuals. The reduced mass-specific metabolic rate and flight performance of larger bees in this species is explained by variation in the relative amount of flight muscle, the primary site of oxygen consumption in flying bees. In this species, larger females have proportionally larger abdomens (Fig. 16.5B) and likely larger ovaries. As a consequence, larger individuals have significantly lower ratios of flight muscle to body mass, lower mass-specific metabolic rate, and reduced scopes for flight performance and metabolic rate (Roberts et al. 2004).

16.3.3 Social insect colonies

Social insect colonies are intriguing organisms from the perspective of MTE because they span multiple levels of biological organization. Individuals within the colony may be expected to exhibit hypometric scaling of metabolic rate with mass, but whole colonies are made up of physically independent individuals at different developmental stages and engaged in a wide variety of different tasks and behaviors. Thus, whole-colony metabolic rate should scale linearly with mass, depending proportionally on the number of individuals in the colony. Surprisingly, social insect colonies exhibit hypometric intraspecific scaling of metabolic rate with colony mass (Fig. 16.6). Intriguingly, while the three social insect datasets illustrated in Figure 16.6 exhibit hypometric metabolic rate scaling consistent with the pattern for individual insects, they are each hypothesized to do so for different reasons. Honeybee clusters maintain a relatively constant core temperature when air temperature falls; mass-specific metabolic rates and mass-specific heat loss from the cluster falls in larger clusters due to a reduced surface area to volume ratio (Southwick et al. 1990). In the polymorphic ant *Pheidole dentata*, lower mass-specific colony metabolic rates arise from larger colonies having a greater fraction of larger "major" workers (Shik 2010). In a harvester ant species with monomorphic workers, *Pogonomyrmex californicus*, it is

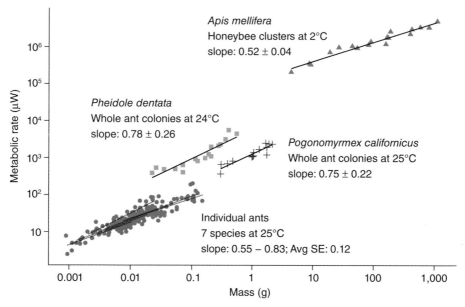

Figure 16.6 Allometry of colony metabolic rate in three colonial species compared with individual metabolic rate in seven solitary species. Social insect colonies, like individual insects, exhibit metabolic rates that scale hypometrically with mass. This figure combines intraspecific data for individual ants (Chown et al. 2007), two functioning whole ant colony species (Shik 2010; Waters et al. 2010), and thermoregulating honeybee clusters (Southwick et al. 1990). The OLS regression results for each group are displayed above (slopes are given with standard errors) and the overall model, which fits a separate slope and intercept for each species, has an $r^2 = 0.99$. The average homogenous slope was 0.62 and the only species that show significantly higher than average scaling slopes are *Eciton hamatum* (slope: 0.83, $p < 0.003$) and *P. dentata* (slope: 0.78, $p < 0.002$).

hypothesized that a lower mass-specific colonial metabolic rate in larger colonies may be due to larger colonies having a lower fraction of active workers (Waters et al. 2010). The similar hypometric scaling patterns with disparate mechanisms do suggest common underlying ecological/evolutionary forces that can be addressed by varied mechanisms in different species. Social insect colonies may be particularly useful for investigating mechanisms responsible for metabolic scaling patterns due to the capacity to more easily manipulate and measure specific components of the superorganism than is possible with individual organisms.

16.3.4 Interspecific allometries

On an interspecific basis, insect metabolic rates scale with $M^{3/4}$ (Chown et al. 2007) as in mammals (Karasov, Chapter 17). Analysis of the intercept (or normaliza-

tion constant) of this relationship indicates that insects have low metabolic rates relative to mammals. After accounting for the effects of mass (the scaling exponent) and temperature (by adjusting insect rates from 25 °C to 37 °C following the activation energy method), insect metabolic rates are approximately half those reported for mammals (Fig. 16.7A). This pattern fits with the ectothermic nature of most insects; ectothermic vertebrates also have lower metabolic rate than mammals at the same body temperature (Hulbert and Else 2000).

Examination of the intercept of metabolic rate allometries (referred to as metabolic coefficient, intensity, elevation or normalization constant) has the potential to reveal evolutionary differences in the metabolic physiology of different animal taxa. Among invertebrates, ticks and scorpions have been shown to exhibit significantly lower metabolic rates than other "typical" arthropods, possibly contributing to the high abundance of these species in some regions (Lighton

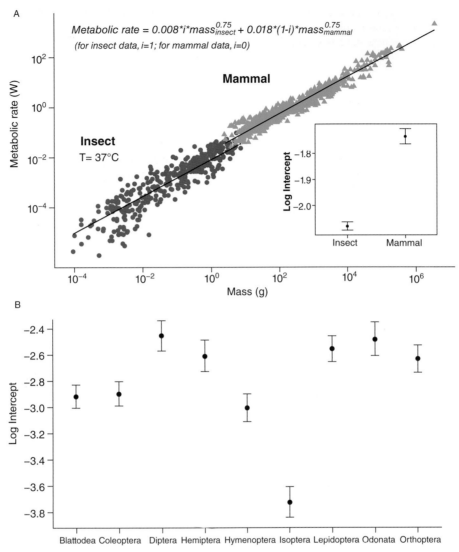

Figure 16.7 The hypometric scaling of inactive insect metabolic rates (Chown 2007) is generally consistent with the pattern observed for mammals (Savage et al. 2004b); both taxa exhibit similar 3/4-power scaling slopes but insects have lower intercepts or normalization constants. In (A) the metabolic rates for 391 insect species have been adjusted using the Arrhenius equation to 37 °C, a standard mammalian body temperature (see Brown and Sibly, Chapter 2). Independent of mass, inactive insects have metabolic rates about two times lower than inactive mammals with the same body temperature (figure inset displays the intercept values from the OLS regression on log-log data). (B) The interspecific data on insect metabolic rates at 25 °C (Chown 2007) can be analyzed by taxonomic order. A linear model that fits unique slopes and intercepts for each order was not a significantly better model than one that preserved a common slope (0.72 ± 0.2 SE) but allowed for variation in intercepts by order. The maximum intercept (Diptera, 3.5 mW) was more than 18 times greater than the minimal intercept (Isoptera, 0.19 mW). Part of this variation may be due to behavioral variation among "inactive" insects, and methodological variation among researchers, but the data suggest substantial order-level variation in inactive metabolic rates among insects.

et al. 2001). Similarly, predatory ant-lion larvae which build pits in which to capture prey exhibit common hypometric metabolic rate scaling exponents but with intercepts depressed lower than insects in general and even lower than similarly sized sit-and-wait predatory spiders (Lucas 1985; Van Zyl et al. 1997). The normalization constant (y-intercept on log-log plots) of the scaling relationship for different insect orders varies by 18-fold (Fig. 16.7B). While some of this variation may be related to methodology and behavioral variation among the taxa, it is likely that these order-level patterns at least partially reflect previously unrecognized evolutionary differences in physiology and life history.

16.4 BROADER IMPLICATIONS

The diversity of insect structure and function provides a powerful tool for testing physiological, ecological, and evolutionary predictions of MTE. While the general equations of MTE seem to fit the modal responses of insects, and thus may be very useful for community and ecosystem ecology, behavioral and physiological divergences of individual species and taxa from the general theory of MTE are considerable. The temperature dependence of insect metabolic rates is highly variable, and as previously discussed, frequently depends on important ecological variables including behavior and thermal preferences. The temperature dependence of insect metabolic rates can also be highly subject to thermal acclimation and adaptation (Chown and Nicolson 2004). All of these factors are critical to developing predictive models for how insect populations will respond to global-scale changes in climate (Dillon et al. 2010).

Applications of MTE may be able to help address some of the great unanswered questions in insect metabolic ecology. What biophysical forces or ecological pressures have driven the evolution of insect endothermy? What are the constraints on behavior and physiological performance imposed by body size? Have the biomechanics of the insect exoskeleton or tracheal system influenced the evolution of insect size, and if so, what is the role of metamorphosis in mediating these potential constraints? How do behavioral and developmental regulation within social insect colonies influence the scaling of supply and demand in these physically independent but functionally integrated systems?

Future mechanistic developments of MTE may also help to explain sources of variation in mass- and temperature-independent metabolic intensity, both among distinct insect taxa and on a larger scale between insects, birds, and mammals. In addition, studying the energetics of insects is tremendously important for ecology and agriculture. As predators, scavengers, detritovores, and herbivores, insects play enormously important roles in ecosystem functioning, so that more energy flows through an ecosystem due to the activity of insects than from the activity of vertebrates (Andersen and Lonsdale 1990). Economic growth and stability may depend on understanding the thermal preferences, metabolic rates, and behaviors of insect pollinators (Potts et al. 2010). By moving beyond broad assumptions and universal characterizations, MTE has the potential not only to integrate fields as diverse as insect ecophysiology and biofluid transport dynamics, but also to reveal questions of basic and fundamental importance to agriculture, biomechanics, ecology, and evolution.

Chapter 17

TERRESTRIAL VERTEBRATES

William Karasov

SUMMARY

1 Terrestrial vertebrates cover a large range in body mass M (eight orders of magnitude) and body temperature T (>45 °C), yet the allometric model of the metabolic theory of ecology (MTE), in which whole-animal metabolism I increases with $M^{3/4}$ and exponentially with temperature, explains much of the variation in their whole-animal metabolism.

2 The large differences in I between terrestrial vertebrate endotherms and ectotherms are due to fundamental differences in body architecture and perhaps cellular function in addition to differences in body temperature.

3 With increasing body size, terrestrial vertebrates are made up of an increasing number of similar-sized cells that have decreasing cell- or mass-specific respiration rates.

4 Metabolic rates of individuals can be linked to population ecology through their energy budgets, which can be summed over all members of a population.

5 Daily food consumption scales with $M^{2/3}$ to $M^{3/4}$ and is much higher in endotherms than ectotherms. The surface area of the gut, where breakdown of substrates and absorption of their monomers occurs, scales with M to the $\approx 3/4$ power, and endotherms exceed the ectotherms by about 4 times. Generally, food assimilation efficiency is relatively independent of M, within a given food type, and is relatively independent of body temperature in ectotherms.

6 MTE provides a mechanistic model of growth based on I/M.

7 The maximum per capita rate of population growth, r_{max}, scales with $\approx -1/4$ power of M. However, the underlying mechanism(s) may not act through metabolism as hypothesized by MTE.

8 MTE predicts that population density (abundance) of a consumer population will decline with $M^{-3/4}$, but this prediction is borne out in only some studies. These relationships can be useful in modeling features of food webs in ecological communities.

9 MTE can highlight special problems of relatively small animals, can advance knowledge about biology and ecology of huge animals that are now extinct, and can advance knowledge about human and ecosystem health.

Metabolic Ecology: A Scaling Approach, First Edition. Edited by Richard M. Sibly, James H. Brown, Astrid Kodric-Brown.
© 2012 John Wiley & Sons, Ltd. Published 2012 by John Wiley & Sons, Ltd.

This chapter focuses on how research with terrestrial vertebrates (mammals, birds, amphibians, and "reptiles" – Testudines, Squamata, Crocodilia) has contributed substantially to progress in major areas of the metabolic theory of ecology (MTE): (1) the mechanistic description of how metabolism varies with body size and temperature, and (2) ecological implications that arise from how patterns govern features at levels of populations, communities, and ecosystems.

17.1 METABOLISM OF TERRESTRIAL VERTEBRATES VARIES PREDICTABLY WITH DIFFERENCES IN BODY SIZE AND TEMPERATURE

Two important features of terrestrial vertebrates that have made them big players in metabolic ecology are a large range in body size and in body temperature (T). The body mass range among extant forms is 6 million-fold, from the smallest amphibian (Brazilian gold frog, *Brachycephalus didactylus*, 7×10^{-2} g) to the largest land mammal (elephant, *Elephas maximus*, 4×10^{6} g). Terrestrial vertebrates include animals that control their body temperatures by relying mainly on endogenous heat production (endotherms; primarily mammals and birds) or by relying mainly on environmental sources of heat (ectotherms; primarily reptiles and amphibians). Representatives from both these metabolic groups can hold their body temperature relatively constant for prolonged periods of time (homeothermy) or allow it to vary with changing environmental heat load (poikilothermy). The full range of possibilities occurs within mammals, which are typically considered homeothermic endotherms during normothermia, but some of which can become ectothermic and poikilothermic during torpor. During normothermia, body temperatures of birds tend to be higher than those of eutherian mammals, which tend (though not always) to be higher than those of marsupials and monotremes. The body temperatures in some groups vary with body size (Clarke and Rothery 2008). The body temperatures experienced by terrestrial vertebrates range from a few degrees *below* 0 °C in some torpid frogs, reptiles, and mammals to around 45 °C in the most heat-loving desert lizards.

Among species of terrestrial vertebrates there have been hundreds of studies of whole-organism metabolic rate (I) measured under standardized conditions, which include a resting and fasted state and, for endo-therms, an air temperature where metabolism is not increased for body temperature regulation (in which case it is called basal metabolic rate; BMR) or an air temperature specified for an ectotherm (in which case it is called a standard metabolic rate; SMR). Based on these measurements, there have been scores of analyses of how I varies with body mass (M) and body temperature (T) (each as a mean value for a particular species). The two factors can be incorporated within a single model described early by Peters (1983) and elaborated upon by Gillooly et al. (2001), who concluded that I increases (or "scales") with $M^{3/4}$ and that it increases exponentially with temperature approximately according to the Arrhenius relation

$$e^{-E/kT} \tag{17.1}$$

where E is the activation energy, k is Boltzmann's constant (8.62×10^{-5} eV K^{-1}), and T is absolute temperature in kelvin. Thus, the joint multiplicative effects of body mass and temperature are expressed in the single model

$$I = i_o M^{3/4} e^{-E/kT} \tag{17.2}$$

where i_o is a normalization constant independent of body size and temperature. There are also alternative expressions for this relationship (Brown et al. 2004). In the plot for normothermic and torpid mammals and birds (Fig. 17.1B, adapted from Gillooly et al. 2001), metabolism standardized for body mass ($IM^{-3/4}$) declines with $1/kT$ in a linear manner on a semi-log plot, as would be expected from equation 17.2. The slope can be used to calculate an estimate of E, which falls within the range of estimates of E from other biological systems (0.2 to 1.2 eV; average $\approx$0.65) (Gillooly et al. 2001). Besides varying among "systems" (including species), the value of E also changes with temperature within an animal (Schmidt-Nielsen 1990), so it is by no means a constant. Indeed, alternative expressions that rely on the van't-Hoff principle (Q_{10}, the rate change for a 10 °C change in temperature) arguably also provide suitable corrections for differences in temperature (Clarke 2004, 2006) (but see Brown and Sibly, Chapter 2). The relationships in Fig. 17.1A,B, based on equation 17.2, suggest (1) that the typically higher values of I in birds compared with similar-sized mammals are largely a result of their typically higher body temperatures, and (2) that a significant fraction of the reduction in metabolism experienced by birds

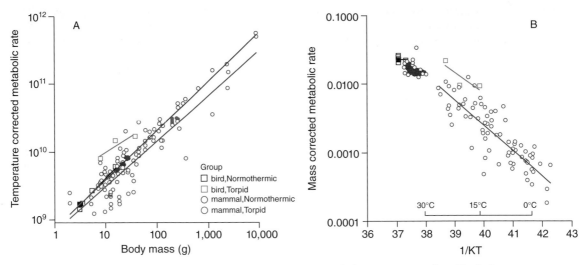

Figure 17.1 Mass and temperature dependence of whole-organism metabolic rate in mammals and birds that were normothermic or torpid. (A) The relationship between temperature-corrected metabolic rate, according to equation 17.2 and $E = 0.65$ eV, and body mass in grams. The slopes did not differ significantly ($p > 0.8$) and the pooled slope was 0.71. Similar-sized normothermic mammals and birds did not differ significantly ($p > 0.9$). (B) Relationship between mass-corrected metabolic rate, measured in W g$^{-3/4}$ and temperature, measured in kelvin. The normothermic and torpid mammals fall along the same line, confirming that most of the reduction in metabolism in torpid mammals is the result of a lowered body temperature. The pooled slope was -0.36, which is taken as an estimate (in eV) of the absolute value of E, and that value is within the range previously observed in Arrhenius plots for other biological processes. Data from Gillooly et al. (2001).

and mammals in torpor is the result of lowered T, but there were still significant differences between normothermic and torpid animals within each of those taxa. Remaining differences after temperature correction may relate to regulated metabolic suppression in torpid animals (Staples and Brown 2008; Armstrong and Staples 2010). Likewise, even after temperature correction by either Boltzmann–Arrhenius or van't Hoff methods, metabolism is still at least 4 times higher in endo- than ectothermic vertebrates (Brown et al. 2004; White et al. 2006).

Researchers have raised many valid concerns about the principles and assumptions underlying the MTE equation 17.2 and have suggested alternative explanations and theories (e.g., Kooijman 2000; Clarke, 2004, 2006; Glazier 2005; van der Meer 2006; Savage et al. 2008; Roberts et al. 2010). As indicated, there have been scores of similar analyses of I vs. M and T that use different datasets of terrestrial vertebrates and even different statistical analyses, and there are plenty that indicate different scaling with M, or heterogeneous scaling among different groups of terrestrial vertebrates (e.g., White and Seymour 2003, 2005; Farrell-Gray and Gotelli 2005; Glazier 2005, 2008, 2009b; White et al. 2006, 2009; Downs et al. 2008; Sieg et al.

2009; Clarke et al. 2010; Isaac and Carbone 2010), and also differences in the relationships between metabolism and temperature (e.g., White et al. 2006; Clarke et al. 2010). Although those findings present challenges to the uniformity suggested by MTE (Glazier 2005; M. P. O'Connor et al. 2007b), the various empirically determined allometric scalings (i.e., slopes of $\log_{10}I$ vs. $\log_{10}M$), as well as those scalings predicted by MTE and competing theories, typically are < 1. This is important when considering robustness of subsequent ecological predictions and implications to be discussed below. Equation 17.2 arguably provides a reasonable starting point for subsequent ecological predictions.

17.2 MECHANISTIC CORRELATES OF METABOLISM VS. SIZE IN TERRESTRIAL VERTEBRATES

17.2.1 Allometry of metabolism at the suborganismal level

Underlying the metabolic scaling at intra- and interspecific levels are fascinating patterns and questions about allometric scaling of intracellular biochemical processes, cellular respiration rates and cell sizes, and body composition (e.g., Calder 1984; Schmidt-Nielsen 1984; Else and Hulbert 1985a; Wang et al. 2001; Glazier 2005; Savage et al. 2007; Raichlen et al. 2010). For example, one key insight helps explain the large differences in metabolism between ecto- and endothermic vertebrates. Recall that whole-organism metabolism (I) is higher in endothermic than ectothermic vertebrates, even after correction for differences in temperature (Brown et al. 2004). A lot of the difference can be accounted for by mammals (including eutherian, marsupial, and monotremes) having larger energy-intensive organs, such as liver, kidney, heart, and brain, with relatively more mitochondria per unit organ volume, when compared with similar-sized reptilian and crocodilian ectotherms (Else and Hulbert 1985b). In general, the total mitochondrial surface area (the product of organ mass and mitochondrial area per unit organ tissue) in mammalian tissues is about 4 times higher than in ectotherms (Fig. 17.2).

This difference is of about the same magnitude as that between terrestrial endotherms and ectotherms in i_o, the normalization constant in equation 17.2 that is independent of body size and temperature (Brown et al. 2004). Additionally, liver cells may also be leakier to sodium in endotherms than ectotherms, which necessitates more ATP generation by mitochondria for pumping out sodium, and hence more oxygen consumption, per unit tissue in the endotherms (Hulbert and Else 1990). Thus, in terrestrial vertebrates, the large differences in I between endotherms and ecto-

therms are due to fundamental differences in body architecture and perhaps cellular function in addition to differences in temperature. In the following paragraphs we will follow a similar analysis for the large changes in I with body size within mammals.

17.2.2 Allometry of metabolism at the organ level

Researchers have evaluated whether the allometry of whole-organism metabolism can be constructed from the allometry of individual organs. For example, Wang et al. (2001) studied allometry of liver, brain, kidneys, and heart, which together account for about 60% of resting metabolism. The basic model is that, for each individual,

$$I = \sum b_{oi}\, m_{oi} \qquad (17.3)$$

where b_{oi} is the mass-specific resting metabolic rate of organ i and m_{oi} is the organ's mass, and the products of these are summed over all organs/tissues. Among five mammalian species ranging in body mass (M) from rats (0.48 kg) to humans (70 kg), b_o, which was measured *in vivo*, scaled with M to the −0.08 to −0.27 power for five organs (−0.27 for liver; Fig. 17.3A) and m_o of organs and tissues scaled with M to the 0.76 to 1.10 power (0.87 for liver) (Wang et al. 2001). Although not all products $b_{oi}\, m_{oi}$ scaled as $M^{3/4}$, the summed model was quite close to the actual scaling of I on M (compare equations 1 and 10 in Wang et al. 2001). These findings illustrate that the allometric scaling of BMR can be accounted for by the scaling of the masses and metabolic rates of these internal organs, and that specific metabolic rate in internal organs declines with body size. Whereas BMR allometry can be accounted for by the scaling of the masses and metabolic rates of internal organs, skeletal muscle is responsible for 90% or more of whole-body metabolism during sustained locomotion at a high velocity (Suarez and Darveau 2005), which is called maximal metabolic rate (MMR). This could explain why the scaling of MMR has been found to differ from that of BMR, being somewhat steeper (Bishop 1999; Glazier 2008, 2009b).

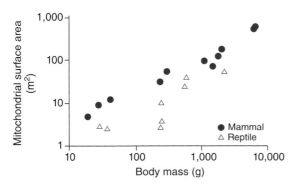

Figure 17.2 Mitochondrial surface area in m² as a function of body mass in grams in mammals (filled circles) and reptiles (unfilled triangles). The slopes did not differ significantly between the two groups, and the fitted slope is 0.76. These data are those reported in Else and Hulbert (1985b).

17.2.3 Allometry of metabolism at the cellular level

Another kind of analysis of the allometry of tissues in relation to body size considers cell size and cell-specific

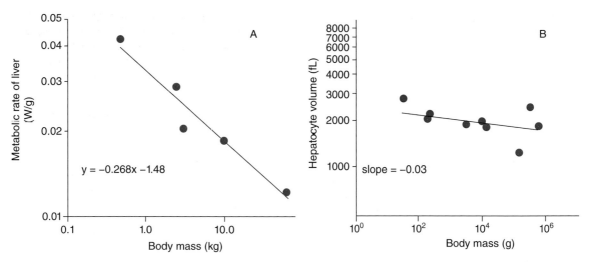

Figure 17.3 Cell size and specific metabolism in the mammalian liver. (A) Specific metabolic rate of liver, measured *in vivo*, in five mammalian species scales approximately with body mass to the −1/4, similar to the scaling for isolated hepatocytes in mammals, which is estimated to be −0.18 (Porter 2001). Data are from Table 1 in Wang et al. (2001). (B) Hepatocyte volume vs. body mass in mammals was invariant (the slope not significantly different from 0) in these nine mammalian species. Replotted from Savage et al. (2007). The view that emerges from these cellular and organ-level allometries is that as mammals increase in size they have larger livers made up of an increasing number of similar-sized cells that have decreasing cell- or mass-specific respiration rates.

respiration among animals increasing in body size. Recent work indicates that two patterns might underlie the increase in whole-animal metabolism as body size increases in terrestrial vertebrates. If $I \propto M^{3/4}$, then: (pattern 1) cell size is fairly invariant, cell number increases ($\propto M$), and presumably respiration per cell declines ($\propto M^{-1/4}$), or (pattern 2) cell size increases with body size and cellular metabolic rate remains unchanged. For 18 cell types in mammals summarized by Savage et al. (2007), 13 (72%) corresponded to pattern 1, where cell size was invariant among mammals that were sized from mice to elephants. Among many of the cell types showing this pattern, which include liver cells (hepatocytes; Fig. 17.3B), respiration per unit cell mass does decline with increasing body mass, and the scaling is estimated to be −0.18 for isolated hepatocytes in mammals (Porter 2001).

The pattern may be similar in birds (Else et al. 2004) and reptiles (Hulbert et al. 2002). Among mammals there were a small number of cell types with cell size allometries different from those in pattern 1 (more like pattern 2), but information about the allometry of their cellular metabolism and how many other cell types might be similar is scanty. Savage et al. (2007) discuss possible reasons for differences in cellular

allometry and some other interesting consequences of such differences.

17.2.4   Allometry of metabolism at the subcellular level

The analysis of the six mammalian species summarized in Else and Hulbert (1985b) revealed that organ masses scaled with body mass in the range of 3/4 to 1 and that mitochondrial surface area per unit organ scales in the range of near 0 to −1/4 (Else and Hulbert 1985a). The result is that their product, which is total mitochondrial surface area and hence capacity for oxygen consumption, scaled with mass to the 0.59 power when considering only internal organs, or 0.76 when skeletal muscle was also included (e.g., the mammals plotted in Fig. 17.2). West et al. (2002) also extended the allometric analysis of whole-animal respiration down to the level of single mitochondria and even respiratory enzymes and concluded that the scaling was approximately 3/4. Thus, these analyses offer some explanation for approximate 3/4 scaling of *I* within mammals (likewise for reptiles; Else and Hulbert 1985b).

17.2.5 The integrated view of metabolic scaling

The general view that emerges from these organ-level, cellular, and subcellular allometries is that as we consider increasing body size within a vertebrate taxon, animals are mainly made up of an increasing number of similar-sized cells that have decreasing cell- or mass-specific respiration rates (Schmidt-Nielsen 1984). Exceptions have been noted for some cell types (e.g., neurons and adipocytes; Savage et al. 2007). This integrated view also underscores how interrelated are the various organs and processes that support whole-organism metabolism. Cellular respiration in different kinds of tissues is functionally and structurally linked to other processes and tissues, such as those that support oxygen delivery (e.g., lung, circulatory system, and blood), removal of waste (e.g., kidney) and supply of nutrients (e.g., the gut; see below, section 17.3.1). A number of ideas consider how the integration of structure and function in such an interacting system works mechanistically and how it came to be through natural selection acting primarily on whole-organism performance. The idea of "symmorphosis" proposes that the capacities of various processes in series are matched to each other and to natural loads (Taylor and Weibel 1981; Weibel 2000). A complementary idea is that each step or process in a series has some degree of reserve capacity, i.e., "enough but not too much" spare capacity (Diamond 1991) or safety factor (Alexander et al. 1998). Darveau et al. (2002) viewed this issue from the perspective of metabolic control analysis, which considers that control of metabolism is vested in both energy supply and energy demand pathways. These and other views provide useful frameworks for understanding the "how and why" of metabolic allometry, which are questions that continue to intrigue biologists.

17.3 MECHANISTIC CORRELATES OF RESOURCE ACQUISITION AND ALLOCATION VS. SIZE IN TERRESTRIAL VERTEBRATES

One way that metabolic rates can be linked to population ecology and life history is through the complete energy budget of individuals, which can be summed over all members of a population (Brown et al. 1993). According to the principle of allocation (Sibly, Chapter 5, Fig. 5.2), the production of individuals, which includes their growth and reproduction, is summarized in the simple energy budget:

$$P = (C \cdot A) - I \tag{17.4}$$

where P is energy allocated to production, C is energy consumed, A is food assimilation efficiency (i.e., the fraction of food energy consumed that is digested, absorbed, and retained for allocation), $(C A)$ is assimilation rate, and I is whole-organism metabolism (all in units of W or kJ/d). Here, we consider these relationships as we begin to explore some of the implications of metabolic scaling for vertebrate population biology.

17.3.1 Consumption and assimilation rates increase with body mass, approximately to the 3/4 power

As R. H. Peters aptly pointed out, "by now, most readers will surely suspect that ingestion rate will rise as $M^{3/4}$ and homeotherms will eat more than poikilotherms" (Peters 1983). Summaries of allometries of C vs. M of hundreds of terrestrial vertebrates provided ample evidence, with mass exponents in the range 0.63–0.82 depending on taxonomic and dietary group (Peters 1983; Reiss 1989). The allometric comparison among carnivorous birds, mammals, and poikilothermic tetrapods indicated that at $M = 1$ kg, endothermic birds and mammals consumed 20 and 14.5 times more energy per day than poikilotherms, respectively. More recently, there have been several reviews of ingestion rates in mammals and birds highly motivated to feed, presumably at near-maximal levels (C_{max}). Typically they involve mammals and birds acclimated to very low temperatures, ideally at their limit of thermal tolerance, to high levels of forced activity, or hyperphagic individuals during lactation (mammals), while storing energy for migration and hibernation, or engaged in rapid growth. As a general rule, C_{max} scales as $M^{2/3}$ to $M^{3/4}$ (Karasov and Martínez del Rio 2007). Because this allometric scaling is similar to the scaling for basal metabolism (I, above), the ratios C_{max}/I cluster together and are in the range of 4–7 (Karasov and Martínez del Rio 2007).

Just as the allometry of cellular respiration is linked to allometries of other processes and tissues that support oxygen transfer, it should not surprise that the allometry of consumption is linked to allometries of

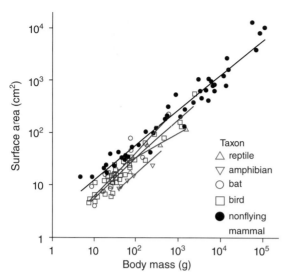

Figure 17.4 Allometry of intestinal surface area. In this plot of nominal small intestine area (area of smooth-bore tube), the slope for all 126 terrestrial vertebrate species is about 3/4 (0.76 ± 0.02), although there are some statistically significant differences among the taxa. For their body size, non-flying mammals (filled black symbols, black line) have the largest nominal surface area, significantly exceeding the other non-flying ectothermic vertebrates (unfilled blue triangles, blue lines). Flying endothermic vertebrates, the bats and birds (red), are intermediate between the non-flying mammals and the ectotherms. Data for endotherms from Caviedes Vidal et al. (2007) and data for ectotherms from Karasov and Hume (1997).

tissues and processes that support assimilation of food energy and nutrients. In terrestrial vertebrates, the surface area of the gut, where breakdown of substrates and absorption of their monomers occurs, scales with body mass to roughly the 3/4 power, and in endothermic mammals and birds it exceeds that in ectothermic tetrapods by about 4 times (Fig. 17.4; Karasov and Martínez del Rio 2007), which is similar to the difference in temperature-corrected minimal metabolism between ectotherms and endotherms (above). These patterns seem appropriate for an organ that delivers nutrients to fuel metabolism, but it is interesting that small flying vertebrates (small birds and bats) have less intestinal surface area than do non-fliers (Fig. 17.4). Caviedes-Vidal et al. (2007) suggested that fliers may rely on an extra pathway of intestinal nutrient absorption as a compensation, namely, passive absorption

between intestinal cells (called paracellular absorption). Another factor is that food processing time, indexed by "mean retention time" (measured with markers), is much faster in endotherms than ectotherms (Karasov and Diamond 1985). Mean retention time itself scales with mass to approximately the 1/4 power, an observation consistent with the allometries of C and total volume of the digestive tract (Karasov and Martínez del Rio 2007; and see below, section 17.5.1). An outcome of the allometric match between digestive features and C is that the assimilation efficiency of the gut is relatively body-size independent, within a given food type. This is apparent in the summaries provided by Peters (1983), although one exception we will discuss below is that very small herbivores are less efficient in fermenting plant material.

You may wonder how temperature affects the digestive processes and overall assimilation. The short answer is that variation in temperature has the effects on digestive rates that one would expect based on Arrhenius principles. It appears that rates of gut motility, enzymatic breakdown, and absorption all have rather similar thermal dependencies, with the net result that assimilation efficiency is relatively independent of changes in temperature within ectotherms (Karasov and Hume 1997).

17.3.2 Metabolic scaling provides a mechanistic model of growth based on individual metabolism

Several MTE studies have developed an explicit, conceptual link between I and growth rates and the timing of life-history events (e.g., generation time) (West et al. 2001, 2004; Gillooly et al. 2002; Hou et al. 2008; Moses et al. 2008a; Zuo et al. 2009). Like many models of growth, including a classic model by von Bertalanffy and Nowinski (1960), the model of West et al. (2001) and its variant by Hou et al. (2008) is a particular form of a general model that estimates the change in mass m per unit time (dm/dt) as a function of m and associated coefficients and mass exponents (see also Kerkhoff, Chapter 4):

$$\text{growth rate} = dm/dt = am^y - bm^z \qquad (17.5)$$

Some investigators using this model simply use statistics to fit their data and derive the coefficients and

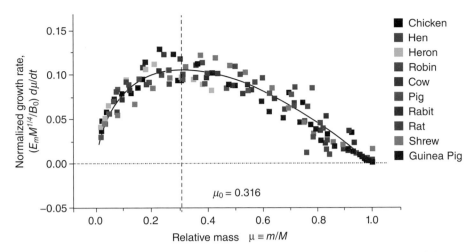

Figure 17.5 Universal growth curve. In this remarkable figure, Hou et al. (2008) plotted their values of dimensionless growth rate vs. dimensional mass ratio for a wide variety of vertebrate endotherms. The curve is universal in the sense that it was generated based on principles of energy allocation and should apply independent of taxon and body size. On the x-axis is the ratio of mass of the growing juvenile (m) relative to the adult mass (M) ($\mu = m/M$). On the y-axis is the normalized growth rate ($d\mu/dt$), which is a function of several variables: E_m, the extra respiration (metabolism) that is associated with depositing a gram of new tissue; B_0, a normalization constant equal to the adult's $I/M^{3/4}$. (From Hou et al. 2008.)

mass exponents, but West et al. directly estimated them from what they considered to be fundamental physiological parameters, some of which relate to I, which they considered scaled with mass$^{3/4}$. In their formulation, in equation 17.5, y = 3/4 and z = 1, and actual growth data for many species fit the theoretical model quite well (Fig. 17.5). Scientific peers who do not necessarily embrace the universality of 3/4 scaling were quick to point out that the original 2/3 scaling proposed by von Bertalanffy (von Bertalanffy and Nowinski 1960) fit the data just as well, and that conceptual models based on different assumptions, such as the dynamic energy budget (DEB) theory of Kooijman (2000), yield similar growth curves (Banavar et al. 2002; Ricklefs 2003; Makarieva et al. 2004; Sousa et al. 2009; Zuo et al. 2009). This point is acknowledged by West et al. (Zuo et al. 2009), who point out that MTE is nonetheless a more general model. The models that link growth to metabolism hinge importantly on the extra respiration (metabolism) that is associated with depositing new tissue (E_m in Fig. 17.5), which has historically been challenging to measure and may vary under different conditions (Karasov and Martínez del Rio 2007).

17.4 IMPLICATIONS OF METABOLIC SCALING FOR VERTEBRATE POPULATION ECOLOGY

17.4.1 Metabolic scaling provides a mechanistic model of growth of populations based on individual metabolism

Peters (1983) showed that the maximum per capita rate of population growth, r_{max}, scaled with approximately the −1/4 power of body mass. A recent review for lizards seems consistent with this (Warne and Charnov 2008), although Sibly and Brown (2007) found the scaling for taxonomic clades of mammals to be significantly steeper (−0.37; see Sibly, Chapter 5, and also results of Hamilton et al. (2011) illustrated there). Near 3/4 scaling seems partly explicable from a consideration that the similar allometries of (C A) and I in equation 17.4 (both $\propto M^{3/4}$) would lead one to predict that P would scale similarly (Reiss 1989), and hence per capita (P/M) the scaling would be near −1/4. Also, taxonomically broader surveys of energy flow through populations have shown that assimilation, respiration, and production scale with body mass in a similar fashion, and that ratios of P/I and P/CA are independent of body mass (Humphreys 1979). An important contribution of MTE has been to provide a more explicit conceptual hypothesis linking allometry of I and r_{max} (Savage et al. 2004a).

Savage et al. developed their model by considering the total metabolic rate of the population, which is the power required for maintenance, replacement, and net new individuals. Modeling maintenance as a function of M was discussed above, but the latter two features required taking into account the new features of lifespan and age at first reproduction (α), both of which empirically scale with $M^{1/4}$ (Peters 1983). Duncan et al. (2007) also allometrically modeled some of these features. We will consider briefly the hypothetical direct link between metabolism and these life-history features.

17.4.2 Is there a direct link between metabolism, lifespan, and age at first reproduction?

A link between metabolism, lifespan, and age at first reproduction (α) has long been suspected. Empirically, the last two scale with $M^{1/4}$ (Peters 1983). As pointed out by Brown et al. (2004), the scaling of mass-specific metabolism (B) is approximately $-1/4$ (because $B = I/M \propto M^{3/4}/M$), and biological times, such as turnover times of metabolic substrates, are the reciprocal of rates and hence exhibit 1/4-power scaling. MTE thus predicted that biological times (t_B), including lifetime and α, scaled as

$$t_B \propto M^{1/4} \, e^{E/kT} \tag{17.6}$$

Munch and Salinas (2009) found that intraspecific variation in lifespan in many ectotherms, including some terrestrial vertebrates, can be explained by temperature in equation 17.6. One mechanism for the 1/4-power body mass scaling of lifespan suggested by Brown et al. was based on a theory of senescence that attributes aging to cumulative damage at the molecular and cellular levels by the free radicals produced as byproducts of aerobic metabolism (see Speakman (2005a) for a brief history of this hypothesis).

Comparative studies among groups of species have been useful for testing this hypothesis, especially after addressing certain technical issues. One was that lifespan and many other traits may be associated with body size, as is metabolism, without their underlying mechanisms necessarily acting through metabolism (the covariation of traits with M; Speakman 2005b). Also, hypothesis testing using comparative data from many species that share phylogenetic history requires special statistical procedures (Speakman 2005b).

de Magalhaes and his colleagues (2007) took account of these issues in their analysis of maximum longevity in 1456 mammals, birds, amphibians, and reptiles. To remove confounding effects of collinearity with M, they regressed body-mass residuals of B against lifespan. They used the method of phylogenetic independent contrasts to address the issue of shared phylogenetic history. Residual B did not correlate significantly with lifespan in either eutherian mammals or birds. Animals with lower B for their body size did not tend to live longer, as might be predicted from MTE. In a somewhat similar analysis, Speakman replaced the basal metabolism with daily energy expenditure of free-living animals and tested for correlations with lifespan in a smaller set of birds and mammals (44 and 48, respectively; Speakman 2005a). In that analysis, there was a significant negative relationship between residual metabolism and lifespan in the mammals, but not in the birds. Both of these studies also underscored a previously known pattern that birds and bats have relatively long lifespans for their size or metabolism, compared with non-flying mammals. So, even though the analyses did not entirely reject the underlying inverse link between metabolism and lifespan hypothesized by MTE, the comparison among flying and non-flying mammals and birds points to a lack of universality in the concept.

Lovegrove (2009) analyzed age at first reproduction (α) in mammals. Like the earlier analysis of de Magalhaes et al. (2007), Lovegrove used various alternative regression models to test for correlations among those variables and I after removing the confounding effects of colinearity, body mass, and phylogeny. In his analysis, α was not significantly partially correlated with either M, I, or growth rate. He thus concluded that there was little evidence that individual metabolism governs the rate at which energy is converted to growth and reproduction at the species level.

Although many population and life-history features scale with body mass to the approximately 3/4 power (or $-1/4$ power if per capita) (above citations; see also Sibly and Brown 2007, 2009), it remains tantalizing but difficult to draw the direct connection with this exponent and metabolism per se. Furthermore, many of the allometric analyses reveal that features of lifestyle (e.g., some unique combination of anatomical, physiological and/or behavioral traits; Sibly and Brown 2009) and phylogenetic association (sometimes these two are correlated) are an additional significant source of variation on top of temperature and size (e.g., Lovegrove 2009; Sibly and Brown 2007). Thus, the difficulty in linking directly the scaling of life-history features to the scaling of metabolism is perhaps predictable as one extends the fundamentals of metabolic allometries to higher levels of biological integration, which seem inevitably more complex. Ricklefs (2010a, 2010b) reviews tests for other factors that influence lifespan and α in some groups. He concludes that the strong vertebrate-wide relationships among size, lifespan, and age at maturity can be explained by evolutionary theories that relate rate of aging (hence lifespan) primarily to extrinsic mortality, rather than to an intrinsic factor such as metabolism.

17.4.3 MTE predicts the allometric scaling of population density (in some cases)

Notwithstanding the challenges of directly linking life-history features to the scaling of metabolism, MTE does yield predictions about vertebrate populations that are partly borne out.

One of the most considered (White et al. 2007) is the expectation that carrying capacity, hence population density (abundance), will decline as $M^{-\alpha}$ (where α is the particular mass scaling relationship for metabolism, e.g., 3/4; Damuth 1987, 2007). This expectation is a direct consequence of the dependence of the metabolic rate of an individual (I) on its body mass (M). Maximum population density for a given species (N_{max}) should be approximately equal to

$$N_{max} = R/I = R/(i_o M^{3/4}) \qquad (17.7)$$

where R represents the rate at which resources become available to the population, and I is the metabolic rate of each individual ($I = i_o M^{3/4}$ as in equation 17.2). This simple model predicts that population density should decrease with a species' body mass to the approximate $-3/4$ power. This pattern, termed energy equivalence by Isaac, Carbone, and McGill in Chapter 7, has been demonstrated in mammals, birds, and reptiles (Marquet et al. 2005; Buckley et al. 2008) (Fig. 17.6). Not surprisingly, densities are lower for secondary consumers

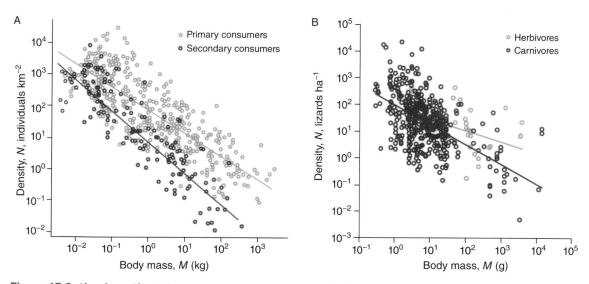

Figure 17.6 Abundance (density) vs. consumer mass in mammals and lizards. (A) Density vs. consumer mass in mammals for primary consumer species and secondary consumer species. The slope of the relationships is −0.73 (not much different from −3/4) for primary consumers and −0.99 (not much different from −1) for secondary consumers. Redrawn from Marquet et al (2005). (B) Density vs. consumer mass in lizard species for herbivores and omnivores (green) and carnivores (red). Plant-eating lizards had significantly higher densities than carnivores, and the scaling slopes did not significantly differ between consumer types or from −3/4. Figure plotted from data provided by Buckley et al. (2008).

(predators) than for primary consumers (herbivores) (e.g., Fig. 17.6A,B) because there is generally greater food abundance for the former than the latter (i.e., larger numerator in equation 17.7). But, what is curious about the dataset in Fig. 17.6A is that predators' scaling seems closer to -1 than to $-3/4$ (Marquet et al. 2005). Assuming that consumer need (the denominator of equation 17.7) indeed scales with M^α where $-2/3 < \alpha < 1$, which is a very reasonable assumption considering the previous discussion, the explanation might have to do with factors that relate to the numerator, which is the food base supporting the consumer.

Carbone and colleagues have shown how the scaling of predator abundance might differ according to characteristics of the prey resource, such as density and size (Carbone and Gittleman 2002; Carbone et al. 2007). One example of this might be instances in which prey size and/or prey density are related to predator size. For mammalian carnivores, for example, large predators tend to eat larger prey, and in fact prey mass tends to scale isometrically with carnviore mass, i.e., $M_{prey} = aM_{pred}^1$ (Peters 1983; Vezina 1985). Consider how incorporating this into the numerator of equation 17.7 changes the predicted scaling of predator density.

For the numerator of equation 17.7, the productivity of a predator's prey base (R_{prey}) can be estimated in terms of mass, and hence energy, based on the premise that

$$R_{prey} = r\, N_{prey}\, M_{prey} \qquad (17.8)$$

where r is the per capita growth rate of the prey population, N_{prey} is the number of individuals in the prey population, and M_{prey} is prey mass. Prey mass tends to scale isometrically with predator mass, i.e., $M_{prey} = aM_{pred}^1$ (Peters 1983; Vezina 1985). Because $r_{max} \propto \text{mass}^{-1/4}$ (previous discussion) and $M_{prey} = aM_{pred}$, we can estimate $r_{prey} = aM_{pred}^{-1/4}$. Hence,

$$R_{prey} = (aM_{pred})^{-1/4}\, N_{prey}(aM_{pred}) \propto (M_{pred})^{3/4} \qquad (17.9)$$

Placing equation 17.9 into the numerator of equation 17.7 yields the conclusion that density of the carnivores is independent of carnivore mass. The problem is that this is not the empirical observation (Fig. 17.6A).

The result of this allometric exercise is perplexing because it is inconsistent with observed patterns, and so this is clearly an area ripe for future work. However, the observed patterns may not be so consistent, as it turns out. Carbone and colleagues (Carbone et al.

2010) found that mammalian carnivore density is best predicted by both the carnivore's mass and the prey's biomass density and their interaction, which means that the allometric model in equation 17.7 is likely too simple. Isaac, Carbone, and McGill (Chapter 7) cite increasing evidence of systems where the energy equivalence rule does not appear to hold true. We agree with Carbone et al. (2007) that to understand patterns of abundance we need to understand the scaling of consumer population density in relation to features of the food resource as well as to consumer need (metabolic rate). In turn, these relationships can be enormously useful in modeling features of food webs in ecological communities (Reuman et al. 2009; Petchey and Dunne, Chapter 8).

17.5 OTHER IMPLICATIONS AND APPLICATIONS OF METABOLIC SCALING IN VERTEBRATE BIOLOGY

17.5.1 MTE can highlight special problems of relatively small animals

Recall that many biological times will be short because they are the reciprocal of rates and hence exhibit 1/4-power scaling. For some processes there may be notable time constraints that lead to interesting ecological implications.

For example, many herbivores rely on their gastrointestinal microbiota to ferment difficult-to-digest plant cell wall material. The microbiota break down cell wall material relatively slowly, and if herbivores retain material in their gut for < 12 hours the extent of cell wall digestion is relatively low (Hummel et al. 2006). Several researchers have derived from allometric considerations the manner in which the food retention time might be coordinated with metabolic demand (Demment and Van Soest 1983; Calder 1984), and retention time goes down with declining body size. That scaling emerges from the observation that energy expenditure and feeding rate at the whole-animal level increase allometrically with body mass (M), approximately as $M^{3/4}$, whereas the volume of the gut (or mass of gut contents) increases isometrically, approximately as M^l (Parra and Montgomery 1978). The turnover (retention) time for contents of a full gut would then be

$$\text{time} \propto \text{amount/rate} \propto M^l/M^{3/4} \propto M^{1/4} \qquad (17.10)$$

Thus, in small herbivores retention time and digestive efficiency on plant material may decline with declining body size (e.g., Hume et al. 1993). Ultimately, this likely imposes a lower size limit on animals that rely on fermentation of difficult-to-digest cell wall material to meet their energy needs. Small herbivores, instead, need to be selective and ingest plant materials that have very low cell wall contents (e.g., new growing rather than senescent leaves, flowers, fruits), which they can efficiently digest without relying on extensive fermentation.

As another example of a possible time constraint, altricial birds have the fastest growth rates among animals and the smallest of them have such rapid development (egg incubation <2 weeks, time from hatch to fledging <2 weeks) that it raises questions about reaching important developmental/immunological milestones. The time to produce B-cell lineages that express functionally different Ig specificities (against possible pathogen antigens) may be at least 4 weeks and independent of body size (Klasing et al. 1999). If this is so, how can small nestlings compensate, or are they fundamentally more vulnerable to infections because of their small size and very rapid development? Other components of the avian immune system may also be size invariant (Banerjee and Moses 2010).

17.5.2 MTE can advance knowledge about biology and ecology of huge animals that are now extinct

The largest flying vertebrates are now extinct (e.g., the bird *Argentavis*, the pterosaur *Quetzalcoatlus*), but their size (<100 kg) was much smaller than that of the largest terrestrial vertebrates. The maximum size for vertebrate fliers is probably limited by flight dynamics such as the aerofoil size-to-weight ratio (Alexander 1998; Sato et al. 2009; Witton and Habib 2010). The largest terrestrial animals were sauropod dinosaurs, whose mass is estimated to be more than 50 000 kg based on allometric relationships between bone linear dimensions and body mass of extant organisms (Packard et al. 2009; Cawley and Janacek 2010) and other methods (Alexander 1998). Supporting such a huge mass would not necessarily have been a problem for a walking animal, but they would have been challenged by the possibility of overheating, which argues against them having higher metabolic rates like modern endotherms (Alexander 1998).

It is fascinating to learn how new technologies and scaling principles are being used to make inferences about body composition (Franz et al. 2009), temperature relations, metabolism, and metabolic ecology of extinct organisms and their communities. Stable isotope ratios in fossilized material can provide evidence about body temperatures, possibly as far back as the Miocene (Eagle et al. 2010). Gillooly et al. (2006b) inferred even earlier dinosaur body temperatures using principles from metabolic ecology. The inference was made using apparent growth rates of dinosaurs, which can be calculated based on morphological and histological features of fossilized bones such as their size and apparent growth lines (e.g., Erickson et al. 2001; Sander et al. 2004; Erickson 2005; Bybee et al. 2006; Lehman and Woodward 2008). With apparent growth rates (G) and estimated body masses (M) of dinosaurs in hand, Gillooly et al. modeled growth rate as a function of mass and temperature using a variant of equation 17.2 and other principles from metabolic ecology that relate to growth and development:

$$G = g_o M^{3/4} e^{0.1T} \tag{17.11}$$

By using the dinosaurian size and growth information, and rearranging to solve for T, they concluded that dinosaur body temperatures increased with body mass from approximately 25 °C at 12 kg to above 40 °C for dinosaurs weighing tens of kilograms. Heat balance models based on first principles of heat exchange between animals and their environment also predict that the very largest of dinosaurs living in warm environments would have had high body temperatures even if they had low, reptilian metabolic rates (Spotila et al. 1973).

Is it possible that some dinosaurs had high metabolic rates, like modern endotherms? Using information from biomechanics (Pontzer et al. 2009) and fossilized evidence of anatomical structures associated with endothermy in extant vertebrates, such as nasal respiratory turbinates and pneumaticity of the skeleton (Hillenius and Ruben 2004; O'Connor and Claessens 2005), researchers have made arguments for and against dinosaur endothermy. Sander et al. (Sander and Clauss 2008; Sander et al. 2011) and McNab (2009) provide good overviews of this topic, as well as discussions of how assumptions about metabolism relate to life history and population sizes of dinosaurs in prehistoric communities (see also Farlow et al. 2010). Similarly, Smith et al. (2010a, 2010b) use metabolic ecology to advance knowledge about prehistoric mammal communities. The study of all these fascinating issues relies on principles, methods, and knowledge about scaling and metabolic ecology in vertebrates.

17.5.3   MTE can advance knowledge about human and ecosystem health

Metabolic ecology, and the scaling principles and data embedded within it, can guide research that advances knowledge and practices relating to health of ourselves and of our ecosystems. For example, one theory of aging views senescence as the gradual accumulation of damage from metabolic byproducts, so longevity would be expected to be inversely related to metabolism (see Speakman (2005a) for a brief history of this hypothesis). But, contradicting this pattern, bats and birds have exceptional longevity despite high metabolic rates (Munshi-South and Wilkinson 2010). Comparative and laboratory studies of mechanisms in bats and birds that resist oxidative damage to DNA and cellular structures could prove informative. As another example, predicting drug and toxin levels in veterinary practice with animals of many different sizes requires

knowledge of the body-size dependence of rates of toxin intake and elimination (Riviere 1999). Finally, as reviewed by Boyer and Jetz (Chapter 22), metabolic ecology can offer prescriptions for protecting the health of our ecosystems through biodiversity conservation.

17.6   CONCLUDING REMARKS

Research with terrestrial vertebrates has contributed substantially to MTE with mechanistic descriptions of how whole-animal metabolism varies with size and temperature. The allometric relationships can be used to make useful predictions for animals of known size and temperature that have not yet been otherwise studied. Furthermore, the allometric relationships can be used to derive new relationships such as those that explore how metabolism might influence features of vertebrate populations and communities. MTE can highlight special problems of relatively small animals, advance knowledge about biology and ecology of huge animals that are now extinct, and advance knowledge about human and ecosystem health.

Chapter 18

SEABIRDS AND MARINE MAMMALS

Daniel P. Costa and Scott A. Shaffer

SUMMARY

1 The metabolic theory of ecology (MTE) provides a baseline for understanding deviations from general patterns. In adapting to life in the marine environment, seabirds and marine mammals have evolved a diverse spectrum of life-history patterns that are fertile ground for examining the importance of metabolic processes and the MTE.

2 We first explore how body size, a fundamental trait in MTE, and foraging mode influence breeding strategies and maternal investment in marine mammals and seabirds.

3 Then we consider the special adaptations that allow albatrosses, among the largest flying birds, to occupy a unique niche in the marine environment.

Although they range widely from the tropics to polar seas, albatrosses occur in the windiest regions because they are specialists at soaring flight, exploiting prevailing winds to travel rapidly over ocean basins with minimal energy expenditure. Consequently, albatrosses have the lowest flight cost of any vertebrate.

4 In general, flight costs decrease with increasing body size, with costs as low as 1.4–$2.0 \times$ BMR reported for wandering albatrosses (*Diomedea exulans*), the largest flying seabird ($10\,$kg). However, the specialization for soaring flight has resulted in evolutionary constraints on physiological performance for powered flight and reproductive output.

18.1 INTRODUCTION

A major criticism of the metabolic theory of ecology (MTE), and allometry in general, is that in an effort to uncover fundamental physiological and ecological constraints, precision is sacrificed for generality. While grand patterns become apparent and fundamental processes emerge, variations from general patterns are often overlooked or simply discounted as "noise." However, deviations from general patterns provide insight into the adaptations of organisms to their spe-cific ecology and/or physiology. Thus MTE provides a useful benchmark for examining the deviations from the general pattern in how organisms expend and acquire energy, and how these differences are related to adaptations for distinctive niches and habitats.

Marine mammals and seabirds are a very diverse group of vertebrates. Their evolution entailed multiple independent invasions of the marine environment by the ancestors of five extant marine mammal taxa (cetaceans, sirenians, pinnipeds, sea otter, and polar bear) and six orders of extant seabirds: (1) penguins;

Metabolic Ecology: A Scaling Approach, First Edition. Edited by Richard M. Sibly, James H. Brown, Astrid Kodric-Brown.

(2) loons; (3) grebes; (4) albatrosses; (5) pelicans, boobies, and cormorants; and (6) the shorebirds, gulls, terns, auks, murres, and puffins.

18.2 GENERAL PATTERNS OF ENERGY INTAKE AND EXPENDITURE, FORAGING, AND REPRODUCTION

Of relevance to the MTE is that these taxa exhibit a remarkable diversity of life-history patterns and body sizes that have resulted in extreme variations in the metabolic intensity and energy allocation to life processes. For example, flying seabirds range in body size from the 42 g Wilson's storm petrel (*Oceanites oceanicus*) to the 10 kg wandering albatross (*Diomedea exulans*), and up to 30 kg for the flightless emperor penguin (*Aptenodytes forsteri*). Many seabird species live 40–50 years, and their reproductive patterns range from rapidly growing altricial hatchling auklets to highly precocial and slowly growing albatrosses (Schreiber and Burger 2001).

Marine mammals are larger and cover an even greater range of body sizes from the 25 kg sea otter (*Enhydra lutris*) to the 200 000 kg blue whale (*Balaenoptera musculus*), the largest animal to have ever

lived! They also tend to be long-lived. Many species live 40–50 years, and data suggest that bowhead whales (*Balaena mysticetus*) live well beyond 100 years (George et al. 1999). With the exception of polar bears (*Ursus maritimus*), all marine mammals are extremely precocial, because they must be capable of surviving in the water or in a crowded colony immediately after birth.

As a group, marine mammals exhibit considerable variation in both the duration and pattern of maternal investment. True seals and baleen whales have extremely short lactation durations relative to their size (Fig. 18.1), made possible by lipid-rich milk (up to 65% lipid; Costa 2009). In the case of the hooded seal (*Cystophora cristata*) a pup can double its body mass prior to weaning in just four days after birth (Bowen et al. 1985)! The reproductive output of blue whales (c. 50–150 tons), which typically produce a calf every two to three years, is also quite astounding. This contrasts markedly with sperm whales (*Physeter macrocephalus*) which are 1/10th the size of a blue whale (females weigh about 15 tons) but produce a calf every five years (Sears and Perrin 2009; Whitehead 2009).

For both marine mammals and seabirds, these diverse life-history patterns have evolved in response to the dynamic nature of marine environments where resource (energy) availability varies dramatically in

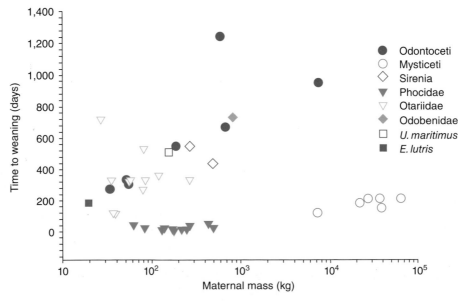

Figure 18.1 Time to weaning plotted as a function of maternal mass for marine mammals. Lactation durations of phocid seals and mysticete whales are shorter than all other marine mammals. Figure adapted from Costa (2009).

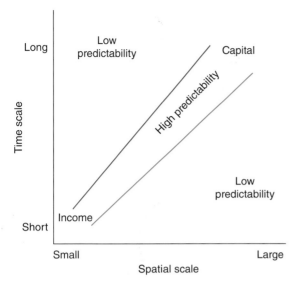

Figure 18.2 The conceptual range of foraging scales that can be effectively utilized by air-breathing marine vertebrates. Income breeders are limited to operating over small spatial scales due to the limited time they can spend away from their young, whereas capital breeders can integrate resources over much larger spatial and temporal scales.

both space and time (Schneider 1994). Reliably finding resources in such a variable environment is limited to a range of foraging patterns (Weimerskirch 2007) (Fig. 18.2). Consequently, marine vertebrates have evolved a suite of life-history traits that allow species to match the spatio-temporal variability in food energy acquisition with energy allocation to meet the demands of reproduction and self-maintenance (Trillmich and Trillmich 1984; Costa 1991, 1993; Boyd 1998; Trillmich and Weissing 2006; Houston et al. 2007).

Since mammals and birds breathe air, reproduction is further constrained by the need to breed on land contrasted by the need to forage at sea (cetaceans and sea snakes, which give birth at sea, are notable exceptions). The separation between breeding and feeding habitats has led to the evolution of two general life-history patterns: (1) income breeders (most seabirds, sea lions, fur seals, toothed whales, and marine reptiles), where the young are provisioned from resources acquired by foraging as they are needed; and (2) capital breeders (true seals and baleen whales) where resources are acquired and stored over a long time period prior to a reproductive event (Costa 1991, 2009). Because

capital breeders obtain all of the resources necessary to provision their offspring during one very long trip to sea prior to parturition (seals) or while residing and feeding in a cool productive habitat before migrating to a warmer oligotrophic environment to calve (baleen whales), they are capable of ranging over vast spatial scales, sometimes with thousands of kilometers separating their foraging and breeding grounds (Fig. 18.3) (Costa 1993, 2009). In contrast, most income breeders return frequently to provision their offspring and are therefore limited to trips lasting hours to a few days. Income breeders are thus limited to foraging at distances of tens to hundreds of kilometers from the rookery (Costa 1993; Trillmich and Weissing 2006; Houston et al. 2007) (Fig. 18.3). Further, there is a premium on maximizing energy gain while minimizing time spent foraging, so income breeders typically occur in oceanic regions where productivity is high, such as upwelling regions (Costa 1993, 2008). To minimize time spent away from offspring, income breeders generally exhibit high levels of foraging effort and more expensive modes of locomotion, resulting in high field metabolic rates (FMR; see Brown and Sibly, Chapter 2 and Karasov, Chapter 17). By contrast, capital breeders are not constrained by the time away from offspring, and can therefore economize on energy expended on foraging (Costa 1991, 1993).

18.3 COMPARISONS AMONG SEABIRDS

Here we use allometric relationships to examine the constraints of powered flight in relation to body size across a broad range of seabirds including the order Procellariiformes, the albatrosses and petrels, an order that contains the largest flying seabirds. We examine how foraging range and method of prey delivery vary across seabirds by comparing the typical foraging range from the colony and method of prey delivery for the four major orders of seabirds (Ashmole and Ashmole 1967; Nelson 1978; Warham 1990; Schreiber and Burger 2001). We then examine how FMR and flight cost vary across different seabird taxa, including flying and flightless seabirds. We compared the FMR and flight costs of 15 species of Procellariiformes (Leach's and Wilson's storm petrels, common, diving, Georgian, snow, Antarctic, Cape, and giant petrels, wedge-tailed shearwaters, Antarctic prions, Northern fulmars, and Laysan, grey-headed,

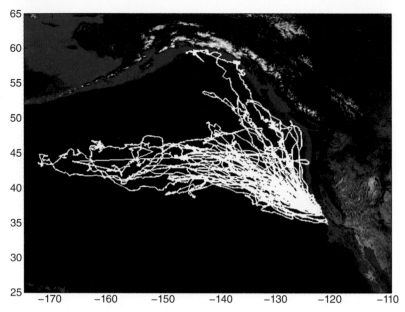

Figure 18.3 ARGOS satellite tracks of foraging trips of capital breeding female northern elephant seals (white) compared to foraging trips of lactating California sea lions (red). Elephant seals were foraging from the Año Nuevo colony (Robinson et al. 2010), California and sea lions from colonies on San Miguel Island, Southern California (Robinson and Costa, unpublished). Figure by P. Robinson.

black-browed, and wandering albatrosses), four species of Pelecaniformes (white-tailed tropicbirds, Cape and Northern gannets, and red-footed boobies), and 14 species of Charadriiformes (common, Arctic, and sooty terns, brown noddies, dovekies, common and thick-billed murres, black guillemots, Cassin's and least auklets, Atlantic puffins, and black-legged kittiwakes) measured with the doubly-labeled water method (Lifson and McClintock 1966). Data were taken from the literature (Ellis and Gabrielsen 2001; Shaffer et al. 2004; Shaffer 2011). FMR was plotted as a function of body mass and compared to the theoretical cost of powered flight as a function of body size (Schmidt-Nielsen 1984). The cost of flight scales linearly with mass while metabolic rate should scale as mass to the exponent of 0.75. First we look at a special example, albatrosses.

18.3.1 A special example, albatrosses

Due to their evolutionary specialization for soaring flight albatrosses can forage over large spatial scales,

often covering thousands of kilometers in a matter of days, and do so with one of the lowest costs of transport measured (Costa and Prince 1987; Weimerskirch et al. 2000; Shaffer et al. 2001, 2004). This efficient mode of locomotion allows them to cover the same distances as a capital breeder, but in a matter of days instead of months, as is the case for an elephant seal (*Mirounga angustirostris*; Fig. 18.4). Furthermore albatrosses, like nearly all Procellariiform seabirds (including shearwaters and petrels), have evolved the ability to concentrate energy from ingested prey into an oil-rich slurry that is fed to chicks (Warham 1990). This key adaptation enhances flight performance by allowing parents to reduce the bulk of transporting solid prey over great distances before returning to the nest. Marine mammals have evolved a similar approach through lactation by producing lipid-rich milk during a foraging trip (Costa 1993). Lactation enables marine mammals to better optimize the delivery of energy to their young as the energy content of the milk is independent of the type or quality of prey consumed, or the distance or time taken to obtain it (Costa 1991).

The economical soaring flight of albatrosses is enabled by several key elements including: (1) long

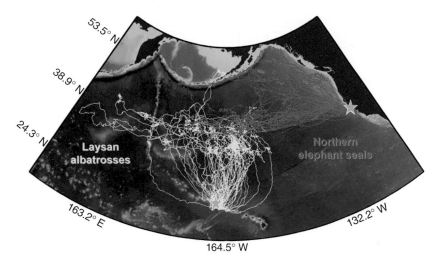

Figure 18.4 The 8- to 9-month foraging tracks of 20 female northern elephant-seals (orange) are compared to the 10- to 30-day foraging tracks (yellow) of 31 Laysan albatrosses. Note that both species range widely across the North Pacific but the time scales used to cover such distances vary dramatically. The background is a relative representation of bathymetry (lighter is shallower, no given scale). Blue stars indicate the positions of the colonies (albatrosses: Tern Island, Hawaii; elephant-seals: Año Nuevo, California coast). (Albatross data from Kappes et al. (2010) and elephant seal data from Robinson et al. (2010). Figure by Y. Tremblay.)

narrow wings (Fig. 18.5) coupled with a wing/shoulder lock mechanism that allows maintenance of an outstretched wing with minimal expenditure of energy (Pennycuick 1982) and (2) a large body size (2–10 kg) that increases wing loading for faster flight and greater penetration through the wind (Pennycuick 1989). Although the economical flight of albatrosses has been described as a major adaptation to the marine environment (Pennycuick 1982), it is also likely this is the only possible niche available for a large flying seabird. In fact the large body size likely prevents albatrosses from utilizing any other form of flight (e.g., sustained flapping, or flap gliding).

ered to nestlings increases compared to the amount of fresh or undigested food. Thus gulls and auks that feed relatively close to the colony return frequently with fresher food (i.e., whole prey) whereas albatrosses return with a higher percentage of stomach oil produced along the foraging trip when ranging far from the colony. Seabirds such as boobies, gannets, and penguins that forage at distances that are intermediate or less than those of albatrosses return with partially digested meals (Fig. 18.6). An analogous pattern is observed in pinnipeds, where the lipid content of the milk is higher for species that make longer trips to sea (Fig. 18.7) (Costa 1991).

18.3.2 Foraging range and method of prey delivery

Across the four main orders of seabirds, there exists a spectrum in the distances at which adults forage from the nest. Overall, these distances typically range from a few tens of kilometers for gulls to several thousand kilometers for albatrosses (Fig. 18.6). It also becomes apparent that when foraging at considerable distances from the nest, the quantity of predigested prey deliv-

18.3.3 Field metabolic rates and flight costs

Further comparison of body size constraints on flight was made by comparing the ratio of FMR to basal metabolic rate (BMR) between flying and non-flying seabirds (i.e., 10 species of penguins; Order, Sphenisciformes), which range in mass from 1 to 35 kg. Like previous analyses (Nagy et al. 1999), allometric comparison of energy expenditures reveals that there was a significant

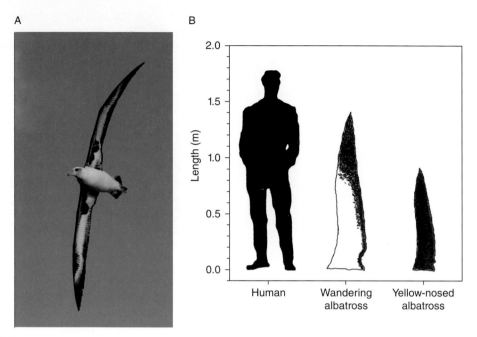

Figure 18.5 (A) Photograph of a Laysan albatross (*Phoebastria immutabilis*) turning on an updraft of wind over Tern Island, Northwest Hawaiian Islands. Photo by D. Costa. (B) Single wing length of the largest and smallest albatross species compared to human height for scale. The length and narrowness give albatross wings a high aspect ratio.

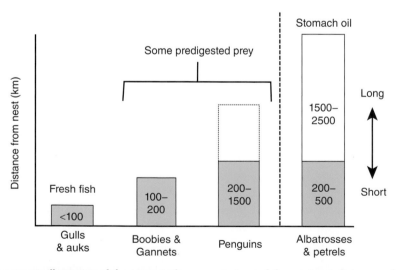

Figure 18.6 Diagrammatic illustration of the amount of prey processing and the maximum distance seabird parents will range from the colony while foraging, in representative seabird groups. The solid portion of each bar represents the contribution of fresh food, whereas the open portion represents the contribution of partially digested food or stomach oil in the diet. The diagram illustrates that as range from a breeding colony increases, the proportion of predigested food or stomach oil becomes greater than the proportion of fresh food in far-ranging species.

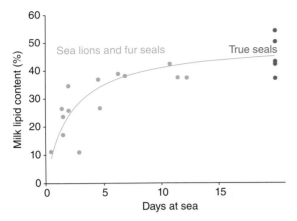

Figure 18.7 The relationship between the relative lipid content (%) of sea lion and fur seal milk (green symbols) is shown relative to the duration of their foraging trips. True seals (blue symbols) are capital breeders and make only one long foraging trip and produce milk of relatively high lipid content. (Data from Costa 1991; Georges et al. 2001; Schulz and Bowen 2004.)

($r^2 = 0.864$, $p < 0.05$) relationship between FMR and body mass for all seabird species combined, described by the equation $y = 0.919M^{0.77}$. While the smaller Procellariiformes such as petrels and shearwaters had FMR similar to other seabirds, albatrosses as a group exhibited FMRs consistently below the FMRs of other seabirds of similar body mass (Fig. 18.8). Further, the capacity for large seabirds to increase their metabolism for powered flight is lower than for small seabirds, giving them a lower metabolic scope (the capacity to increase metabolism above resting levels) (Fig. 18.8). This supports the prediction that birds larger than 10 kg should not be able to sustain powered flight. This prediction was based on the intersection between an assumed maximum sustainable metabolic rate of $10 \times \text{BMR}$ (where BMR scales with mass$^{0.75}$), and the theoretical cost of powered flight, which scales isometrically with mass (Schmidt-Nielsen 1984). A criticism of this prediction is that the cost of powered flight may not scale isometrically as small and large birds would not necessarily employ the same wing design. In fact this is true, as larger Procellariiformes have higher aspect ratio wings than smaller species (Pennycuick et al. 1984). Furthermore, the differences in FMR

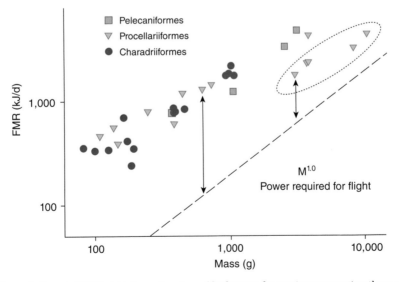

Figure 18.8 Field metabolic rate (FMR) plotted as a function of body mass for species representing three orders of flying seabirds. The dashed line is the theoretical cost of powered flight, which scales isometrically with body mass (Schmidt-Nielsen 1984). Regardless of the placement of the line with respect to the y-intercept, larger species have a lower metabolic capacity. The arrows reinforce the point that the metabolic rate of larger species is closer to the cost of flight than smaller birds. Data for albatrosses fall within the dotted ellipse. Sources for the original FMR data can be found in Shaffer (2011).

between small and large Procellariiformes is consistent with differences in wing geometries, as smaller petrels have shorter wings that allow higher wing beat frequencies and with greater vertical wing displacement stroke, whereas the longer, higher aspect ratio wings of albatrosses result in lower wing beat frequencies and smaller vertical displacement. Thus the albatross wing design trades power for efficiency and an economical cost of flight that restricts them to the windiest regions on earth (Weimerskirch et al. 2000; Suryan et al. 2008).

The above differences in aerodynamic design, flight mechanics, and energetics are consistent with larger flying seabirds having lower ratios of FMR/BMR compared to smaller flying seabirds. Smaller Procellariiform seabirds are more similar in body/wing morphology to other flying seabirds (Pennycuick 1982). More importantly, FMR/BMR ratios for penguins (Sphenisciformes) fall within the range observed for smaller flying seabirds, supporting the idea that, in the absence of flight, there is no intrinsic limitation of body size on metabolic scope for activity (Fig. 18.9).

If we compare seabirds of similar body mass range to the albatrosses, it appears that albatrosses have lower at-sea FMR than a range of most other seabirds. This result is perhaps not all that surprising given that albatrosses are specialists at soaring flight (Pennycuick 1982). Thus, a combination of high aspect ratio wings and anatomical wing lock mechanism (Pennycuick 1982) permits albatrosses to use prevailing winds to travel rapidly over the sea surface with a minimal amount of energy expenditure. Undoubtedly, this evolutionary adaptation is likely one of the primary reasons for the success of albatrosses as pelagic foragers in waters that range from the tropics to polar seas. However, it is also probable that this is the only niche available for a large flying seabird, due to the limitations of powered flight at large body sizes.

Using the comparative approach across a range of flying and flightless seabirds that cover almost four orders of magnitude in body mass, we show that, as body mass increases, the range of foraging patterns and their associated metabolic intensity that are available to flying seabirds are limited. While we chose to examine a very specific trait we view this as a starting point.

18.4 CONCLUDING REMARKS

Given the range of body sizes and diversity of life-history traits and foraging strategies in seabirds and

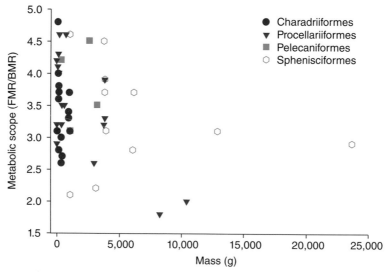

Figure 18.9 The ratio of field metabolic rate to basal metabolic rate (FMR/BMR) is plotted as a function of body mass for four orders of seabirds. Flighted species are solid symbols and flightless species (penguins) are open symbols. Note that the largest Procellariiformes, albatrosses (triangles), have a considerably lower ratio than all other flying or non-flying birds. Original sources for FMR and BMR data can be found in Ellis and Gabrielsen (2001) and Shaffer (2011).

marine mammals, many other questions can be probed. The examination of these relationships across taxa, where the individual life-history patterns are well understood, provides insight into the deviations from the more general patterns generated by the MTE. Thus both approaches provide insight into the adaptive process. MTE provides a general framework that identifies patterns that can be derived from physics, chemistry, and general biology. Observed deviations from these patterns, which occur so dramatically in seabirds and marine mammals, can be used to identify interesting life-history traits and adaptations to the environment that warrant further study. Thus it is not a matter of either or, but of how one uses these insights to understand the process of adaptation.

ACKNOWLEDGMENTS

This work was supported by the TOPP program with funding from the Sloan, Packard and Moore Foundations, as well as from the Office of Naval Research N00014-10-1-0356, the National Science Foundation Office of Polar Programs grant ANT-0838937 and the Oil and Gas Producers E&P Sound and Marine Life Program contract JIP22 07-23.

Chapter 19

PARASITES

*Ryan F. Hechinger, Kevin D. Lafferty,
and Armand M. Kuris*

SUMMARY

1 The metabolic theory of ecology (MTE) can inform parasitology and vice versa. The host–parasite interaction is fundamentally metabolic. Parasites obtain all their energy and materials from hosts. This and other aspects of parasitism consequently impact a range of processes from the performance of individual parasites and hosts, to the organization of food webs in ecosystems. Because parasites comprise a massive component of diversity, considering them can help to test, refine, and expand MTE.

2 Parasites can differ from free-living species in basic attributes of energy metabolism. This might involve maximizing metabolic power instead of efficiency, something not generally realized.

3 Parasite life-history traits seem to scale with parasite body size. The scaling exponents appear to be similar to those characterizing free-living species. Host and parasite allometric scaling can be used to parameterize parasite population models. Measures

of fitness can be extracted from these models to examine how parasite body size and consumer strategy might evolve in relation to host body size.

4 Maximum parasite abundance within hosts might scale with parasite body size to the −3/4 power while total parasite biomass within hosts scales with host body size. However, a thorough MTE framework has not been developed for parasite abundance within hosts.

5 The abundance of parasitic and free-living species in three ecosystems follows the same universal scaling relationship. But this is only apparent after accounting for the flow of energy among trophic levels. The universal −3/4 scaling of abundance implies the invariant production of biomass with body size across all life forms, regardless of functional group: ectotherm or endotherm, vertebrate or invertebrate, free-living or parasitic.

6 There are many, promising avenues for future research, both empirical and theoretical.

19.1 INTRODUCTION

Metabolic ecology uses individual metabolism as the foundation for a conceptual framework of ecology. This is natural, because we can consider metabolism to be the transformation and use of energy and materials by individual organisms, and much of ecology can be expressed as the flow of energy and materials within organisms and between them and their environment (see Brown, Sibly, and Kodric-Brown, Introduction). The metabolic theory of ecology (MTE) (Brown et al. 2004; Brown and Sibly, Chapter 2) puts metabolic

Metabolic Ecology: A Scaling Approach, First Edition. Edited by Richard M. Sibly, James H. Brown, Astrid Kodric-Brown.
© 2012 John Wiley & Sons, Ltd. Published 2012 by John Wiley & Sons, Ltd.

ecology to practice by capitalizing on the empirically documented scaling relationships of individual metabolism with body size and temperature. This enables prediction of broader-scale ecological patterns and processes with the use of few, easily measured variables, in particular, body size and temperature.

Parasitism is a consumer lifestyle and the term applies to all infectious agents (parasites), from viruses to multicellular animal and plant parasites (Kuris and Lafferty 2000; Lafferty and Kuris 2002). We can easily consider parasitism from a metabolic perspective. A parasite feeds on a single individual – the host – for the duration of its life (or life stage, for complex life cycle parasites), while a predator feeds on multiple prey. The durable nature of the parasitic association highlights the possibility of intense and intertwined metabolic interactions between parasites and their hosts above that expected between predatory consumers and their prey. There are several distinctive parasitic consumer strategies. These basic consumer strategies (which are taxon neutral) have different absolute and relative body sizes of parasites and hosts (Fig. 19.1A). The overriding importance of body size for these consumer strategies emphasizes the promise of a metabolic scaling approach to understanding the ecology and evolution of parasitism.

This chapter covers several areas that are relevant to MTE, parasites, and infectious disease. We first expand on the merits of incorporating parasites into MTE. We follow with a discussion of parasite energy metabolism, which is different from that of most free-living species. We then investigate several areas where MTE interfaces with parasitology by covering relevant research and reanalyzing data, spanning topics ranging from individual- to ecosystem-level patterns and processes. We conclude by summarizing and suggesting avenues for future research.

19.2   WHY CONSIDER PARASITES AND INFECTIOUS DISEASE?

19.2.1   MTE should shed light on parasitology

Parasites impact host individuals to varying degrees and can affect populations (e.g., Lafferty 1993; Hudson et al. 1998) and communities (e.g., Dobson and Hudson 1986; Mouritsen and Poulin 2005; Wood et al. 2007). Most aspects of parasite–host interactions are fundamentally metabolic or can be easily considered from a metabolic perspective. Like any consumer, parasites extract energy from their resources. For instance, 21% of an infected isopod's biomass production goes to the growth of a larval acanthocephalan parasite (Lettini and Sukhdeo 2010). Parasites also drain energy from hosts beyond what they consume. Hosts resist parasites in several ways (Rigby et al. 2002) and repair damage caused by parasites. These activities must incur metabolic costs. For example, juvenile chipmunks respond to botfly parasitism by increasing their resting metabolic rate by 8% per fly and they can have up to 6 botflies; the metabolic effects persist after the parasites have left and result in reduced body mass of adult chipmunks (Careau et al. 2010). Also, juvenile damselfish previously infested with a single gnathiid isopod micropredator had a 35% higher oxygen consumption rate than uninfected fish (Grutter et al. 2011). Metabolism also underlies other aspects of parasitism. For example, parasite manipulation of host behavior to alter predation rates (reviewed in Moore 2002) not only involves metabolic processes of the parasite and host, but also involves altering the flow of materials and energy in food webs. Thus, metabolic ecology and MTE can inform the ecology and evolution of parasites and infectious disease at multiple scales. Research covered in this chapter shows that metabolic scaling relationships for hosts and parasites can illuminate optimal parasite consumer strategies, optimal body sizes, and the scaling of abundance of all life forms in ecosystems.

19.2.2   Most species are parasitic and they change the game

Generalizations about life should pertain to most of life. From this standpoint, the increasing attention given parasites by ecological research is warranted because parasites likely comprise over half of species diversity (Price 1980; de Meeûs and Renaud 2002; Dobson et al. 2008). Considering parasites tests the universality of empirical generalizations and theoretical assumptions of research focused on free-living species.

Not only do parasites comprise a major component of diversity, but parasites differ on average from free-living species in ways that directly bear on MTE. For instance, as detailed in section 19.3.1, many parasites primarily rely on anaerobic energy metabolism. This is different than the aerobic metabolism that characterizes most free-living species. Parasites also differ from free-living species in basic ecological attributes directly relevant to MTE. For example, MTE research has been used to help understand the flow of energy among trophic levels in ecological networks of free-living species (e.g., Brown and Gillooly 2003; Jennings and Mackinson 2003; Reuman et al. 2008; Petchey and Dunne, Chapter 8). Much of this work assumes consumer–resource body-size ratios to be constant and larger than 1, or the existence of a positive relationship between trophic level and body size. Although these assumptions may apply to some predator–prey interactions, parasites clearly violate those assumptions (Fig. 19.1). Parasites also greatly distort the relationship between trophic level and body size (section 19.5). As we will see, these typical differences between parasites and free-living species highlight the utility of using parasites not only to test MTE, but also to refine it in a way that enhances its performance for all life forms.

19.3 PARASITE INDIVIDUALS

19.3.1 The scaling of metabolic rate

MTE relies on understanding how metabolic rate scales with body or cell size (Brown et al. 2004; Brown, Sibly, and Kodric-Brown, Introduction). Metabolic rates influence ecological dynamics and patterns. As described in Chapter 2, equation 2.3, whole-organism metabolism, I, scales allometrically with body size, M, and exponentially with temperature as

$$I = I_0 \, M^\alpha \, e^{-E/kT} \qquad (19.1)$$

In equation 19.1, I_0 is a normalization constant that varies for organisms of different physiological types (e.g., plants, endothermic vertebrates, ectothermic vertebrates, invertebrates) and α is the scaling exponent. Across a wide range of multicellular organisms, α has an average value of ~3/4 (e.g., Kleiber 1932; Hemmingsen 1960; Peters 1983). Recent work indicates that metabolic rates in bacteria scale with an exponent larger than 1 and in protists scale with an exponent ~1 (Makarieva et al. 2008; DeLong et al. 2010). Later in the chapter, we will use 3/4 scaling for host metabolic rates, as this applies to most host organisms studied and helps keep clear the identity of exponents. The $e^{-E/kT}$ term in the equation is a formulation of the Arrhenius equation, which can capture the

influence of temperature on metabolic rate (Gillooly et al. 2001; Brown and Sibly, Chapter 2).[1]

Logging both sides of equation 19.1, so that log $I = \log I_0 + \alpha \log M$, gives a linear form that makes data easier to visualize and analyze. Anti-logging the intercept provides the normalization constant and the slope provides the scaling exponent, both of which can be tested by linear regression.

Metabolic scaling studies of parasites have suggested that parasite metabolism scales allometrically, similar to free-living species. For several parasite groups, intraspecific whole-organism oxygen consumption scales with fractional exponents of body size, consistent with observations from free-living species (see Fig. 19.2 for examples). Plotting oxygen consumption against body size across four parasitic nematode species gives a scaling exponent of 0.62 (von Brand 1960). Figure 19.2D indicates that parasitic protists scale with the same exponent as free-living protists, but with a lower normalization constant. Similarly, in his classic comparative work, Hemmingsen (1960) included one parasite, the pig nematode, *Ascaris suum* (as *A. lumbricoides*). This parasite's metabolic scaling fit with the slope characterizing free-living species, with a below-average normalization constant.

However, there may be a major problem with the above scaling studies. They rely on metabolic rates approximated by quantifying oxygen consumption, with the assumption that the organisms rely on standard aerobic mitochondrial respiration.[2] But many parasites, including most of the species in the above scaling evaluation, do not rely primarily on aerobic energy metabolism. Most parasitic worms and protists do not completely oxidize fuel to carbon dioxide, even when in well-oxygenated environments such as the vertebrate bloodstream (von Brand 1973; Barrett 1981, 1984; Saz 1981; Kohler 1985; Bryant et al.

---

[1]In this equation, E is the activation energy for enzymatic reactions (~ 0.63 eV on average for aerobic respiration), k is Boltzmann's constant (8.62×10^{-5} eV K^{-1}), and T is the average operating temperature in kelvin (Gillooly et al. 2001; Brown et al. 2004). The units cancel out, leaving it as a dimensionless modifier.

[2]Aerobic respiration involves the tricarboxylic acid cycle, which generates high energy electrons ("reducing equivalents") by completely oxidizing carbon to CO_2, followed by ATP generation using the electron transport chain and the electrochemical proton gradient. Aerobic respiration occurs in mitochondria for eukaryotic organisms.

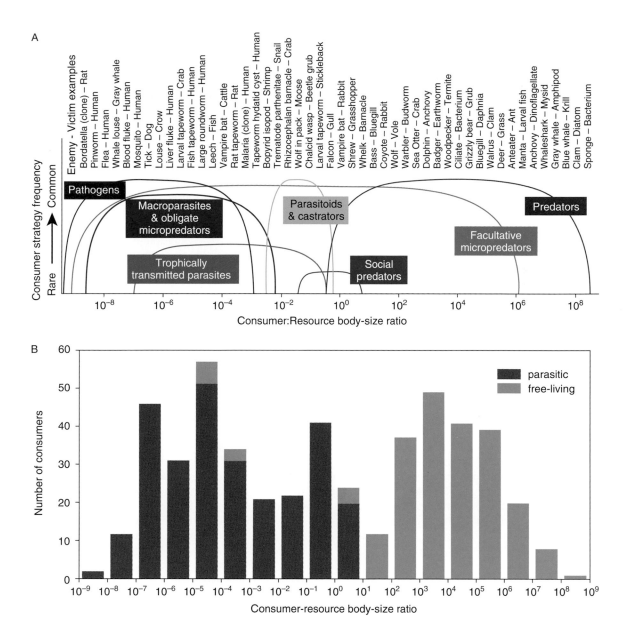

Figure 19.1 Consumer–resource body-size ratios for parasitic and free-living consumers. (A) Conceptual diagram representing the diversity of consumer strategies in nature. The *x*-axis represents consumer–resource body-size ratios on a log scale. The *y*-axis represents the authors' impression of the relative diversity of consumers employing different strategies. Examples of the various interactions are indicated above. Note that "trophically transmitted parasites" also includes trophically transmitted pathogens and castrators, which explains the large size ratios attained. Figure modified from Lafferty and Kuris (2002). (B) Actual frequency distributions of logged consumer–resource body-size ratios from three estuarine food webs. Figure pools the data presented for three estuaries by Hechinger et al. (2011). Values less than 0 are for consumers that are smaller than their resources. Note that body-size ratios vary among both parasitic and free-living species, and when parasites (and micropredators) are considered, body-size ratios are not restricted to being larger than 1. The log consumer–resource body-size ratio is slightly larger than 0 for several parasitic castrator trematode species, because these parasites grow to be 10–40% the mass of their snail hosts (Hechinger et al. 2009), tend to infect larger hosts (Sousa 1983), and consequently can have a slightly larger average size than the average host.

1989).[3] The main source of ATP for parasites is often glycolysis, or slight variants, none of which require oxygen. Even parasites that require oxygen still maintain relatively high rates of glycolysis. A lack of complete oxidation can also characterize parasites of invertebrates, where amino acid catabolism and aerobic respiration can be more important.

The lack of aerobic respiration reflects inefficient energy metabolism in terms of ATP extracted per unit resource. But parasites may gain by the strategy they adopt. Due to an inverse relationship between rate and efficiency, maximum metabolic power (ATP per unit time) is coupled to suboptimal efficiencies (e.g., Angulo-Brown et al. 1995; Waddell et al. 1997). Many readers have experienced this trade-off while driving cars varying in horsepower and fuel efficiency. Easy access to fuel can favor metabolic power gained by glycolysis over the efficiency of aerobic respiration (e.g., see Pfeiffer et al. 2001; Aledo and del Valle 2004; Vazquez et al. 2010).[4] Abundant resources and power maximization can explain the high glycolytic rates of some free-living microbes and fast-growing mammalian cells, including cancer tumor cells.[5] Parasites exist in conditions of abundant resources, potentially explaining why they are both productive and "inefficient" (Barrett 1984).

All this means that empirical scaling relationships based on oxygen consumption underestimate the normalization constants for parasite metabolic rates. This is because parasites generate more ATP than predicted from their oxygen consumption – for the majority of endoparasitic worms potentially about an order of magnitude more (Barrett 1991). Further, if parasites tend to operate under a different metabolic optimization regime than related free-living species, then they may also possess different scaling exponents (similar to how different metabolic constraints may explain the different scaling exponents characterizing bacteria, protists, and animals; DeLong et al. 2010). If parasites scale with different exponents, using standard scaling relationships will lead to inaccurate predictions about how parasite metabolic rates scale with parasite body size. What does this mean about including parasites in MTE analyses? Until high-quality investigations on the scaling of parasite metabolic rates are available, a sensible approach may be to use standard empirical scaling relationships with the expectation that they underestimate parasite metabolic rates relative to free-living species. We should also consider that parasite scaling exponents may differ from free-living species, tempering precise comparisons of parasites of different body sizes until better data and analyses are available.

19.3.2 Parasites and temperature

Just as for free-living species, temperature influences biological and ecological rates for parasites (reviewed in von Brand 1973, 1979). For instance, malaria develops faster in warmer mosquitoes and warmer mosquitoes bite more frequently. But warmer mosquitoes also die faster, and the net result is that there is an optimum temperature for malaria transmission (Anderson and May 1991; Rogers and Randolph 2000). MTE addresses temperature via its influence on metabolic rates (equation 19.1; Brown et al. 2004; Brown, Sibly, and Kodric-Brown, Introduction). Temperature correction in MTE is often carried out in comparative tests of ectothermic organisms from different environments. It also helps explain the faster biological rates characterizing endotherms. With parasites, we have the same concerns, but with fascinating differences. Parasites are ectothermic with body temperatures matching their hosts. Parasites of ectotherms therefore operate at near-ambient environmental temperature. Parasites living in the body of endothermic hosts have a consistently warm environment without paying the high costs of endothermy. Parasites living on the surface of endotherms will approximate the surface temperature of the host, which is often somewhere in-between ambient envi-

---

[3]This pertains to life stages that are parasitic (established on or in a host). Free-living stages of parasites typically use full aerobic respiration. Migrating parasitic larval stages are often somewhere in-between free-living stages and established parasitic stages. Further, almost all the metabolic research pertains to endoparasites; little is known about how ectoparasites metabolize.

[4]Biochemists have only recently appreciated that glycolysis may be thermodynamically optimized for power (e.g., Angulo-Brown et al. 1995; Waddell et al. 1997). We appreciate the power of glycolysis when we sprint and it provides over 100 times more ATP per second than aerobic respiration (Voet and Voet 1995).

[5]The increase in glycolysis characterizing many cancer cells is known as the "Warburg effect." This is also why positron emission tomography (PET) detects cancer cells. The radioactive tracer is a glucose analog, which is disproportionally taken up and trapped in cancer cells.

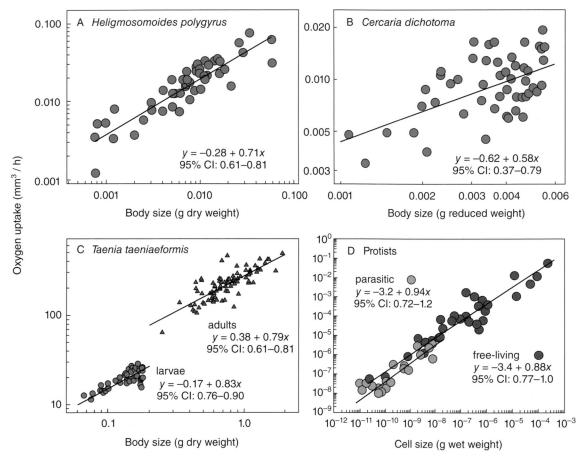

Figure 19.2 Scaling of "standard" or "endogenous" respiration rates (oxygen consumption) versus body size for parasites. (A) An intestinal nematode of mice, *Heligmosomoides polygyrus*. We reanalyzed data from Bryant (1974) Fig. 3a,b, pooling genders. (B) Individual parthenitae (sporocysts) of the trematode *Cercaria dichotoma*. Data from Pascoe et al. (1968). Note, data possesses extra scatter because worms were starved for varying amounts of time (1–19 days). For day 1 worms ($n = 10$), Pascoe estimated an exponent of 0.70 ± 0.03 95% CI. (C) Larval stages (cysticerci) from rat livers and adult stages from cat intestines of a tapeworm, *Taenia taeniaeformis*. Data from von Brand and Alling (1962) Figs 1 and 3. (D) Interspecific scaling of oxygen uptake versus cell size for heterotrophic free-living ($n = 32$) and parasitic protists ($n = 20$). Data from DeLong et al. (2010) (who used data from Makarieva et al. 2008, which itself was mostly compiled from a report by Vladimirova and Zotin 1985). We converted data to O_2 uptake, and used a general linear model on log-log data to test the significance of the different normalization constants for parasites and free-living species ($F_{1,49} = 10.1$, $p = 0.0025$), but then fit separate RMA regressions to each group following DeLong et al.'s analysis. We obtained all other estimates using OLS on log-log data.

ronmental temperature and internal host body temperature. An interesting twist arises for the many parasites with complex life cycles. Such parasites encounter several temperature regimes during their free-living stages and stages parasitic of endothermic and ectothermic hosts. These temperature differences should be appropriately accounted for when applying MTE to parasites. Additionally, the influence of these temperature differences on metabolism likely has interesting ecological and evolutionary ramifications (e.g., differences in rates of growth, interactions, and evolution).

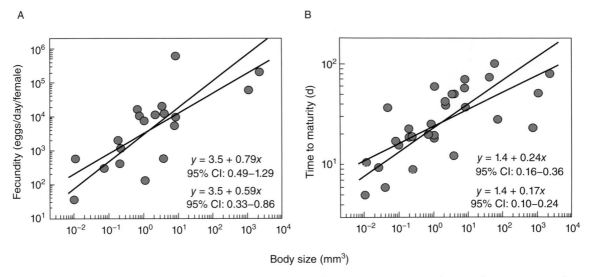

Figure 19.3 The allometric scaling of life-history traits for nematode macroparasite genera of mammal intestines using data from Skorping et al. 1991. (A) Fecundity (eggs/day/female) versus body size. We took data from Skorping et al.'s Fig. 2 fecundity vs. pre-patency plot, matched pre-patency data with data from their Fig. 1, and fit lines. Our OLS estimate matches well with the value that they reported in text. (B) Time to maturity (time post-infection to egg release) versus female body size. We took data from Fig. 1 of Skorping et al. (1991), flipped axes, and fitted lines. In both panels, the upper statistics and steeper lines are from RMA regression, whereas lower statistics and shallower lines are from OLS. 95% CIs are for exponents.

19.3.3 The scaling of life-history traits with parasite body size

Body size affects metabolic rates, which, in turn, set the pace of life histories (Brown et al. 2004; Sibly, Chapter 5). For instance, fecundity and maturation time scale with body size in free-living species (reviewed in Peters 1983; Calder 1984; Schmidt-Nielsen 1984). A few studies indicate that parasite life-history traits vary with body size as they do for free-living species. However, most of these analyses were not performed to examine the allometric scaling of life history, making it difficult to estimate scaling exponents and normalization constants. For instance, Loker (1983) compared life-history relationships among schistosome trematodes. But these blood flukes span less than an order of magnitude in size, limiting our ability to discriminate allometric from isometric scaling. Studies spanning a broader size range have tended to only use phylogenetically independent contrasts, report correlation coefficients instead of slopes and intercepts, and use genus- or family-level data. Additionally, several of these studies used length, or length × width, as a measure of body size for groups that vary considerably in body shape – copepods (Poulin (1995), trematodes (Poulin (1997), nematodes Morand (1996), and platy-

helminths (Trouvé et al. 1998). Hence, although these studies suggest that parasite life histories scale roughly similarly to free-living species, their analyses and data make it difficult to estimate scaling exponents and coefficients.

However, two studies have reported life-history data appropriate for evaluating parasite allometric scaling relationships. Skorping et al. (1991) analyzed several life-history relationships for nematode macroparasites of mammals. They reported that fecundity allometrically scales with an exponent of 0.55 (95% CI: 0.35–0.75). The confidence limits barely reach the predicted exponent of 0.75. However, the slope was estimated using OLS.[6] This might underestimate the slope because error in body-size estimates could be close to the error in fecundity (because the data were means for nematode genera). We found that an RMA slope indicates an exponent of 0.79 (Fig. 19.3A). We were

---

[6] OLS refers to "ordinary least-squares" regression and RMA refers to "reduced major axis" regression (see White, Xiao, Isaac, and Sibly, Chapter 1). RMA provides steeper slope estimates than OLS, and is appropriate when error in estimates of the response (dependent) and predictor (independent) variables are of similar magnitude. OLS provides appropriate slope estimates when error in the response variable is at least three times larger than error in the predictor.

also able to use their data to show that time to maturity may scale as $M^{1/4}$, which is predicted by the allometric exponent for biological times (Fig. 19.3B). Also studying nematode intestinal parasites of mammals, Morand and Poulin (2002) reported $M^{0.76}$ scaling of daily fecundity versus body volume based on reanalyzing data from Morand (1996). These analyses suggest scaling exponents for nematode worms consistent with MTE predictions concerning the timing of life history, but additional studies on a broader range of parasites are needed to assess the generality of these findings for parasites.

19.3.4   The scaling of optimal parasite body size

Measures of parasite fitness are often derived from parasite population models that include both host and parasite dynamics. The scaling relationships between body size and various vital rates and demographic patterns permit us to express model parameters as allometric functions of host and/or parasite body size. Expressing model parameters as allometric functions of body size can enable exploration of parasite evolution, including optimal parasite body size.

Morand and Poulin (2002) explored how optimal parasite body size might vary with host body size for single-host life cycle nematode parasites of mammals. They applied allometric scaling to both hosts and parasites in a macroparasite–host population model. They solved for the body size that maximizes a measure of fitness, R_0, which is the parasite net reproductive rate when the parasite invades an uninfected host population. The equation for R_0 involves several parameters that scale with either parasite body size (parasite fecundity and mortality) or host body size (host mortality and density). They predicted that optimal nematode body length would scale with host mass to the 0.38 power and that optimal nematode body volume would scale with host body mass to the 0.94. Both predictions were supported by estimates derived from data on pinworm nematode parasites in their mammalian hosts: nematode length scaled with host mass to the 0.33 power (95% CI: 0.28–0.39) and nematode volume scaled to the 0.80 power (0.63–1.04). Inputting into their model the body sizes of North American mammals, they generated a predicted distribution of nematode parasite body sizes. The predicted mean was indistinguishable from the observed mean,

and the observed and predicted distribution of lengths was grossly similar. These results imply that the effects of metabolic scaling on host and/or parasite demographic processes affect the evolution of parasite body size.

Kuris and Lafferty (2000) also used allometric scaling to model how R_0 responds to different host and consumer body sizes, comparing a variety of predatory and parasitic consumer strategies. Their modeling indicated that a predator strategy is most profitable on small-bodied resource species, while a parasitic strategy is better when eating large-bodied resources. This outcome was explained by comparing resource lifespan with consumer lifespan. If resource lifespan is short, it does not pay to form a durable feeding relationship and a consumer should immediately consume the entire resource. Relatively large and long-lived resources, however, can be "milked" over time, favoring a parasitic strategy. Parasitoids are able to infect hosts close in size to themselves because they eventually kill the host as part of their development. Parasitic castrators are able to infect hosts that are only a little larger than themselves because they primarily take energy that does not impair the host lifespan. However, the most profitable hosts for castrators tend to be absolutely small because such hosts devote relatively more resources to reproduction. More can be done with such models, including testing with empirical data on consumer–resource size distributions in nature (Fig. 19.1).

19.4   PARASITES IN HOSTS

MTE predicts population abundance by using the allometric scaling of whole-organism metabolism (Brown et al. 2004). If organisms of different body sizes have, on average, equal access to resources, steady-state abundance, N, should decline with body size. This is because a given amount of resources supports fewer large-bodied individuals than small-bodied individuals. Individual resource requirements parallel whole-body metabolic rates, which scale with M^α (see equation 19.1). Hence, holding resource supply constant, abundance will scale with body size as $N \propto M^{-\alpha}$. Recall that temperature effects can also be factored into the relationship (see equation 19.1).

Researchers studying free-living species often measure species abundance as the density of individuals per unit volume or area of habitat. However,

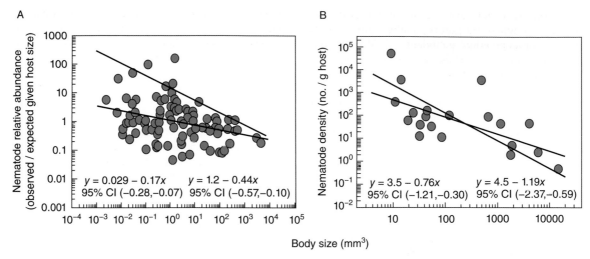

Figure 19.4 Parasitic nematode abundance within infected hosts versus parasite body size. (A) Relative abundance versus female body size for nematode parasites of mammals (data from Fig. 1 of Arneberg et al. 1998). Dependent data represent anti-logged residuals of log mean parasite numbers in infected hosts versus host body size. Arneberg et al. reported an OLS slope of −0.19 (95% CI: −0.31 to −0.07) which agrees well with our OLS estimate (left statistics). We also fitted a 95th percentile quantile regression to approximate the upper bound (right statistics). (B) Mean number of worms per gram of host tissue for nematode parasites of mammalian guts (data from declining part of the abundance distribution in Fig. 2b of Morand and Poulin 2002). Morand and Poulin used RMA to provide a slope of −1.19, which matches with our estimate (right statistics). We found that OLS, which could be more appropriate, provides a slope of −0.76 (left statistics). There is insufficient data for quantile regression to be stable over the 75th percentile, so we did not use it to estimate the upper bound of the data.

ecological parasitologists have traditionally used the natural unit of a single host body as the unit for parasite density (but see section 19.5). A group of conspecific parasites within an individual host is termed an *infrapopulation*, as distinct from the population of parasites in the entire host population or ecosystem (Bush et al. 1997). The suite of parasite species within individual hosts is called an *infracommunity*. Can we apply MTE to parasite infrapopulations and infracommunities?

19.4.1 Parasite infrapopulation abundance vs. parasite body size

Two studies examined the interspecific scaling of infrapopulation abundance. Does parasite abundance decline with $M^{-3/4}$ as it does for free-living animals?[7] Arneberg et al. (1998) examined data for nematode species parasitizing mammals. They used the mean number of parasites per infected host as the raw data for the dependent variable. They showed that OLS

slopes were appropriate, because abundance error was over 11 times greater than body-size error. In a variety of analyses – including those using observed abundance data, abundance residuals corrected for host body size, and phylogenetic contrasts – they documented allometric slopes shallower than −3/4, ranging from −0.12 to −0.21. They also estimated a slope of −0.46 (95% CI: −0.70 to −0.23) for the uppermost boundary of the abundance data. This slope matches our estimation of an upper-bound slope for their global data (Fig. 19.4A). MTE predictions are more likely to apply to upper-bound abundance data because upper bounds are more likely to represent infrapopulations experiencing resource limitation. Morand and Poulin (2002) also examined nematodes of mammals,

[7]The few other comparative studies relating parasite abundance to parasite body size (e.g., Poulin 1999) cannot be used to evaluate hypotheses about MTE because they were performed for other purposes and do not provide the appropriate data (e.g., actual measures of abundance and body mass).

restricting their data to those that live in host intestines. They measured infrapopulation density as the mean infrapopulation abundance divided by host mass. In the analyses using the *maximum* mean infrapopulation density observed for each nematode species among host populations, Morand and Poulin detected a mode where nematode densities peaked at small but not the smallest body sizes.[8] They estimated an RMA slope of −1.19 for the declining part of the distribution (Fig. 19.4B). In analyzing their data, we found that OLS, which could be more appropriate given that density error likely swamps body-size error, provides a slope of −0.76 (Fig. 19.4B).

Although all these analyses document the negative scaling of abundance with parasite body size, findings using maximum mean infrapopulation densities from the second study might be more consistent with MTE predictions, as both RMA and OLS exponent confidence limits overlap −3/4.

19.4.2 Parasite infracommunity biomass vs. host body size

George-Nascimento et al. (2004) and Poulin and George-Nascimento (2007) examined the scaling, with host body size, of the total metazoan parasite assemblage standing-stock biomass (or biovolume) within a wide range of vertebrate host species.[9] In a variety of analyses (using raw data or residuals, with or without temperature correction) on their global dataset, the authors detected that mean total parasite biomass scaled with slopes ranging from 0.54 (95% CI: 0.36–0.71) to 0.82 (0.64–1.0).[10] However, analyses on the highest-quality subset of their data (ecto- and endoparasites of fishes) indicated that infracommunity biomass

scaled isometrically with host mass. This was particularly evident when they incorporated temperature effects and used maximum observed parasite infracommunity biomass: exponent 0.97 (0.80–1.14) (Poulin and George-Nascimento 2007). Isometric scaling is consistent with volumetric space limitations, or with parasites metabolizing at the same rates as host tissues.[11] This research did not factor in parasite body size, which would be important if parasite metabolic rates scale with parasite body size as we expect from the discussion in section 19.3.2. If parasite metabolism scales with parasite body size, and if host metabolism imposes a ceiling on parasite abundance, a particular host would support a smaller biomass of small-bodied parasites than of large-bodied parasites, because smaller parasites require more energy per unit mass. It would be worthwhile to construct and test an MTE framework for within-host total parasite abundance, biomass, and energy flux that explicitly factors in parasite species' body size.

19.5 PARASITES AND FREE-LIVING SPECIES IN ECOSYSTEMS

The above sections consider parasite abundance within hosts. But parasite abundance can also be calculated using the standard spatial units used in ecology. Doing so permits a direct comparison of the abundances of parasites and coexisting free-living species.

The MTE equation for steady-state abundance of multicellular organisms with equal access to resources

[8]This parallels findings observed for some free-living animals that may reflect an optimal body size for the group (reviewed in Brown 1995). As described in section 19.3.4, Morand and Poulin (2002) explored this issue using macroparasite population models.

[9]Also note Muñoz and Cribb (2005), who presented data on total metazoan parasite biomass in 14 individual conspecific fish. They reported a positive correlation for log total parasite biomass and host mass. We estimated allometric exponents of 0.44 (95% CI: 0.21–0.67) using OLS and 0.72 (0.40–1.28) using RMA. However, host size is confounded by host age, and we cannot distinguish longer cumulative exposure time or body size as being the factor driving the positive association.

[10]They also observed the mathematically required reciprocal relationships obtained by dividing parasite biomass by host mass.

[11]The theoretical framework developed in the pioneering papers of George-Nascimento et al. (2004) and Poulin and George-Nascimento (2007) assumes that parasites metabolize at the same rate as host tissues, but makes a faulty prediction that the summed standing-stock parasite biomass will then scale with host whole-body metabolic rate, $M_H^{3/4}$. If parasite tissue metabolized at the same rate as host tissue, we could expect isometric scaling, as is found in the scaling of many host tissues and organs with host body size (Calder 1984). This can be further understood by considering that the amount of tissue supported by a watt scales with $M^{-1/4}$ (the inverse of mass-specific metabolic rate). Consequently the amount of any particular host tissue – or parasite tissue metabolizing like host tissues – supported by host metabolic rate could scale as $M^{1/4}M^{3/4} = M^1$.

captures the decline in abundance with the increase in body size as

$$N = iM^{-3/4} e^{E/kT} \qquad (19.2)$$

where i represents a normalization constant (Brown et al. 2004). Dividing both sides of the equation by the temperature-dependence term gives "temperature-corrected abundance," $N e^{-E/kT}$.

Hechinger et al. (2011) plot abundance versus body size for parasites alongside free-living species. They used data from three estuaries for metazoan parasites and a wide range of free-living species belonging to different physiological groups – invertebrates, ectothermic vertebrates (fishes), and endothermic vertebrates (birds). Plots in log-log space of observed or temperature-corrected abundances versus body size show that equation 19.2 does not predict the scaling exponents and normalization constants for either parasitic or free-living species (Fig. 19.5A). For instance, parasites are less abundant than expected and have slopes that are too shallow.

The authors hypothesized that discrepancies in the data could be explained by trophic transfer efficiency, which is the fractional transfer of energy among trophic levels (Lindeman 1942; E. P. Odum 1971). Some MTE research has factored trophic transfer efficiency into abundance scaling (e.g., Brown and Gillooly 2003; Jennings and Mackinson 2003; Reuman et al. 2008; see also Petchey and Dunne, Chapter 8), but most of this work assumes consumer–resource body-size ratios to be constant and larger than 1, or the existence of a positive relationship between trophic level and body size. These assumptions may apply to some predator–prey interactions. However, parasites (and many micropredators, like mosquitoes) violate the "big consumer" assumption by flipping the typical consumer–resource body-size ratio, and by altering the distribution of these ratios in empirical food webs (Fig. 19.1). Including parasites also changes the trophic level versus body-size relationship in empirical webs (Fig. 19.6), further indicating the need to incorporate into theory the flow of energy among trophic levels independently of body size.

Although rarely done (e.g., Brown et al. 2004; Meehan 2006; McGill 2008), it is possible to incorporate into scaling theory the effects of trophic level and trophic transfer efficiency in a way free of assumptions concerning body-size relationships. Noting that trophic transfer efficiency, ε, can be expressed as a fraction, Hechinger et al. predicted that abundance would exponentially decrease with increasing trophic level, L, as

$$N e^{-E/kT} = iM^{-3/4} \varepsilon^L \qquad (19.3)$$

where the basal level is defined as $L = 0$.

This incorporation of trophic dynamics revealed uniform $-3/4$ ecosystem-wide scaling of abundance with body size across all parasitic and free-living species (Fig. 19.5B). This remarkable finding highlights the utility of using the ε^L term to provide broadly applicable scaling relationships. After accounting for the flow of energy among trophic levels, all species clustered around a single $-3/4$ line, regardless of taxonomic or functional group affiliation: endotherm or ectotherm, invertebrate or vertebrate, free-living or parasite.

Hechinger et al. (2011) also examined an implication of the observed uniform $-3/4$ scaling of abundance with body size – that of the invariant production of biomass with body size. Because a single regression line roughly describes the $M^{3/4}$ scaling of temperature-corrected individual biomass production, P_{ind}, for a wide range of multicellular species (Ernest et al. 2003), total population production, P_{tot}, is $P_{\text{ind}} N$, which scales as $M^{3/4} M^{-3/4} = M^0$. Data from the three estuaries suggest that any species within a specific trophic level can produce biomass at the same rate, whatever their body size or functional group (Fig. 19.5C; Hechinger et al. 2011). This "production equivalence" appears to be a more general rule than the often discussed "energetic equivalence" rule that only applies to metabolically similar organisms (e.g., see Damuth 1981, 1987; Nee et al. 1991; Ernest et al. 2003; Brown et al. 2004). Concerning the importance of parasites to ecosystems, production equivalency reveals the comparable ecological relevance of a parasitic species compared to any free-living species existing at the same trophic level.

19.6 CONCLUSIONS AND FUTURE DIRECTIONS

We have surveyed the literature directly relevant to MTE and parasites, including basic aspects of parasite metabolic scaling. Applying MTE to parasites can inform several aspects of ecological and evolutionary parasitology. Conversely, considering parasites provides a novel way to test, refine, and expand MTE. Our

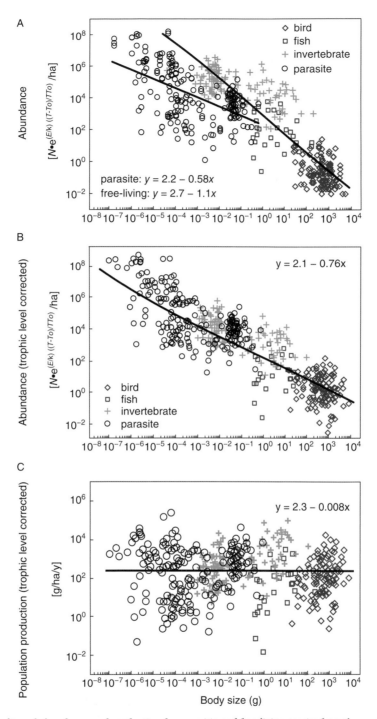

Figure 19.5 The scaling of abundance and production for parasitic and free-living species from three estuaries. (A) Temperature-corrected abundance versus body size for a wide range of parasitic and free-living species. A single line does not describe the relationship: parasites appear to scale less steeply than $-3/4$ and free-living species more steeply. (B) Temperature-corrected abundance decreases with body size to the $-3/4$ power when statistically holding trophic level constant. A single line now describes the various animal groups. (C) Total population production is invariant with respect to body size across all body sizes and animal groups when statistically holding trophic level constant. The data in (A–C) are from Hechinger et al. (2011), but pool the data from three estuaries. The estuary effect and interactions were not apparent and not significant in all cases (all $p > 0.73$). The temperature correction term in (A) and (B) used the Arrhenius equation expressed relative to the ambient estuarine temperatures.

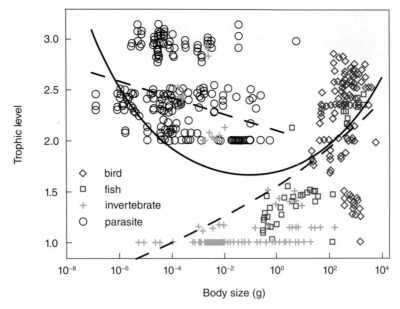

Figure 19.6 The relationship between trophic level and body size for the parasitic and free-living consumer species in three estuarine food webs (A–C). The dashed lines represent Poisson regressions fitted separately to parasites and combined free-living species (birds + fish + invertebrates) and the solid line represents the fit to the pooled data (with a quadratic term for body size). From Hechinger et al. (2011), using data pooled from three estuaries. The estuary effect was very weak and non-significant ($p = 0.60$).

review of the literature (we apologize for any oversights) and new analyses support these contentions. For instance, research that considered parasites simultaneously with free-living species showed how adding trophic dynamics into abundance versus body-size scaling increases the generality of MTE. This extension of MTE adds the flow of energy among trophic levels in a way that appears to perform well for all species. Continued application of MTE to parasites may help further generalize the theory.

There is much more to be done, both in terms of including parasitic species alongside free-living species in MTE and in applying an MTE framework to parasites and infectious disease. We need more, high-quality analyses of basic individual-based scaling relationships for parasites (metabolic rates and life-history traits). A sound understanding of parasite individual metabolic scaling will facilitate the use of MTE to estimate and predict the role of parasitism for hosts and ecosystems.

It would be worthwhile to construct a quantitative MTE framework based on parasite metabolic scaling to estimate and predict within-host parasite abundance, biomass, and energy flux. Such a framework could facilitate estimating the energetic impacts of parasites on hosts. This could help fill a hole in MTE, which currently does not account for the unobserved energetic burden of parasites on free-living species. Such an MTE framework could also provide a perspective on infectious disease previously unavailable to medical and veterinary science. Metabolic scaling may enable prediction of the impacts of infectious agents on hosts. For instance, although we did not cover it here, there is evidence that the allometric scaling of host biological times ($\sim M^{1/4}$) sets the pace of disease symptoms caused by microbial pathogens (Cable et al. 2007). Does disease caused by other sorts of parasites also scale with host biological times? Or does it primarily scale with parasite metabolic rates and population numbers? Expanding this work may help to predict the impacts of diseases when we are lacking data, including new infectious diseases of humans or livestock. Research can also begin to factor in the scaling of host immune responses, which may not simply scale with host body size to the 1/4 power (e.g., see Wiegel and Perelson 2004).

With appropriate extrapolation, an MTE framework for parasites could indicate direct energetic impacts of parasitism for entire populations, communities, and ecosystems. Section 19.3.4 showed that we can incorporate host and parasite allometric scaling into parasite–host population models to ask questions about parasite body-size optimization. However, expressing model parameters as allometric functions of body size also permits exploration of how other model outcomes scale with host or parasite body sizes. For instance, although not covered here, De Leo and Dobson (1996) showed that important rates of parasite transmission may scale with host body size (and see extensions in Bolzoni et al. 2008a, 2008b). One obvious benefit of incorporating MTE into parasite–host population models would be the inclusion of temperature effects. It would also be worthwhile to explore the interface between MTE and parasite diversity within hosts, something we did not investigate here. Finally, the research in section 19.5 shows the promise of MTE for the quantification and prediction of the role of parasites in entire communities and ecosystems. We hope that the scope and content of this chapter convey the promise and excitement of this area of research for ecology, parasitology, and medical and veterinary science.

Chapter 20

HUMAN ECOLOGY

Marcus J. Hamilton, Oskar Burger, and Robert S. Walker

SUMMARY

1 Metabolic ecology provides a robust theoretical framework for understanding how individual metabolic processes constrain flows of energy and materials at all levels of biological organization as organisms interact with their environments. Here we use this framework to examine variation in human ecology across scales, from individual life histories to the ecology of populations, and from hunter-gatherer societies to modern nation-states.
2 Many aspects of human ecology, such as most life-history traits, are remarkably predictable for a mammal of our body size. Deviations from expectations occur primarily at the population level as human population energetics is not simply the linear sum of individual demands.
3 Humans and the complex social systems we create are clearly constrained by the energy fluxes at all scales of social organization. Moreover, the same quantitative scaling relations indicative of economies of scale are found across the spectrum of human socio-economies. Specifically, as human systems grow in size they also increase in per-capita efficiency, due to the benefits of existing within complex social and infrastructure networks where resources, energy, and information can be shared widely and quickly across populations.
4 A metabolic approach to human ecology lends insight into those specific ways in which humans are a predictable mammal species, and the ways in which we differ. Consequently, human metabolic ecology provides anthropology with a theoretical framework to understand the energetic evolution of the human species, and why the human species became the most ecologically dominant species in Earth's history.

20.1 INTRODUCTION

Human ecology is the study of the interactions between humans and their environments. The human species is remarkable in many respects, but surprisingly similar to other species in other respects. The evolutionary history and biogeography of the human species is particularly noteworthy. Beginning around 50 000– 60 000 years ago, humans began their final major biogeographic expansion out of Africa and colonized all of Eurasia, Australia, and the Americas by the end of the Pleistocene some 12 000 years ago (Klein 2009). This expansion had major ecological impacts on Earth's terrestrial environments, including the replacement and/or assimilation of archaic hominid species throughout Eurasia (Klein 2009), and the global

Metabolic Ecology: A Scaling Approach, First Edition. Edited by Richard M. Sibly, James H. Brown, Astrid Kodric-Brown.
© 2012 John Wiley & Sons, Ltd. Published 2012 by John Wiley & Sons, Ltd.

extinction of Pleistocene megafauna (Lyons et al. 2004; Surovell et al. 2005). Beginning about 12 000 years ago, in many parts of the world human societies independently developed methods of cultivation (Bellwood 2004; Purugganan and Fuller 2009) and diets started to shift from a sole reliance on wild plants and animals to one that incorporated domesticates. As a consequence, human societies rapidly diversified along a new socio-economic spectrum of food production.

Today, the diversity of human societies ranges from populations that continue to rely on foraging economies, to large, complex industrial societies that rely on extracting energy sources of fossilized biomass from geologic deposits stored deep in the Earth's crust (Smil 2008). In the twenty-first century, our most developed societies rely on global communication networks that allow rapid transport of energy, materials, and information across vast distances. Currently, the majority of the world's human population, nearing 7 billion, lives in dense urban centers of millions of individuals, and performs specialized functions much removed from the actual task of food production. The human species is clearly the most ecologically dominant species in Earth's history, currently appropriating between 25% and 40% of potential global net primary production for human needs (Vitousek et al. 1997; Haberl et al. 2007). Much of this technological innovation has occurred relatively recently in human evolutionary history, within the last few centuries, and shows no sign of slowing. On the contrary, the pace of technological development is exponential (Arthur 2009), seemingly limited only by our ability to access increasing amounts of energy required to fuel its growth (Brown et al. 2011).

However, our internal biology is quite unremarkable. For example, we share about 99% of our DNA with our closest relative, the chimpanzee. Our bodies metabolize energy at a predictable rate for a mammal of our size (Ulijaszek 1995) and, like all other organisms, our diets must maintain a predictable balance of chemical elements in order to maintain our internal biochemistry (Williams and Frausto da Silva 2006). In comparison to other mammals, our life history is quite predictable for a primate species of our body size (see below), with some unique features, such as a particularly long lifespan, large encephalization, a significant post-reproductive lifespan in human females, and the requirement of large amounts of parental investment in our highly altricial, slow-growing offspring (see

Charnov and Berrigan 1993; Hawkes et al. 1998; Kaplan et al. 2000; Walker et al. 2006).

One of the most striking biological features of the human species is the large size of our brain, a powerful biophysical computer that allows us to absorb, process, interpret, and store large quantities of information about the external world (Eccles 1991; Schoenemann 2006, 2009). These abilities have led to a sophisticated understanding of basic chemical, physical, and biological principles, and have allowed us to use these principles to engineer solutions to many of the energetic and informational constraints faced by other biological species, beyond the genetic pathways shared with other organisms (Williams and Frausto da Silva 2006). The immense human cognitive capacity resulted in the evolution of a key secondary pathway of heritable information some time in our evolutionary past, which has ever since been fundamental to human biocultural evolution. In anthropology, this coevolutionary process is often referred to as dual inheritance theory (Cavalli-Sforza and Feldman 1981; Boyd and Richerson 1985) and has been fundamental to the evolution of culture. Importantly, both biological and social information can now be shared through communication networks that span the entire planet.

A key goal of anthropological science is to develop a core body of theory that can be used to understand human ecology, evolution and diversity in all its dimensions. This theory must be general and consistent with the evolution of all other forms of biological life, and it must follow from the universal laws of physics and chemistry that govern all exchanges of energy, matter, and information in the universe. Such theory is crucial, not only for the basic scientific understanding of human ecological and evolutionary history, but also for the applied aspects of environmental and sustainability science (Brown et al. 2011). Our approach here is to examine how we can quantify energy flows in human systems, across different biological and cultural scales, and from there begin to develop a predictive theory of human energetics, based on first principles of physics, chemistry, and biology.

In this chapter we explore several ways in which metabolic theory (Brown et al. 2004) can be used to model human ecology: (1) in comparison to other mammalian species; (2) across scales of biological organization, from individuals to populations; and (3) across the spectrum of human socio-economic development, from hunter-gatherer societies, to twenty-first-century nation-states.

20.2 COMPARATIVE HUMAN LIFE HISTORY

In Figure 20.1, we compare various human life-history variables with chimpanzees, other primates, and mammals. We exclude bats and non-placental mammals from the analysis because they are known to have unusual life histories (Hamilton et al. 2011). Metabolic theory predicts that across species, life-history parameters scale as power functions of the form $Y = Y_0 M_\alpha$ (see Sibly, Chapter 5), where Y is the life-history parameter of interest, Y_0 is a normalization constant, M is body size, and α is a scaling parameter, predicted to be $-1/4$ for biological rates, and $+1/4$ for biological times and sizes (Brown et al. 2004). The scalings in Figure 20.1 are close to these predicted values, and provide a useful benchmark for comparing different aspects of human life history.

Figure 20.1A,B shows that newborn mass and weaning mass are tightly correlated with adult mass across placental mammals, and while primates in general seem to have slightly higher newborn and weaning masses, human developmental masses are predictable given our adult body mass. In terms of the timing of life-history events, shown in Figure 20.1C–G, primates generally have slow life histories throughout life relative to other mammals (Charnov and Berrigan 1993). While humans have among the slowest life histories of all mammals by most measures, their rates are not exceptional for primates (with the possible exception of maximum lifespan) and, again, seem to follow primarily from the fact that we are large-bodied primates.

Reproductive output can be measured in several ways. Humans normally give birth to a single offspring at a time, as do many other mammals (Fig. 20.1H) and have particularly low litter frequencies (Fig. 20.1I), resulting in low annual fertility rates for a mammal of our size (Fig. 20.1J). Combining some of the above life-history measures to calculate mass-specific production = (weaning mass)/(adult mass) × (litter size) × (litter frequency) (Sibly and Brown 2007), a useful estimate of the overall speed of the life history shows that, again, humans have slow life histories, though not exceptionally slow for a primate of our body size (Fig. 20.1K). Indeed, human hunter-gatherer lifetime reproductive effort is much as predicted (Burger et al. 2010).

This kind of comparative scaling approach to human life history allows us to consider humans in a wider ecological and evolutionary context than is often considered in anthropology. For example, human life histories are often compared to chimpanzee life histories, our closest living genetic relatives, in order to gauge the evolutionary divergence of hominins with reference to our most recent common ancestor (e.g., Kaplan et al. 2000; Kaplan and Robson 2002; Walker et al. 2006). Given the consistent (and thus predictive) scaling relations between body size and life-history components across species shown above, MTE predicts that the slower speed of life history in humans compared to chimpanzees is partially explained by the fact that humans are larger-bodied animals than chimpanzees.

20.3 COMPARATIVE HUMAN POPULATION ECOLOGY

We can use a similar scaling approach to compare human ecology to mammalian ecology at a population level. A key macroecological measure of an animal's spatial energetics is the home range, the area it requires to fuel its metabolic demand (Peters 1983). Across species the size of the home range, H, increases proportionally with adult body size, M, and so scales as $H \propto M^1$. Figure 20.2A demonstrates that hunter-gatherers have the largest home ranges for their body size, and among the largest home ranges of all mammals, irrespective of body size. The one mammal with a larger home range in Figure 20.2A is the orca, a marine species. The large size of hunter-gatherer home ranges is a result of large social group sizes (see below). Sharing a home range with group members allows individuals access to larger areas than most solitary mammals. Chimpanzee home ranges are also larger than the average for a mammal of their body size, but much as predicted for primates.

Hunter-gatherer population densities are low for a mammal of our body size (Fig. 20.2B), likely reflecting the greater need for space with increasingly carnivorous diets across species (Haskell et al. 2002). In comparison, chimpanzee population densities fall on the mammalian scaling slope, suggesting that human population densities exceed those of chimpanzees in part because of the more carnivorous human diet. Figure 20.2B also includes the population density of the current human global population for comparison, showing an increase of about three orders of magnitude in density from that of hunter-gatherers, thus

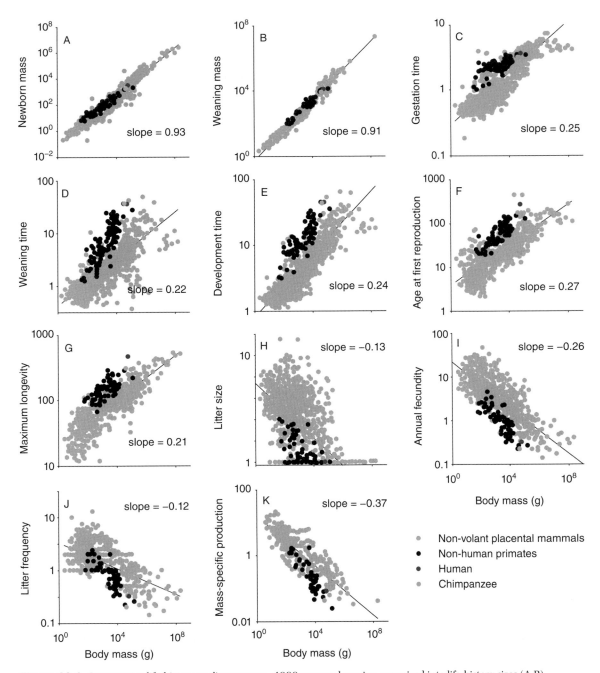

Figure 20.1 Comparative life-history scalings across ~4000 mammal species, organized into life-history sizes (A,B), life-history timings (C–G), and components of reproduction (H–K). Overall, the human life history falls within the range of variation of all other mammals, particularly primates. However, panels D–G show that humans have delayed life-history events after birth, and we have a particularly long maximum lifespan for a mammal of our size. Similarly, panels I–K show that while humans have slow rates of reproduction, they are not exceptionally slow for a large primate and are relatively fast for an ape of our size.

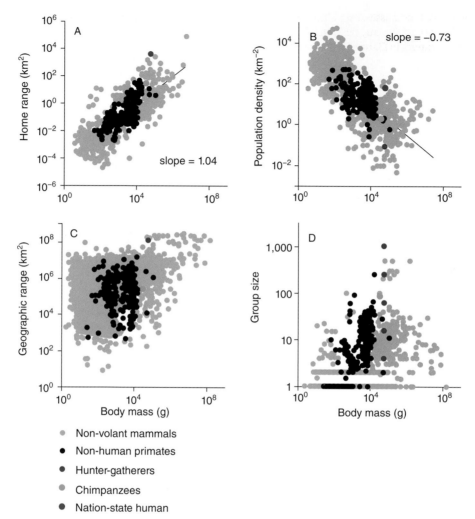

Figure 20.2 Bivariate plots of comparative population ecology across mammal species and human socio-economic conditions. Similar to Figure 20.1, most measures of human ecology fall within the overall range of other mammals, but vary in interesting ways. Hunter-gatherer home ranges are particularly large for our body size (A), but this is primarily due to large hunter-gatherer group sizes (D). Hunter-gatherer population densities are particularly low for a mammal of our body size (B), but densities in modern nation-states have increased from hunter-gatherers by about three orders of magnitude, pushing the upper mammalian bound. We humans have a particularly large species geographic range (C) for terrestrial mammals of our size. Hunter-gatherer group sizes vary throughout the year due to their fission-fusion structure, but interestingly, the average size of total populations is about 1000 individuals, close to the apex of a triangular-shaped distribution of variation of group sizes across mammals (D), suggesting that the large size of human social groups is facilitated by our body mass.

approaching the upper bound for mammals in general, but again, within the overall range of variation for a mammal of our size.

Humans achieved the majority of their current geographic range by the late Pleistocene. Subsequently, while human population densities have increased by about 1000-fold over the last 13 000 years or so, the extent of our geographic range has not changed significantly. The human species has an extremely large geographic range for our body size, effectively occurring on all available landmasses, excluding Antarctica, and among the largest for any terrestrial mammal. Those with larger ranges in Figure 20.2C are all marine mammals.

Group size in humans is particularly interesting. While the definition of group size is often somewhat ambiguous because there are many reasons why social species form groups (e.g., defense, foraging, cooperative breeding, and so on), humans have an adult body size that falls at the apex of the triangular distribution of group sizes across mammals (Fig. 20.2D). The determinants of group size are poorly understood, but clearly the upper constraints to how large social groups can be must be related to the opposing forces of the benefits of cooperation, and the density-dependence of available resources. Some combination of these forces seems to change abruptly at ~60 kg. Below this size, maximum group sizes increase approximately linearly with body size, and above this size, decrease linearly. Multiple group sizes are shown for hunter-gatherers because hunter-gatherer populations consist of a nested hierarchy of social groups that fission and fuse throughout the year as ecological and social conditions vary (Hamilton et al. 2007b). These groups include nuclear families, seasonally varying sizes of foraging groups, local populations, and regional ethnolinguistic populations (Binford 2001). Consequently, hunter-gatherer societies cannot be meaningfully characterized by a single group size estimate. However, the average size of a hunter-gatherer ethnolinguistic population (the largest social group) is about 1000 individuals (Binford 2001; Hamilton et al. 2007b), a value that falls at the apex of the distribution in Figure 20.2D. So, Figure 20.2D suggests that at a mass of ~60 kg, the size of a human, the balance between access to energy and the cost/benefit of cooperation leads to the largest social group sizes in mammals.

Available data suggests that contemporary human populations form a similar nested hierarchy of groups (Arenas et al. 2001, 2004), but it is often difficult to quantify what those group sizes are. If we take the population sizes of current nation-states, the number of speakers of a language, or adherents of a religion, as measures of the largest, most broadly defined social networks, then these values extend into the millions, even billions. Intriguingly, Figure 20.2D suggests that mammals the size of humans would be predicted to have the largest group size, and so perhaps are subject to the weakest constraints on maximum social group size.

20.4 HUMAN ECOLOGICAL AND EVOLUTIONARY ENERGETICS

We now consider the comparative energetics of human populations by reference to the metabolic theory of ecology (MTE), according to which

$$B = B_0 \, M^\alpha e^{-E/kT} \qquad (20.1)$$

where B is mass-specific metabolic rate, B_0 is a normalization constant, independent of body size and temperature, M^α is the power-law scaling with body mass, M, with α an allometric scaling exponent, and $e^{-E/kT}$ is the exponential Arrhenius function, where E is an "activation energy," k is Boltzmann's constant $(8.62 \times 10^{-5} \, \text{eV K}^{-1})$, and T is temperature in kelvin (Brown and Sibly, Chapter 2). A central insight of MTE is that the mass-dependence results from the geometry of the vascular networks that supply nutrients to plants and animals (West et al. 1997, 1999a; Brown et al. 2004; Banavar et al. 2010). In most organisms these networks are hierarchically self-similar, fractal-like structures, an especially efficient design for a distribution network to maximize delivery of nutrients to all points within a three-dimensional organism (West et al. 1997, 1999a). Thus, the metabolic rate of an organism is largely a function of the engineering of its underlying distribution network, and the sublinearity of the mass-dependence captures the fact that, as organisms change in size, the overall rate at which they operate, their metabolism, changes nonlinearly, rather than simply proportionally. Equation 20.1 also shows how mass-specific metabolic rate is affected by the operating temperature of an organism.

While this model for the effects of size and temperature on metabolic demand is constructed at the individual level, it has important implications for understanding the structure and dynamics of ecological systems at all levels of biological organization. Moreover, the model's implications can be generalized straightforwardly to any biophysical system that metabolizes energy in order to maintain structure: the amount of energy a system requires is a function of its size and structure, and the availability of energy is a function of the temperature of the larger system within which that system operates.

This metabolic perspective and the physical and biological principles on which it is based explain much of the observed variation in hunter-gatherer land use (Fig. 20.3). Hunter-gatherer societies are complex adaptive systems that access energy from surrounding

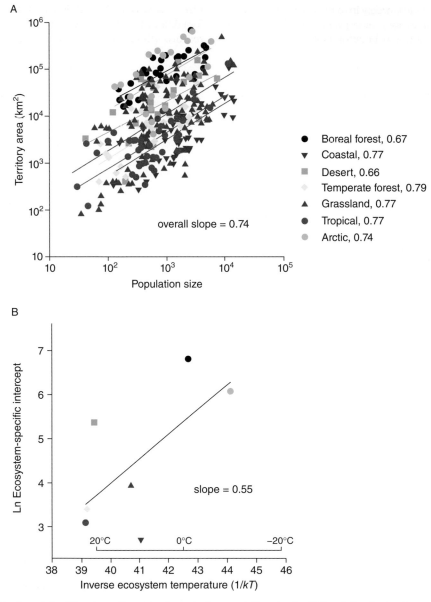

Figure 20.3 Scaling plots of hunter-gatherer territory size by population size for 339 societies, broken into seven primary ecosystem types. (A) shows that populations in all ecosystem types display economies of scale (slopes < 1), with slopes ranging between 0.66 and 0.79, with a consensus GLM slope of 0.74. Generally, intercepts decrease at a consistent rate as ecosystem productivity increases, from Arctic societies, with the highest per-capita space requirements (i.e., highest intercepts) to coastal foragers, reliant primarily on abundant and predictable marine resources, with the lowest per-capita space requirements. (B) plots the intercepts of (A) (i.e., per-capita space requirements) as a function of the average temperature of each ecosystem type. Results demonstrate that much of the change in value of the intercept is explained by ecosystem temperature, a general measure of ecological turnover.

ecosystems that vary widely in temperature, and therefore in the availability of energy. Figure 20.3A shows the territory size of a population plotted as a function of the population size. Territory size estimates the population's total catchment area. While there is considerable variation, the overall scaling relation between territory size, A, and population size, N, is well-described by a power function $A = A_0 N^\beta$, where β is significantly less than 1, and near 3/4 (Hamilton et al. 2007a). Therefore, larger hunter-gatherer populations require proportionally less area to support their energy demands, resulting in an economy of scale in space use where the system increases in efficiency with size. Figure 20.3A shows that when these groups are classified by primary ecosystem type, the same scaling holds within each ecosystem type, and it is only the intercept, A_0, the height of the scaling function, that changes. The slopes vary between 0.66 and 0.79, and have a consensus slope of 0.74 using generalized linear model (GLM) regression methods. This result demonstrates that hunter-gatherer populations exhibit economies of scale independently of ecosystem type, but the absolute amount of area they require to meet their energy demands varies predictably with ecosystem type. This

result accords with intuition that suggests a society foraging in the Arctic requires more space to harvest resources than one in the tropics. However, it is remarkable that the variation in the intercepts in Figure 20.3A that quantify the amount of space needed in each ecosystem seems to be driven by ecosystem temperature, reflecting the effect of metabolic rate on primary production. As such, this leads to the prediction that the ecosystem-specific per-capita space requirements (the intercepts of the scaling relations in Fig. 20.3A) should increase with decreasing environmental temperature with a slope E of between 0.2 and 1.2, the approximate range of activation energies in terrestrial ecosystems. The Arrhenius plot, Figure 20.3B, shows that the slope of this relation is 0.55, which supports the MTE prediction that variation in space use across hunter-gatherers is driven by the effects of temperature on the rate at which energy fluxes through ecosystems.

Do we see similar macroecological scaling relations in other types of human socio-economies? In particular, are similar economies of scale found in societies that process energy more intensively?

In Figure 20.4 we plot territory size as a function of population size for 1030 subsistence-level societies

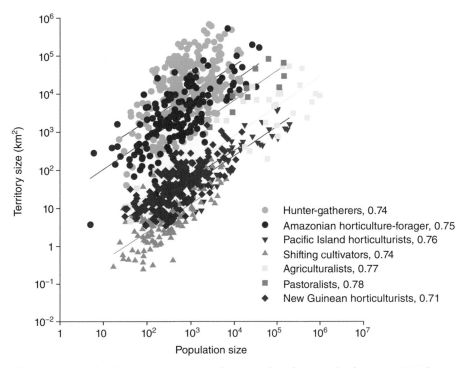

Figure 20.4 Bivariate log-log plot of territory size vs. population size for subsistence-level societies. Sample size is 1030 societies, including ~9 million individuals, covering a total of ~17 million km², ~13% of the planet's land surface.

gathered from the literature, including the 339 hunter-gatherer societies described above, as well as various types of horticulturists, pastoralists, and agriculturists. The slopes range between 0.71 and 0.78, similar to hunter-gatherer societies (Fig. 20.3A), and again, the only quantitative difference between the scaling relations is the change in the intercept, which seems to follow the increasing intensity of land use. The intercepts decrease about three orders of magnitude from hunter-gatherers with the greatest spatial requirements, about 10 km^2 per person on average, and the least intensive energy extraction methods, to shifting cultivators, with the least spatial requirements, about 0.1 km^2 per person on average, and particularly intensive, non-sustainable slash-and-burn agriculture. Thus, empirical data suggests that subsistence-level human societies developed simple yet fundamental economies of scale and maintained them with remarkable consistency. As societies diversified and developed new methods of agriculture that increased the intensity of land use, the energetic outcome was reduced per-capita space requirements. In some cases these methods of resource intensification (i.e., agriculturists, island horticulturists, and pastoralists) increased the total size of populations, from a maximum of a few thousand in hunter-gatherer societies, to hundreds of thousands in the largest agriculturalist societies.

In modern industrialized nation-states, where most individuals are not directly involved in food production, and even fewer exist at a purely subsistence level, there is a fundamentally different relation between population size and area. Clearly, the spatial size of a nation is not determined directly by the subsistence needs of its population, but is determined by a combination of geographic, historical, economic, cultural, and political factors. As such, populations of industrialized nations with complex infrastructure and communication networks have a different set of constraints on how individuals aggregate and fill space than members of subsistence economies. However, this does not mean that the internal structures of industrial societies are not subject to the same overarching energetic constraints that operate on subsistence-scale societies. For instance, Figure 20.5 shows that the total energy consumption of a nation-state scales sublinearly with population size with a slope of about 3/4, similar to the scaling in Figures 20.3 and 20.4. This scaling suggests that individuals within more populous nations benefit energetically from existing within

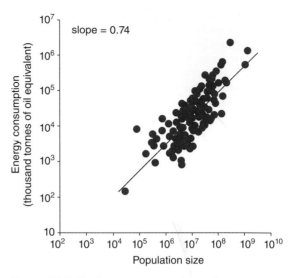

Figure 20.5 Total energy consumption scaling as a function of population size in modern nation-states (data from the World Resources Institute, 2005; http://www.wri.org/). The sublinear scaling between the total energy consumption and population size of a nation of ~0.75 is similar to the economies of scale exhibited by subsistence-level populations in Figures 21.3 and 21.4.

increasingly developed distribution networks, resulting in economies of scale in national-level energy use.

More generally, these results show that the same economies of scale can be traced from hunter-gatherers across the socio-economic spectrum of subsistence societies to industrialized nation-states, suggesting some degree of universality in the energetics of human populations. Moreover, these economies of scale are quantitatively similar to the scaling of individual metabolic rates in mammals, where larger-bodied mammals are more energetically efficient per unit mass than smaller mammals. This consistent scaling suggests that economies of scale can occur in populations just as they do within individual organisms, and for similar reasons. It is interesting to note that similar 3/4-power scaling has recently been reported in studies of the metabolism of social insect colonies (Hou et al. 2010; Waters et al. 2010) (Hayward, Gillooly, and Kodric-Brown, Chapter 6, Fig. 6.6, and Waters and Harrison, Chapter 16, Fig. 16.6), as well as in the energetics of human economies (Brown et al. 2011), suggesting that these scaling relations may be common to social species in general, and quantify the energetic benefits of cooperation.

20.5 CONCLUSIONS

In this chapter we have shown that fundamental components of human ecology, including the speed of life history, rates of reproduction, the energetics of space use, and socio-economic development across populations, are constrained by the same metabolic processes as other mammals, at rates well-predicted by MTE. We have also started to outline a general approach to understanding the ecological and evolutionary energetics of human societies.

MTE provides anthropology with a quantitative body of theory based on the first principles of physics, chemistry, and biology. Because the human species is just another mammalian species, this theoretical perspective has much to offer in the study of humans. Importantly, metabolic theory provides a robust theoretical framework which can be used to both predict relationships, and to understand deviations from those predictions, thus gaining a novel perspective into those biological properties the human species shares with all other mammals, and the ways in which the human species is unique in Earth's history.

Part III

Practical Applications

Chapter 21

MARINE ECOLOGY AND FISHERIES

Simon Jennings, Ken H. Andersen, and Julia L. Blanchard

SUMMARY

1 Variations in metabolic rate with body size and temperature drive patterns of energy acquisition and use. These have consequences for the life histories of marine plants and animals and the structure and function of populations, communities, and ecosystems.

2 Small primary producers and size-based predation characterize marine ecosystems. Since the transfer of energy to predators at each step in the food chain is inefficient, less energy is available for larger individuals at higher trophic levels.

3 Changes in energy supply and demand with size account for observed trends in abundance with size. These are often described with "size spectra," log-log relationships between the total abundance of individuals and their body size. Slopes of size spectra are remarkably consistent among ecosystems that differ in productivity and temperature.

4 Models of the processes structuring size spectra rely on the principles of the metabolic theory of ecology (MTE). Species identities are often ignored in these models, because many of the key processes depend more on body size than on taxonomic identity.

5 The limited parameter demands of size-based models simplify application to diverse systems at many scales. Size-based models have been used to investigate the effects of food chain coupling and the role of prey size selection, as well as to predict fish production, community size composition, and fishing impacts.

6 Accounting for species identity in a size-based framework provides additional insight into life-history trade-offs and food web structure. The approach also supports analysis and advice on management issues that are species focused.

7 Applications of size- and species-based models include the analysis and prediction of life histories in a community context and the assessment of trade-offs between fishing and conservation.

8 Metabolic ecology underpins the assessment and management of fishing impacts on populations, communities, and ecosystems. The drivers for developing these methods remain strong because they are providing new insights into marine ecology and because fishery managers increasingly demand advice that places fishery management in a community and ecosystem context.

Metabolic Ecology: A Scaling Approach, First Edition. Edited by Richard M. Sibly, James H. Brown, Astrid Kodric-Brown.
© 2012 John Wiley & Sons, Ltd. Published 2012 by John Wiley & Sons, Ltd.

21.1 INTRODUCTION

Marine ecosystems cover around two-thirds of the Earth's surface and account for nearly half of global primary production. The body mass of taxa found in these ecosystems spans 20 orders of magnitude, from bacteria to whales, and mean sea temperatures range from less than $0\,°C$ at the poles to almost $30\,°C$ in the tropics. The marine environment provides food for people, with fisheries yielding around 100 million tonnes annually. Metabolic ecology helps us to understand structure and processes in marine ecosystems and to assess and manage the effects of fishing.

Metabolic rate varies with body mass and temperature (see Brown and Sibly, Chapter 2). Variations in metabolic rate drive patterns of energy acquisition and use, with consequences for the life histories of marine plants and animals and the structure and function of their ecosystems. Thus smaller species have higher metabolic rates per unit body mass, faster growth, higher maximum population growth rates, greater annual reproductive output, higher natural mortality, and shorter lifespan. Conversely, larger species have lower metabolic rates, slower growth, lower maximum population growth rates, lower annual reproductive output, lower natural mortality, and greater longevity. These differences in life history affect species' responses to their physical and biological environment and, for those animals exploited by humans, to fishing (Fenchel 1974; Banse and Mosher 1980; Charnov 1993; Brown et al. 2004; see also Sibly, Chapter 5).

We begin this chapter by seeking to understand how metabolism influences the life histories of marine animals and their responses to temperature. Our focus is unashamedly fishy, since metabolic rates (Winberg 1956; Clarke and Johnston 1999) and life histories (Beverton and Holt 1959; Charnov 1993) have been so comprehensively studied in this group of significant ecological, economic, and social importance. The principal link between population processes and the metabolic processes described in other chapters of this book are related to the role of metabolism in setting the "pace of life." The "pace of life" determines other aspects of the life history and responses to mortality, both from other predators in the sea and from humans fishing. In the core of the chapter we consider the life histories of species in the context of communities and ecosystems, and how species interactions lead to the characteristic structuring of marine communities. In the final sections we describe how the understanding of life histories and community interactions has been used for measuring fishing impacts and developing assessment and management tools. The chapter provides examples of the ways in which fundamental advances in the understanding of ecological processes can be used to solve applied problems.

21.2 LIFE HISTORIES

To achieve population persistence, adults that die must be replaced. That replacement can be achieved by many alternate life histories is amply demonstrated by the variety of life forms that persist in the sea. These extant life forms must all achieve $1:1$ replacement despite differences in size, longevity, age at maturity, and reproductive output – a consequence of life-history trade-offs (see Sibly, Chapter 5). If there were no trade-offs every animal would start reproducing at birth, suffer no mortality, and produce large numbers of young at frequent intervals as it got infinitely older: the "Darwinian demon" of Law (1979).

Body size and temperature have a profound influence on observed life histories through their effects on metabolic rate. With metabolic rate driving the "pace of life," other aspects of the life history show compensatory adjustment. These compensatory adjustments serve to maximize lifetime reproductive output and are remarkably consistent within and among taxa. Indeed, relationships among life-history parameters such as reproductive output, size and age at maturity, maximum size and age, natural mortality, and growth rate are often used to predict parameters for taxa where details of the life history are not known (Beverton and Holt 1959; Pauly 1980; Gislason et al. 2010). Both growth and natural mortality have long been used as life-history parameters, although their values and associated trade-offs are more easily understood in a community context (section 21.6).

Temperature effects are manifest across the latitudinal ranges of marine species. With increasing temperature increasing metabolic rate and the "pace of life," aspects of the life history such as age at maturity and annual reproductive output respond to maintain lifetime reproductive output (Beverton 1992). Thus species' populations inhabiting warmer waters have faster growth, early maturity at smaller size, and higher annual reproductive output.

While 1:1 replacement needs to be achieved on average, mortality rates will fluctuate with changes in abiotic conditions and predator abundance. Further, for fished populations, there is additional mortality that must be tolerated for the population to persist. Population responses to increased mortality rates are governed by maximum population growth rate and the strength of compensation. Both are linked to body size.

Maximum population growth rates are faster in smaller species with early maturity and higher annual reproductive output (Myers et al. 1999; Denney et al. 2002) (see also Sibly, Chapter 5). For a given species, they also increase with temperature. Compensation is the capacity for increased population growth as mortality increases. Compensation increases with body size (Goodwin et al. 2006). The balance between compensation and population growth rates explains why populations of larger species often yield large catches as they are fished more heavily but, once they become depleted, their low maximum population growth rates prevent recovery. Small species, conversely, can be significantly depleted as they are fished more heavily but bounce back more quickly once fishing effort is reduced.

We have been referring to body size or mass as convenient correlates of metabolic rate and life history, but this is a generalization that does not apply in a consistent way across all animal groups. The generalization is often considered acceptable for many pelagic species, where differences in the amounts of metabolically active tissue and structural tissue are quite limited, but for some benthic species, and especially for deep-sea species, a large proportion of body mass is metabolically inert or has low metabolic activity (Drazen and Seibel 2007). This includes structural materials such as chitin and shell, and also lipids that provide buoyancy in deep-sea species and other fishes that lack swim bladders. Here, other measures such as energy content of living tissue may be more appropriate. Further, it should be recognized that body sizes of individuals within a fish species can vary by many orders of magnitude, from egg sizes around 1 mg in most species to maximum body sizes of several hundred kg in whale sharks, marine sunfish, marlins, and tunas. It is therefore important to distinguish between individual-level rates, such as metabolic rate, which scale with individual size, and population rates, such as fecundity, mortality, and population growth rates, which scale with a characteristic size at maturity or maximum size.

21.3 FOOD WEBS

Metabolism creates an energy demand and consumers acquire the energy to meet this demand by feeding on primary producers and/or other consumers. This establishes a food web (Petchey and Dunne, Chapter 8; Hechinger, Lafferty, and Kuris, Chapter 19), where most energy passes from smaller to larger individuals to provide the resources needed to metabolize, grow, and reproduce, and ultimately to achieve the 1:1 replacement needed for persistence. The transfer of energy through the food web is inefficient, so the aggregate production of predators is always less than the production of their prey.

Trophic levels are used to measure the number of steps in a food chain that lead to a given consumer, with primary producers typically assigned a trophic level of 1. When multiple feeding pathways support a consumer, as is usually the case, the trophic level is likely to be fractional. For this reason there is a trophic continuum in the sea, with trophic level rising almost continuously as a function of body mass (Fig. 21.1). The relationship tends to be strongest in pelagic systems and weakest in coastal systems, where phytoplankton production accounts for a smaller proportion of total primary production and larger primary producers such as macroalgae, sea grasses, or corals may be grazed directly by large herbivores (invertebrates, fish, reptiles, and mammals).

Globally, phytoplankton account for approximately 90% of total marine primary production (Duarte and Cebrián 1996). Consequently most grazers are also small, and energy is transported from the phytoplankton to the macrofauna through predation by larger organisms on smaller ones. Several predation events may occur before energy reaches the largest fish, which is why marine food chains often extend to 4 or 5 trophic levels (Vander Zanden and Fetzer 2007). With metabolic rate setting the pace of life for individuals, the smaller individuals and species in a food web turn over faster and consume more than the larger ones, so rates of energy flux per unit mass are higher. These differential rates affect community interactions and the balance of species abundances (Fig. 21.1.).

Marine food webs have been conceptualized in three principal ways, reflecting interactions among species, interactions among size-classes, and interactions among species and size-classes. Abstractions based purely on body mass of individuals are rarely used in

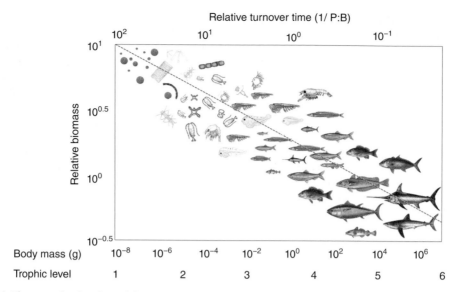

Figure 21.1 The generalized scaling of the total biomass of individuals in a marine food web sorted into logarithmic size bins (e.g., 1 g–10 g, 10 g–100 g), illustrating the biomass "size spectrum." Body mass spans many orders of magnitude from primary producers to top predators and trophic level increases with body mass owing to size-based predation. The scaling exponent is approximately −0.1, so the total biomass of smaller organisms tends to be slightly higher than the biomass of larger ones; but smaller organisms have relatively faster turnover times so their production is much higher. The inefficient transfer of energy from prey to predators means that production falls by 80–90% at each step in the food chain. Owing to their considerable scope for growth, the same species are found in several size-classes and feed at several trophic levels. Drawings and photographs of organisms courtesy of R. Beckett, S. R. Jennings, and J. H. Nichols.

terrestrial food web analysis (Woodward et al. 2005a) but are especially appropriate in the marine environment for at least three reasons. First, detailed aspects of the life history and population dynamics are not well known, especially in less accessible environments such as the open ocean and deep sea. Second, the entire food web in the sea is structured by size as much as by species identity, with tiny phytoplankton accounting for around 90% of primary production and the body size of consumers broadly linked to their position in the food chain (Duarte and Cebrián 1996). Third, many marine species grow by 5–6 orders of magnitude in mass during their life cycle, performing a series of trophic roles that are driven as much by size as their taxonomic identity (Cushing 1975). Indeed, studies of fish communities have shown that increases in trophic level with body size are primarily driven by increases in the trophic level of component species as they grow (Jennings et al. 2001).

Studies of the size composition of individuals in marine food webs have long demonstrated an approxi-

mately linear negative relationship between the logarithm of abundance and the logarithm of body mass (Sheldon et al. 1972, 1977). The slopes of these so-called size spectra (Fig. 21.1) are remarkably consistent in environments ranging from the tropics to temperate latitudes and the poles (Boudreau and Dickie 1992). The description and analysis of these patterns by Sheldon and others was a great example of the approach that was later described as "macroecology." When the size spectra are represented as the total biomass in logarithmic size groups (e.g., placing all individuals from 1 g to 10 g in one bin, from 10 g to 100 g in the next, and so on), the biomass in each group is roughly constant or decreases weakly with body size (Sheldon et al. 1972). This pattern is referred to as the "Sheldon hypothesis," and it implies that the total biomass of predators is the same as, or a consistent but relatively large fraction of, the total biomass of their prey. Note that size spectra shown here are similar to the ones commonly plotted for trees in forests, which also consider size but not species identity (Enquist and

Bentley, Chapter 14). They differ from most of the abundance–body mass relationships described elsewhere in this book (Isaac, Carbone, and McGill, Chapter 7; Petchey and Dunne, Chapter 8; Hechinger, Lafferty, and Kuris, Chapter 19), which are relationships between log mean abundance and log mean body mass where the data points represent single species or populations (e.g., of birds or mammals).

Explanations for the slopes of size spectra, where slope reflects the rate of change in abundance with size, have been based on detailed process-based models of predator–prey interactions and more simplistic models based on fundamental ecological principles. Almost all models are underpinned by the recognition that the scaling of metabolism with body size accounts for differences in the energy requirements of animals in different size-classes.

In general terms, the slope of the size spectrum is a function of the efficiency of energy transfer from prey to predators, dubbed "trophic transfer efficiency," and the predator–prey size ratio (Borgmann 1987; Hechinger, Lafferty, and Kuris, Chapter 19). Trophic transfer efficiency can be estimated directly or predicted by modeling the processes that account for energy transfer, including the probability of encountering prey, the probability of prey capture, and the gross growth efficiency (Andersen et al. 2009; Petchey and Dunne, Chapter 8). The scaling of maintenance metabolism with size (Brown and Sibly, Chapter 2) does not affect the relative abundance of animals with different maximum sizes in the size spectrum, and it can be shown analytically that a change in the scaling from $M^{3/4}$ to $M^{2/3}$ changes the slope of the resulting size spectrum by only <5%, an effect that is unlikely to be detectable in data (Andersen and Beyer 2006).

Measurements of trophic transfer efficiency in different marine environments suggest that it is not affected by changes in temperature and productivity among ecosystems and typically ranges from 10% to 20% (Ware 2000). Mean predator–prey mass ratios are similarly unaffected and often range from 10^2 to $10^4:1$ (Barnes et al. 2010). For this reason, patterns of energy transfer from primary producers to fish are broadly comparable in marine ecosystems, and high rates of primary production tend to be translated into high rates of fish production in all areas (Fig. 21.2).

There is some evidence that trophic transfer efficiency decreases at higher trophic levels and predator–prey mass ratios increase, but these changes appear to counter one another as they do not have a discernible impact on the slope of the size spectrum (Barnes et al. 2010). The relative constancy of predator–prey mass ratios and transfer efficiency and the absence of relationships with temperature or primary production help explain why the slope of the size spectrum is also relatively constant among ecosystems. However, the "height" of the size spectrum does change with productivity, reflecting changes in the total numbers of animals present with changes in the energy to support them.

21.4 FOOD WEB COMPLEXITY

The broad characterizations of marine food webs based on size spectra provide an appealing synthesis that links process and structure, but the acceptable level of abstraction depends on the questions being addressed. For example, knowledge of processes governing bulk energy flux is important for understanding and predicting community structures and ecological processes in ecosystems that vary widely in productivity, biodiversity, and physical characteristics, and where knowledge of the biota may not support consistently detailed analyses. For these ecosystems, simple size-based rules have been used to estimate global fish biomass and the role of fishes in biogeochemical processes, analyses that would not have been tractable if approached on a species-by-species basis (Jennings et al. 2008; Wilson et al. 2009). Conversely, when questions about species groups and species need to be addressed, for example in regional and ecosystem-scale assessments, then additional information has to be included.

Of course, the range of trophic ecologies in marine food webs is more complex than implied by simple size-based abstractions. Size spectra most accurately describe communities of the open ocean where phytoplankton are the primary producers and almost all energy transfer occurs in the pelagic environment. By contrast, coastal and shelf communities are typically more complex, with much of the energy captured and processed by communities on the seafloor. Seafloor animals can be herbivores and detritivores, which obtain energy from benthic primary producers and detritus exported from the pelagic food web. Owing to the relatively shallow depths in coastal and shelf seas, the production of seafloor detritivores is accessible to predators that may also forage in the open sea.

As well as being used for system-wide analysis, size spectra can be compiled for different components of a

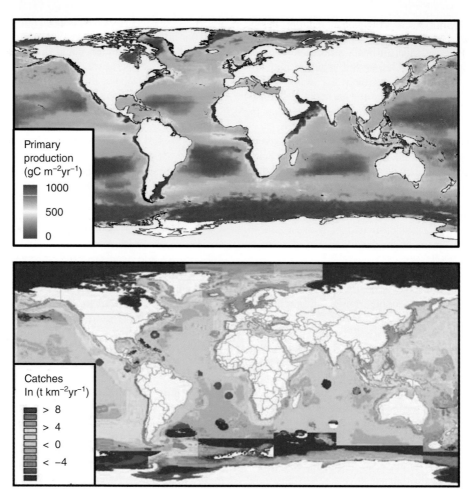

Figure 21.2 The global distribution of phytoplankton primary production (upper panel) broadly reflects the global distribution of fisheries catches (lower panel). Global phytoplankton production of 45×10^8 tonnes C per year is estimated to be ≈90% of total marine primary production. From SeaWIFS data processed by Frédéric Mélin and Rodney Forster (A) and from Worm et al. (2009), based on an analysis by Reg Watson (B).

community. Differences in size-spectra slopes for different components of the community reflect differences in feeding interactions. Several investigators have identified differences between the slope of a size spectrum in a predation-based community and in a community in which energy is shared (Brown and Gillooly 2003; Blanchard et al. 2009). In communities that share energy (such as shellfish of different sizes that all eat phytoplankton and detritus) the slope of the relationship between log numbers and log body size tends to be shallower than in predation-based communities and approximates −0.75 (equivalent to +0.25 for log

biomass and log body size). Given that the rate of metabolism scales as approximately $M^{3/4}$, energy use by animals in such communities is hypothesized to be independent of body mass. This is the "energy equivalence hypothesis" as first proposed by terrestrial ecologists based on the scaling relationship between mean abundance and mean size of different animals (Damuth 1981; see also Isaac, Carbone, and McGill, Chapter 7, and Hechinger, Lafferty, and Kuris, Chapter 19). Their argument is not entirely consistent with the argument we are making here for a marine community that shares the same energy source, but leads to the same hypothesis (Jennings et al. 2007). In a predation-based community, available energy supply decreases with increasing body size, because the transfer of energy from prey to predators is inefficient. This accounts for the steepening of the size spectrum to slopes of −1 to −1.2 or so (equivalent to 0 to −0.2 for log biomass and log body size in Fig. 21.1), rather than −0.75.

If the food web is divided into food chains that are based on energy sharing and predation then these will be coupled to some extent. When coupling is stronger, more energy from one food chain is transferred to the other, either by predation or by consumption of detritus. Models of the dynamic coupling between food chains have shown that the strength of coupling affects food web resilience. Resilience is a measure of the rate at which a community returns to a "steady state" following a change in primary production or predation (Blanchard et al. 2011). Although just a model-based analysis at present, the results imply that deep pelagic oceans, where there is low coupling, are more likely to be vulnerable to human and environmental impacts than shallow coastal seas.

Species are a key focus of ecology, conservation, and fisheries but their identities are ignored in most treatments of the size spectrum. Incorporating species in the size spectrum is challenging when most fish eggs have a size of 1 mg but fish species reach maturity at sizes ranging from 1 g to 100 kg. In terrestrial ecology, species are often characterized by a representative body size, such as the size of an adult, but this approach oversimplifies marine food webs where individuals of the same species can fulfill many roles in the food web as they grow. From an analytical perspective, the large scope for growth in fish requires that the whole size structure of a species' population is resolved in food web analysis, and the community size spectrum can be decomposed into the size spectra for species with different maximum size (Andersen and Beyer et al. 2006; Pope et al. 2006; Fig. 21.3). Such approaches have now been adopted to assess the effects of fishing on marine communities and ecosystems and are addressed in section 21.5. Species-specific differences in feeding strategies and behavior can lead to interesting and informative departures from the average tendency. For example, filter-feeding sharks and whales, that can "feed down the food chain" on smaller and more productive size-classes of prey, have evolved a strategy that gives them access to a greater resource base by using their mobility to follow areas of abundant zooplankton at cross-ecosystem scales.

21.5 FISHING IMPACTS

Fishing has wide-ranging impacts on populations, communities, and ecosystems. The capacity of a species' population to withstand fishing mortality depends on the rate of mortality and the life history of the population. The faster life histories of smaller species typically confer more resilience to fishing mortality. Fishing affects communities and ecosystems by changing the relative abundance of species, with knock-on impacts on food web interactions. The impacts of fishing on communities and ecosystems have been an increasing focus of recent research, reflecting societal desire and political commitments to meet environmental targets for ecosystems as well as sustainable fisheries. Analysis of these impacts poses many new challenges for a scientific community that has largely focused on fishing impacts on a few well-studied populations.

Susceptibilities of populations to fishing mortality are body-size dependent because smaller species have greater capacity to sustain additional mortality (Fig. 21.3). Rates of fishing mortality also tend to be size-dependent because different fishing gears target different size-classes in populations and communities and because the management system often defines regulations for minimum mesh sizes to allow smaller individuals to escape. Consequently, in real fisheries, larger species are usually subject to higher mortality rates and less able to sustain them. This leads to the differential depletion of larger individuals and species in many communities, modifying the slopes of size spectra that would be expected in unexploited systems (section 21.3). There are many examples of slopes of size spectra becoming steeper with increased fishing rates (Rice and Gislason 1996; Bianchi et al. 2000). Changes in the slope of the spectrum are due to the depletion of large fish and the proliferation of small fish as their larger predators are depleted (Dulvy et al. 2004; Daan et al. 2005).

The relative impacts of fishing on a community can be quantified by comparing the slopes of observed size spectra with the predicted slopes of the spectra in the absence of fishing. In the North Sea, this approach was used to predict that the biomass of large fishes weighing 4–16 kg and 16–66 kg, respectively, was 97.4% and 99.2% lower than would be expected in the absence of fisheries exploitation. In addition, because the smaller fishes that now dominate the size spectrum have faster turnover times, the mean turnover time of

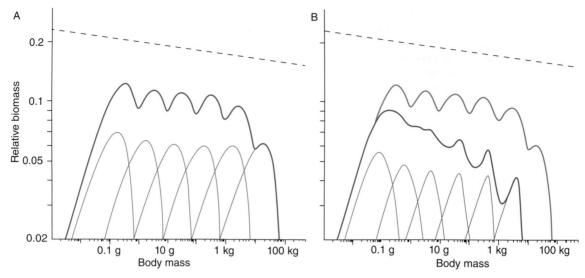

Figure 21.3 Biomass size spectra for a modeled community of six species spanning a range of maximum body sizes (A) that is not fished and (B) where each species is fished with a trawl-type of size selection pattern. Both panels show the biomass of all fish in the fish community (magenta lines) and of six species with varying asymptotic sizes (red lines) as a function of individual weight. The dashed line is the theoretical spectrum (elevated for clarity), and the green line in (B) is the community spectrum from (A). The calculations are based on the assumption that the total metabolism of an individual scales as $M^{3/4}$, and fishing impacts each species from 5% of the asymptotic weight and onwards with a fishing mortality of 0.75 per year. Fishing on all species with the same fishing mortality has the largest impact on the bigger species in the community, as the smaller species have a higher rate of production than the bigger species. Modeling approach based on Anderson and Beyer (2006).

the fished community fell from 3.5 to 1.9 years (Jennings and Blanchard 2004; Fig. 21.4).

As well as the size-selective effects of fishing on fish communities, the physical contact between seabed fauna and towed fishing gears also results in size-selective mortality. Towed gears may differentially kill larger animals because smaller ones can be pushed aside by the pressure wave in front of the gear (Gilkinson et al. 1998). Consequently, benthic communities in trawled areas tend to be dominated by smaller individuals and species, and the slopes of benthic invertebrate size spectra become steeper in more heavily trawled areas (Duplisea et al. 2002; Hiddink et al. 2006).

21.6 FISHERY ASSESSMENT AND MANAGEMENT

Fisheries are managed because unregulated fisheries have consequences that society deems undesirable. These consequences include collapses of fisheries that result in long-term loss of yield, with costly social and

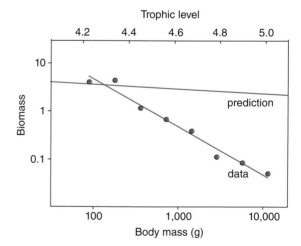

Figure 21.4 The predicted slope of a size spectrum for an unexploited North Sea fish community and the slope of the size spectrum as determined from data for the exploited North Sea fish community in 2001. The steeper slope for the exploited community revealed the extent of decline in the abundance of large fishes following intensive fishing.

economic impacts, as well as unwanted changes to marine habitats, species, and ecosystems that are seen as a conservation issue.

To support management and to avoid the unwanted impacts of fishing, assessments of the rates of fishing that lead to high and sustainable catches have been provided for decades. Developing processes to conduct and improve these assessments is a central focus of fisheries science. Broadly, the assessment process involves estimating population size and sustainable rates of fishing by accounting for inputs (growth, recruitment) to and outputs (natural mortality, fishing mortality) from the population. The assessment process is currently focused on individual populations, and is data intensive. Natural mortality is a parameter that provides a link between single-species population dynamics and the influence of the wider food web, and may be estimated from life-history invariants and community models (Pope 1979; Andersen et al. 2009; Gislason et al. 2010).

Despite the management focus on population assessment, populations are embedded in communities and ecosystems, and there is growing interest in understanding interactions among small groups of species caught in the same fisheries. To some extent there has always been a push–pull in fisheries and marine environmental management between tractable abstractions of the ecosystem characterized by single-species population analysis and community-based analyses where mortality and growth are treated explicitly. However, there has been a recent and sustained resurgence of interest in community-based analysis, not least because management targets for community properties such as size composition are now being considered in some jurisdictions. Understanding the response of the interacting populations to fishing allows explicit examination of the trade-offs between the status of fished populations and aspects of the community such as size composition and trophic level. The disadvantages of community-based analysis are the greater parameter demands and difficulties of formalizing transient and complex interactions between species.

We have already discussed the development of size-based models that incorporate species identity. This is achieved by using maximum body size as a proxy for species identity and has provided a generalized understanding of the processes structuring communities and responses to fishing (Fig. 21.3; Pope et al. 2006; Andersen et al. 2008). However, real species matter in fisheries since they are known and desired by consumers, targeted by fishers, and regulated by managers. Accordingly, fish communities have also been modeled to predict the abundance of particular species in defined size-classes. Such species-specific models are based on the assumptions that growth dynamics can be described by known parameters describing growth rate, asymptotic size, size at maturity, and size-dependent rates of reproduction. Feeding interactions can be characterized by a binary diet matrix that defines which species eat which other species, as in Petchey and Dunne (Chapter 8). Mortality rates are the sum of non-predation mortality, predation mortality, and fishing mortality. Predation mortality depends on

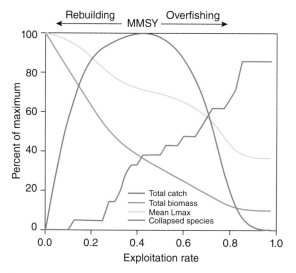

Figure 21.5 Modeled relationships between fishery catches, properties of a fish community, and the exploitation rate. Lines denote the total catch from a fish community (blue), the biomass of fish in the community (green), the mean maximum size of fishes in the community (a metric of the life-history composition, yellow), and the number of collapsed species (species are defined as collapsed when biomass falls to <10% of predicted biomass in the absence of fishing, red). MMSY is the maximum multispecies sustainable yield that can be sustained by the community, but achieving MSSY comes at costs to the larger, more vulnerable species which are expected to collapse at these exploitation rates. In this example, managers that seek to reconcile demands for fishing and conservation might aim for exploitation rates around 0.2. At this exploitation rate there is a small loss of potential yield to the fishery but changes to the community and the collapses of vulnerable species are minimized. Simulations by J. S. Collie using the size-based model of Hall et al. (2006); from Worm et al. (2009).

the predator–prey size ratio and predator abundance, while fishing mortality depends on size, species identity, and the selectivity of fishermen (Hall et al. 2006). The strength of these models is that they allow explicit consideration of the trade-offs between objectives for fisheries (e.g., high and sustainable yields or profits) and conservation (e.g., maintaining viable populations of vulnerable species and community properties) (Fig. 21.5).

Size- and species-based models have since been elaborated and extended to include dynamic food-dependent growth, species-specific feeding behavior, and an interaction matrix based on species spatial co-occurrence). The new methods allow for both population and overall community dynamics to be tracked through time and for trade-offs between fisheries yields and environmental "health" indicators to be evaluated across different species and size-specific harvesting strategies. Alternate species and size-structured models are also being used as part of simulation frameworks to evaluate the performance of indicators in supporting achievement of conservation and management objectives (Fulton et al. 2008). Size- and species-based models can also be used to support single-species assessments by better predicting natural mortality, one of the most challenging parameters to measure directly. Mortality rates can thus be seen as a consequence of predators in the community rather than as intrinsic life-history parameters related to lifespan (Andersen et al. 2009). The models "explain" natural mortality as a consequence of predator–prey interactions. In other words, for one predator to satisfy its metabolic demands, a corresponding number of prey have to die. Interestingly, the balance between growth of predators, determined by their metabolic demands, and the rate of death of smaller prey organisms, leads to mortality scaling as $M^{-1/4}$ where M is the body mass of an individual (Andersen et al. 2009).

21.7 CONCLUSIONS

The application of metabolic scaling theory to infer population and ecosystem properties has a long history in marine science, from the empirical determination of relations between metabolism and population growth rate to the role of metabolic scaling in predicting the community size spectrum. Variations in metabolic rate with body size and temperature drive patterns of energy acquisition and have been shown to have consequences for the life histories of marine plants and animals and the structure and function of populations, communities, and ecosystems. These insights into the processes influencing marine communities have led to present-day size-based analyses and dynamical community models to describe the effect of fishing, climate, and changes in primary production on the marine ecosystem. Further, the growing capacity to account for species identity in size-based community models has provided additional insights into life-history trade-offs and food web structure, while supporting analysis and advice on fisheries management issues that are species focused. Based on such insights we conclude that the understanding of the importance of metabolism in driving some of the main population, community, and ecosystem processes in marine ecosystems is maturing. The next steps are likely to involve moving beyond the study of broad scaling patterns, to understand differences between metabolic predictions and observations, at both the process and system levels. For example, an intriguing new analysis of empirical data shows that the scaling of natural mortality with body size varies systematically with maximum size (Gislason et al. 2010). Drivers for the next steps in this research remain strong, as fishery managers increasingly demand advice that places fishery management in a community and ecosystem context.

Chapter 22

CONSERVATION BIOLOGY

Alison G. Boyer and Walter Jetz

SUMMARY

1 Modern conservation biology is faced with the challenge of maintaining the biodiversity and function of global ecosystems despite escalating human impacts on the environment. Conservation efforts are often limited by the lack of detailed population-level data required for a formal evaluation of extinction risk.

2 The metabolic theory of ecology (MTE) provides a quantitative link between metabolism and many organismal traits known to be associated with extinction risk. The MTE facilitates predictions of broad-scale patterns of extinction risk for understudied groups, based only on the limited data available.

3 We demonstrate several applications of MTE to understanding and predicting species-level extinction risk using case studies of mammals and birds

of the world. MTE provides clear linkages between extinction risk and species' life-history traits and energy and space requirements. We also discuss the broad-scale conservation applications of the MTE, including the design of conservation reserves and efforts to identify key taxa, specifically ectotherms, that due to their metabolic dependence on ambient temperatures may be most affected by global warming.

4 While the application of the MTE to conservation biology has been limited in the past, we argue that broad-scale MTE-based models have the potential to aid conservation efforts. Quantitative predictions produced by such models may help conservationists to estimate future risk and to focus effort on the species and habitats that need it most.

22.1 INTRODUCTION

Biodiversity is faced with an extinction crisis caused by escalating human environmental impacts (Ceballos and Ehrlich 2002; Butchart et al. 2010; Rands et al. 2010). According to estimates by the International Union for the Conservation of Nature's *Red List of Threatened Species* ("Red List"), more than one of every five vertebrate species is now threatened with extinction (Hilton-Taylor et al. 2009). The increasing scale

of modern biodiversity loss requires that conservationists look beyond the population level, which has characterized most of conservation biology to date, to examine broad patterns in extinction risk among taxa and habitats and to identify key regions and taxa that require greater conservation attention.

A recent and growing body of work has found that intrinsic biology has a significant influence on extinction risk (Pimm et al. 1988; Laurance 1991; McKinney 1997; Purvis et al. 2000a; Reynolds 2003; O'Grady

---

Metabolic Ecology: A Scaling Approach, First Edition. Edited by Richard M. Sibly, James H. Brown, Astrid Kodric-Brown.
© 2012 John Wiley & Sons, Ltd. Published 2012 by John Wiley & Sons, Ltd.

Table 22.1 Metabolically linked traits associated with elevated extinction risk.

Taxon	Traits linked to extinction risk	Reference
Mammals	Large body size, small geographic range, slow life history, high trophic level	Purvis et al. 2000c; Jones et al. 2003; Cardillo 2003; Cardillo et al. 2008; Davidson et al. 2009
Amphibians	Restricted geographic range, large body size, low fecundity	Lips et al. 2003; Bielby et al. 2008; Cooper et al. 2008; Sodhi et al. 2008
Birds	Large body size, long generation time, low fecundity, small geographic range	Owens et al. 1999; Owens and Bennett 2000; Boyer 2010; Lee and Jetz 2011
Reptiles	Foraging strategy	Reed and Shine 2002
Fish	Large body size, age at maturity, slow life history, small geographic range	Reynolds et al. 2005; Olden et al. 2007, 2008
Crayfish	Low fecundity, habitat specialization	Larson and Olden 2010
Insects	Habitat specialization, small geographic range, large body size	Koh et al. 2004; Williams et al. 2009
Plants	Local abundance, body size, diaspore mass, restricted geographic range, dispersal ability, habitat specialization	Duncan and Young 2000; Freville et al. 2007; Bradshaw et al. 2008; Sutton and Morgan 2009; Polidoro et al. 2010

et al. 2004; Davies et al. 2008; Foden et al. 2009; Lee and Jetz 2011; see Table 22.1). As Brown (1995) emphasized, extinction may be predictable. However, despite increases in the total number of species assessed by the Red List in recent years, the conservation status of most of the world's species remains poorly known (Pennisi 2010). Even in well-studied groups, there is a high proportion of species for which current data are insufficient to allow a full evaluation of extinction risk, including 15% of mammals, 14% of reptiles, 25% of amphibians, and 22% of fishes (Butchart and Bird 2010). The metabolic theory of ecology (MTE) provides a strong, quantitative link between metabolism and many organismal constraints and traits associated with extinction risk, including body size (Brown et al. 2004), home range size (Jetz et al. 2004), trophic level (Arim et al. 2007), population density (Silva and Downing 1994; Allen et al. 2002; Wilman and Jetz, in preparation), population growth rate (Savage et al. 2004a), life history (Ernest et al. 2003; Sibly and Brown 2007), overheating, water loss, and activity time. MTE may offer a powerful predictive framework to help evaluate and project extinction risk for understudied groups for which detailed population-level data is lacking. Despite these considerable implications for conservation management and policy, to date the application of MTE to conservation biology has been relatively limited.

22.2 METABOLIC LINKAGES TO EXTINCTION RISK

Understanding the factors which predispose a species to become endangered and extinct is one of the major challenges for conservation biologists. The impacts of external threats, such as habitat loss, overexploitation, and climate change, can vary depending on an animal or plant's intrinsic vulnerability, as determined by traits such as life history, body size, and trophic or niche characteristics (McKinney 1997; Reynolds 2003; Foden et al. 2009). Here we use case studies of the world's mammal and bird species to discuss a few of the predictions of MTE that influence species' intrinsic vulnerability to human environmental impacts, including life history, energy use, and space requirements of animal populations.

22.2.1 Body size, life history, and extinction risk

Within mammals, large-bodied species have a shorter fossil "lifespan" and generally suffer a higher risk of extinction than do small-bodied species (Van Valen 1975; Diamond 1984; Martin 1984; Van Valkenburgh 1999). The hypothesized mechanisms underlying this relationship include the lack of "sleep-or-hide"

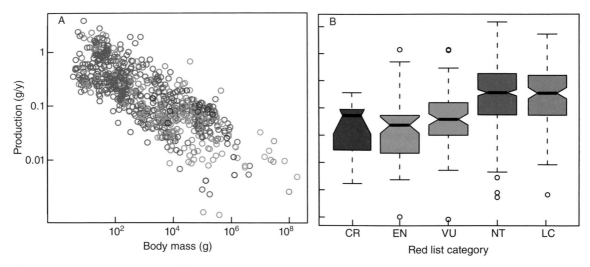

Figure 22.1 Body size and speed of life history are related to extinction risk in the world's mammals. (A) Scaling of mass-specific production with body size (see Sibly, Chapter 5) among species in five Red List categories of extinction risk (data from Davidson et al. 2009). (B) Distribution of mass-specific production within each Red List category (in order of decreasing extinction risk: CR, critically endangered; EN, endangered; VU, vulnerable; NT, near threatened; LC, least concern). Boxplots encompass the lower (25%) and upper (75%) quartiles, with a heavy line at the median, and whiskers extend to 1.5 times the interquartile range. The most endangered species have on average lower rates of production and hence slower life histories and slower recovery from population reductions.

strategies in large mammals which could buffer against environmental change (Liow et al. 2008), targeted hunting of larger mammal species by humans both in the past and in the present day (Lyons et al. 2004; Price and Gittleman 2007), and lower reproductive rates and slower population growth in large mammals (Johnson 2002; Purvis et al. 2000b; Cardillo 2003; Cardillo et al. 2005; Charnov and Zuo 2011). Life-history variation is increasingly recognized as a key predictor of extinction risk in mammals (Cardillo et al. 2006; Davidson et al. 2009). Even though geographic range size remains the strongest predictor, in birds life history explains a substantial amount of the variation in extinction risk across species, above and beyond other effects such as human encroachment (Lee and Jetz 2011). Together with development mode, body mass emerges as a key factor behind the role of life history in extinction risk.

One critical aspect of mammalian life history is mean total biomass of offspring produced per year, which can be estimated from measures of mean offspring mass, litter size, and number of litters per year. When normalized by adult body size, this value represents the annual amount of biomass produced per unit of adult body size, in units of grams per year, known

as the mass-specific production rate (see Sibly, Chapter 5). Mass-specific production scales negatively with mass, so that larger-bodied species have lower production per year (Fig. 22.1A; Sibly and Brown 2007; Hamilton et al. 2011; see also Sibly, Chapter 5), although there is significant variation around this central tendency (see Bielby et al. 2007).

We illustrate the tight linkage between the thus-quantified "speed" of life history and extinction risk using a global database of mammalian life histories (data from Davidson et al. 2009) in conjunction with species-level estimates of extinction risk from the Red List. Red List evaluations are based on five independent criteria relating to observed population loss and geographic range contraction, and species are assigned to risk categories based on the probability of extinction within the next 50–100 years (Mace et al. 2008). Using a global dataset of mammal species, we found that species in high-risk Red List categories (vulnerable, endangered, and critically endangered) had slower rates of mass-specific production than species of lower conservation concern (Fig. 22.1B; ANOVA, $p < 0.001$, $F = 28.65$, df = 4, 702). Consistent with previous studies (Cardillo et al. 2005; Davidson et al. 2009) life history explained more variation in extinction risk than body size across all mammals (general linear model, $N = 704$, adjusted $r^2 = 0.06$ for $\log_{10}$ body mass and 0.14 for $\log_{10}$ mass-specific production).

This observed linkage between life-history characters and extinction risk supports the idea that species metabolism and resulting relative speed of reproduction affects species' ability to rebound from human-caused population declines. Arguably, most human environmental impacts serve, either directly or indirectly, to increase the mortality rate experienced by species. Logically then, such an "across-the-board" increase in mortality rate will have greatest impact on those species with "slow" reproductive rates. The metabolic theory of ecology provides a powerful framework linking reproductive rates (or mass-specific production rates; Charnov and Zou 2011) to metabolic rates and body size (Brown and Sibly 2006). Such relationships, based on first principles of physiological ecology, may provide valuable predictions about species' ability to respond to future environmental threats.

22.2.2 Energy requirements and extinction risk

Another commonly cited correlate of extinction risk in vertebrates is trophic level, where predatory species at

higher trophic levels are thought to be at higher risk of extinction (Bennett and Owens 1997; McKinney 1997; Purvis et al. 2000b; Cardillo et al. 2004; Lee and Jetz 2011). One prominent explanation for this pattern is that species at higher trophic levels are more vulnerable to the cumulative effects of disturbance on species lower down the food chain (Diamond 1984; Crooks and Soulé 1999; Petchey et al. 2008b). However, others have emphasized the tendency for predators to maintain lower population densities and require larger habitat areas (Cardillo et al. 2004; see also Petchey and Dunne, Chapter 8 and Karasov, Chapter 17) than other species due to energy requirements (Damuth 1987; Carbone et al. 1999; Jetz et al. 2004).

Primary production required, or PPR, represents the amount of energy (kJ/day) in the form of net primary productivity that is required to support the energy needs (metabolism) of a single individual (Pauly and Christensen 1995; Jetz and Wilman, in preparation). Because predators rely on populations of prey at lower trophic levels, and due to the inefficiency of energy transfer between trophic levels, much more primary productivity is required to support a species at the top of the food chain. Thus, PPR increases with trophic level, and distinct trophic levels clump together, appearing as parallel lines in Figure 22.2, with apex predators represented by the topmost group, and herbivores at the

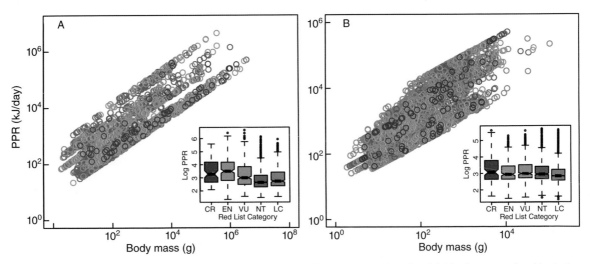

Figure 22.2 Primary productivity required (PPR, kJ/day) and body size in (A) mammal and (B) bird species colored by Red List categories of extinction risk (in order of decreasing extinction risk: CR, critically endangered; EN, endangered; VU, vulnerable; NT, near threatened; LC, least concern). Insets show the distribution of PPR within each Red List category (data from Jetz and Wilman, in preparation). Species at greatest risk tend to be at higher trophic levels and/or to require greater ecosystem primary production per individual, putting them at greater risk when faced with habitat reduction and other kinds of environmental change.

bottom. Within trophic levels, PPR increases with body size due to the positive allometric scaling of metabolic rate (Brown et al. 2004). PPR strongly constrains the number of individuals of a species that can survive in a given area, such that local abundance scales with PPR^{-1} (Jetz and Wilman, in preparation). Because low population density and low local abundance is linked to extinction risk (Gilpin and Soule 1986; Silva and Downing 1994; Davidson et al. 2009), we may expect species that require more energy to sustain a population (i.e., higher PPR) to be at higher risk of extinction.

We used the estimates of PPR for 8226 birds and 3210 terrestrial mammal species gathered by Jetz and Wilman (in preparation) and tested their relationship to extinction risk as measured by the Red List categories. PPR was strongly related to Red List category, and to extinction risk, in both terrestrial mammals and birds. Mammal species in high-risk categories (critically endangered and endangered) had significantly higher PPR than species in lower-risk Red List categories (Fig. 22.2A; ANOVA, $p < 0.001$, $F = 57.28$, df = 4, 3208). However, many critically endangered species were located on the lower edge of points within Figure 22.2A because the majority of critically endangered mammals are herbivorous (i.e., at the lowest trophic level). A similar relationship held for birds, where species of least concern had significantly lower PPR than species in the higher-risk categories (Fig. 22.2B; ANOVA, $p < 0.001$, $F = 24.07$, df = 4, 8224). Controlling for the effect of body mass on PPR, we found that PPR remained a significant predictor of extinction risk (ANOVA, mammals: $p < 0.001$, $F = 22.56$, df = 4, 3208; birds: $p < 0.001$, $F = 5.535$, df = 4, 8224).

The strong linkage between PPR and extinction risk demonstrates the direct relevance of metabolic ecology for conservation. PPR measures the relative energetic needs of each species. Species with high PPR may meet these needs by inhabiting small home ranges within a high-productivity habitat or by utilizing large areas of lower-productivity habitat. However, over 40% of global net primary productivity is now being used by a single species – humans (Rojstaczer et al. 2001; Haberl et al. 2007). Species with high energetic requirements, whether they live in high-productivity tropical forests or lower-productivity habitats, are increasingly impacted by encroaching human populations. Because of the unavoidable trade-off in energy use between human and "natural" systems, one may expect future extinctions to selectively remove species with high energetic requirements as human populations dominate more and more of the Earth's productivity.

22.2.3 The energetics of space use, geographic range size, and extinction risk

Species with large body size, high metabolic rate, and large PPR tend to have low population densities and larger space needs (Damuth 1981; Marquet et al. 1990; Jetz et al 2004; Wilman and Jetz, in preparation). While the drivers of geographic range occupancy (Gaston et al. 2000; Freckleton et al. 2006; Jetz et al. 2008) represent a critical additional factor and these links require further study, it has been suggested that large local space needs alone may contribute strongly to larger geographic ranges in order to maintain a viable, range-wide population size (Brown and Maurer 1989). According to this idea, the lower boundary of the relationship between body size and geographic range size may represent some minimum geographic range size required to support a species of a given body size.

Small geographical range size is one of the strongest, and most consistently observed, predictors of extinction risk across taxa (Table 22.1; Jones et al. 2003; Cardillo et al. 2008; Cooper et al. 2008; Davies et al. 2008; Davidson et al. 2009; Lee and Jetz 2011). Large geographic ranges are associated with resistance to extinction both during deep-time extinctions in the fossil record (Jablonski 2005; Jablonski and Hunt 2006; Payne and Finnegan 2007) and in more recent human-mediated extinctions (Harris and Pimm 2008; Boyer 2010). Geographic range size and recent declines are also a primary criterion that the IUCN uses for classifying threat (Mace et al. 2008). Any factor that caused a contraction of the geographic range, and thus a decline in population size, would be expected to detrimentally affect species that already had a small geographic range extent. Empirically, macroecologists and biogeographers have observed a "triangular" relationship between body size and geographic range area in terrestrial species (Brown and Maurer 1987). While species of all body sizes may have large geographic ranges, the minimum range size exhibited by species tends to increase with body size, such that large-bodied species typically have only large ranges (Fig. 22.3; Gaston and Blackburn 1996), providing some support for the aforementioned connections.

We examined the relationship between body size, range size, and extinction risk using global databases of mammalian and bird species' geographic range areas (data from Davidson et al. 2009; Jones et al. 2009; Lee and Jetz 2011). We found evidence of a lower bound on the relationship between body size and geographic

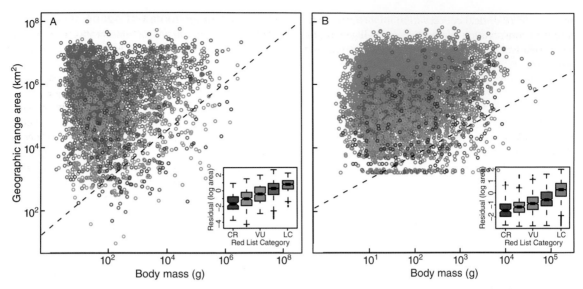

Figure 22.3 Body size and geographic range area in (A) mammal and (B) bird species colored by Red List categories of extinction risk. Dashed lines show the hypothesized scaling of minimum geographic range area, $A \times M^{3/4}$, placed at an arbitrary y-intercept for comparison. Insets show the distribution of the residuals of range area within each Red List category after controlling for the effects of body mass. CR, critically endangered; EN, endangered; VU, vulnerable; NT, near threatened; LC, least concern. Mammals: data from Davidson et al. (2009) and Jones et al. (2009); birds: data from Lee and Jetz (2011). The most endangered species consistently have relatively small geographic ranges relative to their body size.

range size in birds and terrestrial mammals. While small range size and large body size are each independently associated with extinction risk in this dataset (ANOVA, mammals: range size $p < 0.001$, $F = 218.455$, body mass $p < 0.001$, $F = 17.408$, df $= 1, 3362$; birds: range size $p < 0.001$, $F = 576.132$, body mass $p < 0.001$, $F = 80.171$, df $= 1, 8961$), threatened species (identified by Red List categories critically endangered, endangered, and vulnerable) tend to be found near the lower edge of the body size–range size space due to a significant interaction between the body size and range size (Fig. 22.3; ANOVA, mammals: $p < 0.001$, $F = 7.441$, df $= 4, 3362$; birds: $p = 0.03$, $F = 2.665$, df $= 4, 8961$; see also Davidson et al. 2009).

We briefly elaborate on the potential connection between body size-dependent space needs and geographic range size in determining extinction risk. Following MTE, the area A required to support an individual I of mass M scales with body mass as $A/I \times M^b$ (Jetz et al. 2004; see also Hamilton, Burger, and Walker, Chapter 20), where the exponent b represents the body-mass scaling exponent of metabolic rate B, $B \times M^b$. Assuming that the minimum viable number of

individuals, N_{min} (minimum viable population size; Traill et al. 2007), does not vary systematically with regard to body size, then the area necessary to accommodate the total minimum viable population size, N_{min}, should be approximated by $A \times N_{min} M^b$ (Boyer and Jetz 2010). If b is approximately 0.75 (Savage et al. 2004b; Anderson and Jetz 2005; White et al. 2009), we expect the lower threshold of the range size–body size relationship to scale as $M^{3/4}$. However, the lower boundary of body size–range size space occupied by non-threatened species appears to scale somewhat steeper than $M^{3/4}$ (dashed lines in Fig. 22.3). This suggests that either (1) the minimum viable population size, N_{min}, is not independent of body size (Brook et al. 2006; Traill et al. 2007), or (2) species occupy only a subset of their geographic range (Hurlbert and White 2007; Jetz et al. 2008), or (3) the scaling of space use with body size is better characterized by the home range area ($H \times M^1$; Jetz et al. 2004) rather than population density, or some combination of these and other possibilities. However, given that there seems to be a robust scaling relationship between body size and minimum geographical range size, the exact slope of the relationship may not be of critical conservation importance. From a species conservation perspective, a small geographic range relative to body size, and conversely a large body size within a given range size category, may be much more important than whether the species is large or small or widespread or narrowly distributed in absolute terms. Habitat loss and the resulting contraction and fragmentation of geographic ranges is a concern, not only for large-bodied species, but for species of all sizes if their range size approaches the lower bound (Gaston and Blackburn 1996).

22.3 BROAD-SCALE CONSERVATION APPLICATIONS OF MTE

From its inception, two core goals of conservation science have been, not only to maintain viable populations of the world's species, but also to preserve functioning samples of global ecosystems (Convention on Biological Diversity 1993). Part of the challenge to conservation biologists has been to use scientific principles to select and manage wildlife reserves that meet these two goals. To meet this challenge, a systematic conservation planning framework has been developed so as to optimize the allocation of scarce conservation funding by prioritizing areas for protection (Margules and Pressey 2000). Over the past decade, this approach has been applied at the global scale to identify areas of high conservation priority, such as the biodiversity hotspots (Myers et al. 2000; Olson and Dinerstein 2002; Mittermeier et al. 2005; Cardillo et al. 2006; Ceballos and Ehrlich 2006; Conservation International 2009). Just as MTE has relevance for determining the extinction risk of a single species when faced with anthropogenic threats, MTE may be informative for identifying species and, through their geographic distribution, regions and habitats for conservation attention. Here we discuss two areas in which MTE has particular application: the selection and placement of reserves, and conservation strategies to deal with climatic warming.

22.3.1 Design and placement of conservation reserves

Habitat destruction threatens biodiversity at all levels, from ecosystems to individuals, and is considered the most important cause of species extinction worldwide

(Pimm and Raven 2000; WWF 2004; Hoekstra et al. 2005). One common goal of conservation reserves and protected areas is to preserve functioning ecosystems in the face of ongoing habitat loss and human encroachment. As we saw above, the MTE provides an energetic framework linking metabolism and body size to energy and habitat area requirements. MTE-based assessments of habitat and energy requirements, such as PPR (Jetz and Wilman, in preparation), can help to refine predictions of the effects of habitat loss on biodiversity, and may also provide insight into the design of effective conservation reserves.

Determining the area and placement of conservation reserves to support biodiversity is a central question in conservation biology. Ever since the theory of island biogeography (MacArthur and Wilson 1967), there has been an ongoing debate over the so-called SLOSS question: Is a single large reserve preferable to several small reserves of the same total area (Diamond 1975; Simberloff and Abele 1976; Terborgh 1976; Wilcox and Murphy 1985; Boecklen 1997; Etienne and Heesterbeek 2000)? Population dynamic models suggest that the design that minimizes the risk of extinction of species is case-specific, with the optimal number of reserves ranging between one and very many (McCarthy et al. 2009). However, as we saw above, MTE and the allometry of space use can give an approximate lower bound on the area of viable reserves, suggesting that several small reserves will only be effective if each is larger than some minimum area.

Given the strong links between local abundance and reserve areas required for species persistence ("minimum dynamic areas"; Pickett and Thompson 1978; Leroux et al. 2007), MTE-based predictions of individual space needs can aid conservation planning, even in the absence of resources for local population-level surveys. Metabolism and PPR-based statistical models of abundance may be applied across species and regions (Wilman and Jetz, in preparation).

22.3.2 Impacts of global climate change

Anthropogenic climate change, and specifically global warming, are having observable and likely detrimental effects on organisms (Parmesan and Yohe 2003; Colwell et al. 2008), and global temperature is projected to increase by an additional 1 to 6 °C over the next century, including more variable weather conditions and more frequent heat waves (IPCC 2007a).

Conservation scientists are faced with the challenge of predicting which species will be most threatened by climate change, and determining how best to mitigate these impacts (Brooke 2008; Foden et al. 2009). In this context, there is a growing focus on using "macrophysiology" to predict the responses of organisms to climate change (Chown et al. 2004; Chown and Gaston 2008), especially for vertebrate ectotherms (Deutsch et al. 2008; Kearney et al. 2009; Buckley et al. 2010; Sinervo et al. 2010).

Ectotherms are especially likely to be vulnerable to warming because, depending on the amount of behavioral thermoregulation, their internal temperature, and thus metabolic rate, are often strongly influenced by environmental temperature (Dillon et al. 2010; Buckley et al. in press; see also O'Connor and Bruno, Chapter 15 and Waters and Harrison, Chapter 16). These findings are consistent with MTE, where temperature mediates the scaling of metabolic rate with body size, such that metabolic rate increases approximately exponentially with body temperature (Gillooly et al. 2001). Environmental temperature may also constrain the rate at which ectotherms can assimilate energy, forming a hump-shaped relationship with temperature (i.e., performance curve; Huey 1982). However, the biological impact of rising temperatures depends on the degree of behavioral thermoregulation and the physiological sensitivity of organisms to temperature change (Deutsch et al. 2008). Because tropical organisms tend to have narrower ranges of thermal tolerance (Ghalambor et al. 2004), and may be at or near their thermal optima already, some authors suggest the greatest extinction risks from global warming may be in the tropics (Deutsch et al. 2008; Tewksbury et al. 2008; Dillon et al. 2010). Recent physiological models suggest that climate change in already warm regions may reduce potential activity time while also increasing energetic maintenance costs of ectotherms (Kearney et al. 2009; Dillon et al. 2010), and may result in substantial extinctions (Huey et al. 2010; Sinervo et al. 2010).

In endotherms, the relationships between environmental temperature and metabolism, or other biological parameters affecting population growth rate, are more complex (Anderson and Jetz 2005; La Sorte and Jetz 2010). In wintering birds, a metabolic framework has been shown to capture the variation in abundance due to temperature-dependence of food availability and body-size related thermal constraints (Meehan et al. 2004). Short-term population declines and mass

die-offs have been observed as a result of extreme heat events (e.g., Jiguet et al. 2006). For example, in 2009, the death of thousands of small birds was reported in Western Australia during a severe heat wave (McKechnie and Wolf 2009). In hot environments, endotherms evaporate water from their bodies in order to maintain body temperatures below lethal limits (Wolf and Walsberg 1996). Because increasing temperatures may also lead to changes in the physiological water balance, especially in arid environments where access to water is limited, water loss rather than overheating is likely to be the main physiological constraint of global warming on endotherms (McKechnie and Wolf 2009).

The increasing emphasis on physiology as a key predictor of climate change effects on organisms brings additional relevance of MTE for conservation science. Ectotherms constitute the vast majority of terrestrial biodiversity (Wilson 1992), and physiological differences between endo- and ectotherms may result in highly disparate responses to projected temperature changes (Buckley et al. in press). While many more physiological details will be crucial to predicting the exact responses of particular organisms, the majority of evidence points to relatively large potential effects of temperature change on the metabolism and survival of tropical ectotherms (Dillon et al. 2010). This result is quite troubling, given that it places the greatest biological risks of climate change in the tropics where biodiversity is greatest. Recognizing the physiological differences in sensitivity to temperature change among organisms and habitats, conservationists may be able to identify those species most at risk from climate-related threats. In terms of conservation strategies, these results highlight the need to provide connectivity between and buffer zones around established reserves, perhaps orienting corridors pole-ward and upslope to allow for temperature-related shifts (Halpin 1997).

22.4   CONCLUSIONS

Given the pressing need to slow down the extinction crisis and the lack of detailed demographic information for the majority of endangered species, there are many opportunities for MTE to inform conservation biology (Calder 2000). MTE-based models can be informative even when based on simple biological traits, and models may prove useful for prioritizing data collection itself. However, perhaps because of its pursuit of generalizable insights rather than detailed predictions, the application of MTE to conservation biology has been limited. In many cases, the application of general ecological "principles" to specific conservation and management problems requires tedious and often difficult elaboration of details (Simberloff 2004). While broad-scale MTE-based models are no substitute for detailed studies of population declines and habitat loss, we argue that quantitative predictions from such models may help conservationists to focus research and conservation attention on the species and habitats that need it most. Currently, some models of extinction risk use allometric scaling to estimate population-level parameters (such as RAMAS® Ecorisk; Hajagos and Ferson 2001); however, the application of more sophisticated, MTE-based models at geographic scales could help identify species and higher taxa that are likely to be at imminent risk and in need of conservation action. Examining the energy (PPR) and space requirements of species may help determine the habitat requirements of threatened species and lead to more effective reserve design and allocation. Finally, understanding the physiological mechanisms of species' response to climate change will increase our ability to predict the effects of projected changes and will aid efforts to mediate these effects on biodiversity. Predictive models, including MTE, will be a vital tool as ecologists and conservationists anticipate future challenges.

Chapter 23

CLIMATE CHANGE

Kristina J. Anderson-Teixeira, Felisa A. Smith, and S. K. Morgan Ernest

SUMMARY

1 Climate change is a reality that will profoundly impact ecological systems at all levels of organization.
2 Changes in climate are expected to impact individual metabolism, either directly through temperature effects on metabolism, or indirectly through body size changes driven by evolutionary responses to increases in temperature.
3 Temperature-driven changes in metabolic rate, as well as changes in body size, will affect community properties such as numbers of individuals and total biomass.
4 Changes in temperature, moisture, and carbon dioxide will impact ecosystem energetics, carbon cycling, and organization.
5 As a theory linking multiple levels of ecological organization, metabolic ecology has great potential to contribute to our understanding of how ecological systems will respond to the novel environmental conditions produced by climate change.

23.1 INTRODUCTION

By the end of the century, it is predicted that atmospheric carbon dioxide concentrations will have doubled or tripled, global mean temperature will have increased by 1.8–4.0 °C, and the amount and timing of precipitation will have been altered substantially in many parts of the world (IPCC 2007b). While these values are disturbing enough, global averages mask the true magnitude of the problem. Climate shifts are larger over continents (particularly in their interiors), at high latitudes, in more arid regions, and in the Northern Hemisphere. Moreover, the rate of climate change may be even more critical than its magnitude and duration (Davis et al. 2005; Loarie et al. 2009).These anthropogenic environmental perturbations will have profound effects on ecological systems at all levels of organization.

Understanding how ecological systems will respond to interacting elements of climate change is one of the great challenges in science today – one that requires mechanistic understanding in order to accurately predict responses to novel environmental conditions. By providing a mechanistic framework grounded in the energetics of individuals and spanning multiple levels of organization, metabolic ecology has great potential

Metabolic Ecology: A Scaling Approach, First Edition. Edited by Richard M. Sibly, James H. Brown, Astrid Kodric-Brown.
© 2012 John Wiley & Sons, Ltd. Published 2012 by John Wiley & Sons, Ltd.

to contribute to our understanding of the response of ecological systems to climate change.

In this chapter, we discuss the impacts of climate change at three levels of ecological organization: individuals, communities, and ecosystems. We review key concepts for understanding the likely impacts of climate change and highlight some interesting applications of metabolic ecology.

23.2 INDIVIDUALS

In recent decades, considerable effort has gone into predicting how species will respond to the large anthropogenic climate shifts expected over the next few centuries. The temperature increases that have already occurred (IPCC 2007b) have had detectable impacts on species (Hughes 2000; Davis and Shaw 2001; McCarty 2001; Stenseth et al. 2002; Walther et al. 2002; Parmesan and Yohe 2003; Root et al. 2003). While some studies have documented changes in the morphology, phenology, genetics, and community interactions of organisms (Smith et al. 1998; Hughes 2000; Bradshaw and Holzapfel 2001; Davis and Shaw 2001), most have focused on shifts in the distribution and abundance of species (reviewed in Parmesan and Yohe 2003; Root et al. 2003; Ackerly et al. 2010). A variety of detailed, spatially explicit models have been developed to determine whether individual species can keep pace with climate change. These are largely based on characterizing the abiotic environmental niche space now occupied by species and projecting forward, although more recent iterations include biotic interactions as an important aspect (Berry et al. 2002; Peterson et al. 2002; Iverson et al. 2004; Thuiller et al. 2004; Araújo et al. 2006; Araújo and Luoto 2007). However, these models of "species velocity" ignore other possible responses to climate, particularly those relating to physiology and adaptation.

23.2.1 Metabolism and temperature

All organisms require energy for survival, reproduction, and growth (see Sibly, Chapter 5). The total energy required is not only a function of size, but is also dependent on whether the organism maintains a constant body temperature. How organisms obtain that energy, and how they allocate it, is directly related to environmental temperature (Brown et al. 2004).

The rate at which energy is acquired, transformed, and used drives the rate of all biological activities of the organism and, moreover, sets its energetic demands or footprint on the environment. The occupation of novel environments or abrupt environmental shifts – such as those predicted under even the best-case scenarios of climate change – can radically alter the crucial pattern of energetic allocation between survival, reproduction, and growth. Such shifts often represent a stress, but could also benefit some organisms; for example, if winters become warmer, animals may be able to divert energy from maintenance to growth or reproduction. Moreover, metabolism will be indirectly influenced by energy availability in the environment; for example, increases in high-latitude productivity could result in larger body sizes, increased fecundity, and higher abundance (Yom-Tov and Yom-Tov 2004, 2005). Consequently, knowledge of energetics is central to an understanding of the selective forces that shape an organism's physiology, natural history, and evolution, and central to a synoptic understanding of how organisms will respond to climate change.

23.2.1.1 Ectotherms

For ectotherms in particular, many of the biotic impacts of climate shifts are mediated through physiology (Fig. 23.1). As discussed in previous chapters, the rate of biological activity increases approximately exponentially with temperature in the range of approximately 0–40 °C (Gillooly et al. 2001). This means that metabolism changes more in response to a shift in high temperature than low temperature. The implications of an exponential relationship between metabolism and temperature are striking: not only can temperature shifts influence ontogeny through the influence on growth rates or development times (Gillooly et al. 2001; West et al. 2001; Brown et al. 2004; Hou et al. 2008), they can lead to differences in realized adult body size (Atkinson et al. 1994; Partridge et al. 1994; Sibly and Atkinson 1994; Ashton 2004). Moreover, there are other feedbacks since many fundamental biological rates also demonstrate temperature and body-size dependence (Fig. 23.1), as has been discussed throughout this volume. For example, for many ectotherms, maximal population growth rates scale positively with temperature and negatively with mass (Savage et al. 2004; see also Sibly, Chapter 5 and Isaac, Carbone, and McGill, Chapter 7).

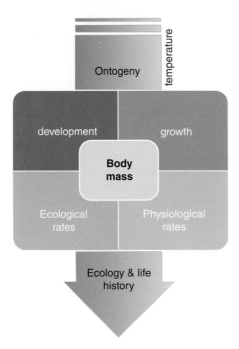

Figure 23.1 Temperature can influence ontogeny (of ectotherms in particular), through changes in development or growth rate, which can result in altered adult body mass. Similarly, changing temperature regimes can result in differential mortality of adults of varying body mass. In turn, metabolism and other physiological and ecological rates are influenced by body mass.

How will climate change influence the physiology of organisms? A recent study used a metabolic ecology approach to estimate the effects of temperature shifts on the metabolism of terrestrial ectotherms over the last 30 years (Dillon et al. 2010). Although high latitudes have experienced a greater rise in temperature (Fig. 23.2A) (IPCC 2007b), the exponential temperature dependence of metabolism means that the impact on metabolic rate should be greater in warmer areas. The calculations by Dillon et al. (2010) suggest that metabolic rates have increased more in the tropics and north temperate zones than in the Arctic (Fig. 23.2). This is particularly striking because the increase in temperature in the tropics has been modest to date, leading to the perception that little biotic change has occurred. The exponential relationship between metabolism and temperature, coupled with the narrower temperature tolerances of tropical species, suggests that future warming will impact these high-biodiversity areas much more than previously thought. Higher metabolic rates for ectotherms likely lead to reallocation of resources between growth, maintenance, and reproduction (Hou et al. 2008). Moreover, greater energetic demands could potentially lead to less "discretionary energy" for reproduction, which could ultimately influence population abundance and dynamics in these regions. One aspect not addressed by the Dillon et al. (2010) study was body size; increases in temperature could also influence ontogeny and/or lead to differential survival of juveniles or adults (Fig. 23.1). If higher temperatures select for smaller body size, for example, then the effects Dillon et al. (2010) found could be confounded by the allometric scaling of metabolic rate with mass.

23.2.1.2 Endotherms

Generally, endotherms are considered to be buffered against environmental fluctuations since they usually maintain a constant core temperature of 36–40 °C. However, while many physiological rates and processes are not directly influenced by environmental temperatures, metabolism is. If ambient air temperature is lowered significantly, for example, animals begin to shiver to maintain core body temperature, which raises their metabolic rate. Similarly, when ambient temperature exceeds a certain level, organisms somewhat paradoxically raise their metabolic rate to cool the core body temperature. Several studies have indicated that foraging and other activities are constrained by temperature, presumably because the metabolic cost outweighs the potential gain (Huey and Slatkin 1976; Huey and Pianka 1981; Traniello et al. 1984; Fraser et al. 1993). If climate shifts result in endotherms restricting activity and/or expending more energy regulating body temperature, then there is a tangible physiological cost to the individual. Interestingly, there are several studies that suggest endotherms occupying high-latitude habitats tend to have elevated metabolic rates relative to those in lower latitudes (Portner 2002; Anderson and Jetz 2005; Clarke 2006).

23.2.2 Body size and temperature

Body size is of great physiological and ecological significance, as shown throughout this book (see also Peters 1983). How animals interact with their envi-

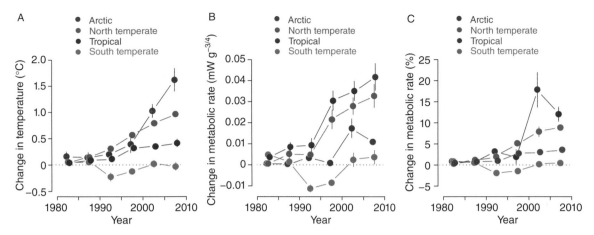

Figure 23.2 Global changes in temperatures and metabolic rates since 1980. (A) Changes in mean temperature (5-year averages); (B) predicted absolute changes in mass-normalized metabolic rates by geographic region; (C) predicted relative changes in mass-normalized metabolic rates. Both temperature and metabolic rate are expressed as differences from the standard reference period (1961–1990) and calculated on the basis of E and b_0 for an average ectotherm. This study is the only one to date that has explicitly computed changes in metabolism that should have occurred with the documented warming. Reprinted by permission from Macmillan Publishers Ltd: Dillon, M. E., Wang, G. and Huey, R. B. (2010) Global metabolic impacts of recent climate warming. *Nature*, 467, 704–706. © 2010.

ronment is strongly mediated by their body mass. Regular patterns of morphological variation with abiotic factors (especially temperature) have been observed repeatedly over time and space (see Millien et al. 2006). The existence of such ecogeographic patterns demonstrates the ability of species to adapt to fluctuating abiotic conditions, as well as highlighting the strong selection imposed by the environment.

One of the best-supported ecogeographic patterns is Bergmann's rule: the principle that within a broadly distributed taxon, representatives of larger size are found in colder environments, and those of smaller size are found in warmer areas (Bergmann 1847; Mayr 1956). Although originally formulated in terms of species within a genus, Bergmann's rule is often recast in terms of populations within a species. The rule holds for the majority of vertebrates examined to date (Table 23.1). Moreover, many ectotherms also appear to demonstrate Bergmann clines, including bacteria, protists, plants, insects, marine organisms, and turtles (Huey et al. 2000; Ashton 2002; Belk and Houston 2002; Ashton and Feldman 2003; Blanckenhorn and Demont 2004). Large-bodied animals may more tightly conform, perhaps because they are less able to avoid stressful thermal environments by diapause, torpor, burrowing, nests, or other means (Freckleton et al.

Table 23.1 Percent of various vertebrate groups that conform to Bergmann's rule. Although endothermic vertebrates demonstrate a strong body-size cline with temperature, ectotherms are much more variable in their adherence to the rule (Millien et al. 2006).

	Conformation to Bergmann's rule
Endotherms	
Birds	76%
Mammals	71%
Ectotherms	
Lizards	30%
Snakes	21%
Turtles	83%
Salamanders	72%
Frogs	62%
Fish	28%

2003). This suggests that larger-bodied animals may be more sensitive to climatic fluctuations.

Bergmann's rule has generally been interpreted as a direct selective response to temperature (Birch 1957; Brown and Lee 1969; Dawson 1992). When organisms increase in body mass, their surface area increases more slowly than does their volume (surface area $\propto$ length^2, versus volume $\propto$ length^3). Thus the ratio of surface area to volume scales as ~$M^{2/3}$. Because heat is dissipated from surfaces, this relationship means that larger animals tend to lose less heat per unit mass than do smaller animals and are at an advantage under cold environmental conditions (e.g., polar bears). Conversely, smaller animals have a greater surface to volume ratio and are more capable of dissipating heat under thermally stressful conditions (e.g., black bears). While a number of other mechanisms for Bergmann's rule have been postulated (e.g., productivity gradients, selection on life-history characteristics, development rates; Scholander 1955; Rosensweig 1968; Burnett 1983; Yom-Tov and Nix 1986), most of them are usually factors related to thermal characteristics of the environment.

While most studies have focused on spatial gradients of body size and temperature, Bergmann's rule has been observed in animal populations over both historical and evolutionary time (e.g., Smith et al. 1995, 1998, 2009; Yom-Tov 2001; Hunt and Roy 2006; Millien et al. 2006; Yom-Tov et al. 2006). Body size decreases have been documented in several species of passerine birds over the last 20–50 years in both Israel and England (Yom-Tov 2001; Yom-Tov et al. 2006). Several investigators have been able to tie evolutionary change in the populations to underlying temperature changes. For example, studies of woodrats (*Neotoma*) over both historic and millennial timescales find that these small rodents rapidly adapt to climate shifts by adjusting body size (Smith et al. 1995, 1998; Smith and Betancourt 1998, 2003, 2006). The response is in the direction predicted by Bergmann's rule: woodrats are larger during cold intervals, and smaller during warmer episodes. In some cases, the proximate mechanism is even clear. In a 10-year study of these rodents in southern New Mexico, investigators found that smaller-bodied animals were able to successfully overwinter when winter temperatures were milder, and larger-bodied animals suffered differential mortality as summers warmed (Smith et al. 1998). This led to significant shifts in the mean adult body size of the population. Bergmann's rule has also been observed in deep time; it has been proposed as a mechanism for the size increases documented in deep-sea ostracods over the past 40 million years (Hunt and Roy 2006).

While the ability of endotherms to adapt to recent anthropogenic climate change has only recently been seriously considered (e.g., Smith et al. 1995; Bradshaw and Holzapfel 2001; Smith and Betancourt 2006), this may be a possible response to climate change. Most likely to be affected will be small-bodied species with short generation times and medium to large population sizes because climate shifts may be too rapid for species with longer generation times to adapt.

The significance of body-size shifts as an adaptive response to climate change stems from the fundamental role of size in mediating most fundamental ecological, life-history, and physiological processes, as has been shown throughout this book (see also Peters 1983). Thermal selection on the body mass of organisms directly influences many biological processes including, but not restricted to, metabolism. Moreover, because body size so strongly constrains the energetic demands of organisms and their interaction with the environment, it is not surprising that it also influences space and home range requirements, population densities, and other important population- and community-level characteristics (Peters 1983; Calder 1996; Isaac, Carbone, and McGill, Chapter 7; Karasov, Chapter 17; Jennings, Andersen, and Blanchard, Chapter 21). Below, we discuss how these shifts in size and metabolic rate of individuals may influence higher levels of ecological organization.

23.3 COMMUNITIES

As discussed above, changes in climate are expected to impact individual metabolism either directly through temperature effects on metabolism or indirectly through body size changes driven by evolutionary and/or plastic responses to increases in temperature. These changes in individual metabolism will likely lead to reorganization of communities. Understanding how changes in individual metabolic rate will impact total abundance and biomass requires models that explicitly link metabolism with these community-level properties.

23.3.1 Metabolic models of community abundance and biomass

There have been several papers proposing models for using metabolic scaling to link individual metabolic rate with community-level abundance and/or biomass (Ernest and Brown 2001; Enquist et al. 2003; Meehan et al. 2004; White et al. 2004; Ernest et al. 2009; see also Isaac, Carbone, and McGill, Chapter 7). While each of these models has some unique components, they all have the same essential core (Brown et al. 2004): the community property reflects an interplay between the rate of resource supply and individual metabolic rate. The exact structure of the relationship varies depending on whether the variable of interest is total abundance (N) or standing biomass (W). For abundance, the core relationship is:

$$N = \frac{[R]}{I} = \frac{[R]}{i_0 \overline{M}^{3/4} e^{-E/kT}} \tag{23.1}$$

And for standing biomass:

$$W = N\overline{M} = \frac{[R]\overline{M}}{i_o \overline{M}^{3/4} e^{-E/}} \tag{23.2}$$

where $[R]$ is the rate of supply or concentration of resource, i_o is a normalization constant, and $\overline{M}^{3/4}$ and $\overline{M}$ are averages taken across all individuals within a community or assemblage.[1]

Depending upon the taxa and the question being addressed, modifications can be made to equations 23.1 and 23.2 to improve both predictive ability and biological accuracy. For example, inclusion of the specific gravity of a tree species' wood is important for accurate biomass estimates for forest ecosystems (Enquist et al. 1999; Chave et al. 2005). Additionally, it may be necessary to use different scaling relationships between mass and metabolic rate when asking questions about endotherms operating at temperatures beyond their thermoneutral zone (Meehan et al. 2004).

23.3.2 Possible impacts of climate change on community properties

The relationships outlined above make it clear that shifts in average individual size, temperature, and resource concentration or rate of supply should all impact community-level properties directly through their relationship with metabolism. Using equations

23.1 and 23.2, we can predict the responses of total abundance and standing biomass to shifts in temperature and average individual mass for a hypothetical ectotherm community under three different climate change scenarios (no change and 1.8 °C and 4 °C increases in ambient temperature) (Fig. 23.3). With no change in temperature, as average size of an individual shifts from 80 g to 30 g, we predict an increase in total abundance and a decrease in standing biomass, demonstrating both the potential importance of shifting size on community-level properties and the different effect of changing metabolic rate on community-level total abundance and standing biomass. If temperature alone changes (i.e., no corresponding change in average size of an individual), then increasing temperature is expected to cause a decrease in both abundance and biomass. Finally, if temperature increases *and* average size decreases, as might initially be expected from biogeographic patterns such as Bergmann's rule (see section 23.2.2), expected impacts on metabolic rate can be more complicated. As the community shifts from an average mass of 80 g to 45 g concurrently with a 4 °C increase in temperature, we predict very little change in total abundance, but standing biomass should decline by approximately 40%. The magnitudes and broad ranges of possible responses of communities to climate change highlight the importance of better understanding not only the impacts of increasing temperature on metabolic rate but also how changes in temperature and average individual size might interact.

Should increases in temperature impact the average size of an individual in communities? Based on Bergmann's rule, we might expect shifts in temperature to cause shifts in the average size of an individual through selection for smaller individuals within species, or smaller species, based on physiological and life-history responses to temperature (Atkinson and Sibly 1997). However, one study examining both species-level and community-level changes in average size showed that a community can exhibit an overall decrease in average size even though the dominant species increased in size over the same time period (White et al. 2004). These different responses at the species and community level emerge, because at the community level the average size of an individual is influenced by the distribution of abundance across sizes. If individual species are being selected to become larger or smaller in size, they can concurrently also become more or less abundant within the community.

---

[1] Brown et al. (2004) simplify the relationship for standing biomass to yield: $W [R]M^{1/4}e^{E/kT}$. However, because $\overline{M}^{3/4}$ does not generally equal $\overline{M}$ (Savage 2004), equation 23.2 is more generally appropriate.

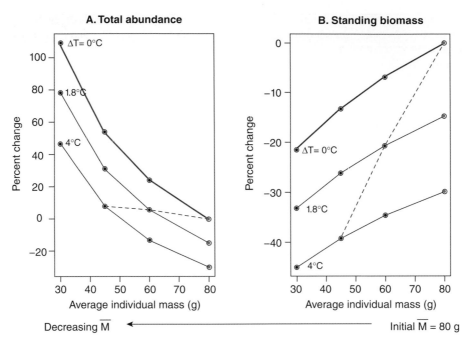

Figure 23.3 Predicted responses, as expected from MTE, to shifts in average individual size and temperature of community-level abundance (A) and standing biomass (B) using equations 23.1 and 23.2. Black and red lines indicate the percent expected change in total abundance or standing biomass under a specific climate change scenario, as average individual mass decreases. Blue dots (at 80 g mass) denote the percent expected change if only temperature is changing. Dashed lines denote expected changes if both size and temperature are changing concurrently. The initial community was modeled as having a normally distributed individual size distribution (mean = 80 g, standard deviation = 5), under an average environmental temperature of 11.8 °C (contiguous US average for 1980–1999; data from www.ncdc.noaa.gov). Expected percent change in abundance and biomass were calculated using the initial community as a baseline and comparing to expectations for three different shifts in average individual mass (mean = 60 g, 45 g, and 30 g) and three different temperature scenarios (no change in average temperature, and 1.8 °C and 4 °C increases). 1.8 °C and 4 °C increases in temperature were chosen based on the minimum and maximum temperature changes projected in the most recent IPCC report (IPCC 2007b).

Therefore small increases in size at the species level can be offset or overwhelmed by large shifts in community composition. Answering the question of whether increases in temperature will impact average individual size in communities requires an understanding of how the individual size distribution (i.e., the frequency distribution of individual masses; *sensu* White et al. 2007) for an entire community is affected by temperature. The limited number of studies comparing communities along environmental gradients – as well as those specifically considering the effects of warming (e.g., Daufresne et al. 2009; Morán et al. 2010; Yvon-Durocher et al. 2011b) – have generally found that average body size within a community decreases with increasing temperature. However, most of these studies are on aquatic systems (e.g., fish: Blanchard et al.

2005; Meerhoff et al. 2007; Brucet et al. 2010; plankton: Bays and Crisman 1983; Duncan 1984; Saito et al. 2011; Sommer and Lewandowska 2011; but see Meehan et al. 2004 for an example with birds). Obviously, more work examining temperature impacts on the individual size distribution in terrestrial ecosystems is required before we can confidently predict the metabolic implications of climate change for community-level properties.

23.4   ECOSYSTEMS

We have already seen how two important climatic elements, temperature and water, shape ecosystem-level processes through their effects on metabolic rate

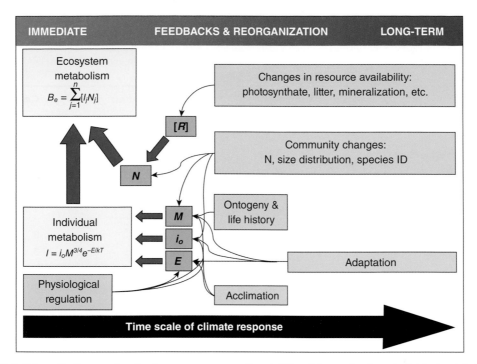

Figure 23.4 Schematic diagram showing how processes operating at different timescales shape individual metabolism (I; equation 1.1) and thereby ecosystem metabolism (B_e). Individual metabolism: I, individual metabolic rate; i_o, normalization constant; M, body mass; E, effective activation energy (eV); k, Boltzmann's constant (8.62×10^{-5} eV K^{-1}); T, temperature (Kelvin); N, number of individuals; $[R]$, rate of supply or concentration of a limiting resource; j, size class; n, number of size-classes represented in a community.

(Anderson-Teixeira and Vitousek, Chapter 9). Here, we consider how *changes* in temperature, precipitation, and atmospheric carbon dioxide will affect ecosystems. One of the most fascinating – and challenging – aspects of understanding these changes is that they will affect many intimately linked ecosystem processes in different ways. Understanding how ecosystems will ultimately respond to these changes is challenging, yet of utmost importance. Climate change will affect not only the structure and internal dynamics of ecosystems, but also the global climate. Any shift in the balance between CO_2 uptake through photosynthesis and CO_2 release through respiration will feed back to affect atmospheric CO_2 concentration and climate. Therefore, it is important to understand how production and respiration will differ in their responses to various elements of climate change and, thereby, how climate change will affect the carbon balance of ecosystems.

On the surface, using metabolic theory to predict how ecosystems will respond to climate change seems

straightforward. Knowing the temperature dependence of soil respiration, for example, we should be able to calculate the expected soil respiration rate at a new temperature. Although this approach works on very short timescales, responses are quickly complicated by acclimation, adaptation, feedbacks, shifts in community composition, and ecosystem state changes (Fig. 23.4) (e.g., Luo 2007; Reich 2010). The challenge to predicting ecosystem responses to climate change lies in understanding the different responses of multiple interacting processes to various elements of climate change, as well as how these play out across trophic levels and through time (e.g., Heimann and Reichstein 2008; Walther 2010; Woodward et al. 2010c). Ultimately, of course, fundamental biophysical constraints shape ecosystem properties, and understanding how to integrate these basic constraints with the complexity at the ecosystem level will help us to better understand the role of metabolism in determining ecosystem responses to climate change.

Table 23.2 Expectations for short-term responses of terrestrial production and respiration to increases in temperature, precipitation, and atmospheric carbon dioxide (+ indicates increase; − indicates decrease; 0 indicates no change). The respective effective temperature sensitivities (ε; Anderson-Teixeira and Vitousek, Chapter 9) for these processes are also shown. Based on Campbell and Norman 1998; Bernacchi et al. 2001; Gillooly et al. 2001; Atkin and Tjoelker 2003; Reichstein et al. 2003; Allen et al. 2005; Atkin et al. 2007; Reich 2010.

	Temperature	Precipitation[a]	Carbon dioxide
Terrestrial C$_3$ photosynthesis[b]			
Rate	+[c] ($E \approx 0.3$ eV)	+	+
Temperature sensitivity (ε)	−	+	+
Plant respiration[d]			
Rate	+ ($E \approx 0.3$ eV)[d]	0/+	0/+
Temperature sensitivity (ε)	−	0/+	0/+
Heterotrophic respiration			
Rate	+ ($E \approx 0.65$ eV)	+	0
Temperature sensitivity (ε)	0/−	+	0

[a] Response asymptotes at high precipitation levels.
[b] Approximation for terrestrial Rubisco-limited C$_3$ photosynthesis over the temperature range 0–30 °C.
[c] Response becomes negative at higher temperatures.
[d] Assumes acclimation to photosynthate supply.

Metabolic ecology is helpful for predicting responses of ecosystems at the short timescale (Table 23.2), understanding the mechanisms underlying the complexity at intermediate timescales, and also for understanding how ecosystems are ultimately constrained by climate across broad climatic gradients (Fig. 23.4). Currently, however, the vast majority of research on this topic does not employ the metabolic scaling approach. Here, we relate to metabolic theory the current understanding of how changes in temperature, precipitation, and CO$_2$ will affect ecosystem metabolism. We give special attention to the differential responses of ecosystem production and respiration, as such differences will drive feedbacks in resource availability and the long-term carbon balance.

23.4.1 Temperature change

Ecological rates, including those at the ecosystem level (Anderson-Teixeira and Vitousek, Chapter 9), generally increase with temperature – at least in the short term and below stress-inducing temperatures. Based on the inherent temperature dependence of physiological rates (Table 23.2), we expect warming to increase rates of energy and material flow in ecosystems. Indeed, experimental increases in temperature often

stimulate production and respiration in both terrestrial ecosystems (Fig. 23.5) (Rustad et al. 2001; Wu et al. 2011) and aquatic mesocosms (Yvon-Durocher et al. 2010). Moreover, there is evidence that climatic warming has increased both net primary productivity and soil respiration globally (Nemani et al. 2003; Bond-Lamberty and Thomson 2010). Thus, climate change is likely to accelerate ecosystem dynamics.

23.4.1.1 Modification of temperature response by acclimation and feedbacks

Observed increases in productivity and ecosystem respiration often do not quantitatively match the increase that would be expected based on the intrinsic temperature dependence of physiological rates under current conditions (Table 23.2). Rather, biotic acclimation, changes in resource supply, changes in community composition, and interactions with other environmental variables modify the response expected from metabolic theory (Table 23.2; Figs 23.4 and 23.6) (Luo 2007; Reich 2010).

The relationship between productivity and temperature will almost certainly be affected by climate change (Table 23.2). The effective temperature dependence of photosynthesis depends upon the photosynthetic pathway (C$_3$, C$_4$, or CAM), ambient CO$_2$ concentration,

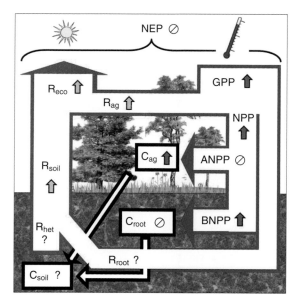

Figure 23.5 Typical responses (as characterized by meta-analysis) of the metabolic fluxes in terrestrial ecosystems to increases in temperature. Carbon fluxes into the ecosystem are represented by green arrows, and those out by red arrows. Warming-induced changes are symbolized as follows: up or down arrow indicates increase or decrease, respectively; Ø indicates no change, ? indicates unknown. Blue arrows indicate a change that tends to cool the climate by removing CO_2 from the atmosphere; orange arrows indicate a change that tends to warm the climate by adding CO_2 to the atmosphere. NEP, net ecosystem production (GPP minus R_{eco}); GPP, gross primary production; NPP, net primary production; ANPP, above-ground NPP; BNPP, below-ground NPP; C_{ag}, carbon storage in above-ground biomass; C_{root}, carbon storage in root biomass; C_{soil}, carbon storage in soil organic matter; R_{ag}, above-ground plant respiration; R_{root}, root respiration; R_{het}, heterotrophic respiration (below-ground); R_{soil}, total soil (below-ground) respiration ($R_{het} + R_{root}$); R_{eco}, ecosystem respiration. Based on Wu et al. (2011).

water availability, and the temperature range over which it is characterized (Campbell and Norman 1998). Therefore, changes in temperature, precipitation regimes, atmospheric CO_2 concentrations, and species composition will cause the effective temperature dependence of photosynthesis to deviate from metabolic theory's "canonical" temperature dependence of terrestrial production ($E = 0.3\,eV$; Table 23.2) (Allen et al. 2005; Anderson-Teixeira and Vitousek, Chapter 9).

Likewise, the temperature response of ecosystem respiration will be modified by numerous feedbacks including changes in resource supply (e.g., photosynthate), biotic acclimation, changes in the microbial community, and environmental constraints (Table 23.2; Figs 23.4 and 23.6) (Davidson and Janssens 2006; Davidson et al. 2006; Allison et al. 2010; Reich 2010). This is particularly true of plants, which acclimate to warming by downregulating respiration so as to maintain homeostasis (Atkin and Tjoelker 2003; King et al. 2006; Atkin et al. 2007). Given that warming will typically cause feedbacks and acclimation (Fig. 23.4), the long-term responses of respiration and production to warming are unlikely to match the temperature sensitivity expected from metabolic theory.

23.4.1.2 Response of the carbon balance

In the absence of feedbacks and acclimation, respiration generally responds more strongly to temperature than does production (Figure 10.3). This is worrisome because, when sustained over time, this difference would cause larger increases in CO_2 release through respiration than in CO_2 uptake through primary production, which would imply a positive feedback to climate change. Indeed, there are situations in which this appears to occur (e.g., Piao et al. 2008; Yvon-Durocher et al. 2010). However, the net carbon balance of terrestrial ecosystems does not respond consistently to warming (Fig. 23.5) (Wu et al. 2011). Rather, on average, production and respiration have responded with similar temperature sensitivity in warming experiments (Wu et al. 2011).

Insights into the long-term responses to warming may be gained by considering how the carbon balance changes across climatic gradients. In terrestrial ecosystems, biomass does not vary systematically with temperature across global climatic gradients (Allen et al. 2005; Anderson-Teixeira et al. 2011; Stegen et al. 2011); however, carbon storage in detritus and soil generally decreases with increasing temperature (Jobbágy and Jackson, 2000a; Allen et al. 2005), indicating that steady-state carbon storage may decline with increases in temperature. In epipelagic oceans, the ratio of production to respiration decreases with increasing temperature globally, implying that increases in temperature will reduce the ability of oceans to sequester carbon (López-Urrutia et al. 2006). While such evidence lends support to the idea that

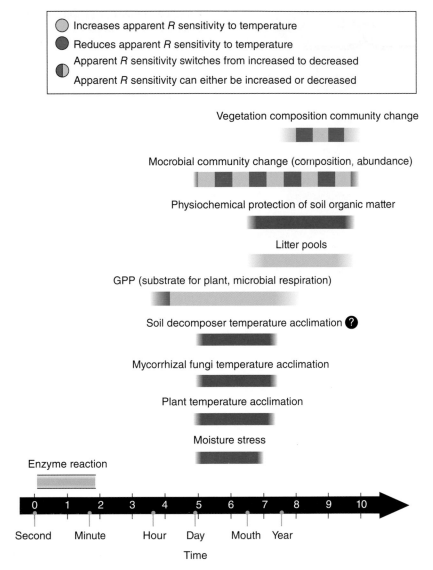

Figure 23.6 In terrestrial ecosystems, the apparent sensitivity of plant and microbe respiration to temperature is modified by a wide range of processes acting over a range of timescales (represented logarithmically by numbers on the arrow). From Reich (2010).

warming may eventually shift the metabolic balances of many ecosystems in favor of respiration, we have much to learn about how the myriad of interacting temperature-dependent processes in ecosystems will combine to affect their metabolic balances in future climates (Luo 2007).

23.4.2 Precipitation changes

Precipitation is another important driver of ecosystem processes – one that affects both production and respiration strongly but in somewhat different ways (Anderson-Teixeira and Vitousek, Chapter 9). Both

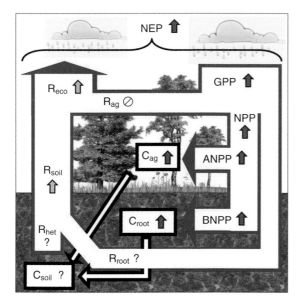

Figure 23.7 Typical responses (as characterized by meta-analysis) of the metabolic fluxes in terrestrial ecosystems to increases in precipitation. Carbon fluxes into the ecosystem are represented by green arrows, and those out by red arrows. Warming-induced changes are symbolized as follows: up or down arrow indicates increase or decrease, respectively; Ø indicates no change, ? indicates unknown. Blue arrows indicate a change that tends to cool the climate by removing CO_2 from the atmosphere; orange arrows indicate a change that tends to warm the climate by adding CO_2 to the atmosphere. NEP, net ecosystem production (GPP minus R_{eco}); GPP, gross primary production; NPP, net primary production; ANPP, above-ground NPP; BNPP, below-ground NPP; C_{ag}, carbon storage in above-ground biomass; C_{root}, carbon storage in root biomass; C_{soil}, carbon storage in soil organic matter; R_{ag}, above-ground plant respiration; R_{root}, root respiration; R_{het}, heterotrophic respiration (below-ground); R_{soil}, total soil (below-ground) respiration ($R_{het} + R_{root}$); R_{eco}, ecosystem respiration. Based on Wu et al. (2011) by permission of John Wiley & Sons, Ltd.

production and respiration increase with precipitation, at least up to some level of rainfall. These responses have been confirmed by experimental addition of precipitation to terrestrial ecosystems (Fig. 23.7) (Wu et al. 2011). On balance, increases in precipitation have a stronger effect on production than on respiration – even at high levels of precipitation (Austin 2002; Chen et al. 2009; Anderson-Teixeira et al. 2011). Therefore, increased precipitation generally results in

net uptake of carbon by the ecosystem (Fig. 23.7) (Wu et al. 2011). Global patterns in net ecosystem productivity, biomass, and soil carbon suggest that this response will persist over the long term (Jobbágy and Jackson 2000b; Anderson-Teixeira et al. 2011).

The timing of precipitation is also important to an ecosystem's carbon balance. Whereas small pulses of precipitation trigger pulses of CO_2 releases through soil respiration, more sustained and deeper increases in soil moisture increase carbon uptake by ecosystems by stimulating plant growth (Huxman et al. 2004b; Jenerette et al. 2008; Chen et al. 2009). These transient effects can significantly affect an ecosystem's long-term metabolic balance; for example, in a Chihuahuan desert grassland, plots receiving single large precipitation additions had greater productivity than those receiving the same amount in multiple smaller additions (Thomey et al. 2011). We can therefore expect that changes in the timing – and not just the amount – of precipitation will be very important in shaping future ecosystems.

23.4.3 Carbon dioxide fertilization

Increases in ambient CO_2 concentrations fundamentally alter metabolism on both the individual and ecosystem level (Table 23.2). Free-air CO_2 enrichment (FACE) experiments show that photosynthesis, net primary production, yield of crop ecosystems, and ecosystem respiration all increase – at least temporarily – under elevated CO_2 (DeLucia et al. 1999; King et al. 2004; Ainsworth and Long 2005). Increases in productivity are at least partially offset, however, by the acclimation of photosynthetic capacity, the upregulation of leaf respiration, and nutrient limitation (Oren et al. 2001; Leakey et al. 2009). Increased productivity in turn stimulates soil respiration (King et al. 2004; Drake et al. 2011) such that CO_2 fertilization only sometimes supports sustained increases in carbon uptake (Oren et al. 2001; Menge and Field 2007; Langley et al. 2009; McCarthy et al. 2010). The question of whether increased atmospheric CO_2 will ultimately increase total carbon storage in ecosystems remains an important question in global change research.

Beyond its net effects on the carbon balance, higher atmospheric CO_2 alters concentrations of metabolic structures and nutrients, changes the temperature dependence of photosynthesis, and affects nutrient cycling (Table 23.2). Increased CO_2 alters the concentrations of terminal metabolic units in plants (*sensu* West et al. 1997, 1999a, 1999b), decreasing Rubisco and chlorophyll concentrations (Ainsworth and Long 2005) and increasing numbers of mitochondria (Griffin et al. 2001). Moreover, increased CO_2 increases the temperature dependence of photosynthesis (Long et al. 2004), alters nutrient cycling and stoichiometry (Leakey et al. 2009; Drake et al. 2011), mitigates moisture stress (Leakey et al. 2009), and alters trophic interactions (DeLucia et al. 2008). Thus, while some physiological rates may acclimate close to their original level, increasing CO_2 concentrations will have important implications for ecosystems – many of which we do not yet understand.

23.4.4 Interactive and indirect effects

Of course, ecosystems will not react to each element of global change in isolation; rather, they will be faced with simultaneous changes in multiple variables. The overall response cannot necessarily be predicted based on individual responses because effects are interactive. In a California grassland, for example, productivity responses to multiple interacting global change elements differed greatly from simple combinations of single-factor responses (Shaw et al. 2002). Considering this complexity, much remains to be learned about how the metabolic balance of ecosystems will respond to the full suite of global change elements.

Further complexity is added when we consider that changes in energy and carbon cycling will profoundly affect the organization of ecosystems. Climate change will alter body size distribution, biotic interactions, and trophic structure (Petchey et al. 1999; Schmitz et al. 2003; Voigt et al. 2003; O'Connor et al. 2009;

Sarmento et al. 2010; Woodward et al. 2010c; Blankinship et al. 2011; Yvon-Durocher et al. 2011b). For example, herbivore–plant interactions respond to climate manipulations in both terrestrial (e.g., DeLucia et al. 2008) and aquatic systems (e.g., O'Connor 2009; O'Connor and Bruno, Chapter 15). In some cases, ecosystems may suddenly shift into an alternate state. Gradual climate changes may have minimal perceivable effects on ecosystems but reduce resilience so that a small perturbation can trigger a sudden and drastic shift to an alternate state (Scheffer et al. 2001). Thus, there is a great deal of complexity to the effects of climate change on the organization of ecosystems. Biotic interactions and feedbacks will mediate direct responses to abiotic factors (Fig. 23.4), likely triggering nonlinear and sometimes abrupt responses of ecosystems to climate change (Walther 2010).

23.5 CONCLUSIONS

Climate changes are already impacting ecological systems at all levels of organization – from the concentrations of metabolic structures in cells and the temperature sensitivity of biochemical reactions to community structure and ecosystem dynamics. Such perturbations will only increase into the foreseeable future. By providing mechanistic linkages across all of these levels of organization, metabolic ecology has great potential to contribute to our understanding of how ecological systems respond to novel environmental conditions. Research on the effects of climate change remains a frontier for metabolic ecology – one that holds both great challenge and opportunity but also considerable uncertainty as we confront a future of human-caused environmental change.

Chapter 24

BEYOND BIOLOGY

Melanie E. Moses and Stephanie Forrest

SUMMARY

1 Many of the greatest challenges facing science and engineering concern the flow of information, energy, and materials through networks. Examples include the spread of disease in increasingly interconnected human populations, the impact of fossil-fueled transportation on global climate, and advances in computation and telecommunication. The model of West et al. (1997, WBE) provides a unifying framework for understanding fundamental constraints, improving design, and predicting behavior in all of these complex systems.

2 Just as the cardiovascular network supplies energy to cells in organisms, so networks of transistors in computer chips support the information revolution, and road networks are the conduits through which people and goods move and interact to create vibrant modern cities. WBE offers insights into how networks govern the dynamics of these human-constructed systems.

3 Applying WBE in these settings reveals important commonalities and differences between biological and human-designed systems. The differences have led to important extensions and refinements of network scaling theory to account for issues such as: decentralized networks where resources do not flow from a single source; systems that become more densely populated as they increase in size; and modeling more carefully how resources travel from the terminus of a network to the components they service – the so-called "last mile."

4 These model enhancements have allowed us to apply WBE to human ecology and engineered systems, and they may lead to wider application of the theory in biology.

24.1 INTRODUCTION

In 1997 West, Brown, and Enquist (WBE: West et al. 1997) demonstrated how the branching architecture of the cardiovascular network generates the canonical metabolic scaling relationship, $B \sim M^{3/4}$ where B is the metabolic rate of an organism and M is its mass. The 3/4 exponent results from networks evolved to simultaneously maximize energy and resources delivered to cells, minimize the cost of transporting those resources, and minimize the cost of constructing and maintaining the network itself. The paper has been influential, not just because it proposed a mechanism to explain $M^{3/4}$ scaling, but also because it demonstrated how much of biology (described in previous chapters of this book) can be explained by understanding the flow of energy and other resources through networks. In this chapter, we show that the WBE approach can be extended even further to explain how networks constrain the design and growth of human-constructed systems, and in turn, how the topology and dynamics of engineered networks broadly affects how those systems function.

Metabolic Ecology: A Scaling Approach, First Edition. Edited by Richard M. Sibly, James H. Brown, Astrid Kodric-Brown.
© 2012 John Wiley & Sons, Ltd. Published 2012 by John Wiley & Sons, Ltd.

We take as examples two different human-constructed systems: cities and computers. Cities are central to modern human ecology, with more than half of the human population living in urban areas. By even greater majorities, people in cities dominate the consumption of energy and materials and the production of new ideas, research, and inventions (Bettencourt et al. 2007). Computers and information systems are also central to modern human ecology, increasingly dominating our time, interactions with each other, and ways of solving problems. How do metabolic networks in biology relate to the flow of energy, information, and materials in cities and computers? In this chapter we discuss how the topology and dynamics of these networks constrain the way that humans move and process information, energy, and materials.

Although the examples of road networks and computer chips illustrate how WBE provides a unifying framework for understanding scaling in human-engineered systems, certain properties of these systems differ from cardiovascular networks and other biological resource distribution networks described by the WBE (West et al. 1997) model. Thus, applying WBE to these systems requires extensions and refinements to the original theory.

1 *Distributed networks.* Computer and road networks are less centralized than the cardiovascular system – there is not necessarily a single central source, like the heart, for all flow through the network. The theory can be corrected to account for multiple sources and destinations of information or resources within a network.

2 *Density dependence.* To a first approximation, the size and density of cells do not change with organism size within a taxonomic group. However, the density of transistors on computer chips has increased exponentially over time from thousands to millions of transistors per square millimeter, a phenomenon described by Moore's Law. Similarly, the density of people and businesses in cities increases with larger population size such that cities with a larger population have a higher density as well as a larger spatial extent (Samaniego and Moses 2008).

3 *The last mile.* Many networks deliver resources, energy, or information to a local service unit, and from this terminus the resource is transported to its final destination by other means, for example, by diffusion in the case of cells, or by wireless signals in the case of the Internet. We refer to this terminal service unit as "the last mile" by analogy with telecommunication

networks. Work by Banavar and colleagues (2010) revised WBE to show how quarter-powers arise not from the fractal structure of the cardiovascular network, but instead from scaling of velocity to match the characteristic length of the capillary, or more generally, the length of the last mile.

4 *Accommodating superlinear scaling.* WBE demonstrates that, in centralized networks, the volume of a network that could deliver resources to each cell in a large animal as fast as they are delivered to a small animal must increase superlinearly with the organism volume (Fig. 24.1). This would require that the volume of the cardiovascular network grows faster than the volume of the animal. WBE assumes that this does not happen, consistent with the observation that mammals from mice to elephants are all approximately 8% blood (Peters 1983). Further, it is logically impossible for superlinear scaling to hold over large ranges in size: if

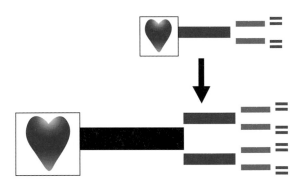

Figure 24.1 The top network shows a single aorta branching into two arteries that each branch into two capillaries. Following WBE each successive branch is shorter (in this case, 1/2 the length of the parent branch) and narrower (such that the summed cross-sectional areas of the daughter branches equals the area of the parent branch). In order to double the number of capillaries (from four in the upper network to eight in the lower network), the number of green arteries is doubled and the single blue aorta is duplicated. In order to obey the constraint that the blood flows through a single aorta, an *additional* branch must be added to the network. The black aorta in the larger network is both longer and wider than the other branches. Thus, the volume of red, green, and blue branches is doubled, but the volume of the larger network is more than doubled due to the addition of the black branch. In this way, the volume of the network grows faster than the number of capillaries that deliver resources: the network volume scales superlinearly with the network delivery rate.

the cardiovascular system in a mouse were 8% of the volume of the mouse, superlinear scaling would result in the cardiovascular system of an elephant being larger than the elephant itself. This is a key constraint in deriving the 3/4-power scaling exponent: because the network volume is constrained to scale linearly with the organism volume or mass, blood flows through the network at a sublinear rate. However, engineers have found several ways to accommodate superlinear network scaling in order to maintain fast delivery rates in large systems. These strategies, described below, accommodate superlinear network scaling over some range of system sizes.

To summarize, the scaling behavior of human-constructed systems poses a challenge for WBE because these systems have distributed networks, components whose densities vary with system size, different transport mechanisms to cross the last mile, and different strategies for overcoming the problem of superlinear network scaling. In the following, we give examples of how these challenges have been met in two quite different systems, cities and computer chips, and we then explore how these extensions of the theory may apply to biological systems, particularly ant colonies and other social animals.

24.2 CITIES

24.2.1 Scaling of road networks

Cities do not exist without roads. Just as the cardiovascular network distributes energy and materials to cells in an organism, urban roads are the arteries that transport people and goods to make the activities of businesses, households, and communities possible. Understanding the topology of urban networks that connect people and places provides insight into how cities are organized and may provide clues to how cities might be better designed to reduce traffic and increase interactions and innovation.

There are some similarities between urban road arteries and biological arteries. Cities have large highways and multi-lane surface streets that branch into successively smaller and slower boulevards that eventually deliver cars to surface streets, driveways, and parking lots. Roads should be designed so that surface streets that connect to highways have enough capacity to accommodate all of the exiting cars. When this ideal

is not met, the result is a traffic jam. But traffic could be much worse than it actually is. If the topology of city roads matched that of the cardiovascular network, every city would have a central intersection downtown that every car passed through on every trip (Fig. 24.2A,B). In a sprawling city such as Los Angeles, each car would have to travel a substantial fraction of the radius of the city to drive to an enormous central intersection on every trip. Fortunately, the geometry of urban roadways and how drivers use them is not quite so simple. One usually does not require a trip across town to buy a gallon of milk or gasoline. In this sense, urban road networks are *less centralized* than biological vascular networks. There is not a single central place through which all traffic flows.

At the opposite extreme, we can imagine a completely decentralized city. In such a city, destinations are distributed evenly through the city, and no one would ever have to travel farther than the distance to their nearest grocery store, school, or restaurant (Fig. 24.2C,D). In such a city, the length of a trip would be determined not by the radius of the city (frequently tens of kilometers), but rather by the distance between destinations, perhaps a few blocks between coffee shops, or a few kilometers between shopping malls. In the latter case, per-capita transportation would depend entirely on density, and in dense cities like New York, people would hardly have to travel any distance at all.

Analysis of travel distances and road capacities (Samaniego and Moses 2008) indicates that cities are neither completely centralized (like the cardiovascular network and Fig. 24.2A,B) nor completely decentralized (as in Fig. 24.2C,D); they exist somewhere in between. Figure 24.3 shows that the average trip length is affected by both area and population density. Many people in large, densely populated cities travel very short distances to buy gasoline or groceries, and very long distances to commute downtown. Interestingly, in US cities, city area and density are correlated. That is, as cities (defined functionally as Metropolitan Statistical Areas) increase in population, they have both larger areas and more people and businesses per unit area. The greater area of large cities tends to increase travel distances, but this effect is somewhat mitigated by the increased density of large cities which reduces average driving distances. The scaling approach also points to a potential explanation for why traffic is more congested in large cities: empirically, per-capita driving distances are influenced by

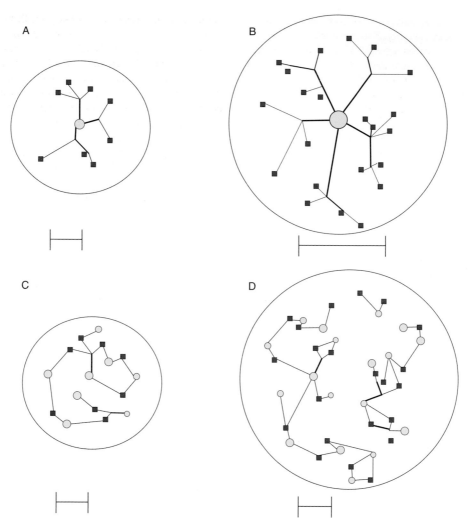

Figure 24.2 (A,B) In a completely centralized city every trip would originate at some location inside the city area (red squares) and end in a single downtown location (blue circles). The average trip distance would depend entirely on the area of the city. Since the city in (B) has twice the area of city A, average travel distances (shown below each figure) would be longer by a factor of the square root of 2. In a completely decentralized city (C,D), each trip would again originate from some location in the city area, but each trip would end at the nearest blue circle. Each blue circle represents a type of destination – a business or residence or park or other destination. In such a city, the area of the city has no impact on travel distances, only the density of destinations affects per-capita transport distances. Because city density increases with city size, the average distance traveled in city D is smaller than in city C. (Figure from Samaniego and Moses 2008.)

both city area and city density, but road capacity scales only with city density. The systematic deviation between road capacity and miles driven in large cities results in more congestion in larger cities. More generally, this approach offers a macroscopic perspective on the differences between small and large cities and on how road infrastructure and traffic might change as cities grow, and it gives urban planners a quantitative tool to understand the impact on traffic of different urban designs.

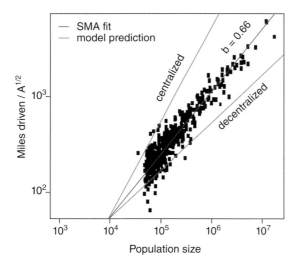

Figure 24.3 Predicted and observed relationships between total miles driven, population size, and area (A) of a city. The predicted relationship for centralized networks (as in Fig. 24.2A,B) is that the total number of miles driven is proportional to the population size multiplied by the radius of the city. This is equivalent to predicting that the number of miles driven divided by city radius ($A^{1/2}$) is linearly proportional to population size, as shown by the upper blue line with a slope equal to 1. The predicted relationship for decentralized networks (as in Fig. 24.2C,D) is that the total number of miles driven is proportional to population size divided by the square root of population density. Rearranging terms to show the decentralized prediction on the same axis as the centralized prediction, miles driven divided by city radius is predicted to be proportional to the square root of population size, shown by the lower blue line with a slope equal to 1/2. The empirical data from US urban areas shows that traffic follows an intermediate scaling that is influenced by both city area and population density.

24.2.2   Beyond infrastructure: scaling of human interactions in cities

Bettencourt and colleagues (2007) proposed that the size of a city systematically affects not just physical networks such as roads and power grids, but also the virtual networks of interactions between people. They consider how a wide range of city characteristics, such as rate of patent production, number of crimes, length of electrical cables, and total wealth of a city, relate to the size of the city population. They represent these relationships as scaling equations of the form $Y \sim N^{\beta}$

where Y is the city characteristic of interest, N is the city population size, and β is the scaling exponent. They find that wealth creation and innovation have superlinear scaling exponents, meaning exponents greater than 1, and often these exponents are near 1.15. Consequently, "larger cities are disproportionally the centers of innovation, wealth and crime, all to approximately the same degree" (Bettencourt et al. 2010). In contrast, scaling exponents related to physical infrastructure have sublinear exponents less than 1, and often close to the 3/4-power exponents seen in biology (Bettencourt et al. 2007).

These observations suggest that, as people are concentrated in larger cities, they interact more, and any phenomenon that occurs as a result of interaction happens more often for everyone. Both larger numbers and higher densities of people in larger cities may contribute to increased interactions. Each person in a large city has a greater chance to build on the ideas of others to generate a new idea, create a new business, and make more money. Since larger cities have more people and each person has more opportunity for innovation and wealth, there is a multiplicative effect, and innovation and wealth increase faster than population size. On the other hand, each person also has a greater chance to be involved in a crime or to spread a disease, and these phenomena are also disproportionally greater in larger cities. The theory allows predictions of how police forces, road maintenance costs, and other infrastructure and services should be expected to grow as population size grows in a city.

The Bettencourt et al. analysis provides a quantitative understanding of the human ecology of cities and is particularly relevant to understanding paths to sustainability. Because city size affects the rate at which people consume resources and generate creative new solutions and technologies, scaling in urban environments plays an important role in determining strategies for sustainable human populations (Moses 2009).

The authors also consider the effect of superlinear scaling exponents on urban growth. Metabolic growth equations for organisms show how sublinear exponents lead to asymptotic growth of animals to some maximum size (West et al. 2001; see also Brown and Sibly, Chapter 2). Systems with superlinear exponents could, in theory, grow indefinitely to infinite size. Given that cities and people living in them require resources that are limited, the theory predicts boom and bust cycles for cities, as are evident in empirical data. The theory also predicts that to avoid economic collapse, cities require increasingly faster cycles of innovation that increase the wealth and resource base of the city. In other words, as population increases, cities have to produce increasingly more wealth and innovation to avoid collapse.

This work also provides a way to quantify which cities are the most creative, the most violent, or most effective at generating wealth. Scaling exponents provide a baseline of expectation for a city given its size, so that individual characteristics of cities can be meaningfully compared. Just as allometric plots tell us that humans have unusually long lifespans for our size because they appear as outliers on an log-log plot, plots that take into account the expected increase of wealth and violence and innovation allow us to see which cities are particularly poor or innovative or free from crime. So for example Bridgeport, CT and San Francisco are particularly wealthy, and Corvallis, OR and San Jose, CA produce unusually large numbers of patents, given their size (Bettencourt et al. 2010).

24.3 COMPUTERS

24.3.1 Wire scaling on computer chips

Modern computer chips contain billions of transistors networked together in a few square centimeters of surface area. There are several distinct networks on these chips, which deliver power to the individual components, deliver a synchronizing clock signal, connect memory components and input and output, and implement the logic operations comprising the chip's functionality. As a concrete example, we will focus on the network known as the "clock tree," which delivers a timing signal to individual transistors, and we will show how clock trees and cardiovascular networks have similar topologies (Moses et al. 2008b). The clock tree is important because it consumes up to 40% of the chip's power, and it is a good example of the network scaling predictions from WBE once corrections are made to account for the clock tree being a two-dimensional rather than a three-dimensional structure. A common clock tree design known as the "H-tree" is shown in Figure 24.4. It is designed to deliver a timing signal from a clock to every location on the chip simultaneously. The topology of the H-tree precisely follows two branching rules proposed by WBE to minimize the resistance of

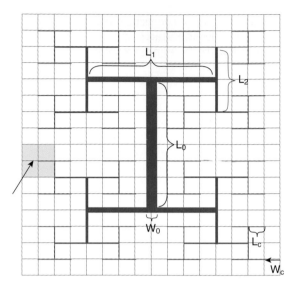

Figure 24.4 The H-tree is a fractal branching network that delivers a timing signal to all areas of a computer chip. Each successive branch is regularly shorter and narrower. The final branch delivers a timing signal to a service unit called an "isochronic region," shaded in blue.

flow through biological networks. When translated from three to two dimensions, those rules are:

1 *Wire width preserving.* When one wire splits into two daughter wires, their summed cross-sectional areas equal that of the parent.

2 *Area-filling branch lengths.* Each daughter wire is systematically shorter than its parent. In Fig. 24.4, the longest wire extends from the center of the chip, half way to the edge; the next wire goes from that point to half of the remaining distance to the edge, and so on. The length of each branch is the radius of the area to which it delivers a signal. The lengths are area-filling in that signal is delivered to all regions of the chip.

However, there is an additional assumption about cardiovascular networks in the WBE model which does not apply to H-trees. WBE assumed that capillaries have the same lengths across species, but the lengths of terminal wires of H-trees are different for large and small computer chips. This is because, empirically, the density of transistors on large chips is higher than the density on small chips: when transistors are closer together, the wires connecting them are shorter.[1] When this difference is accounted for, rules 1 and 2 precisely describe the branching geometry of the original H-tree design, shown in Figure 24.4, and are surprisingly similar to the depiction of the WBE model created by Etienne and colleagues (2006). These rules, along with other design constraints on computer chips, allow us to predict how much power is consumed by the H-tree and the whole computer chip. Additionally, a detailed analysis of the H-tree network sheds light on a debate about assumptions in the WBE model, leading to a more general model of biological network scaling (Banavar et al. 2010).

24.3.2   Scaling in information networks

The H-tree example shows that beneath some obvious differences (e.g., computer chips are measured in square millimeters, while animals are measured in grams or kilograms), there are striking similarities in the topology of distribution networks. There are also similarities in the function of these networks. Cardiovascular systems and H-trees are both infrastructure networks that connect components into a functioning system. Cells do not function without oxygen and nutrients, and transistors cannot compute without receiving electrons and timing signals from the clock tree. Just as the cardiovascular network dictates the pace of life, computational speed is constrained by H-trees and the logical "interconnect" that sends electronic ones and zeros to transistors.

Understanding this analogy requires understanding certain differences between computational networks and the resource distribution networks described by WBE. These differences include decentralized network flow, variation in the density of components, the role of the last mile, and accommodating superlinear network scaling. Once we account for the differences, we can use WBE network scaling principles to describe how power consumption, latency, and the physical

footprint of a network scale as functions of system size, the number of components, and the degree of centralization.

Like the idealized cardiovascular network (Fig. 24.1), the size of the H-tree scales superlinearly: the footprint of the H-tree grows faster than the number of terminal wires that deliver timing signals to each isochronic region[2] (Fig. 24.4). However, there is an important difference in how the cardiovascular network and the H-tree accommodate superlinear network scaling. Because the original H-tree design (Fig. 24.4) required so much power to drive the long wide wires, engineers developed a way to minimize wire widths. Repeaters amplify signals so that even long wires have a small footprint. As the signal splits off at branch points, it is repeated, or amplified, to make up for losses. In that sense, there is no conservation of signal the way that there is conservation of blood. In this way, modern H-tree designs are modified to scale up more efficiently. It is noteworthy that amplifying a signal is possible in an information network, but amplifying energy (i.e., producing new blood cells or oxygen at intermediate points through the cardiovascular network) is not possible. Information can be amplified, but energy and materials cannot.

Engineers developed a second innovation to address superlinear wire scaling of the "interconnect," the network that forms electrical connections between logical elements on the chip. The footprint of the interconnect is simply allowed to consume a larger fraction of the surface of larger chips by adding metal layers of wire on top of the two-dimensional surface that holds the transistors (Moses et al. 2008b). Chips with more transistors have more metal layers to accommodate excess wires. Thus, while biological and computational networks face the same trade-off – the output of the network scales sublinearly with network size – the

---

[1] The exponential increase in transistor density has occurred because technological advances have made smaller transistors and thinner wires, allowing transistors to be packed more closely together. As the distances between transistors shrinks, wires connecting those transistors are shorter, so the density of components affects the size of the network connecting them.

[2] H-trees do not directly connect to every transistor. Rather, each terminal wire of an H-tree delivers a timing signal to an "isochronic region," the service unit that contains some number of transistors all receiving the same timing signal. When the isochronic regions are smaller and more dense, the frequency of the timing signal increases, and the chip can process data faster (this is the chip frequency that has increased from kHz to GHz and is often used to market computer chips). The density of this isochronic region affects the lengths of the H-tree network segments.

trade-off is managed differently. In biology, where superlinear network volume is not possible, the output is slowed in larger organisms. In two-dimensional computer chips, where economic pressures maximize speed (output), wire footprints that grow more quickly than chip surface areas are accommodated on additional surfaces.

These engineering innovations of moving wire onto metal layers and using repeaters to amplify signals have met market pressures that drive the design of faster and faster chips, with more and more transistors. However, those additional metal layers and repeaters consume power, and power has become a fundamental limit on modern chip design. By early in the twenty-first century, most PC chips consumed in the order of 100 watts (roughly the same metabolism as a human being) and more power was consumed in wires than in the transistors they connect. Miniaturization in transistor sizes has led to power-efficient transistors, but the wires that connect them scale in the opposite direction – their power consumption increases (Ho 2003).

Wire scaling became the fundamental problem in producing chips that conformed to Moore's Law – performing more computations with less power becomes impossible when wire scaling dominates power consumption. This problem is solved by decentralization. Just as decentralization allows cities to scale up more efficiently, exploiting decentralized networks improves scaling properties of computational networks. Our analysis suggests that much of the power consumed on chips can be predicted by assuming decentralized flow over the interconnect (Fig. 24.5) as indicated by the slope of the regression line that equals 1. This decentralization is being extended even further by the recent innovation of multi-core architectures; the process of placing increasing numbers of centralized processing units (CPUs) on a chip allows the interconnect to benefit from the same kind of locality that reduces traffic volume in decentralized cities (Fig. 24.2C,D).

Study of these decentralized networks suggests that wire scaling will remain an issue: as more cores are added to a chip, wire scaling *between* cores will dominate communication and power. There are ways to mitigate this problem by clever programming and architectures that maximize localized use of cores (Bezerra et al. 2010; Zarkesh-Ha et al. 2010). Just as people do not have to drive across town to buy a gallon of milk, electrons usually do not have to travel across an entire chip to compute.

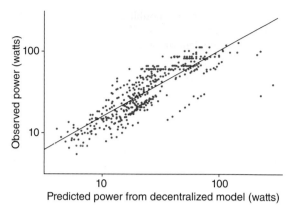

Figure 24.5 Measured versus predicted power consumed by microprocessors. Power is predicted assuming that wire length is determined by transistor density rather than chip area. Excluding the outliers on the right (diamonds representing chips with an unusually large number of transistors in the memory cache) the scaling exponent (slope of the regression line) is indistinguishable from 1 (Moses et al. 2008b).

The transition to multi-core chips mirrors the evolutionary transition to multicellular animals. The interconnect and H-tree on a single core becomes analogous to the transport of energy, materials, and signals within a cell, and the wires that connect cores become analogous to the cardiovascular network between cells. Interestingly, recent work in biology shows that the metabolic scaling exponent shifts in the transition from unicellular to multicellular organisms, the former having a nearly linear exponent and the latter having an exponent near 3/4 (deLong et al. 2010). Whether a similar shift in the scaling exponent will be seen in multi-core systems depends on whether engineers can continue to develop innovations to escape scaling constraints.

24.3.3   The last mile

Although the density of cells in animals is roughly constant, the density of transistors on computer chips varies by many orders of magnitude, from thousands to millions of transistors per mm^2. The enormous increase in transistor density over the last 40 years explains much (but not all) of "Moore's Law," an empirical observation that computing power has increased exponentially, doubling every 18–24 months over this period.

The absolute length of the isochronic region to which timing signals are delivered in computer chips is measured in nanometers, but despite its minuscule size, it determines the frequency of the clock and thus the information-processing power of the chip (Bakoglu 1990). Therefore, smaller isochronic regions and denser packing of both transistors and the terminal wire of the network lead to much faster speeds through the network. This is important because it means that the density of components alters network scaling.

The service unit has its analog in other transportation systems that distribute energy, materials, and information from a central source to dispersed locations and that have been designed to maximize performance. Examples in engineered networks are the last mile that connects individual consumers to global infrastructure networks such as the Internet or telephone networks or the electrical power grid. The service unit is also where a package flown across the globe at a speed of hundreds of miles per hour is walked to a door by a mail carrier, and where passengers exit high-speed planes and trains to take slower modes of transport home. Banavar et al. (2010) propose that the length of the service volume in an animal, proportional to the length of a capillary, also determines the speed of oxygen delivery and ultimately the scaling of metabolism. This model shows that the 3/4-power exponent is generated in networks without fractal structure because, in three dimensions, the radius of the organism scales as $M^{1/3}$, but the velocity is determined by the length of the last mile, which turns out to scale as $M^{1/12}$. This additional factor of $M^{1/12}$ transforms biological rates and times into "quarter"-powers of mass. The length and speed of transport over this last mile ultimately constrains delivery rates in a variety of biological and engineered networks.

24.4   BACK TO BIOLOGY

By applying network analysis and the MTE approach to human-constructed systems, we gain insights into how decentralization, density of components, and differences between networks of information and energy affect network scaling. These insights can be applied back to biology. Humans are not the only animals concentrated in dense populations with complex social structures and complex systems of information exchange. Recent work on ant societies shows that the size of an ant colony is related to its life history in the same way that an animal's body size determines its life history: a colony's metabolism, lifespan, and allocations to reproduction are all quarter-powers of colony size (Hou et al. 2010). This is particularly surprising because ant foraging networks are primarily two-dimensional, and so the scaling exponent might be expected to be lower (i.e., 2/3) if a two-dimensional network topology were governing colony metabolism in the same way that three-dimensional network topology governs organism metabolism. Evidence for "quarter"-power scaling in human socio-economic networks (Hamilton, Burger, and Walker, Chapter 20) suggests the same conundrum.

The surprising similarity in scaling patterns of organisms and "superorganisms" causes us to re-examine whether the topology of energetic networks is sufficient to explain 3/4-power scaling. The decentralized networks connecting ants in a colony are very much like roads in a city – they govern the rates of interaction among ants as well as the rates that ants collect resources and bring them back to a central nest. Larger ant colonies appear to have more interactions and communicate faster, and have more coordinated group foraging strategies, and more elaborate social structures than small colonies (Beckers et al. 1989). One hypothesis is that ants in larger colonies benefit from increasing returns in information exchange – that is, each ant in a larger colony is more informed and better able to share information about food locations, much as humans in cities are more able to exchange information to innovate and build wealth. The increased interactions and information exchange in large colonies might mitigate the diminishing returns of foraging networks that bring food to a central nest (Jun et al. 2003). However, a preliminary study shows that large and small colonies use information to improve foraging equally well (Flanagan et al. 2011). Why the networks in ant colonies that exchange information as well as transport energy and resources should lead to the same quarter-power scaling relationships seen in unitary organisms remains a mystery.

Understanding scaling properties of societies, including humans, ants, and other social organisms, requires a theory of how energy and information flow through networks. This has the potential to unify the study of organisms that are characterized as open energetic systems built upon networks that exchange both energy and information.

Chapter 25

SYNTHESIS AND PROSPECT

James H. Brown, Richard M. Sibly, and Astrid Kodric-Brown

SYNTHESIS

The picture that emerges in this book is of sunlight flooding the Earth. Dotted about the landscape are organisms: plants, animals, and microbes. Living things are found nearly everywhere in this landscape, on land, the oceans, and fresh waters, from the equator to the poles, from the tops of the highest mountains to the deepest ocean depths.

Organisms are packed into this global landscape as tightly as possible. The abundance and distribution of each species is limited by the extrinsic environmental conditions that comprise its ecological niche, and simultaneously by intrinsic traits that allow it to obtain energy and the materials – water and nutrients – that comprise its tissues. When material resources are abundant, all the energy is used, as in a forest where all the usable sunlight is filtered out before reaching the forest floor. When material resources are limiting, much of the sunlight goes unused by living things, as in deserts where water is in short supply and in the oceans where nutrients such as nitrogen, phosphorus, and iron limit metabolism and biomass.

In order to survive and reproduce, each organism needs to get enough of the available energy to power its biological activities – its metabolism. The energy ultimately comes from the Sun. Green plants use pho-

tosynthesis to convert the Sun's energy into the organic molecules that comprise their bodies. Animals and most microbes obtain their energy by feeding on the plants or on each other. The rate at which an organism uses energy depends largely on its body size and temperature. These two factors determine not only its requirements for food, water, and nutrients, but also its pace of life – how long it lives, how fast it grows, and how much it reproduces.

This book is built on the premise that the amount of energy an organism needs each day – its metabolic rate – is mechanistically constrained by its size and body temperature, according to quantitative scaling relationships. Brown et al. (2004; Brown and Sibly, Chapter 2) introduced the Metabolic Theory of Ecology (MTE), explored the mechanisms, and introduced MTE's pivotal proposal that the "central equation" encapsulates many of the relevant scaling relationships. One additional input is needed to make the link from individuals to populations and higher ecological levels. This is the principle of "energy equivalence," discussed further below. In essence this is the idea that if the energy flux per unit area is constant, then it can be split between many small or a few large organisms, if all are packed in as tightly as possible. This allows deduction of population densities from individual scaling relationships. So knowledge of the scaling relationships allows deduction of how ecological processes work at levels above the individual, in populations, communities, and ecosystems. This is the framework developed in the chapters of this book. It holds out the promise of a unified theory of ecology.

So ecology is largely about how resources are filtered and apportioned among species according to their body size and temperature. Many precise details of the filtering mechanisms remain to be worked out. But enough is known to erect a scaffold of how metabolism – the uptake of energy and materials from the environment, the transformations of these commodities within organisms, and the allocation of resources to survival, growth, and reproduction – affects all aspects of ecology at all levels of organization from the individual to the biosphere. This scaffold identifies both the general themes that hold across most kinds of organisms and environments as well as the special cases that can be identified as departures from the general pattern.

This book highlights the success to date in using metabolism to erect a theoretical scaffold for ecology. Metabolic ecology uses some basic tenets of physics, chemistry, and biology and a little math to understand the division of resources among living things, and the distribution, abundance, and diversity of life on Earth. It aims to provide a general synthetic explanation for why there are more small organisms than large ones and why the pace of life and the diversity of organisms are so much higher in the tropics than at the poles. It sheds new light on some long-recognized patterns: (1) Kleiber's rule for the 3/4-power scaling of metabolic rate with body mass; (2) life histories, wherein small organisms live fast and die young while large ones live slow and long; (3) Bergmann's rule for larger organisms in colder environments; (4) the energy equivalence rule, wherein energy use of populations and species is approximately independent of body size; (5) assembly rules for the organization of food webs; and (6) latitudinal, altitudinal, and depth gradients of increasing biodiversity with increasing temperature.

On the one hand, metabolism provides the fundamental mechanisms to understand much of the diversity of life. As pointed out by Peters (1983) in "The Ecological Implications of Body Size," nearly all characteristics of organisms and their ecology vary with size. Empirically and descriptively, knowing body size and temperature allows many orders of magnitude in ecological variables (food requirements, life-history characteristics, abundance, interaction strength, species diversity) to be collapsed to about one order of magnitude. Theoretically and mechanistically, the basic processes are being revealed: metabolic rate scales with body mass and temperature and most ecological processes are influenced, directly or indirectly, by metabolic rate.

On the other hand, metabolic ecology provides a framework for focusing on the variation, and especially the extreme cases. Some of these may reveal inconsistencies and limitations of the theoretical scaffold. Examples are the superlinear scaling of metabolic rate in heterotrophic prokaryotes (DeLong et al. 2010; Okie, Chapter 12) and the lower population densities of the very smallest birds and mammals compared to somewhat larger species, so an exception to energy equivalence (Brown and Maurer 1989; Brown 1995; Isaac, Carbone, and McGill, Chapter 7, Fig. 7.3). Other extreme cases may be "exceptions that prove the rule." One example is the extremely high temperature dependence of primary succession, which appears to reflect positive feedbacks among multiple biogeochemical processes rather than a single process with a single activation energy such as respiration (Anderson-Teixeira et al. 2008; Anderson-Teixeira and Vitousek, Chapter 9). Another example is the energetics of flight, foraging, and reproduction in albatrosses, which illustrate how adaptation for an exceptional lifestyle can result in extreme values, but still within the fundamental constraints of avian metabolic ecology (Costa and Shaffer, Chapter 18).

PROSPECT

Many chapters in this book contain examples of how metabolic ecology has provided a useful unifying conceptual scaffold not only for interpreting and synthesizing existing information, but also for motivating new research. There is enormous opportunity to apply MTE and less specific versions of metabolic ecology to answer long-standing questions and to explore new puzzles. Collectively, the chapters indicate that there is considerable variation among the different subdisciplines of ecology and taxa of organisms, both in the extent to which metabolic perspectives have been applied to date and in the opportunities for future work. So, for example, starting long before MTE, metabolism provided a unifying framework, well-grounded in physics and chemistry, for understanding resource requirements, metabolic rates, and population growth rates of microbes (Okie, Chapter 12). Ever since the seminal work of Lindeman (1942) and Sheldon and Parsons (1967) energetics have provided mechanistic explanations for food web organization (reviewed in Kerr and Dickie 2001; Pascual and Dunne 2006; Petchey and Dunne, Chapter 8) and macroecological patterns (e.g., Brown 1981, 1995; Damuth 1981; Brown and Maurer 1987; Gaston and Blackburn 2000). On the other hand, with the conspicuous exception of optimal foraging theory, energy and metabolic rates have not been widely used currencies in traditional community (Isaac, Carbone, and McGill, Chapter 7) and behavioral ecology (Hayward, Gil,looly, and Kodric-Brown, Chapter 6).

We are reluctant to single out areas to recommend for future research. There are so many of them that to highlight a few might be seen as reflecting our personal biases rather than any objective criteria. Nevertheless, there are a few areas that seem so promising as to deserve comment. One is species diversity. The major geographic patterns of biodiversity have been known for decades and are increasingly accurately quantified using GIS and electronic databases (Lomolino et al. 2010; Storch, Chapter 11). There are strong signatures of temperature and also body size (e.g., May 1978, 1986), suggesting mechanistic underpinnings in metabolism. But a unified theory of biodiversity remains elusive, and David Storch correctly concludes that efforts to apply MTE are at best a work in progress.

One theme that arises repeatedly throughout the book is the idea of energy equivalence. This was first suggested by John Damuth (1981) who noticed that empirically population density or number of individuals per unit area of mammals scales as $M^{-3/4}$. Since metabolic rate or rate of energy use per individual scales as $M^{3/4}$, the rate of energy use per unit area by mammals is invariant with body size: $M^{-3/4} M^{3/4} = M^0$. So, to a first approximation, all species use energy at the same rate. Damuth subsequently noted that trophic level accounted for much of the residual variation, with herbivores consistently having higher densities than carnivores. Ryan Hechinger, Kevin Lafferty, and Armand Kuris (Hechinger et al. 2011; Chapter 19) have applied MTE to incorporate variation in energy supply due to trophic level and derived a related concept of production equivalence: to a first approximation all populations convert a constant fraction of available or assimilated energy into biomass production. Nevertheless there is still much to be done. In particular, it is still not clear why coexisting species of different body sizes

should have equivalent energy availability which they then use at equivalent rates (see Isaac, Carbone, and McGill, Chapter 7).

Another exciting prospect is to extend concepts of metabolic ecology beyond typical organisms and ecology to explore applications to human activities. We present three examples. Hamilton, Burger, and Walker (Chapter 20; see also Burnside et al. 2011) use metabolic ecology as a conceptual framework for understanding trends in energy and space use, demography, and economics as humans expanded out of Africa to colonize the entire Earth and developed first agricultural and then industrial-techological societies. One pervasive change is the increasing reliance on non-human metabolism, initially burning biofuels and now mostly fossil fuels, to supplement human biological metabolism fueled by food consumption. The implications for economics, sociology, and environmental relationships of modern humans are profound and only just beginning to be explored (e.g., Hubbert 1981; H. T. Odum 1971, 2007; Hall et al. 2001; Brown et al. 2011). Another promising area is the metabolic ecology of societies. Waters and Harrison (Chapter 16) and Hayward, Gilooly, and Kodric-Brown (Chapter 6) discuss recent studies of social insects, which show that they appear to act like superorganisms and to exhibit sublinear scaling of metabolic rate with colony mass. Many other organisms, including humans, aggregate into social groups and a metabolic and scaling perspective could guide studies of the energetic and fitness consequences of these organizations. A final example is application of concepts of metabolic ecology to areas outside biology. Moses and Forrest present examples of "biologically inspired designs" applied to engineering. Examples to date range from computer chips to highway networks, but the possibilities are endless. For example, we have an example for a near-optimal design for a distributed array of solar panels – it is called a tree.

We conclude by emphasizing the enormous opportunity to use experimental manipulations to address big, interesting questions in metabolic ecology. Much of the work to date has come from a macroecological perspective. Most of the empirical studies have compiled and analyzed existing data. This is not surprising, given the recent emergence of macroecology, and the fact that many mechanistic hypotheses to account for macroecological patterns are metabolic. This approach has been valuable, but scientists delight in experiments. This is because good experiments allow powerful testing of theories and detection of inadequacies, so enabling science to progress. So many readers will enjoy spotting cases where experiments are particularly feasible.

We are struck by the suitability of microbes (Okie, Chapter 12; Litchman, Chapter 13) and insects (Waters and Harrison, Chapter 16) for experimental work. The small size, fast life histories, and diverse niches make the former well suited for experiments that combine perspectives of metabolic ecology and traditional microbiology. Examples of possibilities range from the experimental microcosms used by Peter Morin and colleagues (e.g., Beckerman et al. 2010) to address questions of community and ecosystem ecology to the "long-term" (20 000 generations in two years) experiments of Lenski, Bennett, and colleagues (e.g., Leroi et al. 1994; Bennett and Lenski 2007). Insects offer the potential to address different questions. Comparisons of structure and function over ontogeny can address the constraints and opportunities that come with

changes in body size and design. The ability to rear insects on controlled diets and environmental conditions (e.g., temperature and partial pressure of oxygen) allows for experiments to address big questions at the intersection of physiology, developmental biology, and ecology (Kerkhoff, Chapter 4; Waters and Harrison, Chapter 16). And even entire colonies of social insects can be manipulated experimentally to address fascinating questions about how the behavior and physiology of members contribute to the attributes of the "superorganism."

The central question of metabolic ecology is: How do the salient features of populations, communities, and ecosystems emerge from scaling relations and other constraints on morphology, physiology, and behavior of individual organisms? Ecology is all about organism–environment interactions. Asking how energy and materials are exchanged in these interactions and how these fluxes affect both the organism and the environment opens new windows on ecology.

GLOSSARY

Note: The original papers referred to in this book sometimes vary in the terms they use to describe particular ideas, variables or equations. This is because metabolic ecology draws on several areas, including biological scaling, allometry, and comparative biochemistry and physiology in addition to all subdisciplines of ecology. In this book we have tried to standardize terminology. The principal terms and symbols are defined below, together with some of the variants employed in source papers.

abundance Number of individuals, usually expressed per unit area, or per standardized sample, and given the symbol N.

Arrhenius equation Widely used equation giving the rate of a chemical reaction as a function of absolute temperature. The expression is $= A\,e^{-E/kT}$, where A is a normalization constant, E is the activation energy, k is Boltzmann's constant and T is absolute temperature in kelvin ($= °C + 273.15$).

assimilation efficiency and rate The efficiency is the fraction of food energy consumed that is digested, absorbed, and retained for allocation; the rate is the product of efficiency and food intake rate. See Karasov, Chapter 17.

autotrophs Organisms that obtain carbon from non-living sources, so photosynthesis or chemosynthesis (e.g., algae, plants, and some prokaryotes).

B Mass-specific metabolic rate in watts/gram, $= I/M$.

basal metabolic rate, BMR Minimum metabolic rate of a resting, fasting endotherm within its thermoneutral zone. See Brown and Sibly, Chapter 2, Karasov, Chapter 17.

Boltzmann's constant A constant in the Arrhenius equation with value $8.617 \times 10^{-5}\,\text{eV}\,\text{K}^{-1}$.

Boltzmann factor Means the same as the Arrhenius equation.

D Diameter of a plant stem or tree trunk.

density dependence The effect of population density on a life-history trait or population growth rate. Generally the effect is to decrease population growth rate (though some cases of "positive density dependence," i.e. Allee effect, are known).

Droop equation Equation relating population growth rate, μ, to the concentration of an element or compound within a cell, where $\mu = \mu_\infty\,(1 - Q_{min}/Q)$, where μ is the biomass growth rate (day^{-1}); μ_∞ is the maximum growth rate; Q_{min} is the minimum concentration of the element necessary to just support life (moles per cell at zero growth); and Q is the observed intracellular concentration (moles per liter). See Kaspari, Chapter 3, Litchman, Chapter 13.

Metabolic Ecology: A Scaling Approach, First Edition. Edited by Richard M. Sibly, James H. Brown, Astrid Kodric-Brown.
© 2012 John Wiley & Sons, Ltd. Published 2012 by John Wiley & Sons, Ltd.

e   2.718. . . . Magic number discovered by mathematicians.

E   Activation energy in the Arrhenius equation in electron volts. MTE assumes that $E \approx 0.65\,eV$ for processes governed by respiration, and $E \approx 0.30\,eV$ for processes governed by photosynthesis. See Brown and Sibly, Chapter 2.

EER   Energy equivalence rule. See below.

energy equivalence   Idea that the energy use of populations and species per unit area is approximately independent of body size. See Karasov, Chapter 17, Isaac, Carbone, and McGill, Chapter 7, Storch, Chapter 11, Enquist and Bentley, Chapter 14, Brown, Sibly, and Kodric-Brown, Chapter 25.

F   Rate of energy flux through a population, community, or ecosystem expressed in watts per unit ground area.

field metabolic rate, FMR   Metabolic rate of a non-reproducing, free-ranging animal in its natural habitat. See Brown and Sibly, Chapter 2, Karasov, Chapter 17.

heterotrophs   Organisms that obtain carbon from feeding on organic matter produced by other organisms (e.g., fungi, animals, and many prokaryotes).

I   Whole-organism metabolic rate, in watts.

I_0 Normalization constant for *I*. Equivalent to antilog of the intercept on a log-log plot of *I* vs. *M*.

intrinsic rate of increase   Means the same as maximum population growth rate, r_{max}.

isometric scaling   Proportionate scaling of body dimensions. Thus volumes scale as *M*, areas as $M^{2/3}$, and lengths as $M^{1/3}$.

k   Used generally as a constant, and specifically as Boltzmann's constant in the Arrhenius equation, when $k = 8.617 \times 10^{-5}\,eV\,K^{-1}$.

K   Carrying capacity of an environment, i.e., number of individuals per unit area that a specified environment can support long term when the population growth rate is zero.

kelvin, K   A unit of absolute temperature, $= °C + 273.15$.

linear   Sometimes used to refer to a power law with an exponent of 1, so a scaling relationship with a slope of 1 on a log-log plot.

ln   Abbreviation for natural logarithm, also written $\log_e$. $\log_e(x) = 2.30259 \log_{10}(x)$.

M   Body mass, in grams unless otherwise stated.

maximal metabolic rate, MMR   Maximal metabolic rate during sustained aerobic activity. Sometimes referred to as $VO_{2\,max}$. See Brown and Sibly, Chapter 2, Karasov, Chapter 17.

metabolic rate   Rate of energy turnover of an organism. Technically best measured as heat production by calorimetry but often measured as rate of O_2 consumption or CO_2 production in animals or as CO_2 consumption in plants.

MTE   Metabolic theory of ecology, based on the central equation $B = B_0\,M^{-1/4}e^{-0.65/kT}$, see Brown and Sibly, Chapter 2.

Michaelis–Menten equation   Equation describing the kinetics of a reaction, $V = V_{max}[S]/K_m + [S]$, where *V* is the reaction rate, V_{max} is the reaction rate when substrate is not limiting, [*S*] is substrate concentration, and K_m is the Michaelis constant, the substrate concentration at which $V = V_{max}/2$. This expression is sometimes used as an alternative to the Droop equation to model population growth of microbes.

μ Symbol for population growth rate in microbiology (e.g., in the Droop equation).

N   Abundance, sometimes measured simply as number of individuals and sometimes as population density in individuals per unit area. N_t refers to the number at time *t*.

P   Rate of whole-organism biomass production, in grams per unit time.

parasite   A consumer that interacts with only one resource individual for the duration of its life (or life stage, for complex life cycle organisms). There is a diversity of parasitic strategies, but they are all distinguished from predators (e.g., goshawks, scorpions, clams) and micropredators (e.g., mosquitoes, fish lice, cicadellid sharpshooter insects).

population growth rate   The per capita rate at which abundance increases with time. Given the symbol *r* except in microbiology where the symbol μ is used. *r* ~ birth rate minus death rate in the absence of migration. *r* may be expressed as % increase per year and can be measured as $\log_e(N_{t+1}/N_t)$. Population growth rate is also sometimes (though not in this book) expressed as λ, the multiplication factor by which the population increases per year. $\lambda = e^r$.

production equivalence The idea that biomass production of species and populations is invariant with body size. See Hechinger, Lafferty, and Kuris, Chapter 19.

Q_{10} A measure of temperature dependence. The factor by which a biological rate increases for a $10\,°C$ rise in temperature. See Brown and Sibly, Chapter 2.

r Population growth rate, expressed as % increase per unit time.

r_{max} Maximum population growth rate under ideal conditions. In theoretical population studies this is taken to be the growth rate in the absence of density dependence.

R Rate of resource supply, usually expressed per unit area, for a population, community, or ecosystem. The resource may be sunlight, an element, or chemical reactant.

[R] Concentration of resource R. More generally, square brackets denote concentration, so [S] is the substrate concentration in the Michaelis–Menten equation.

resting metabolic rate Similar to BMR but with the animal not necessarily fasting or in a postabsorptive state. See Karasov, Chapter 17.

standard metabolic rate Similar to BMR but for ectotherms, when it is measured at a specified temperature.

sublinear scaling Scaling with body mass as a power function. Given the symbol r except in microbiology where the symbol μ is used. with exponent <1, so with slope <1 on a log-log plot.

superlinear scaling Scaling with body mass as a power function with exponent > 1, so with slope > 1 on a log-log plot.

t Time.

T Temperature in kelvin, $= 273.15 +$ temperature in degrees Celsius.

trophic level Used to enumerate links in a food chain, so = 1 for primary producers, = 2 for herbivores, and so on.

Van't Hoff-Arrhenius equation Means the same as Arrhenius equation.

W Biomass, usually measured as standing stock in grams per unit area.

WBE The mechanistic model for 3/4-power scaling of metabolic rate as published in West, Brown, and Enquist (1997).

WBE2 The mechanistic model for scaling of metabolic rate in plants as published in West, Brown, and Enquist (1999b).

Z Per-capita death rate per unit time, also called mortality rate.

REFERENCES

Ackerly, D. (2009) Conservatism and diversification of plant functional traits: Evolutionary rates versus phylogenetic signal. *Proceedings of the National Academy of Sciences of the USA* 106, 19699–19706.

Ackerly, D. D., Loarie, S. R., Cornwell, W. K., et al. (2010) The geography of climate change: implications for conservation biogeography. *Diversity and Distributions* 16, 476–487.

Adams, H. D., Guardiola-Claramonte, M., Barron-Gafford, G. A., et al. (2009) Temperature sensitivity of drought-induced tree mortality portends increased regional die-off under global-change-type drought. *Proceedings of the National Academy of Sciences of the USA* 106, 7063–7066.

Addo-Bediako, A., Chown, S. L., and Gaston, K. J. (2002) Metabolic cold adaptation in insects: a large-scale perspective. *Functional Ecology* 16, 332–338.

Agustí, S. (1991) Light environment within dense algal populations: cell size influences on self-shading. *Journal of Plankton Research* 13, 863–871.

Agustí, S. and Kalff, J. (1989) The influence of growth conditions on the size dependence of maximal algal density and biomass. *Limnology and Oceanography* 34, 1104–1108.

Agustí, S., Duarte, C. M., and Kalff, J. (1987) Algal cell size and the maximum density and biomass of phytoplankton. *Limnology and Oceanography* 32, 983–986.

Ainsworth, E. A. and Long, S. P. (2005) What have we learned from 15 years of free-air CO_2 enrichment (FACE)? A meta-analytic review of the responses of photosynthesis, canopy properties and plant production to rising CO_2. *New Phytologist* 165, 351.

Aksnes, D. L. and Egge, J. K. (1991) A theoretical model for nutrient uptake in phytoplankton. *Marine Ecology Progress Series* 70, 65–72.

Albery, W. and Knowles, J. (1976) Free-energy profile for the reaction catalyzed by triosephosphate isomerase. *Biochemistry* 15, 5627–5631.

Aledo, J. C. and Del Valle, A. E. (2004) The ATP paradox is the expression of an economizing fuel mechanism. *Journal of Biological Chemistry* 279, 55372–55375.

Alerstam, T., Hedenström, A., and Åkesson, S. (2003) Long-distance migration: evolution and determinants. *Oikos* 103, 247–260.

Alexander, R. D. (1974) The evolution of social behavior. *Annual Review of Ecology and Systematics* 5, 325–383.

Alexander, R. M. (1998) All-time giants: the largest animals and their problems. *Palaeontology* 41, 1231–1245.

Alexander, R. M., Weibel, E. R., Taylor, C. R., and Bolis, L. (1998) Symmorphosis and safety factors. In: Weibel, E. R., Taylor, C. R., and Bolis, L. (Ed.) *Principles of Animal Design.* Cambridge, Cambridge University Press.

Algar, A. C., Kerr, J. T., and Currie, D. J. (2007) A test of metabolic theory as the mechanism underlying broad-scale species-richness gradients. *Global Ecology and Biogeography* 16, 170–178.

Algar, A. C., Kerr, J. T., and Currie, D. J. (2009) Evolutionary constraints on regional faunas: whom, but not how many. *Ecology Letters* 12, 57–65.

Aljetlawi, A. A., Sparrevik, E., and Leonardsson, K. (2004) Prey-predator size-dependent functional response:

derivation and rescaling to the real world. *Journal of Animal Ecology* 73, 239–252.

Allen, A. P. and Gillooly, J. F. (2007) The mechanistic basis of the metabolic theory of ecology. *Oikos* 116, 1073–1077.

Allen, A. P. and Gillooly, J. F. (2009) Towards an integration of ecological stoichiometry and the metabolic theory of ecology to better understand nutrient cycling. *Ecology Letters* 12, 369–384.

Allen, A. P., Brown, J. H., and Gillooly, J. F. (2002) Global biodiversity, biochemical kinetics, and the energetic-equivalence rule. *Science* 297, 1545–1548.

Allen, A. P., Gillooly, J. F., and Brown, J. H. (2005) Linking the global carbon cycle to individual metabolism. *Functional Ecology* 19, 202–213.

Allen, A. P., Gillooly, J. F., Savage, V. M., and Brown, J. H. (2006) Kinetic effects of temperature on rates of genetic divergence and speciation. *Proceedings of the National Academy of Sciences of the USA* 103, 9130–9135.

Allen, A. P., Gillooly, J. F., and Brown, J. H. (2007) Recasting the species-energy hypothesis: the different roles of kinetic and potential energy in regulating biodiversity. In: Storch, D., Marquet, P. A., and Brown, J. H. (Eds), *Scaling Biodiversity*. Cambridge: Cambridge University Press, pp. 283–299.

Allesina, S. (2011) Predicting trophic relations in ecological networks: a test of the Allometric Diet Breadth Model. *Journal of Theoretical Biology* 279, 161–168.

Alleyne, M., Chappell, M. A., Gelman, D. B., and Beckage, N. E. (1997) Effects of parasitism by the braconid wasp *Cotesia congregata* on metabolic rate in host larvae of the tobacco hornworm, *Manduca sexta*. *Journal of Insect Physiology* 43, 143–154.

Allison, S. D., Wallenstein, M. D., and Bradford, M. A. (2010) Soil-carbon response to warming dependent on microbial physiology. *Nature Geoscience* 3, 336–340.

Alonso, J. C., Magaña, M., Martin, C. A., and Palacin, C. (2010) Sexual traits as quality indicators in lekking male Great Bustards. *Ethology* 116, 1084–1098.

Alperin, M. J. and Hoehler, T. M. (2009) Anaerobic methane oxidation by archaea/sulfate-reducing bacteria aggregates: 1. Thermodynamic and physical constraints. *American Journal of Science* 309, 869–957.

Altmann, S. A. (1987) The impact of locomotor energetics on mammalian foraging. *Journal of Zoology* 211, 215–225.

Amend, J. P. and Shock, E. L. (2001) Energetics of overall metabolic reactions of thermophilic and hyperthermophilic Archaea and Bacteria. *FEMS Microbiology Reviews* 25, 175–243.

Andersen, A. N. and Lonsdale, W. M. (1990) Herbivory by insects in Australian tropical savannas: a review. *Journal of Biogeography* 17, 433–444.

Andersen, K. H. and Beyer, J. E. (2006) Asymptotic size determines species abundance in the marine size spectrum. *American Naturalist* 168, 54–61.

Andersen, K. H., Beyer, J. E., and Lundberg, P. (2008) Trophic and individual efficiencies of size-structured communities.

Proceedings of the Royal Society B: Biological Sciences 276, 109–114.

Andersen, K. H., Farnsworth, K. D., Pedersen, M., Gislason, H., and Beyer, J. E. (2009) How community ecology links natural mortality, growth and production of fish populations. *ICES Journal of Marine Science* 66, 1978–1984.

Anderson, K. J. (2007) Temporal patterns in rates of community change during succession. *American Naturalist* 169, 780–793.

Anderson, K. J. and Jetz, W. (2005) The broad-scale ecology of energy expenditure of endotherms. *Ecology Letters* 8, 310–318.

Anderson, K. J., Allen, A. P., Gillooly, J. F. and Brown, J. H. (2006) Temperature-dependence of biomass accumulation rates during secondary succession. *Ecology Letters* 9, 673–682.

Anderson, R. and May, R. (1991) *Infectious Diseases of Humans*. Oxford: Oxford University Press.

Anderson-Teixeira, K. J., Vitousek, P. M., and Brown, J. H. (2008) Amplified temperature dependence in ecosystems developing on the lava flows of Mauna Loa, Hawai'i. *Proceedings of the National Academy of Sciences of the USA* 105, 228–233.

Anderson-Teixeira, K. J., DeLong, J. P., Fox, A. M., Brese, D. A., and Litvak, M. E. (2011) Differential responses of production and respiration to temperature and moisture drive the carbon balance across a climatic gradient in New Mexico. *Global Change Biology* 17, 410–424.

Anfodillo, T., Carraro, V., Carrer, M., Fior, C., and Rossi, S. (2006) Convergent tapering of xylem conduits in different woody species. *New Phytologist* 169, 279–290.

Angulo-Brown, F., Santillan, M., and Calleja-Quevedo, E. (1995) Thermodynamic optimality in some biochemical reactions. *Il Nuovo Cimento* 17D, 87–90.

Antoniou, P., Hamilton, J., Koopman, B., et al. (1990) Effect of temperature and pH on the effective maximum specific growth rate of nitrifying bacteria. *Water Research* 24, 97–101.

Apol, M. E. F., Etienne, R. S., and Olff, H. (2008) Revisiting the evolutionary origin of allometric metabolic scaling in biology. *Functional Ecology* 22, 1070–1080.

Araújo, M. B. and Luoto, M. (2007) The importance of biotic interactions for modelling species distributions under climate change. *Global Ecology and Biogeography* 16, 743–753.

Araújo, M. B., Thuiller, W., and Pearson, R. G. (2006) Climate warming and the decline of amphibians and reptiles in Europe. *Journal of Biogeography* 33, 1712–1728.

Arenas, A., Díaz-Guilera, A., and Guimerà, R. (2001) Communication in networks with hierarchical branching. *Physical Review Letters* 86, 3196.

Arenas, A., Danon, L., Díaz-Guilera, A., Gleiser, P. M., and Guimerá, R. (2004) Community analysis in social networks. *European Physical Journal B* 38, 373–380.

Arim, M., Bozinovic, F., and Marquet, P. A. (2007) On the relationship between trophic position, body mass and temperature: reformulating the energy limitation hypothesis. *Oikos* 116, 1524–1530.

Arim, M., Abades, S. R., Laufer, G., Loureiro, M., and Marquet, P. A. (2010) Food web structure and body size: trophic position and resource acquisition. *Oikos* 119, 147–153.

Armstrong, C. and Staples, J. F. (2010) The role of succinate dehydrogenase and oxaloacetate in metabolic suppression during hibernation and arousal. *Journal of Comparative Physiology B: Biochemical Systemic and Environmental Physiology* 180, 775–783.

Armstrong, R. A. (1994) Grazing limitation and nutrient limitation in marine ecosystems: steady-state solutions of an ecosystem model with multiple food chains. *Limnology and Oceanography* 39, 597–608.

Armstrong, R. A. (1999) Stable model structures for representing biogeochemical diversity and size spectra for plankton communities. *Journal of Plankton Research* 21, 445–464.

Arneberg, P., Skorping, A., and Read, A. F. (1998) Parasite abundance, body size, life histories, and the energetic equivalence rule. *American Naturalist* 151, 497–513.

Arnone III, J. A., Verburg, P. S. J., Johnson, D. W., et al. (2008) Prolonged suppression of ecosystem carbon dioxide uptake after an anomalously warm year. *Nature* 455, 383–386.

Arrhenius, O. (1920) Distribution of the species over the area. Meddelanden Fran K. *Vetenskapsakademiens Nobelinstitut* 4, 1–6.

Arrhenius, S. (1889) Uber die Reaktionsgeschwindigkeit bei der Inversion von Rohrzucker durch Sauren. *Zeitschrift für Physik Chemie* 4, 226–248.

Arthur, W. B. (2009) *The Nature of Technology: What it is and How it Evolves.* New York: Free Press.

Ashmole, N. P. and Ashmole, M. J. (1967) Comparative feeding ecology of sea birds of a tropical oceanic island. *Peabody Museum of Natural History Bulletin* No. 24.

Ashton, K. G. (2002) Do amphibians follow Bergmann's rule? *Canadian Journal of Zoology* 80, 708–716.

Ashton, K. G. (2004) Sensitivity of intraspecific latitudinal clines of body size for tetrapods to sampling, latitude and body Size. *Integrative and Comparative Biology* 44, 403–412.

Ashton, K. G. and Feldman, C. R. (2003) Bergmann's rule in nonavian reptiles: turtles follow it, lizards and snakes reverse it. *Evolution* 57, 1151–1163.

Atchley, W. R. (1984) The effect of selection on brain and body size associations. *Genetical Research* 43, 289–298.

Atkin, O. K. and Tjoelker, M. G. (2003) Thermal acclimation and the dynamic response of plant respiration to temperature. *Trends in Plant Science* 8, 343–351.

Atkin, O. K., Scheurwater, I., and Pons, T. L. (2007) Respiration as a percentage of daily photosynthesis in whole plants is homeostatic at moderate, but not high, growth temperatures. *New Phytologist* 174, 367–380.

Atkins, P. and De Paula, J. (2009) *Elements of Physical Chemistry.* WH Freeman & Co.

Atkinson, D. (1994) Temperature and organism size – a biological law for ectotherms. *Advances in Ecological Research* 25, 1–58.

Atkinson, D. and Sibly, R. M. (1997) Why are organisms usually bigger in colder environments? Making sense of a life history puzzle. *Trends in Ecology and Evolution* 12, 235–239.

Atkinson, D., Begon, M., and Fitter, A. H. (1994) Temperature and organism size – A biological law for ectotherms? *Advances in Ecological Research.* Academic Press.

Auerswald, L. and Gäde, G. (2000) Metabolic changes in the African fruit beetle, *Pachnoda sinuata,* during starvation. *Journal of Insect Physiology* 46, 343–351.

Austin, A. T. (2002) Differential effects of precipitation on production and decomposition along a rainfall gradient in Hawaii. *Ecology* 83, 328–338.

Badgerow, J. P., and Hainsworth, F. R. (1981) Energy savings through formation flight? A re-examination of the vee formation. *Journal of Theoretical Biology* 93, 41–52.

Bailey, J. E. and Ollis, D. F. (1986) *Biochemical Engineering Fundamentals.* McGraw-Hill Science, Engineering and Mathematics.

Bairoch, A. (2000) The enzyme database in 2000. *Nucleic Acids Research* 28, 304.

Bakoglu, H. B. (1990) *Circuits, Interconnections, and Packaging for VLSI.* New York: Addison-Wesley.

Balderrama, N. M., Almeida, L. O. B., and Núñez, J. A. (1992) Metabolic rate during foraging in the honeybee. *Journal of Comparative Physiology B: Biochemical Systemic and Environmental Physiology* 162, 440–447.

Baldwin, B. G., and Sanderson, M. J. (1998) Age and rate of diversification of the Hawaiian silversword alliance (Compositae). *Proceedings of the National Academy of Sciences of the USA* 95, 9402–9406.

Banavar, J. R., Damuth, J., Maritan, A., and Rinaldo, A. (2002) Ontogenetic growth – modelling universality and scaling. *Nature* 420, 626–626.

Banavar, J. R., Moses, M. E., Brown, J. H., et al. (2010) A general basis for quarter-power scaling in animals. *Proceedings of the National Academy of Sciences of the USA* 107, 15816–15820.

Banerjee, S. and Moses, M. (2010) Scale invariance of immune system response rates and times: perspectives on immune system architecture and implications for artificial immune systems. *Swarm Intelligence* 4, 301–318.

Banse, K. (1976) Rates of growth, respiration and photosynthesis of unicellular algae as related to cell size: review. *Journal of Phycology* 12, 135–140.

Banse, K. (1982) Cell volumes, maximal growth rates of unicellular algae and ciliates, and the role of ciliates in marine pelagial. *Limnology and Oceanography* 26, 1059–1071.

Banse, K. and Mosher, S. (1980) Adult body mass and annual production/biomass relationships of field populations. *Ecological Monographs* 50, 355–379.

Barnes, C., Maxwell, D. L., Reuman, D. C., and Jennings, S. (2010) Global patterns in predator-prey size relationships reveal size-dependency of trophic transfer efficiency. *Ecology* 91, 222–232.

Barraclough, T. G. and Savolainen, V. (2001) Evolutionary rates and species diversity in flowering plants. *Evolution* 55, 677–683.

Barrett, J. (1981) *Biochemistry of Parasitic Helminths.* London: Macmillan.

Barrett, J. (1984) The anaerobic end-products of helminths. *Parasitology* 88, 179–198.

Barrett, J. (1991). Parasitic helminths. In: Bryant, C. (Ed.), *Metazoan Life Without Oxygen.* London, New York: Chapman & Hall, pp. 146–164.

Barron, A., Wurzburger, N., Bellenger, J., Wright, S., Kraepiel, A., and Hedin, L. (2008) Molybdenum limitation of asymbiotic nitrogen fixation in tropical forest soils. *Nature Geoscience* 2, 42–45.

Bartholomew, G. A. and Casey, T. M. (1977) Body temperature and oxygen consumption during rest and activity in relation to body size in some tropical beetles. *Journal of Thermal Biology* 2, 173–176.

Barton, K. A. and Zalewski, A. (2007) Winter severity limits red fox populations in Eurasia. *Global Ecology and Biogeography* 16, 281–289.

Basolo, A. L., and Alcaraz, G. (2003) The turn of the sword: length increases male swimming costs in swordtails. *Proceedings of the Royal Society B: Biological Sciences* 270, 1631–1636.

Basolo, A. L., and Wagner, W. E. (2004) Covariation between predation risk, body size and fin elaboration in the green swordtail, *Xiphophorus helleri. Biological Journal of the Linnean Society* 83, 87–100.

Bays, J. S. and Crisman, T. L. (1983) Zooplankton and trophic state relationships in Florida lakes. *Canadian Journal of Fisheries and Aquatic Sciences* 40, 1813–1819.

Bazzaz, F. A. and Grace, J. (1997). *Plant Resource Allocation.* London: Academic Press.

Beardall, J., Allen, D., Bragg, J., et al. (2009) Allometry and stoichiometry of unicellular, colonial and multicellular phytoplankton. *New Phytologist* 181, 295–309.

Beckerman, A. P., Petchey, O. L., and Morin, P. J. (2010) Adaptive foragers and community ecology: linking individuals to communities and ecosystems. *Functional Ecology* 24, 1–6.

Beckers, R., Goss, S., Deneubourg, J., and Pasteels, J. (1989) Colony size, communication, and ant foraging strategy. *Psyche: A Journal of Entomology* 96, 239–256.

Beer, C., Reichstein, M., Tomelleri, E., et al. (2010) Terrestrial gross carbon dioxide uptake: global distribution and covariation with climate. *Science* 329, 834–838.

Belgrano, A., Allen, A. P., Enquist, B. J., and Gillooly, J. F. (2002) Allometric scaling of maximum population density: a common rule for marine phytoplankton and terrestrial plants. *Ecology Letters* 5, 611–613.

Belk, M. C. and Houston, D. D. (2002) *Bergmann's Rule in Ectotherms: A Test Using Freshwater Fishes.* University of Chicago Press.

Bellwood, P. (2004) *First Farmers: The Origins of Agricultural Societies.* Oxford: Wiley-Blackwell.

Belmaker, J. and Jetz, W. (2011) Cross-scale variation in species richness-environment associations. *Global Ecology and Biogeography* DOI: 10.1111/j.1466-8238.2010.00615.x.

Bennett, A. F. and Lenski, R. E. (2007) An experimental test of evolutionary trade-offs during temperature adaptation. *Proceedings of the National Academy of Sciences of the USA* 104, 8649–8654.

Bennett, P. M. and Owens, I.P.F. (1997) Variation in extinction risk among birds: chance or evolutionary predisposition? *Proceedings of the Royal Society B: Biological Sciences* 264, 401–408.

Berg, S., Christianou, M., Jonsson, T., and Ebenman, B. (2011) Using sensitivity analysis to identify keystone species and keystone links in size-based food webs. *Oikos* 120, 510–519.

Bergmann, C. (1847) Über die Verhältnisse der Wärmeökonomie der Thiere zu ihrer Grösse. *Göttinger Studien* 3, 595–708.

Berlow, E., Dunne, J., Martinez, N., Stark, P. B., Williams, R., and Brose, U. (2009) Simple prediction of interaction strengths in complex food webs. *Proceedings of the National Academy of Sciences of the USA* 106, 187–191.

Bernacchi, C. J., Singsaas, E. L., Pimentel, C., Portis, A. R., and Long, S. P. (2001) Improved temperature response functions for models of Rubisco-limited photosynthesis. *Plant Cell and Environment* 24, 253–259.

Bernacchi, C. J., Portis, A. R., Nakano, H., von Caemmerer, S., and Long, S. P. (2002) Temperature response of mesophyll conductance. Implications for the determination of Rubisco enzyme kinetics and for limitations to photosynthesis in vivo. *Plant Physiology* 130, 1992–1998.

Berry, H. (2002) Monte Carlo simulations of enzyme reactions in two dimensions: fractal kinetics and spatial segregation. *Biophysical Journal* 83, 1891–1901.

Berry, P. M., Dawson, T. P., Harrison, P. A., and Pearson, R. G. (2002) Modelling potential impacts of climate change on the bioclimatic envelope of species in Britain and Ireland. *Climate Change and Conservation* 11 (Special Issue), 453–462.

Bersier, L. F. (2007) A history of the study of ecological networks. In: Képès, F. (Ed.) *Complex Systems and Interdisciplinary Science.* World Scientific.

Bettencourt, L. M. A., Lobo, J., Helbing, D., Kühnert, C., and West, G. B. (2007) Growth, innovation, scaling, and the pace of life in cities. *Proceedings of the National Academy of Sciences of the USA* 104, 7301–7306.

Bettencourt, L. M. A., Lobo, J., Strumsky, D., and West, G. B. (2010) Urban scaling and its deviations: revealing the structure of wealth, innovation and crime across cities. *PLoS ONE* 5(11): e13541. doi:10.1371/journal.pone.0013541

Beverton, R. J. H. (1992) Patterns of reproductive strategy parameters in some marine teleost fishes. *Journal of Fish Biology* 41 (Supplement B), 137–160.

Beverton, R. J. H. and Holt, S. J. (1959) A review of the lifespans and mortality rates of fish in nature, and their relationship to growth and other physiological characteristics. *Ciba Foundation Colloquim on Ageing* 5, 142–180.

Bezerra, G., Forrest, S., Davis, A. L., Moses, M. E. (2010) Rent's Rule traffic patterns and NoC topology scaling. In: *Proceedings of the 12th ACM/IEEE international Workshop on System Level Interconnect Prediction* (Anaheim, California, June 13, 2010). SLIP '10. ACM, New York, pp. 3–8.

Bianchi, G., Gislason, H., Graham, K., et al. (2000) Impact of fishing on size composition and diversity of demersal fish communities. *ICES Journal of Marine Science* 57, 558–571.

Bielby, J., Mace, G. M., Bininda-Emonds, O.R.P., et al. (2007) The fast-slow continuum in mammalian life history: An empirical reevaluation. *American Naturalist* 169, 748–757.

Bielby, J., Cooper, N., Cunningham, A. A., Garner, T. W. J., and Purvis, A. (2008) Predicting susceptibility to future declines in the world's frogs. *Conservation Letters* 1, 82–90.

Binford, L. R. (2001) *Constructing Frames of Reference.* Berkeley: University of California Press.

Biondi, F., Myers, D. E., and Avery, C. C. (1994) Geostatistically modeling stem size and increment in an old-growth forest. *Canadian Journal of Forestry Research* 24, 1354–1368.

Birch, L. C. (1957) The role of weather in determining the distribution and abundance of animals. Population studies: animal ecology and demography. *Cold Spring Harbor Symposia on Quantitative Biology* 22, 203–218.

Bishop, C. M. (1999) The maximum oxygen consumption and aerobic scope of birds and mammals: getting to the heart of the matter. *Proceedings of the Royal Society of London Series B: Biological Sciences* 266, 2275–2281.

Bissinger, J. E., Montagnes, D. J. S., Sharples, J., and Atkinson, D. (2008) Predicting marine phytoplankton maximum growth rates from temperature: improving on the Eppley curve using quantile regression. *Limnology and Oceanography* 53, 487–493.

Blackburn, T. M. and Gaston, K. J. (2001) Linking patterns in macroecology. *Journal of Animal Ecology* 70, 338–352.

Blackburn, T. M., Lawton, J. H., and Pimm, S. L. (1993) Nonmetabolic explanations for the relationship between body-size and animal abundance. *Journal of Animal Ecology* 62, 694–702.

Blackman, V. H. (1919) The compound interest law and plant growth. *Annals of Botany* 33, 353–360.

Blanchard, J. L., Dulvy, N. K., Jennings, S., Ellis, J. R., Pinnegar, J. K., Tidd, A., and Kell, L. T. (2005) Do climate and fishing influence size-based indicators of Celtic Sea fish community structure? *ICES Journal of Marine Science: Journal du Conseil* 62, 405–411.

Blanchard, J. L., Jennings, S., Law, R., et al. (2009) How does abundance scale with body size in coupled size-structured food webs. *Journal of Animal Ecology* 78, 270–280.

Blanchard, J. L., Law, R., Castle, M. D., and Jennings, S. (2011) Coupled energy pathways and the resilience of size-structured food webs. *Theoretical Ecology* 4, 289–300.

Blanckenhorn, W. U. and Demont, M. (2004) Bergmann and converse Bergmann latitudinal clines in arthropods: Two ends of a continuum? *Integrative and Comparative Biology* 44, 413–424.

Blankinship, J., Niklaus, P., and Hungate, B. (2011) A meta-analysis of responses of soil biota to global change. *Oecologia* 165, 553.

Blatt, J. and Roces, F. (2001) Haemolymph sugar levels in foraging honeybees (*Apis mellifera carnica*): dependence on metabolic rate and *in vivo* measurement of maximal rates of trehalose synthesis. *Journal of Experimental Biology* 204, 2709–2716.

Bleiweiss, R. (1998) Slow rate of molecular evolution in highelevation hummingbirds. *Proceedings of the National Academy of Sciences of the USA* 95, 612–616.

Blomberg, S. P., Theodore Garland, J. R., and A. R. Ives, A. R. (2003) Testing for phylogenetic signal in comparative data: behavioral traits are more labile. *Evolution* 57, 717–745.

Blueweiss, L., Fox, H., Kudzma, V., Nakashima, D., Peters, R., and Sams, S. (1978) Relationships between body size and some life history parameters. *Oecologia* 37, 257–272.

Blum, J. J. (1977) On the geometry of four-dimensions and the relationship between metabolism and body mass. *Journal of Theoretical Biology* 64, 599–601.

Boecklen, W. J. (1997) Nestedness, biogeographic theory, and the design of nature reserves. *Oecologia* 112, 123–142.

Bokma, F. (2004) Evidence against universal metabolic allometry. *Functional Ecology* 18, 184–187.

Boltzmann, L. (1905) *Populare Schriften.* Leipzig: J. A. Barth.

Bolzoni, L., De Leo, G. A., Gatto, M., and Dobson, A. P. (2008a) Body-size scaling in an SEI model of wildlife diseases. *Theoretical Population Biology* 73, 374–382.

Bolzoni, L., Dobson, A. P., Gatto, M., and De Leo, G. A. (2008b) Allometric scaling and seasonality in the epidemics of wildlife diseases. *American Naturalist* 172, 818–828.

Bond-Lamberty, B. and Thomson, A. (2010) Temperature-associated increases in the global soil respiration record. *Nature* 464, 579.

Borgmann, U. (1987) Models of the slope of, and biomass flow up, the biomass size spectrum. *Canadian Journal of Fisheries and Aquatic Sciences* 44 (Supplement 2), 136–140.

Bosma, T. N., Middeldorp, P. J. M., Schraa, G., and Zehnder, A. J. B. (1996) Mass transfer limitation of biotransformation: quantifying bioavailability. *Environmental Science and Technology* 31, 248–252.

Botkin, D. B., Jordan, P. A., Dominski, A. S., Lowendorf, H. S., and Hutchinson, G. E. (1973) Sodium dynamics in a northern ecosystem. *Proceedings of the National Academy of Sciences of the USA* 70, 2745–2748.

Boudreau, P. R. and Dickie, L. M. (1992) Biomass spectra of aquatic ecosystems in relation to fisheries yield. *Canadian Journal of Fisheries and Aquatic Science* 49, 1528–1538.

Bouwma, A., Nordheim, E. and Jeanne, R. (2006) Per-capita productivity in a social wasp: no evidence for a negative effect of colony size. *Insectes Sociaux* 53, 412–419.

Bowen, W. D., Oftedal, O. T., and Boness, D. J. (1985) Birth to weaning in 4 days: remarkable growth in the hooded seal, *Cystophora cristata*. *Canadian Journal of Zoology* 63, 2841–2846.

Bower, F. O. (1930). *Size and Form in Plants*. London: Macmillan.

Boyd, I. L. (1998) Time and energy constraints in pinniped lactation. *American Naturalist* 152, 717–728.

Boyd, P., Watson, A., Law, C., et al. (2000) A mesoscale phytoplankton bloom in the polar Southern Ocean stimulated by iron fertilization. *Nature* 407, 695–702.

Boyd, R. and Richerson, P. (1985) *Culture and the Evolutionary Process*. Chicago: University of Chicago Press.

Boyer, A. G. (2010) Consistent ecological selectivity through time in Pacific island avian extinctions. *Conservation Biology* 24, 511–519.

Boyer, A. G. and Jetz, W. (2010) Biogeography of body size in Pacific island birds. *Ecography* 33, 369–379.

Bradley, T. J., Brethorst, L., Robinson, S., and Hetz, S. (2003) Changes in the rate of CO_2 release following feeding in the insect *Rhodnius prolixus*. *Physiological And Biochemical Zoology* 76, 302–309.

Bradshaw, C. J. A., Giam, X., Tan, H. T. W., Brook, B. W., and Sodhi, N. S. (2008) Threat or invasive status in legumes is related to opposite extremes of the same ecological and life-history attributes. *Journal of Ecology* 96, 869–883.

Bradshaw, W. E. and Holzapfel, C. M. (2001) Genetic shift in photoperiodic response correlated with global warming. *Proceedings of the National Academy of Sciences of the USA* 98, 14509–14511.

Bragg, J. G., Thomas, D., and Baudouin-Cornu, P. (2006) Variation among species in proteomic sulphur content is related to environmental conditions. *Proceedings of the Royal Society B: Biological Sciences* 273, 1293.

Branda, S. S. and Kolter, R. (2004) Multicellularity and biofilms. In: Ghannoum, M. and O'Toole, G. A. (Eds), *Microbial Biofilms*. Washington, DC: American Society for Microbiology, pp. 20–29.

Breashears, D., Myers, O., Meyer, C. W., et al. (2009) Tree die-off in response to global change-type drought: mortality insights from a decade of plant water potential measurements. *Frontiers in Ecology and the Environment* 7, 185–189.

Brischoux, F., Bonnet, X., Cook, T. R., and Shine, R. (2008) Allometry of diving capacities: ectothermy vs. endothermy. *Journal of Evolutionary Biology* 21, 324–329.

Brock, D. A., Douglas, T. E., Queller, D. C., and Strassmann, J. E. (2011) Primitive agriculture in a social amoeba. *Nature* 469, 393–396.

Brock, W. H. (2002) *Justus von Liebig: The Chemical Gatekeeper*. Cambridge: Cambridge University Press.

Brody, S. and Proctor, R. C. (1932) Growth and development, with special reference to domestic animals. XXIII. Relation between basal metabolism and mature body weight in different species of mammals and birds. *Missouri University Agricultural Experiment Station Research Bulletin* 166, 89–101.

Bromham, L. (2002) Molecular clocks in reptiles: life history influences rate of molecular evolution. *Molecular Biology and Evolution* 19, 302–309.

Bromham, L., and Penny. D. (2003) The modern molecular clock. *Nature Reviews Genetics* 4, 216–224.

Bromham, L., Rambaut, A., and Harvey, P. H. (1996) Determinants of rate variation in mammalian DNA sequence evolution. *Journal of Molecular Evolution* 43, 610–621.

Brook, B. W., Traill, L. W., and Bradshaw, C. J. A. (2006) Minimum viable population sizes and global extinction risk are unrelated. *Ecology Letters* 9, 375–382.

Brooke, C. (2008) Conservation and adaptation to climate change. *Conservation Biology* 22, 1471–1476.

Brose, U. (2010) Body-mass constraints on foraging behaviour determine population and food-web dynamics. *Functional Ecology* 24, 28–34.

Brose, U., Williams, R. J., and Martinez, N. D. (2003) Comment on "Foraging adaptation and the relationship between food-web complexity and stability". *Science* 301, 918b.

Brose, U., Berlow, E. L., and Martinez, N. D. (2005) Scaling up keystone effects from simple to complex ecological networks. *Ecology Letters* 8, 1317–1325.

Brose, U., Jonsson, T., Berlow, E. L., et al. (2006a) Consumer-resource body-size relationships in natural food webs. *Ecology* 87, 2411–2417.

Brose, U., Williams, R. J., and Martinez, N. D. (2006b) Allometric scaling enhances stability in complex food webs. *Ecology Letters* 9, 1228–1236.

Brose, U., Ehnes, R., Rall, B., Vucic-Pestic, O., Berlow, E., and Scheu, S. (2008) Foraging theory predicts predator-prey energy fluxes. *Journal of Animal Ecology* 77, 1072–1078.

Brown, J. H. (1981) Two decades of homage to Santa Rosalia: toward a general-theory of diversity. *American Zoologist* 21, 877–888.

Brown, J. H. (1995). *Macroecology*. Chicago: University of Chicago Press.

Brown, J. H. and Gillooly, J. F. (2003) Ecological food webs: High-quality data facilitate theoretical unification. *Proceedings of the National Academy of Sciences of the USA* 100, 1467–1468.

Brown, J. H. and Lee, A. K. (1969) Bergmann's rule and climatic adaptation in woodrats (*Neotoma*). *Evolution* 23, 329–338.

Brown, J. H. and Maurer, B. A. (1987) Evolution of species assemblages: effects of energetic constraints and species dynamics on the diversification of the North American avifauna. *American Naturalist* 130, 1–17.

Brown, J. H., and Maurer, B. A. (1989) Macroecology: the division of food and space among species on continents. *Science* 243, 1145–1150.

Brown, J. H. and Sibly, R. M. (2006) Life-history evolution under a production constraint. *Proceedings of the National Academy of Sciences of the USA* 103, 17595–17599.

Brown, J. H., Marquet, P. A., and Taper, M. L. (1993) Evolution of body size – consequences of an energetic definition of fitness. *American Naturalist* 142, 573–584.

Brown, J. H., Gillooly, J. F., Allen, A. P., Savage, V. M., and West, G. B. (2004) Toward a metabolic theory of ecology. *Ecology* 85, 1771–1789.

Brown, J. H., Burnside, W. R., Davidson, A. D., et al. (2011) Energetic limits to economic growth. *BioScience* 61, 19–26.

Brucet, S., Boix, D., Quintana, X., et al. (2010) Factors influencing zooplankton size structure at contrasting temperatures in coastal shallow lakes: implications for effects of climate change. *Limnology and Oceanography* 55, 1697–1711.

Bryant, C., Behm, C. A., and Howell, M. J. (1989) *Biochemical Adaptation in Parasites*. London; New York: Chapman & Hall.

Bryant, V. (1974). Growth and respiration throughout life-cycle of Nematospiroides-Dubius Baylis (1926) (Nematoda-Heligmosomidae): parasitic stages. *Parasitology* 69, 97–106.

Buck, J. (1962) Some physical aspects of insect respiration. *Annual Reviews of Entomology* 7, 27–56.

Buckley, L. B. and Jetz, W. (2007): Environmental and historical constraints on global patterns of amphibian richness. *Proceedings of the Royal Society B: Biological Sciences* 274, 1167–1173.

Buckley, L. B., Rodda, G. H., and Jetz, W. (2008) Thermal and energetic constraints on ectotherm abundance: a global test using lizards. *Ecology* 89, 48–55.

Buckley, L. B., Urban, M. C., Angilletta, M. J., Crozier, L. G., Rissler, L. J., and Sears, M. W. (2010) Can mechanism inform species' distribution models? *Ecology Letters* 13, 1041–1054.

Buckley, L. B., Hurlbert, A. H., and Jetz, W. (in press) Broad-scale ecological implications of ectothermy and endothermy in changing environments. *Global Ecology and Biogeography*.

Bukovinszky, T., van Veen, F. J. F., Jongema, Y., and Dicke, M. (2008) Direct and indirect effects of resource quality on food web structure. *Science* 319, 804–807.

Burger, O., Walker, R., and Hamilton, M. J. (2010) Lifetime reproductive effort in humans. *Proceedings of the Royal Society B: Biological Sciences* 277, 773–777.

Burnett, C. D. (1983) Geographic and climatic correlates of morphological variation in *Eptesicus fuscus*. *Journal of Mammalogy* 64, 437–444.

Burnham, K. P. and Anderson, D. R. (2002) *Model Selection and Multimodel Inference: A Practical Information–Theoretic Approach*. 2nd Edition. New York: Springer-Verlag.

Burnside, W. R., Brown, J. H., Burger, O., Hamilton, M. J., Moses, M., and Bettencourt, L. M. A. (2011) Human macroecology: linking pattern and process in big-picture human ecology. *Biological Reviews*. doi: 10.1111/j.1469-185X.2011.00192.x.

Bush, A. O., Lafferty, K. D., Lotz, J. M., and Shostak, A. W. (1997) Parasitology meets ecology on its own terms: Margolis et al revisited. *Journal of Parasitology* 83, 575–583.

Butchart, S. H. M. and Bird, J. P. (2010) Data deficient birds on the IUCN Red List: What don't we know and why does it matter? *Biological Conservation* 143, 239–247.

Butchart, S. H. M., Walpole, M., Collen, B., et al. (2010) Global biodiversity: Indicators of recent declines. *Science* 328, 1164–1168.

Butler, M. A. and King, A. A. (2004) Phylogenetic comparative analysis: a modeling approach for adaptive evolution. *American Naturalist* 164, 683–695.

Button, D. K. (1978) On the theory of control of microbial growth kinetics by limiting nutrient concentrations. *Deep Sea Research* 25, 1163–1177.

Button, D. K. (1985) Kinetics of nutrient-limited transport and microbial growth. *Microbiology and Molecular Biology Reviews* 49, 270–297.

Button, D. K. (1998) Nutrient uptake by microorganisms according to kinetic parameters from theory as related to cytoarchitecture. *Microbiology and Molecular Biology Reviews* 62, 636–645.

Bybee, P. J., Lee, A. H., and Lamm, E. T. (2006) Sizing the Jurassic theropod dinosaur Allosaurus: assessing growth strategy and evolution of ontogenetic scaling of limbs. *Journal of Morphology* 267, 347–359.

Cable, J. M., Enquist, B. J., and Moses, M. E. (2007) The allometry of host–pathogen interactions. *PLoS ONE* 2, e1130.

Calder, W. A. (1984) *Size, Function and Life History*. Cambridge, MA: Harvard University Press.

Calder, W. (1996) *Size, Function, and Life History*, Mineola, NY: Dover.

Calder, W. A. III (2000) Diversity and convergence: scaling for conservation. In: Brown, J. H. and West, G.B. (Eds), *Scaling in Biology*. Oxford University Press, New York, pp. 297–323.

Caley, M. J. and Schluter, D. (1997) The relationship between local and regional diversity. *Ecology* 78, 70–80.

Campbell, G. and Norman, J. (1998) *An Introduction to Environmental Biophysics*. New York: Springer.

Canfield, D. E., Habicht, K. S., and Thamdrup, B. (2000) The Archean sulfur cycle and the early history of atmospheric oxygen. *Science* 288, 658–661.

Carbone, C. and Gittleman, J. L. (2002) A common rule for the scaling of carnivore density. *Science* 295, 2273–2276.

Carbone, C., Mace, G. M., Roberts, S. C., and Macdonald, D. W. (1999) Energetic constraints on the diet of terrestrial carnivores. *Nature* 402, 286–288.

Carbone, C., Cowlishaw, G., Isaac, N. J. B., and Rowcliffe, J. M. (2005) How far do animals go? Determinants of day range in mammals. *American Naturalist* 165, 290–297.

Carbone, C., Cowlishaw, G., Rowcliffe, J. M., and Isaac, N. J. B. (2007) The scaling of abundance in consumers and their resources: implications for the Energy Equivalence Rule. *American Naturalist* 170, 479–484.

Carbone, C., Pettorelli, N., and Stephens, P. A. (2010) The bigger they come, the harder they fall: body size and prey abundance influence predatory–prey ratios. *Biology Letters* 7, 312–315.

Cardillo, M. (2003) Biological determinants of extinction risk: why are smaller species less vulnerable? *Animal Conservation* 6, 63–69.

Cardillo, M., Purvis, A., Sechrest, W., Gittleman, J. L., Bielby, J., and Mace, G.M. (2004) Human population density and extinction risk in the world's carnivores. *PLOS Biology* 2, 909–914.

Cardillo, M., Orme, C. D. L., and Owens, I. P. F. (2005a) Testing for latitudinal bias in diversification rates: an example using new world birds. *Ecology* 86, 2278–2287.

Cardillo, M., Mace, G. M., Jones, K. E., et al. (2005b) Multiple causes of high extinction risk in large mammal species. *Science* 309, 1239–1241.

Cardillo, M., Mace, G. M., Gittleman, J. L., and Purvis, A. (2006) Latent extinction risk and the future battlegrounds of mammal conservation. *Proceedings of the National Academy of Sciences of the USA* 103, 4157–4161.

Cardillo, M., Mace, G. M., Gittleman, J. L., Jones, K. E., Bielby, J., and Purvis, A. (2008) The predictability of extinction: biological and external correlates of decline in mammals. *Proceedings of the Royal Society B: Biological Sciences* 275, 1441–1448.

Cardinale, B. J., Duffy, J. E., Srivastava, D. S., Loreau, M., Thomas, M., and Emmerson, M. (2008) Towards a food web perspective on biodiversity and ecosystem functioning. In: Naeem, S., Bunker, D. E., Hector, A., Loreau, M., and Perrings, C. (Eds), *Biodiversity, Ecosystem Functioning, and Human Wellbeing: An Ecological and Economic Perspective.* Oxford: Oxford University Press.

Careau, V., Thomas, D. W., and Humphries, M. M. (2010) Energetic cost of bot fly parasitism in free-ranging eastern chipmunks. *Oecologia* 162, 303–312.

Carpenter, S. R. and Kitchell, J. F. (1984) Plankton community structure and limnetic primary production. *American Naturalist* 124, 159–172.

Casey, T. (1977) Physiological responses to temperature of caterpillars of a desert population of *Manduca sexta* (Lepid: Sphingidae). *Comparative Biochemistry and Physiology Part A: Physiology* 57, 53–58.

Casey, T. and Ellington, C. (1989) Energetics of insect flight. In: *Energy Transformations in Cells and Organisms.* Stuttgart, Germany: Thieme, pp. 258–272.

Casey, T. M. and Hegel-Little, J. R. (1987) Instantaneous oxygen consumption and muscle stroke work in *Malacosoma americanum* during pre-flight warm-up. *Journal of Experimental Biology* 127, 389–400.

Casey, T. M. and Knapp, R. (1987) Caterpillar thermal adaptation: behavioral differences reflect metabolic thermal sensitivities. *Comparative Biochemical Physiology* 86, 679–682.

Cassemiro, F. A. S. and Diniz-Filho, J. A. F. (2010) Deviations from predictions of the metabolic theory of ecology can be explained by violations of assumptions. *Ecology* 91, 3729–3738.

Cattin, M.-F., Bersier, L.-F., Banašek-Richter, C., Baltensperger, R., and Gabriel, J.-P. (2004) Phylogenetic constraints and adaptation explain food-web structure. *Nature* 427, 835–839.

Cavalieri, R. (1997) Iodine metabolism and thyroid physiology: current concepts. *Thyroid* 7, 177–181.

Cavalli-Sforza, L. L. and Feldman, M. (1981) *Cultural Transmission and Evolution.* Princeton, NJ: Princeton University Press.

Caviedes-Vidal, E., McWhorter, T. J., Lavin, S. R., Chediack, J. G., Tracy, C. R., and Karasov, W. H. (2007) The digestive adaptation of flying vertebrates: high intestinal paracellular absorption compensates for smaller guts. *Proceedings of the National Academy of Sciences of the USA* 104, 19132–19136.

Cawley, G. C. and Janacek, G. J. (2010) On allometric equations for predicting body mass of dinosaurs. *Journal of Zoology* 280, 355–361.

Ceballos, G. and Ehrlich, P. R. (2002) Mammal population losses and the extinction crisis. *Science* 296, 904–907.

Ceballos, G. and Ehrlich, P. R. (2006) Global mammal distributions, biodiversity hotspots, and conservation. *Proceedings of the National Academy of Sciences of the USA* 103, 19374–19379.

Cermeño, P., Maranon, E., Rodriguez, J., and Fernandez, E. (2005) Size dependence of coastal phytoplankton photosynthesis under vertical mixing conditions. *Journal of Plankton Research* 27, 473–483.

Cermeño, P., Marañón, E., Harbour, D., and Harris, R. P. (2006) Invariant scaling of phytoplankton abundance and cell size in contrasting marine environments. *Ecology Letters* 9, 1210–1215.

Ceuterick, F., Peeters, J., Heremans, K., De Smedt, H., and Olbrechts, H. (1978) Effect of high pressure, detergents and phospholipase on break in Arrhenius plot of *Azotobacter nitrogenase*. *European Journal of Biochemistry* 87, 401–407.

Chadwick, O. A., Derry, L. A., Vitousek, P. M., Huebert, B. J., and Hedin, L. O. (1999) Changing sources of nutrients during four millions years of ecosystem development. *Nature* 397, 491–497.

Chapin, F. S., Matson, P. A., and Mooney, H. A. (2002) *Principles of Ecosystem Ecology*, New York: Springer Verlag.

Chappell, M. A. (1983) Metabolism and thermoregulation in desert and montane grasshoppers. *Oecologia* 56, 126–131.

Chappell, M. A. (1984) Temperature regulation and energetics of the solitary bee *Centris pallida* during foraging and intermale mate competition. *Physiological Zoology* 57, 215–225.

Chappell, M. and Rogowitz, G. (2000) Mass, temperature and metabolic effects on discontinuous gas exchange cycles in eucalyptus-boring beetles (Coleoptera: cerambycidae). *Journal of Experimental Biology* 203, 3809–3820.

Charnov, E. L. (1991) Evolution of life history variation among female mammals. *Proceedings of the National Academy of Sciences of the USA* 88, 1134–1137.

Charnov, E. L. (1993) *Life History Invariants: Some Explorations of Symmetry in Evolutionary Ecology.* Oxford: Oxford University Press.

Charnov, E. L. and Berrigan, D. (1993) Why do female primates have such long lifespans and so few babies? Or life in the slow lane. *Evolutionary Anthropology* 1, 191–194.

Charnov, E. L. and Gillooly, J. F. (2004) Size and temperature in the evolution of fish life histories. *Integrative and Comparative Biology* 44, 494–497.

Charnov, E. L. and Zuo, W. (2011) Human hunting mortality threshold rules for extinction in mammals (and fish). *Evolutionary Ecology Research* 13, 1–7.

Chase, J. M. and Leibold, M. A. (2003) *Ecological Niches: Linking Classical and Contemporary Approaches.* Chicago: University of Chicago Press.

Chave, J., Andalo, C., Brown, S., et al. (2005) Tree allometry and improved estimation of carbon stocks and balance in tropical forests. *Oecologia* 145, 87–99.

Chen, S., Lin, G., Huang, J., and Jenerette, G. D. (2009) Dependence of carbon sequestration on the differential responses of ecosystem photosynthesis and respiration to rain pulses in a semiarid steppe. *Global Change Biology* 15, 2450.

Cheung, W. W. L., Lam, V. W. Y., Sarmiento, J. L., et al. (2010) Large-scale redistribution of maximum fisheries catch potential in the global ocean under climate change. *Global Change Biology* 16, 24–35.

Childress, J. J. (1971) Respiratory rate and depth of occurrence of midwater animals. *Limnology and Oceanography* 16, 104–106.

Childress, J. J. (1995) Are there physiological and biochemical adaptations of metabolism in deep-sea animals? *Trends in Ecology and Evolution* 10, 30–37.

Childress, J. J. and Seibel, B. A. (1998) Life at stable low oxygen: adaptations o aniamls to oceanic oxygen minimum layers. *Journal of Experimental Biology* 201, 1223–1232.

Chisholm, S. W. (1992) Phytoplankton size. In: Falkowski, P. G. and Woodhead, A. D. (Eds) *Primary Productivity and Biogeochemical Cycles in the Sea.* New York: Plenum Press.

Chown, S. L. and Gaston, K. J. (2008) Macrophysiology for a changing world. *Proceedings of the Royal Society B: Biological Sciences* 275, 1469.

Chown, S. and Nicolson, S. W. (2004) *Insect Physiological Ecology: Mechanisms and Patterns.* Oxford; New York: Oxford University Press.

Chown, S. L., Addo-Bediako, A., and Gaston, K. J. (2003) Physiological diversity: listening to the large-scale signal. *Functional Ecology* 17, 568–572.

Chown, S. L., Gaston, K. J., and Robinson, D. (2004) Macrophysiology: large-scale patterns in physiological traits and their ecological implications. *Functional Ecology* 18, 159–167.

Chown, S. L., Marais, E., Terblanche, J. S., Klok, C. J., Lighton, J. R. B., and Blackburn, T. M. (2007) Scaling of insect metabolic rate is inconsistent with the nutrient supply network model. *Functional Ecology* 21, 282–290.

Ciais, P., Reichstein, M., Viovy, N., et al. (2005) Europe-wide reduction in primary productivity caused by the heat and drought in 2003. *Nature* 437, 529–533.

Clapham, D. (2007) Calcium signaling. *Cell* 131, 1047–1058.

Clark, J. S. (1990) Integration of ecological levels: individual plant-growth; population mortality and ecosystem processes. *Journal of Ecology* 78, 275–299.

Clark, J. S. (1992) Density-independent mortality; density compensation; gap formation; and self-thinning in plant-populations. *Theoretical Population Biology* 42, 172–198.

Clark, R. M., Cox, S. J. D., and Laslett, G. M. (1999) Generalizations of power-law distributions applicable to sampled fault-trace lengths: model choice, parameter estimation and caveats. *Geophysical Journal International* 136, 357–372.

Clarke, A. (1983) Life in cold water: the physiological ecology of polar marine ectotherms. *Oceanography and Marine Biology: An Annual Review* 21, 341–453.

Clarke, A. (1992) Reproduction in the cold: Thorson revisited. *Invertebrate Reproduction and Development* 22, 175–184.

Clarke, A. (2004) Is there a universal temperature dependence of metabolism? *Functional Ecology* 18, 252–256.

Clarke, A. (2006) Temperature and the metabolic theory of ecology. *Functional Ecology* 20, 405–412.

Clarke, A. and Fraser, K. P. P. (2004) Why does metabolism scale with temperature? *Functional Ecology* 18, 243–251.

Clarke, A. and Gaston, K. J. (2006) Climate, energy and diversity. *Proceedings of the Royal Society of London B* 273, 2257–2266.

Clarke, A. and Johnston, N. M. (1999) Scaling of metabolic rate with body mass and temperature in teleost fish. *Journal of Animal Ecology* 68, 893–905.

Clarke, A. and Rothery, P. (2008) Scaling of body temperature in mammals and birds. *Functional Ecology* 22, 58–67.

Clarke, A., Rothery, P., and Isaac, N. J. B. (2010) Scaling of basal metabolic rate with body mass and temperature in mammals. *Journal of Animal Ecology* 79, 610–619.

Clauset, A., Shalizi, C. R., and Newman, M. E. J. (2009) Power-law distributions in empirical data. *SIAM Review* 51, 661–703.

Claussen, D., Gerald, G., Kotcher, J., and Miskell, C. (2008) Pinching forces in crayfish and fiddler crabs, and comparisons with the closing forces of other animals. *Journal of Comparative Physiology B: Biochemical, Systemic, and Environmental Physiology* 178, 333–342.

Clutton-Brock, T., West, S., Ratnieks, F., and Foley, R. (2009) The evolution of society. *Philosophical Transactions of the Royal Society B: Biological Sciences* 364, 3127–3133.

Cody, M. L. (1966) A general theory of clutch size. *Evolution* 20, 174–184.

Cohen, A. C. (2004) *Insect Diets: Science and Technology*. Boca Raton, FL: CRC Press.

Cohen, J. E. (1977) Ratio of prey to predators in community food webs. *Nature* 270, 165–167.

Cohen, J. E., Briand, F., and Newman, C. M. (1990) *Community Food Webs*. Berlin: Springer-Verlag.

Cohen, J. E., Pimm, S. L., Yodzis, P., and Saldana, J. (1993) Body sizes of animal predators and animal prey in food webs. *Journal of Animal Ecology* 62, 67–78.

Cohen, J. E., Jonsson, T., Muller, C. B., Godfray, H. C. J., and Savage, V. M. (2005) Body sizes of hosts and parasitoids in individual feeding relationships. *Proceedings of the National Academy of Sciences of the USA* 102, 684–689.

Coleman, J., McConnaughay, K., and Ackerly D. (1994) Interpreting phenotypic variation inplants. *Trends in Ecology and Evolution* 9, 187–191.

Collin, R. (2003) Worldwide patterns in mode of development in calyptraeid gastropods. *Marine Ecology: Progress Series* 247, 103–122.

Colwell, R. K., Brehm, G., Cardelus, C. L., Gilman, A. C., and Longino, J. T. (2008) Global warming, elevational range shifts, and lowland biotic attrition in the wet tropics. *Science* 322, 258–261.

Comstock, J. P. and Sperry, J. S. (2000) Theoretical considerations of optimal conduit length for water transport in vascular plants. *New Phytologist* 148, 195–218.

Connolly, S. R., Hughes, T. P., Bellwood, D. R., and Karlson, R. H. (2005) Community structure of corals and reef fishes at multiple scales. *Science* 309, 1363–1365.

Conservation International (CI) (2009) *The Biodiversity Hotspots*. Available at: http://www.conservation.org/explore/priority_areas/hotspots/pages/hotspots_main.aspx

Contreras, H. L. and Bradley, T. J. (2009) Metabolic rate controls respiratory pattern in insects. *Journal of Experimental Biology* 212, 424–428.

Convention on Biological Diversity (CBD) (1993) *CBD Handbook* (3rd Edition). Available at http://www.cbd.int/convention/refrhandbook.shtml

Coomes, D. A. (2006) Challenges to the generality of WBE theory. *Trends in Ecology and Evolution* 21, 593–596.

Coomes, D. A. and Allen, R. B. (2007) Effects of size, competition and altitude on tree growth. *Journal of Ecology* 95, 1084–1097.

Coomes, D. A. and Allen, R. B. (2009) Testing the metabolic scaling theory of tree growth. *Journal of Ecology* 97, 1369–1373.

Coomes, D. A., Duncan, R. P., Allen, R. B., and Truscott, J. (2003) Disturbances prevent stem size-density distributions in natural forests from following scaling relationships. *Ecology Letters* 6, 980–989.

Coomes, D. A., Lines, E. R., and Allen, R. B. (2011) Moving on from Metabolic Scaling Theory: hierarchical models of tree growth and asymmetric competition for light. *Journal of Ecology* 99, 748–756.

Cooper, N., and Purvis, A. (2009) What factors shape rates of phenotypic evolution? A comparative study of cranial morphology of four mammalian clades. *Journal of Evolutionary Biology* 22, 1024–1035.

Cooper, N., Bielby, J., Thomas, G. H., and Purvis, A. (2008) Macroecology and extinction risk correlates of frogs. *Global Ecology and Biogeography* 17, 211–221.

Costa, D. P. (1991) Reproductive and foraging energetics of high latitude penguins, albatrosses and pinnipeds: implications for life history patterns. *American Zoologist* 31, 111–130.

Costa, D. P. (1993) The relationship between reproductive and foraging energetics and the evolution of the Pinnipedia. In: Boyd, I. L. (Ed.), *Marine Mammals: Advances in Behavioural and Population Biology*. Oxford: Oxford University Press, Symposium Zoological Society of London, pp. 293–314.

Costa, D. P. (2008) A conceptual model of the variation in parental attendance in response to environmental fluctuation: foraging energetics of lactating sea lions and fur seals. *Aquatic Conservation: Marine and Freshwater Ecosystems* 17, S44–S52.

Costa, D. P. (2009) Energetics. In: Perrin, W. F., Thewissen, J. G. M., and Wursig, B. (Eds), *Encyclopedia of Marine Mammals*. San Diego, CA: Academic Press, pp. 791–796.

Costa, D. P. and Prince, P. A. (1987) Foraging energetics of gray-headed albatrosses Diomedea-Chrysostoma at Bird Island, South Georgia, South Atlantic Ocean. *Ibis* 129, 149–158.

Costa, G. C. (2009) Predator size, prey size, and dietary niche breadth relationships in marine predators. *Ecology* 90, 2014–2019.

Costerton, J. W., Lewandowski, Z., Caldwell, D. E., Korber, D. R., and Lappin-Scott, H. M. (1995) Microbial biofilms. *Annual Reviews in Microbiology* 49, 711–745.

Cotgreave, P. (1993) The relationship between body-size and population abundance in animals. *Trends in Ecology and Evolution* 8, 244–248.

Cotgreave, P. and Harvey, P. H. (1992) Relationships between body size, abundance and phylogeny in bird communities. *Functional Ecology* 6, 248–256.

Courties, C., Vaquer, A., Troussellier, M., et al. (1994) Smallest eukaryotic organism. *Nature* 370, 255.

Cowles, D. L., Childress, J. J., and Wells, M. E. (1991) Metabolic rates of midwater crustaceans as a function of depth of occurance off the Hawaiian Islands: food availability as a selective factor? *Marine Biology* 110, 75–83.

Creel, S. (1997) Cooperative hunting and group size: assumptions and currencies. *Animal Behaviour* 54, 1319–1324.

Crooks, K. R. and Soulé, M. E. (1999) Mesopredator release and avifaunal extinctions in a fragmented system. *Nature* 400, 563–566.

Cummings, M., and Gelineau-Kattner, R. (2009) The energetic costs of alternative male reproductive strategies in *Xiphophorus nigrensis*. *Journal of Comparative Physiology A: Neuroethology, Sensory, Neural, and Behavioral Physiology* 195, 935–946.

Curds, C. R. (1975) A guide to the species of the genus Euplotes (Hypotrichida, Ciliatea). *Bulletin of the British Museum (Natural History) Zoology* 28, 1–61.

Currie, D. J. (1991) Energy and large-scale patterns of animal-species and plant-species richness. *American Naturalist* 137, 27–49.

Currie, D. J. and Fritz, J. T. (1993) Global patterns of animal abundance and species energy use. *Oikos* 67, 56–68.

Currie, D. J., Mittelbach, G. G., Cornell, H. V., et al. (2004) Predictions and tests of climate-based hypotheses of broad-scale variation in taxonomic richness. *Ecology Letters* 7, 1121–1134.

Cushing, D. H. (1975) *Marine Ecology and Fisheries*. Cambridge: Cambridge University Press.

Cusimano, N., and Renner, S. S. (2010) Slowdowns in diversification rates from real phylogenies may not be real. *Systematic Biology* 59, 458–464.

Cyr, H., Peters, R. H., and Downing, J. A. (1997) Population density and community size structure: comparison of aquatic and terrestrial systems. *Oikos* 80, 139–149.

Da Silva, F. A., Cassemiro, M., Barreto, B.da S., Rangel, T. F. L. V. B., and Diniz-Filho, J. A. F. (2007) Non-stationarity, diversity gradients, and the metabolic theory of ecology. *Global Ecology and Biogeography* 16, 820–822.

Daan, N., Gislason, H., Pope, J. G., and Rice, J. C. (2005) Changes in the North Sea fish community: evidence of the indirect effects of fishing? *ICES Journal of Marine Science* 62, 177–188.

Damuth, J. (1981) Population density and body size in mammals. *Nature* 290, 699–700.

Damuth, J. (1987) Interspecific allometry of population density in mammals and other animals: the independence of body mass and population energy use. *Biological Journal of the Linnean Society* 31, 193–246.

Damuth, J. (2007) A macroevolutionary explanation for energy equivalence in the scaling of body size and population density. *American Naturalist* 169, 621–631.

Daniel, R. M. and Danson, M. J. (2010) A new understanding of how temperature affects the catalytic activity of enzymes. *Trends in Biochemical Sciences* 35, 584–591.

Darveau, C. A., Suarez, M. D., Andrews, R. D., and Hochachka, P. W. (2002) Allometric cascade as a unifying principle of body mass effects on metabolism. *Nature* 417, 166–170.

Darwin, C. (1859) *On the Origin of Species by Means of Natural Selection, or the preservation of favored races in the struggle for life*. London: John Murray.

Daufresne, M., Lengfellner, K., and Sommer, U. (2009) Global warming benefits the small in aquatic ecosystems. *Proceedings of the National Academy of Sciences of the USA* 106, 12788–12793.

Davey, K. R. (1989) A predictive model for combined temperature and water activity on microbial growth during the growth phase. *Journal of Applied Microbiology* 67, 483–488.

Davey, M. E. and O'Toole, G. A. (2000) Microbial biofilms: from ecology to molecular genetics. *Microbiology and Molecular Biology Reviews* 64, 847–867.

Davidson, A. D., Hamilton, M. J., Boyer, A. G., Brown, J. H., and Ceballos, G. (2009) Multiple ecological pathways to extinction in mammals. *Proceedings of the National Academy of Sciences of the USA* 106, 10702–10705.

Davidson, E. and Janssens, I. (2006) Temperature sensitivity of soil carbon decomposition and feedbacks to climate change. *Nature* 440, 165–173.

Davidson, E. A., Janssens, I. A., and Luo, Y. (2006) On the variability of respiration in terrestrial ecosystems: moving beyond Q10. *Global Change Biology* 12, 154–164.

Davies, R. D., Orme, C. D. L., Storch, D., et al. (2007) Topography, temperature and the global distribution of bird species richness. *Proceedings of the Royal Society of London B* 274, 1189–1197.

Davies, T. J., Savolainen, V., Chase, M. W., Moat, J., and Barraclough, T. G. (2004) Environmental energy and evolutionary rates in flowering plants. *Proceedings of the Royal Society of London B* 271, 2195–2200.

Davies, T. J., Fritz, S. A., Grenyer, R., et al. (2008) Phylogenetic trees and the future of mammalian biodiversity. *Proceedings of the National Academy of Sciences of the USA* 105 (Supplement 1), 11556–11563.

Davis, A. J., Lawton, J. H., Shorrocks, B., and Jenkinson, L. S. (1998) Individualistic species responses invalidate simple physiological models of community dynamics under global environmental change. *Journal of Animal Ecology* 67, 600–612.

Davis, A. L. V., Chown, S. L., and Scholtz, C. H. (1999) Discontinuous gas-exchange cycles in Scarabaeus dung beetles (Coleoptera : Scarabaeidae): mass-scaling and temperature dependence. *Physiological and Biochemical Zoology* 72, 555–565.

Davis, A. L. V., Chown, S. L., McGeoch, M. A., and Scholtz, C. H. (2000) A comparative analysis of metabolic rate in six *Scarabaeus* species (Coleptera: Scarabaeidae) from southern Africa: further caveats when inferring adaptation. *Journal of Insect Physiology* 46, 553–562.

Davis, M. B. and Shaw, R. G. (2001) Range shifts and adaptive responses to quaternary climate change. *Science* 292, 673–679.

Davis, M. B., Shaw, R. G., and Etterson, J. R. (2005) Evolutionary responses to changing climate. *Ecology* 86, 1704–1714.

Dawkins, M. S. (1993) Are there general principles of signal design? *Philosophical Transactions: Biological Sciences* 340, 251–255.

Dawson, P. S. (1974) *Microbial Growth*. New York: Dowden, Hutchinson & Ross; distributed by Halsted Press, Stroudsburg, PA.

Dawson, T. H. (2003) Scaling laws for capillary vessels of mammals at rest and in exercise. *Proceedings of the Royal Society of London Series B: Biological Sciences* 270, 755–763.

Dawson, W. R. (1992) Physiological responses of animals to higher temperatures. In: Peters, R. L. and Lovejoy, T. E. (Eds), *Global Warming and Biological Diversity*. New Haven, CT: Yale University Press, pp. 158–170.

De Castro, F. and Gaedke, U. (2008) The metabolism of lake plankton does not support the metabolic theory of ecology. *Oikos* 117, 1218–1226.

De Leo, G. A. and Dobson, A. P. (1996) Allometry and simple epidemic models for microparasites. *Nature* 379, 720–722.

De Magalhaes, J. P., Costa, J., and Church, G. M. (2007) An analysis of the relationship between metabolism, developmental schedules, and longevity using phylogenetic independent contrasts. *Journals of Gerontology Series A: Biological Sciences and Medical Sciences* 62, 149–160.

De Mazancourt, C., Johnson, E., and Barraclough, T. G. (2008) Biodiversity inhibits species' evolutionary responses to changing environments. *Ecology Letters* 11, 380–388.

De Meeûs, T. and Renaud, F. (2002) Parasites within the new phylogeny of eukaryotes. *Trends in Parasitology* 18, 247–251.

De Roos, A. M., Diekmann, O., and Metz, J. A. J. (1992) Studying the dynamics of structured population models: a versatile technique and its application to *Daphnia*. *American Naturalist* 139, 123–147.

De Ruiter, P. C., Neutel, A., and Moore, J. C. (1994) Modelling food webs and nutrient cycling in agro-ecosystems. *Trends in Ecology and Evolution* 9, 378–383.

Dekas, A. E., Poretsky, R. S., and Orphan, V. J. (2009) Deep-sea archaea fix and share nitrogen in methane-consuming microbial consortia. *Science* 326, 422–426.

Dell, A. I., Pawar, S., and Savage, V. M. (2011) Systematic variation in the temperature dependence of physiological and ecological traits. *Proceedings of the National Academy of Sciences of the USA* 108, 10591–10596.

DeLong, J. P., Okie, J. G., Moses, M. E., Sibly, R. M., and Brown, J. H. (2010) Shifts in metabolic scaling, production, and efficiency across major evolutionary transitions of life. *Proceedings of the National Academy of Sciences of the USA* 107, 12941–12945.

DeLucia, E. H., Hamilton, J. G., Naidu, S. L., et al. (1999) Net primary production of a forest ecosystem with experimental CO_2 enrichment. *Science* 284, 1177–1179.

DeLucia, E. H., Casteel, C. L., Nabity, P. D., and O'Neill, B. F. (2008) Insects take a bigger bite out of plants in a warmer, higher carbon dioxide world. *Proceedings of the National Academy of Sciences of the USA* 105, 1781–1782.

Demment, M. W. and Van Soest, P. J. (1983) *Body Size, Digestive Capacity, and Feeding Strategies of Herbivores*. Morrilton, AR: Winrock International.

Deng, J.-M., Wang, G.-X., Morris, E.C., et al. (2006). Plant mass–density relationship along a moisture gradient in north-west China. *Journal of Ecology* 94, 953–958.

Denney, N. H., Jennings, S., and Reynolds, J. D. (2002) Life history correlates of maximum population growth rates in marine fishes. *Proceedings of the Royal Society: Biological Sciences* 269, 2229–2237.

Deutsch, C. A., Tewksbury, J. J., Huey, R. B., et al. (2008) Impacts of climate warming on terrestrial ectotherms across latitude. *Proceedings of the National Academy of Sciences of the USA* 105, 6668–6672.

Dial, K. P., Greene, E., and Irschick, D. J. (2008) Allometry of behavior. *Trends in Ecology and Evolution* 23, 394–401.

Diamond, J. M. (1975) The island dilemma: lessons of modern biogeographic studies for the design of natural reserves. *Biological Conservation* 7, 129–146.

Diamond, J. M. (1984) "Normal" extinctions of isolated populations. In: Nitecki, M. H. (Ed.), *Extinctions*. Chicago: University of Chicago Press, pp. 191–246.

Diamond, J. (1991) Evolutionary design of intestinal nutrient absorption: enough but not too much. *News in Physiological Science* 6, 92–96.

Dietze, M., Wolosin, M., and Clark, J. (2008) Capturing diversity and interspecific variability in allometries: a hierarchical approach. *Forest Ecology and Management* 256, 1939–1948.

Digel, C., Riede, J. O., and Brose, U. (2011) Body sizes, cumulative and allometric degree distributions across natural food webs. *Oikos* 120, 503–509.

Dillon, M. E., Wang, G., and Huey, R. B. (2010) Global metabolic impacts of recent climate warming. *Nature* 467, 704–706.

Dixon, P. M. and Pechmann, J. H. K. (2005) A statistical test to show negligible trend. *Ecology* 86, 1751–1756.

Djawdan, M., Rose, M. R., and Bradley, T. J. (1997) Does selection for stress resistance lower metabolic rate? *Ecology* 78, 828–837.

Dobson, A. P. and Hudson, P. J. (1986) Parasites, disease and the structure of ecological communities. *Trends in Ecology and Evolution* 1, 11–15.

Dobson, A. P., Lafferty, K. D., Kuris, A. M., Hechinger, R. F., and Jetz, W. (2008) Homage to Linneaus: How many parasites? How many hosts? *Proceedings of the National Academy of Sciences of the USA* 105, 11482–11489.

Dodds, P. S. (2010) Optimal form of branching supply and collection networks. *Physical Review Letters* 104, 048702.

Doemel, W. N. and Brock, T.D. (1977) Structure, growth, and decomposition of laminated algal-bacterial mats in alkaline hot springs. *Applied and Environmental Microbiology* 34, 433–452.

Doi, H., Cherif, M., Iwabuchi, T., Katano, I., Stegen, J. C., and Striebel, M. (2010) Integrating elements and energy through the metabolic dependencies of gross growth efficiency and the threshold elemental ratio. *Oikos* 119, 752–765.

Douglas, A. E. (2010) *The Symbiotic Habit*. Princeton, NJ: Princeton University Press.

Downs, C. J., Hayes, J. P., and Tracy, C. R. (2008) Scaling metabolic rate with body mass and inverse body temperature: a test of the Arrhenius fractal supply model. *Functional Ecology* 22, 239–244.

Drake, J. E., Gallet-Budynek, A., Hofmockel, K. S., et al. (2011) Increases in the flux of carbon belowground stimulate nitrogen uptake and sustain the long-term enhancement of forest productivity under elevated CO_2. *Ecology Letters* 14, 349–357.

Drazen, J. C. and Seibel, B. A. (2007) Depth-related trends in metabolism of benthic and benthopelagic deep-sea fishes. *Limnology and Oceanography* 52, 2306–2316.

Droop, M. R. (1973) Some thoughts on nutrient limitation in algae. *Journal of Phycology* 9, 264–272.

Droop, M. (1974) The nutrient status of algal cells in continuous culture. *Journal of the Marine Biological Association of the UK* 54, 825–855.

Drummond, A. J. and Rambaut, A. (2007) BEAST: Bayesian evolutionary analysis by sampling trees. *BMC Evolutionary Biology* 7, 214.

Duarte, C. M. (2007). Marine ecology warms up to theory. *Trends in Ecology and Evolution* 22, 331–333.

Duarte, C. M. and Cebrián, J. (1996) The fate of marine autotrophic production. *Limnology and Oceanography* 41, 1758–1766.

Dulvy, N. K., Polunin, N. V. C., Mill, A. C., and Graham, N. A. J. (2004) Size structural change in lightly exploited coral reef fish communities: evidence for weak indirect effects. *Canadian Journal of Fisheries and Aquatic Sciences* 61, 466–475.

Duncan, A. (1984) Assessment of factors influencing the composition, body size and turnover rate of zooplankton in Parakrama Samudra, an irrigation reservoir in Sri Lanka. *Hydrobiologia* 113, 201–215.

Duncan, R. and Young, J. (2000) Determinants of plant extinction and rarity 145 years after European settlement of Auckland, New Zealand. *Ecology* 81, 3048–3061.

Duncan, R. P., Forsyth, D. M., and Hone, J. (2007) Testing the metabolic theory of ecology: allometric scaling exponents in mammals. *Ecology* 88, 324–333.

Dunne, J. and Williams, R. (2009) Cascading extinctions and community collapse in model food webs. *Philosoophical Transactions of the Royal Society B* 364, 1711–1723.

Dunne, J. A. (2005) The network structure of food webs. In: Pascual, M. and Dunne, J. (Eds), *Ecological Networks: Linking Structure to Function*. Oxford: Oxford University Press.

Dunne, J. A., Williams, R. J., and Martinez, N. D. (2002) Food-web structure and network theory: the role of connectance and size. *Proceedings of the National Academy of Sciences of the USA* 99, 12917–12922.

Dunne, J. A., Brose, U., Williams, R. J., and Martinez, N. D. (2005) Modeling food web dynamics: complexity-stability implications. In: Belgrano, A., Scharler, U. M., Dunne, J., and Ulanowicz, R. E. (Eds), *Aquatic Food Webs: An Ecosystem Approach*. Oxford: Oxford University Press.

Duplisea, D. E., Jennings, S., Warr, K. J., and Dinmore, T. A. (2002) A size-based model to predict the impacts of bottom trawling on benthic community structure. *Canadian Journal of Fisheries and Aquatic Science* 59, 1785–1795.

Dussutour, A. and Simpson, S. J. (2009) Communal nutrition in ants. *Current Biology* 19, 740–744.

Eagle, R. A., Schauble, E. A., Tripati, A. K., Tutken, T., Hulbert, R. C., and Eiler, J. M. (2010) Body temperatures of modern and extinct vertebrates from C-13-O-18 bond abundances in bioapatite. *Proceedings of the National Academy of Sciences of the USA* 107, 10377–10382.

Eccles, J. C. (1991) *Evolution of the Brain: Creation of the Self*. London; New York: Routledge.

Economo, E. P., Kerkhoff, A. J., and Enquist, B. J. (2005) Allometric growth, life-history invariants and population energetics. *Ecology Letters* 8, 353–360.

Edwards, A. M. (2008) Using likelihood to test for Lévy flight search patterns and for general power-law distributions in nature. *Journal of Animal Ecology* 77, 1212–1222.

Edwards, A. M., Phillips, R. A., Watkins, N. W., et al. (2007) Revisiting Lévy flight search patterns of wandering albatrosses, bumblebees and deer. *Nature* 449, 1044–1048.

Edwards, K. F., Klausmeier, C. A., and Litchman, E. (2011) Evidence for a three-way tradeoff between nitrogen and phosphorus competitive abilities and cell size in phytoplankton. *Ecology* 92, 2085–2095.

Eldredge, N. and Gould, S. J. (1972) Punctuated equilibria: an alternative to phyletic gradualism. In: Schopf, T. J. M. (Ed.), *Models in Paleobiology*. San Francisco, CA: Freeman, pp. 82–115.

Ellis, H. I. and Gabrielsen, G. W. (2001) Energetics of freeranging seabirds. In: Schreiber, E. A. and Burger, J. (Eds), *Biology of Marine Birds*. Boca Raton, FL: CRC Press, pp. 359–408.

Else, P. L. and Hulbert, A. J. (1985a) Mammals – an allometric study of metabolism at tissue and mitochondrial level. *American Journal of Physiology: Regulatory Integrative and Comparative Physiology* 248, R415–R421.

Else, P. L. and Hulbert, A. J. (1985b) An allometric comparison of the mitochondria of mammalian and reptilian tissues – the implications for the evolution of endothermy.

Journal of Comparative Physiology B: Biochemical Systemic and Environmental Physiology 156, 3–11.

Else, P. L., Brand, M. D., Turner, N., and Hulbert, A. J. (2004) Respiration rate of hepatocytes varies with body mass in birds. *Journal of Experimental Biology* 207, 2305–2311.

Elser, J., Dobberfuhl, D., Mackay, N., and Schampel, J. (1996) Organism size, life history, and N:P stoichiometry. *Bioscience* 46, 674–684.

Elser, J. J., Acharya, K., Kyle, M., et al. (2003) Gorwth rate-stoichiometry couplings in diverse biota. *Ecology Letters* 6, 936–943.

Elser, J. J., Fagan, W. F., Kerkhoff, A. J., Swenson, N. G., and Enquist, B. J. (2010) Biological stoichiometry of plant production: metabolism, scaling and ecological response to global change. *New Phytologist* 186, 593–608.

Elton, C. (1927) *Animal Ecology*. London: Sidgwick & Jackson.

Emlen, J. M. (1966) The role of time and energy in food preference. *American Naturalist* 100, 611–617.

Emlet, R. B., Mcedward, L. R., and Strathmann, R. R. (1987) Echinoderm larval ecology viewed from the egg. In: Jangoux, M. and Lawrence, J. M. (Eds), *Echinoderm Studies*. Rotterdam: A.A. Balkema, pp. 55–136.

Emmerson, M. C. and Raffaelli, D. (2004) Predator-prey body size, interaction strength and the stability of a real food web. *Journal of Animal Ecology* 73, 399–409.

Emmerson, M. C., Montoya, J. M., and Woodward, G. (2005) Body size, interaction strength, and food web dynamics. In: de Ruiter, P., Moore, J. C., and Wolters, V. (Eds), *Dynamic Food Webs: Multispecies Assemblages, Ecosystem Development and Environmental Change*. Amsterdam: Elsevier.

Enquist, B. J. (2002) Universal scaling in tree and vascular plant allometry: toward a general quantitative theory linking plant form and function from cells to ecosystems. *Tree Physiology* 22, 1045–1064.

Enquist, B. J. (2003) Cope's Rule and the evolution of long-distance transport in vascular plants: allometric scaling, biomass partitioning and optimization. *Plant Cell and Environment* 26, 151–161.

Enquist, B. J. (2011) Forest annual carbon cost: a global-scale analysis of autotrophic respiration: Comment. *Ecology* 92, 1994–1998.

Enquist, B. J. and Niklas, K. J. (2001) Invariant scaling relations across tree-dominated communities. *Nature* 410, 655–660.

Enquist, B. J. and Niklas, K. J. (2002a) Global allocation rules for patterns of biomass partitioning in seed plants. *Science* 295, 1517–1520.

Enquist, B. J. and Niklas, K. J. (2002b) Global allocation rules for patterns of biomass partitioning – Response. *Science* 296, U1–U2.

Enquist, B. J., Brown, J. H., and West, G. B. (1998) Allometric scaling of plant energetics and population density. *Nature* 395, 163–165.

Enquist, B. J., West, G. B., Charnov, E. L., and Brown, J. H. (1999) Allometric scaling of production and life-history variation in vascular plants. *Nature* 401, 907–911.

Enquist, B. J., Economo, E. P., Huxman, T. E., Allen, A. P., Ignace, D. D., and Gillooly, J. F. (2003) Scaling metabolism from organisms to ecosystems. *Nature* 423, 639–642.

Enquist, B. J., Kerkhoff, A. J., Huxman, T. E. and Economo, E. P. (2007a) Adaptive differences in plant physiology and ecosystem paradoxes: insights from metabolic scaling theory. *Global Change Biology* 13, 591–609.

Enquist, B. J., Allen, A. P., Brown, J. H., et al. (2007b) Biological scaling: Does the exception prove the rule? *Nature* 445, E9–E10.

Enquist, B. J., Kerkhoff, A. J., Stark, S. C., Swenson, N. G., McCarthy, M. C., and Price, C. A. (2007c) A general integrative model for scaling plant growth, carbon flux, and functional trait spectra. *Nature* 449, 218–222.

Enquist, B. J., West, G. B., and Brown, J. H. (2009) Extensions and evaluations of a general quantitative theory of forest structure and dynamics. *Proceedings of the National Academy of Sciences of the USA* 106, 7046–7051.

Eppley, R. W. (1972) Temperature and phytoplankton growth in the sea. *Fishery Bulletin of the National Oceanographic and Atmospheric Administration* 70, 1063–1085.

Erecinska, M., Cherian, S., and Silver, I. A. (2004) Energy metabolism in mammalian brain during development. *Progress in Neurobiology* 73, 397–445.

Erickson, G. M. (2005) Assessing dinosaur growth patterns: a microscopic revolution. *Trends in Ecology and Evolution* 20, 677–684.

Erickson, G. M., Rogers, K. C., and Yerby, S. A. (2001) Dinosaurian growth patterns and rapid avian growth rates. *Nature* 412, 429–433.

Ernest, S. K. M. (2005) Body size, energy use, and community structure of small mammals. *Ecology* 86, 1407–1413.

Ernest, S. K. M. and Brown, J. H. (2001) Homeostasis and compensation: the role of species and resources in ecosystem stability. *Ecology* 82, 2118–2132.

Ernest, S. K. M., Enquist, B. J., Brown, J. H., et al. (2003) Thermodynamic and metabolic effects on the scaling of production and population energy use. *Ecology Letters* 6, 990–995.

Ernest, S. K. M., Brown, J. H., Thibault, K. M., White, E. P., and Goheen, J. R. (2008) Zero sum, the niche, and metacommunities: long-term dynamics of community assembly. *American Naturalist* 172, E257–E269.

Ernest, S. K. M., White, E. P., and Brown, J. H. (2009) Changes in a tropical forest support metabolic zero-sum dynamics. *Ecology Letters* 12, 507–515.

Etienne, R. S. and Heesterbeek, J. A. P. (2000) On optimal size and number of reserves for metapopulation persistence. *Journal of Theoretical Biology* 203, 33–50.

Etienne, R. S., Apol, M. E. F., and Olff, H. (2006) Demystifying the West, Brown and Enquist model of the allometry of metabolism. *Functiional Ecology* 20, 394–399.

Etter, R. J., Rex, M. A., Chase, M. R., and Quattro, J. M. (2005) Population differentiation decreases with depth in deep-sea bivalves. *Evolution* 59, 1479–1491.

Evans, K. L., Warren, P. H., and Gaston, K. J. (2005) Species-energy relationships at the macroecological scale: a review of the mechanisms. *Biological Reviews* 80, 1–25.

Evans, K. L., Newson, S. E., Storch, D., Greenwood, J. J. D., and Gaston, K. J. (2008) Spatial scale, abundance and the species-energy relationship in British birds. *Journal of Animal Ecology* 77, 395–405.

Falster, D. S., Brännström, Å., Dieckmann, U., and Westoby, M. (2011) Influence of four major plant traits on average height, leaf-area cover, net primary productivity, and biomass density in single-species forests: a theoretical investigation. *Journal of Ecology* 99, 148–164.

Farlow, J. O., Coroian, I. D., and Foster, J. R. (2010) Giants on the landscape: modelling the abundance of megaherbivorous dinosaurs of the Morrison Formation (Late Jurassic, western USA). *Historical Biology* 22, 403–429.

Farnsworth, K. D. and Niklas, K. J. (1995) Theories of opitimization, form and function in branching architecture in plants. *Functional Ecology* 9, 355–363.

Farquhar, G. D., Von Caemerrer, S., and Berry, J. A. (1980) A biochemical model of photosynthetic CO_2 assimilation in leaves of C_3 plants. *Planta* 149, 78–90.

Farrell-Gray, C. C. and Gotelli, N. J. (2005) Allometric exponents support a 3/4-power scaling law. *Ecology* 86, 2083–2087.

Fauchon, M., Lagniel, G., Aude, J., et al. (2002) Sulfur sparing in the yeast proteome in response to sulfur demand. *Molecular Cell* 9, 713–723.

Feldman, H. A. and McMahon, T. A. (1983) The 3/4 mass exponent for energy metabolism is not a statistical artifact. *Respiratory Physiology* 52, 149–163.

Felsenstein, J. (1985) Phylogenies and the comparative method. *American Naturalist* 125, 1–15.

Felsenstein, J. (1988) Phylogenies and quantitative characters. *Annual Review of Ecology and Systematics* 19, 445–471.

Fenchel, T. (1974) Intrinsic rate of natural increase: the relationship with body size. *Oecologia* 14, 317–326.

Fenchel, T. and Finlay, B.J. (1983) Respiration rates in heterotrophic, free-living protozoa. *Microbial Ecology* 9, 99–122.

Field, C. B. and Mooney, H. A. (1986) The photosynthesis-nitrogen relationship in wild plants. In: Givnish, T. J. (Ed.), *The Economy of Plant Form and Function*. Cambridge: Cambridge University Press.

Field, C. B., Behrenfeld, M. J., Randerson, J. T., and Falkowski, P. G. (1998) Primary production of the biosphere: integrating terrestrial and oceanic components. *Science* 281, 237–240.

Fielden, L. J., Krasnov, B. R., Khokhlova, I. S., and Arakelyan, M. S. (2004) Respiratory gas exchange in the desert flea *Xenopsylla ramesis* (Siphonaptera: Pulicidae): response to temperature and blood-feeding. *Comparative Biochemistry and Physiology A: Molecular and Integrative Physiology* 137, 557–565.

Fierer, N. and Jackson, R. B. (2006) The diversity and biogeography of soil bacterial communities. *Proceedings of the National Academy of Sciences of the USA* 103, 626–631.

Finkel, Z. V. (2001) Light absorption and size scaling of light-limited metabolism in marine diatoms. *Limnology and Oceanography* 46, 86–94.

Finkel, Z. V., Irwin, A. J., and Schofield, O. (2004) Resource limitation alters the 3/4 size scaling of metabolic rates in phytoplankton. *Marine Ecology: Progress Series* 273, 269–279.

Finkel, Z. V., Sebbo, J., Feist-Burkhardt, S., et al. (2007) A universal driver of macroevolutionary change in the size of marine phytoplankton over the Cenozoic. *Proceedings of the National Academy of Sciences of the USA* 104, 20416–20420.

Fittkau, E. J. and Klinge, H. (1973) On biomass and trophic structure of the central Amazonian rain forest ecosystem. *Biotropica* 5, 2–14.

Flanagan, T. P., Letendre, K., Burnside, W. R., Fricke, G. M., and Moses, M. E. (2011) How ants turn information into food. *Proceedings of the IEEE conference on Artificial Life* (Paris, France April 11–15: 178–185).

Flores, E. and Herrero, A. (2009) Compartmentalized function through cell differentiation in filamentous cyanobacteria. *Nature Reviews Microbiology* 8, 39–50.

Foden, W.B., Mace, G.M., Vié, J.-C., et al. (2009) Species susceptibility to climate change impacts. In: Vié, J.-C., Hilton-Taylor, C., and Stuart, S. N. (Eds), *Wildlife in a Changing World – An analysis of the 2008 IUCN Red List of Threatened Species*. Gland, Switzerland: IUCN, pp. 77–87.

Forsman, A. (2000) Some like it hot: intra-population variation in behavioral thermoregulation in color-polymorphic pygmy grasshoppers. *Evolutionary Ecology* 14, 25–38.

Foster, R. C. (1988) Microenvironments of soil microorganisms. *Biology and Fertility of Soils* 6, 189–203.

Fox, J. W. and Morin, P. J. (2001) Effects of intra- and interspecific interactions on species responses to environmental change. *Journal of Animal Ecology* 70, 80–90.

Francis, A. P. and Currie, D. J. (2003) A globally-consistent richness-climate relationship for angiosperms. *American Naturalist* 161, 523–536.

Franz, R., Hummel, J., Kienzle, E., Kolle, P., Gunga, H. C., and Clauss, M. (2009) Allometry of visceral organs in living amniotes and its implications for sauropod dinosaurs. *Proceedings of the Royal Society B: Biological Sciences* 276, 1731–1736.

Fraser, N. H. C., Metcalfe, N. B., and Thorpe, J. E. (1993) Temperature-dependent switch between diurnal and nocturnal foraging in salmon. *Proceedings of the Royal Society of London B: Biological Sciences* 252, 135–139.

Frausto Da Silva, J. J. R. and Williams, R. J. P. (2001) *The Biological Chemistry of the Elements: The Inorganic Chemistry of Life*. Oxford: Oxford University Press.

Freckleton, R. P. (2002) On the misuse of residuals in ecology: regression of residuals vs. multiple regression. *Journal of Animal Ecology* 71, 542–545.

Freckleton, R. P. (2009) The seven deadly sins of comparative analysis. *Journal of Evolutionary Biology* 22, 1367–1375.

Freckleton, R. and Harvey, P. (2006) Detecting non-Brownian trait evolution in adaptive radiations. *PLoS Biology* 4, 2104–2111.

Freckleton, R. P., Noble, D., and Webb, T. J. (2006) Distributions of habitat suitability and the abundance-occupancy relationship. *American Naturalist* 167, 260–275.

Freckleton, R. P., Harvey, P. H., and Pagel, M. (2003) Bergmann's Rule and body size in mammals. *American Naturalist* 161, 821–825.

Freville, H., McConway, K., Dodd, M., Silvertown, J. (2007) Prediction of extinction in plants: Interaction of extrinsic threats and life history traits. *Ecology* 88, 2662–2672.

Fuchs, H. L. and Franks, P. J. S. (2010) Plankton community properties determined by nutrients and size-selective feeding. *Marine Ecology: Progress Series* 413, 1–15.

Fuhrman, J. A., Steele, J. A., Hewson, I., et al. (2008a) A latitudinal diversity gradient in planktonic marine bacteria. *Proceedings of the National Academy of Sciences of the USA* 105, 7774–7778.

Fuhrman, J. A., Schwalbach, M. S., and Stingl, U. (2008b) Proteorhodopsins: an array of physiological roles? *Nature Reviews Microbiology* 6, 488–494.

Full, R. J. (1997) Invertebrate locomotor systems. In: Dantzler, W. H. (Ed.), *Handbook of Physiology. Section 13: Comparative Physiology*. New York: Oxford University Press.

Fuller, W. A. (1987) *Measurement Error Models*. New York: John Wiley & Sons.

Fulton, E. A., Smith, A. D. M., and Smith, D. C. (2008) Alternative management strategies for Southeastern Australian Commonwealth Fisheries. Stage 2. Quantitative management strategy evaluation. *Report to the Australian Fisheries Management Authority*. Hobart, Australia: CSIRO.

Gallardo, A. and Schlesinger, W. H. (1994) Factors limiting microbial biomass in the mineral soil and forest floor of a warm-temperate forest. *Soil Biology & Biochemistry* 26, 1409–1415.

Garland, T. and Ives, A. R. (2000) Using the past to predict the present: confidence intervals for regression equations in phylogenetic comparative methods. *American Naturalist* 155, 346–364.

Garland, T. (1983a) Scaling the ecological cost of transport to body-mass in terrestrial mammals. *American Naturalist* 121, 571–587.

Garland, T. (1983b) The relation between maximal running speed and body-mass in terrestrial mammals. *Journal of Zoology* 199, 157–170.

Gaston, K. (1988) The intrinsic rates of increase of insects of different sizes. *Ecological Entomology* 13, 399–409.

Gaston, K. J. (1991) The magnitude of global insect species richness. *Conservation Biology* 5, 283–296.

Gaston, K. J. (2000): Global patterns in biodiversity. *Nature* 405, 220–227.

Gaston, K. J. and Blackburn, T. M. (1996) Conservation implications of geographic range size–body size relationships. *Conservation Biology* 10, 638–646.

Gaston, K. J. and Blackburn, T. M. (2000) *Pattern and Process in Macroecology*. Oxford: Blackwell Science.

Gaston, K. J., Blackburn, T. M., Greenwood, J. J. D., Gregory, R. D., Quinn, R. M., and Lawton, J. H. (2000) Abundance–occupancy relationships. *Journal of Applied Ecology* 37, 39–59.

Gause, G. F. (1931) The influence of ecological factors on the size of populations. *American Naturalist* 65, 70–76.

Gause, G. F. (1932) Ecology of populations. *Quarterly Review of Biology* VII, 27–46.

Geider, R. J., Platt, T., and Raven, J. A. (1986) Size dependence of growth and photosynthesis in diatoms: a synthesis. *Marine Ecology: Progress Series* 30, 93–104.

George, J. C., Bada, J., Zeh, J. E., et al. (1999) Age and growth estimates of bowhead whales (Balaena mysticetus) via aspartic acid racemization. *Canadian Journal of Zoology* 77, 571–580.

George-Nascimento, M., Munoz, G., Marquet, P. A., and Poulin, R. (2004) Testing the energetic equivalence rule with helminth endoparasites of vertebrates. *Ecology Letters* 7, 527–531.

Georges, J. Y., Groscolas, R., Guinet, C., and Robin, J. P. (2001) Milking strategy in subantarctic fur seals *Arctocephalus tropicalis* breeding on Amsterdam Island: evidence from changes in milk Composition. *Physiological and Biochemical Zoology* 74, 548–559.

Gergs, A. and Ratte, H. T. (2009) Predicting functional response and size selectivity of juvenile *Notonecta maculata* foraging on *Daphnia magna*. *Ecological Modelling* 220, 3331–3341.

Geritz, S. A. H., Kisdi, E., Meszena, G., and Metz, J. A. J. (1998) Evolutionary singular strategies and the adaptive growth and branching of the evolutionary tree. *Evolutionary Ecology* 12, 35–57.

Gerstner, G. E. and Gerstein, J. B. (2008) Chewing rate allometry among mammals. *Journal of Mammalogy* 89, 1020–1030.

Ghalambor, C. K., Huey, R. B., and Martin, P. R. (2004) Why mountain passes are higher in the tropics – revisited. *Integrative and Comparative Biology* 44, 558–558.

Ghannoum, M. A. and O'Toole, G. A. (2004) *Microbial Biofilms*. Washington, DC: American Society for Microbiology.

Gholz, H. L., Wedin, D. A., Smitherman, S. M., Harmon, M. E., and Parton, W. J. (2000) Long-term dynamics of pine and hardwood litter in contrasting environments: toward a global model of decomposition. *Global Change Biology* 6, 751–765.

Gifford, R. M. (2003) Plant respiration in productivity models: conceptualisation, representation and issues for global terrestrial carbon-cycle research. *Functional Plant Biology* 30, 171–186.

Gilkinson, K., Paulin, M., Hurley, S., and Schwinghamer, P. (1998) Impacts of trawl door scouring on infaunal bivalves: results of a physical trawl door model/dense sand interaction. *Journal of Experimantal Marine Biology and Ecology* 224, 291–312.

Gillooly, J. F. and Allen, A. P. (2007) Linking global patterns in biodiversity to evolutionary dynamics using metabolic theory. *Ecology* 88, 1890–1894.

Gillooly, J. F. and Ophir, A. G. (2010) The energetic basis of acoustic communication. *Proceedings of the Royal Society B: Biological Sciences* 277, 1325–1331.

Gillooly, J. F., Brown, J. H., West, G. B., Savage, V. M., and Charnov, E. L. (2001) Effects of size and temperature on metabolic rate. *Science* 293, 2248–2251.

Gillooly, J. F., Charnov, E. L., West, G. B., Savage, V. M., and Brown, J. H. (2002) Effects of size and temperature on developmental time. *Nature* 417, 70–73.

Gillooly, J. F., Allen, A. P., Brown, J. H., et al. (2005a) The metabolic basis of whole-organism RNA and phosphorus content. *Proceedings of the National Academy of Sciences of the USA* 102, 11923–11927.

Gillooly, J. F., Allen, A. P., West, G. B.,and Brown, J. H. (2005b) The rate of DNA evolution: Effect of body size and temperature on the molecular clock. *Proceedings of the National Academy of Sciences of the USA* 102, 140–145.

Gillooly, J. F., Allen, A. P., Savage, V. M., Charnov, E. L., West, G. B., and Brown, J. H. (2006a) Response to Clarke and Fraser: effects of temperature on metabolic rate. *Functional Ecology* 20, 400–404.

Gillooly, J. F., Allen, A. P., and Charnov, E. L. (2006b) Dinosaur fossils predict body temperatures. *PLOS Biology* 4, 1467–1469.

Gillooly, J. F., Hou, C., and Kaspari, M. (2010) Eusocial insects as superorganisms: Insights from metabolic theory. *Communicative and Integrative Biology* 3, 360–362.

Gilman, S. E., Wethey, D. S., and Helmuth, B. (2006) Variation in the sensitivity of organismal body temperature to climate change over local and geographic scales. *Proceedings of the National Academy of Sciences of the USA* 103, 9560–9565.

Gilpin, M. E., and Soulé, M. E. (1986) Minimum viable populations: processes of species extinction. In: Soulé, M. E. (Ed.), *Conservation Biology*. Sunderland, MA: Sinauer, pp. 19–34.

Gingerich, P. D. (1993) Quantification and comparison of evolutionary rates. *American Journal of Science* 293, 453–478.

Gingerich, P. D. (2009) Rates of evolution. *Annual Review of Ecology, Evolution, and Systematics* 40, 657–675.

Gingerich, P. D. (2000) Arithmeic or geometric normality of biological variation: an empirical test of theory. *Journal of Theoretical Biology* 204, 201–221.

Ginzburg, L. and Damuth, J. (2008) The space-lifetime hypothesis: viewing organisms in four dimensions, literally. *American Naturalist* 171, 125–131.

Giraud, T., Pedersen, J., and Keller, L. (2002) Evolution of supercolonies: the Argentine ants of southern Europe. *Proceedings of the National Academy of Sciences of the USA* 99, 6075.

Gislason, H., Daan, N., Rice, J. C., and Pope, J. G. (2010) Size, growth, temperature and the natural mortality of marine fish. *Fish and Fisheries* 11, 149–158.

Gittleman, J. L. (1986) Carnivore life-history patterns – allometric, phylogenetic, and ecological associations. *American Naturalist* 127, 744–771.

Gittleman, J. L., Anderson, C. G., Kot, M., and Luh, H. K. (1996) Phylogenetic lability and rates of evolution: a comparison of behavioral, morphological and life history traits. In: Martins, E. P. (Ed.), *Phylogenies and the Comparative Method in Animal Behavior*. New York; Oxford: Oxford University Press, pp. 166–205.

Glazier, D. S. (2005) Beyond the "3/4-power law": variation in the intra- and interspecific scaling of metabolic rate in animals. *Biological Reviews* 80, 611–662.

Glazier, D. S. (2006) The 3/4-power law is not universal: evolution of isometric, ontogenetic metabolic scaling in pelagic animals. *Bioscience* 56, 325–332.

Glazier, D. S. (2008) Effects of metabolic level on the body size scaling of metabolic rate in birds and mammals. *Proceedings of the Royal Society B: Biological Sciences* 275, 1405–1410.

Glazier, D. S. (2009a) Metabolic level and size scaling of rates of respiration and growth in unicellular organisms. *Functional Ecology* 23, 963–968.

Glazier, D. S. (2009b) Activity affects intraspecific body-size scaling of metabolic rate in ectothermic animals. *Journal of Comparative Physiology B: Biochemical Systemic and Environmental Physiology* 179, 821–828.

Glazier, D. S. (2010) A unifying explanation for diverse metabolic scaling in animals and plants. *Biological Reviews* 85, 111–138.

Gnaiger, E. and Kuznetsov, A. V. (2002) Mitochondrial respiration at low levels of oxygen and cytochrome c. *Biochemical Society Transactions* 30, 252–258.

Goldman, J. C. and Carpenter, E. J. (1974) A kinetic approach to the effect of temperature on algal growth. *Limnology and Oceanography* 19, 756–766.

Goldsworthy, G. and Coupland, A. (1974) The influence of the corpora cardiaca and substrate availability on flight speed and wing beat frequency inLocusta. *Journal of Comparative Physiology A: Neuroethology, Sensory, Neural, and Behavioral Physiology* 89, 359–368.

Gooding, R. A., Harley, C. D. G., and Tang, E. (2009) Elevated water temperature and carbon dioxide concentration increase the growth of a keystone echinoderm. *Proceedings of the National Academy of Sciences of the USA* 106, 9316–9321.

Goodwin, N. B., Grant, A., Perry, A. L., Dulvy, N. K., and Reynolds, J. D. (2006) Life history correlates of density-dependent recruitment in marine fishes. *Canadian Journal of Fisheries and Aquatic Sciences* 63, 494–509.

Gould, S. J. (1974) The evolutionary significance of "bizarre" structures: antler size and skull size in the Irish Elk. *Evolution* 28, 191–220.

Gould, S. J. (1983) *Ontogeny and Phylogeny*. Cambridge, MA: Harvard University Press.

Gouveia, S. M., Simpson, S., Raubenheimer, D., and Zanotto, F. P. (2000) Patterns of respiration in *Locusta migratoria* nymphs when feeding. *Physiological Entomology* 25, 88–93.

Greenlee, K. and Harrison, J. (2004a) Development of respiratory function in the American locust *Schistocerca americana* I. Across-instar effects. *Journal of Experimental Biology* 207, 497–508.

Greenlee, K. and Harrison, J. (2004b) Development of respiratory function in the American locust *Schistocerca americana* II. Within-instar effects. *Journal of Experimental Biology* 207, 509–517.

Greenlee, K. J. and Harrison, J. F. (2005) Respiratory changes throughout ontogeny in the tobacco hornworm caterpillar, *Manduca sexta*. *Journal of Experimental Biology* 208, 1385–1392.

Greenlee, K., Nebeker, C., and Harrison, J. (2007) Body size-independent safety margins for gas exchange across grasshopper species. *Journal of Experimental Biology* 210, 1288.

Greenlee, K. J., Henry, J. R., Kirkton, S. D., et al. (2009) Synchrotron imaging of the grasshopper tracheal system: morphological and physiological components of tracheal hypermetry. *American Journal of Physiology: Regulatory, Integrative and Comparative Physiology* 297, R1343.

Griffin, K. L., Anderson, O. R., Gastrich, M. D., et al. (2001) Plant growth in elevated CO_2 alters mitochondrial number and chloroplast fine structure. *Proceedings of the National Academy of Sciences of the USA* 98, 2473–2478.

Grimaldi, D. and Engle, M. S. (2005) *Evolution of the Insects*. New York: Cambridge University Press.

Groenendijk, M., Van Der Molen, M. K., and Dolman, A. J. (2009) Seasonal variation in ecosystem parameters derived from FLUXNET data. *Biogeosciences Discussions* 6, 2863–2912.

Grover, J. P. (1991) Resource competition in a variable environment: phytoplankton growing according to the variable-internal-stores model. *American Naturalist* 138, 811–835.

Grutter, A. S., Crean, A. J., Curtis, L. M., Kuris, A. M., Warner, R. R., and McCormick, M. I. (2011) Indirect effects of an ectoparasite reduce successful establishment of a damselfish at settlement. *Functional Ecology* 25, 586–594.

Guill, C. and Drossel, B. (2008) Emergence of complexity in evolving niche-model food webs. *Journal of Theoretical Biology* 251, 108–120.

Guillou, L., Chrétiennot-Dinet, M. J., Boulben, S., Moon-Van Der Staay, S. Y., and Vaulot, D. (1999) *Symbiomonas scintillans* gen. et sp. nov. and *Picophagus flagellatus* gen. et sp. nov. (Heterokonta): two new heterotrophic flagellates of picoplanktonic size. *Protist* 150, 383–398.

Gunther, B. and Leon De La Barra, B. (1966) Physiometry of the mammalian circulatory system. *Acta Physiologica Latino Americana* 16, 32–42.

Guppy, M. and Withers, P. (1999) Metabolic depression in animals: physiological perspectives and biochemical generalizations. *Biological Reviews* 74, 1–40.

Habeck, C. W. and Meehan, T. D. (2008) Mass invariance of population nitrogen flux by terrestrial mammalian herbivores: an extension of the energetic equivalence rule. *Ecology Letters* 11, 898–903.

Haberl, H., Erb, K.-H., Krausmann, F., et al. (2007) Quantifying and mapping the human appropriation of net primary production in Earth's terrestrial ecosystems. *Proceedings of the National Academy of Sciences of the USA* 104, 12942–12947.

Hacke, U. G., Sperry, J. S., Pockman, W. T., Davis, S. D., and McCulloh, K. A. (2001) Trends in wood density and structure are linked to prevention of xylem implosion by negative pressure. *Oecologia* 126, 457–461.

Hahn, D. A. and Denlinger, D. L. (2010) Energetics of insect diapause. *Annual Review of Entomology* 56, 103–121.

Hajagos, J. and Ferson, S. (2001). *RAMAS® Ecorisk: Software for Rapid Ecological Risk Analysis*. Setauket, New York: Applied Biomathematics.

Haldane, J. B. S. (1949) Suggestions as to quantitative measurement of rates of evolution. *Evolution* 3, 51–56.

Hall, C., Lindenberger, D., Kümmel, R., Kroeger, T., and Eichhorn, W. (2001) The need to reintegrate the natural sciences with economics. *BioScience* 51, 663–673.

Hall, J. R., Mitchell, K. R., Jackson-Weaver, O., et al. (2008) Molecular Characterization of the diversity and distribution of a thermal spring microbial community using rRNA and metabolic genes. *Applied and Environmental Microbiology* 74, 4910–4922.

Hall, S. J., Collie, J. S., Duplisea, D. E., Jennings, S., Bravington, M., and Link, J. (2006) A length-based multi-species model for evaluating community responses to fishing. *Canadian Journal of Fisheries and Aquatic Sciences* 63, 1344–1359.

Hallberg, K. B. and Lindstrom, E. B. (1994) Characterization of *Thiobacillus caldus* sp. nov., a moderately thermophilic acidophile. *Microbiology* 140, 3451–3456.

Hallé, F., Oldeman, R. A. A., and Tomlinson, P. B. (1978) *Tropical Trees and Forests: An Architectural Analysis*. New York: Springer-Verlag.

Halpin, P.N. (1997) Global climate change and natural-area protection: management responses and research directions. *Ecological Applications* 7, 828–843.

Halsey, L. G., Blackburn, T. M., and Butler, P. J. (2006a) A comparative analysis of the diving behaviour of birds and mammals. *Functional Ecology* 20, 889–899.

Halsey, L. G., Butler, P. J., and Blackburn, T. M. (2006b) A phylogenetic analysis of the allometry of diving. *American Naturalist* 167, 276–287.

Hamilton, M. J., Milne, B. T., Walker, R. S., and Brown, J. H. (2007a) Nonlinear scaling of space use in human hunter-gatherers. *Proceedings of the National Academy of Sciences of the USA* 104, 4765–4769.

Hamilton, M. J., Milne, B. T., Walker, R. S., Burger, O., and Brown, J. H. (2007b) The complex structure of hunter–gatherer social networks. *Proceedings of the Royal Society B: Biological Sciences* 274, 2195–2202.

Hamilton, M. J., Davidson, A. D., Sibly, R.M., and Brown, J. H. (2011) Universal scaling of production rates across mammalian lineages. *Proceedings of the Royal Society B: Biological Sciences* 278, 560–566.

Hamilton, W. D. (1964) The genetical evolution of social behaviour. I. *Journal of Theoretical Biology* 7, 1–16.

Hammes, G. G. (2007) *Physical Chemistry for the Biological Sciences*. Hoboken, NJ: Wiley-Interscience.

Hansen, T. and Martins, E. (1996) Translating between microevolutionary process and macroevolutionary patterns: the correlation structure of interspecific data. *Evolution* 50, 1404–1417.

Harley, C. D. G. (2003) Abiotic stress and herbivory interact to set range limits across a two-dimensional stress gradient. *Ecology* 84, 1477–1488.

Harmon, L. J., Melville, J., Larson, A., and Losos, J. B. (2008) The role of geography and ecological opportunity in the diversification of day geckos (Phelsuma). *Systematic Biology* 57, 562–573.

Harmon, J. P., Moran, N. A., and Ives, A. R. (2009) Species response to environmental change: impacts of food web interactions and evolution. *Science* 323, 1347–1350.

Harper, J. L. (1977) *The Population Biology of Plants*. London: Academic Press.

Harris, G. and Pimm, S. L. (2008) Range size and extinction risk in forest birds. *Conservation Biology* 22, 163–171.

Harrison, J. F. (1997) Ventilatory mechanism and control in grasshoppers. *American Zoologist* 37, 73–81.

Harrison, J. F. and Fewell, J. H. (2002) Environmental and genetic influences on flight metabolic rate in the honey bee, *Apis mellifera*. *Comparative Biochemistry and Physiology* 133, 323–333.

Harrison, J. F. and Lighton, J. R. B. (1998) Oxygen-sensitive flight metabolism in the dragonfly *Erythemis simplicicollis*. *Journal of Experimental Biology* 201, 1739–1744.

Harrison, J. F. and Roberts, S. P. (2000) Flight respiration and energetics. *Annual Review of Physiology* 62, 179–205.

Harrison, J. F., Lafreniere, J. J., and Greenlee, K. J. (2005) Ontogeny of tracheal dimensions and gas exchange capacities in the grasshopper, *Schistocerca americana*. *Comparative Biochemistry and Physiology Part A: Molecular and Integrative Physiology* 141, 372–380.

Harrison, J. F., Fewell, J. H., Anderson, K. E., and Loper, G. M. (2006a) Environmental physiology of the invasion of the Americas by Africanized honeybees. *Integrative and Comparative Biology* 46, 1110–1122.

Harrison, J. F., Frazier, M. R., Henry, J. R., Kaiser, A., Klok, C. J., and Rascón, B. (2006b) Responses of terrestrial insects to hypoxia or hyperoxia. *Respiratory Physiology and Neurobiology* 2006, 4–17.

Harrison, J. F., Kaiser, A., and Vandenbrooks, J. M. (2010) Atmospheric oxygen level and the evolution of insect body size. *Proceedings of the Royal Society B: Biological Sciences* 277, 1937.

Harrison, P. M. and Arosio, P. (1996) The ferritins: molecular properties, iron storage function and cellular regulation. *Biochimica et Biophysica Acta (BBA)–Bioenergetics* 1275, 161–203.

Harshman, L. G., Hoffmann, A. A., and Clark, A. G. (1999) Selection for starvation resistance in *Drosophila melanogaster*?: physiological correlates, enzyme activities and multiple stress responses. *Journal of Evolutionary Biology* 12, 370–379.

Hart, M. W. (1995) What are the costs of small egg size for a marine invertebrate with feeding planktonic larvae? *American Naturalist* 146, 415–426.

Harte, J. (2002) Towards a synthesis of the Newtonian and Darwinian world views. *Physics Today* 55, 29–37.

Harte, J. (2004) The value of null theories in ecology. *Ecology* 85, 1792–1794.

Harte, J., Zillio, T., Conlisk, E., and Smith, A. B. (2008) Maximum entropy and the state-variable approach to macroecology. *Ecology* 89, 2700–2711.

Harte, J., Smith, A. B., and Storch, D. (2009) Biodiversity scales from plots to biomes with a universal species–area curve. *Ecology Letters* 12, 789–797.

Hartvig, M., Andersen, K. H., and Beyer, J. E. (2011) Food web framework for size-structured populations. *Journal of Theoretical Biology* 272, 113–122.

Harvey, P. H. and Pagel, M. (1991) *The Comparative Method in Evolutionary Biology*. Oxford: Oxford University Press.

Haskell, J. P., Ritchie, M. E., and Olff, H. (2002) Fractal geometry predicts varying body size scaling relationships for mammal and bird home ranges. *Nature* 418, 527–530.

Hassell, M. P., Lawton, J. H., and Beddington, J. R. (1976) The components of arthropod predation. I. The prey death-rate. *Journal of Animal Ecology* 45, 135–164.

Hassler, C.S., Schoemann, V., Nichols, C. M., Butler, E. C., and Boyd, P. W. (2011) Saccharides enhance iron bioavailability to Southern Ocean phytoplankton. *Proceedings of the National Academy of Sciences of the USA* 108, 1076–1081.

Hasson, O. (1997) Towards a general theory of biological signaling. *Journal of Theoretical Biology* 185, 139–156.

Havenhand, J. N. (1993) Egg to juvenile period, generation time, and the evolution of larval type in marine-invertebrates. *Marine Ecology: Progress Series* 97, 247–260.

Hawkes, K., O'Connell, J. F., Jones, N. G. B., Alvarez, H., and Charnov, E. L. (1998) Grandmothering, menopause, and the evolution of human life-histories. *Proceedings of the National Academy of Sciences of the USA* 95, 1336–1339.

Hawkins, B. A. (2001) Ecology's oldest pattern? *Trends in Ecology and Evolution* 16, 470.

Hawkins, B. A. (2010) Multiregional comparison of the ecological and phylogenetic structure of butterfly species richness gradients. *Journal of Biogeography* 37, 647–656.

Hawkins, B. A., Field, R., Cornell, H. V., et al. (2003) Energy, water, and broad-scale geographic patterns of species richness. *Ecology* 84, 3105–3117.

Hawkins, B. A., Albuquerque, F. S., Araújo, M. B., et al. (2007) A global evaluation of metabolic theory as an explanation for terrestrial species richness gradients. *Ecology* 88, 1877–1888.

Hayssen, V. and Lacy, R. C. (1985) Basal metabolic rates in mammals – taxonomic differences in the allometry of BMR and body-mass. *Comparative Biochemistry and Physiology A: Physiology* 81, 741–754.

Hayward, A. and Gillooly, J. F. (2011) The cost of sex: quantifying energetic investment in gamete production by males and females. *PLoS ONE* 6, e16557.

Hayward, A., Khalid, M., and Kolasa, J. (2009) Population energy use scales positively with body size in natural aquatic microcosms. *Global Ecology and Biogeography* 18, 553–562.

Hayward, A., Kolasa, J., and Stone, J. R. (2010) The scale-dependence of population density-body mass allometry: statistical artefact or biological mechanism? *Ecological Complexity* 7, 115–124.

Hechinger, R. F., Lafferty, K. D., Mancini III, F. T., Warner, R. R., and Kuris, A. M. (2009) How large is the hand in the puppet? Ecological and evolutionary factors affecting body mass of 15 trematode parasitic castrators in their snail host. *Evolutionary Ecology* 23, 651–667.

Hechinger, R. F., Lafferty, K. D., Dobson, A. P., Brown, J. H., and Kuris, A. M. (2011) A common scaling rule for the abundance and energetics of parasitic and free-living species. *Science* 333, 445–448.

Heckmann, K., Hagen, R. T., and Görtz, H. D. (1983) Freshwater *Euplotes* species with a 9 type 1 cirrus pattern depend upon endosymbionts. *Journal of Eukaryotic Microbiology* 30, 284–289.

Heimann, M. and Reichstein, M. (2008) Terrestrial ecosystem carbon dynamics and climate feedbacks. *Nature* 451, 289.

Heinrich, B. (1980) Mechanisms of body temperature regulation in honeybees, *Apis mellifera* II. Regulation of thoracic temperature at high air temperatures. *Journal of Experimental Biology* 85, 73–87.

Heinrich, B. (1981) Energetics of honeybee swarm thermoregulation. *Science* 212, 565–566.

Heinrich, B. (1992) *The Hot-Blooded Insects*. Cambridge, MA: Harvard University Press.

Heinze, J. and Schrempf, A. (2008) Aging and reproduction in social insects: a mini-review. *Gerontology* 54, 160–167.

Hemmingsen, A. M. (1950) The relation of standard (basal) energy metabolism to total fresh weight of living organisms. *Reports of the Steno Memorial Hospital and the Nordisk Insulinlaboratorium* 4, 7–51.

Hemmingsen, A. M. (1960) Energy metabolism as related to body size and respiratory surfaces, and its evolution. *Reports of the Steno Memorial Hospital Copenhagen* 9, 1–110.

Hempleman, S. C., Kilgore, D. L. Jr, Colby, C., Bavis, R. W., and Powell, F. L. (2005) Spike firing allometry in avian intrapulmonary chemoreceptors: matching neural code to body size. *Journal of Experimental Biology* 208, 3065–3073.

Henderiks, J. and Pagani, M. (2008) Coccolithophore cell size and the Paleogene decline in atmospheric CO_2. *Earth and Planetary Science Letters* 269, 576–584.

Henderson, P. A. and Magurran, A. E. (2010) Linking species abundance distributions in numerical abundance and biomass through simple assumptions about community structure. *Proceedings of the Royal Society B: Biological Sciences* 277, 1561–1570.

Hendriks, A. J. (2007) The power of size: a meta-analysis reveals consistency of allometric regressions. *Ecological Modelling* 205, 196–208.

Hennemann, W. W. (1983) Relationship among body mass, metabolic rate and the intrinsic rate of natural increase in mammals. *Oecologia* 56, 104–108.

Herreid, C. F., Full, R. J., and Prawel, D. A. (1981) Energetics of cockroach locomotion. *Journal of Experimental Biology* 94, 189–202.

Hetz, S. K. and Bradley, T. J. (2005) Insects breathe discontinuously to avoid oxygen toxicity. *Nature* 433, 516–519.

Heusner, A. A. (1982) Energy-metabolism and body size. 1. Is the 0.75 mass exponent of Kleiber's equation a statistical artifact? *Respiration Physiology* 48, 1–12.

Heusner, A. A. (1983a) Body size, energy-metabolism, and the lungs. *Journal of Applied Physiology* 54, 867–873.

Heusner, A. A. (1983b) Mathematical expression of the effects of changes in body size on pulmonary-function and structure. *American Review of Respiratory Disease* 128, S72–S74.

Heusner, A. A. (1991) Size and power in mammals. *Journal of Experimental Biology* 160, 25–54.

Hiddink, J. G., Jennings, S., Kaiser, M. J., Queirós, A. M., Duplisea, D. E., and Piet, G. J. (2006) Cumulative impacts of seabed trawl disturbance on benthic biomass, production and species richness in different habitats. *Canadian Journal of Fisheries and Aquatic Science*, 63, 721–736.

Hilborn, R. and Mangel, M. (1997) *The Ecological Detective: Confronting Models with Data*. Princeton, NJ: Princeton University Press.

Hillebrand, H., Borer, E., Bracken, M., et al. 2009. Herbivore metabolism and stoichiometry each constrain herbivory at different organizational scales across ecosystems. *Ecology Letters* 12, 516–527.

Hillenius, W. J. and Ruben, J. A. (2004) The evolution of endothermy in terrestrial vertebrates: Who? when? why? *Physiological and Biochemical Zoology* 77, 1019–1042.

Hilton-Taylor, C., Pollock, C. M., Chanson, J. S., Butchart, S. H. M., Oldfield, T. E. E., and Katariya, V. (2009) State of the world's species. In: Vié, J.-C., Hilton-Taylor, C., and Stuart, S. N. (Eds) *Wildlife in a Changing World: An analysis of the 2008 IUCN Red List of Threatened Species*. Gland, Switzerland: IUCN, pp. 15–42.

Hirst, A. and Lopez-Urrutia, A. (2006) Effects of evolution on egg development time. *Marine Ecology: Progress Series* 326, 29–35.

Ho, R. (2003) On-chip wires: scaling and efficiency. PhD thesis, Stanford University.

Hochachka, P. W. and Somero, G. N. (2002) *Biochemical Adaptation: Mechanism and Process in Physiological Evolution*. Oxford: Oxford University Press.

Hodges, C. M. and Wolf, L. L. (1981) Optimal foraging in bumblebees: Why is nectar left behind in flowers? *Behavioral Ecology and Sociobiology* 9, 41–44.

Hoehler, T.M. (2004) Biological energy requirements as quantitative boundary conditions for life in the subsurface. *Geobiology* 2, 205–215.

Hoehler, T.M. (2007) An energy balance concept for habitability. *Astrobiology* 7, 824–838.

Hoehler, T. M., Amend, J. P., and Shock, E. L. (2007) A "follow the energy" approach for astrobiology. *Astrobiology* 7, 819–823.

Hoekstra, J. M., Boucher, T. M., Ricketts, T. H., and Roberts, C. (2005) Confronting a biome crisis: global disparities of habitat loss and protection. *Ecology Letters* 8, 23–29.

Hölldobler, B. and Wilson, E. O. (1990) *The Ants*. Cambridge, MA: Belknap Press.

Hölldobler, B. and Wilson, E. O. (2008) *The Superorganism: The Beauty, Elegance, and Strangeness of Insect Societies*. New York: W. W. Norton & Co.

Holling, C. S. (1959) Some characteristics of simple types of predation and parasitism. *Canadian Entomologist* 91, 385.

Holsinger, K. E. and Roughgarden, J. (1985) A model for the dynamics of an annual plant population. *Theoretical Population Biology* 28, 288–313.

Holt, R. D. (1977) Predation, apparent competition, and the structure of prey communities. *Theoretical Population Biology* 12, 197–229.

Horn H. (1971) *The Adaptive Geometry of Trees*. Princeton, NJ: Princeton University Press.

Horn, H. S. (2000) Twigs, trees, and the dynamics of carbon in the landscape. In: Brown, J. H. and West, G. B. (Eds.), *Scaling in Biology*. Oxford: Oxford University Press, pp. 199–210.

Horner, J., Gosz, J., and Cates, R. (1988) The role of carbon-based plant secondary metabolites in decomposition in terrestrial ecosystems. *American Naturalist* 132, 869–883.

Hortal, J., Rodriguez, J., Nieto-Diaz, M., and Lobo, J. M. (2008) Regional and environmental effects on the species richness of mammal assemblages. *Journal of Biogeography* 35, 1202–1214.

Hou, C., Zuo, W., Moses, M. E., Woodruff, W. H., Brown, J. H., and West, G. B. (2008) Energy uptake and allocation during ontogeny. *Science* 322, 736–739.

Hou, C., Kaspari, M., Vander Zanden, H. B., and Gillooly, J. F. (2010) Energetic basis of colonial living in social insects. *Proceedings of the National Academy of Sciences of the USA* 107, 3634–3638.

Houlton, B. Z., Wang, Y.-P., Vitousek, P. M., and Field, C. B. (2008) A unifying framework for dinitrogen fixation in the terrestrial biosphere. *Nature* 454, 327–330.

Houston, A. I., Stephens, P. A., Boyd, I. L., Harding, K. C., and McNamara, J. M. (2007) Capital or income breeding? A theoretical model of female reproductive strategies. *Behavioral Ecology* 18, 241–250.

Hubbell, S. P. (2001) *The Unified Theory of Biodiversity and Biogeography*. Princeton, NJ: Princeton University Press.

Hubbell, S. P. and Foster, R. B. (1990) Structure, dynamics, and equilibrium status of old-growth forest on Barro Colorado Island. In: Gentry, A. H. (Ed.), *Four Neotropical Rainforests*. New Haven, CT: Yale University Press, pp. 522–541.

Hubbert, M. K. (1981) The world's evolving energy system. *American Journal of Physics* 49, 1007–1029.

Huber, B. (1932) Observation and measurement of plant sap streams. *Berichte Deutsche Botanische Gesellschaft* 50, 89–109.

Huber, B. and Schmidt, E. (1936) Weitere thermo-elektrische untersuchungen uber den transpirationsstrom der baume. *Tharandt Forst Jb* 87, 369–412.

Hudson, P. J., Dobson, A. P., and Newborn, D. (1998) Prevention of population cycles by parasite removal. *Science* 282, 2256–2258.

Huey, R. B. (1982) Temperature, physiology, and the ecology of reptiles. *Biology of the Reptilia* 12, 25–91.

Huey, R. B. and Bennett, A. F. (1987) Phylogenetic studies of coadaptation: preferred temperatures versus optimal performance temperatures of lizards. *Evolution* 41, 1098–1115.

Huey, R. B. and Berrigan, D. (2001) Temperature, demography, and ectotherm fitness. *American Naturalist* 158, 204–210.

Huey, R. B. and Pianka, E. R. (1981) Ecological Consequences of foraging mode. *Ecology* 62, 991–999.

Huey, R. and Slatkin, M. (1976) Cost and benefits of lizard thermoregulation. *Quarterly Review of Biology* 51, 363–384.

Huey, R. B. and Stevenson, R. (1979) Integrating thermal physiology and ecology of ectotherms: a discussion of approaches. *Integrative and Comparative Biology* 19, 357.

Huey, R. B., Gilchrist, G. W., Carlson, M. L., Berrigan, D., and Serra, Â. L. Ì. S. (2000) Rapid evolution of a geographic cline in size in an introduced fly. *Science* 287, 308–309.

Huey, R. B., Losos, J. B., and Moritz, C. (2010). Are lizards toast? *Science* 328, 832.

Hughes, D. T. and Sperandio, V. (2008) Inter-kingdom signalling: communication between bacteria and their hosts. *Nature Reviews Microbiology* 6, 111–120.

Hughes, L. (2000) Biological consequences of global warming: is the signal already apparent? *Trends in Ecology and Evolution* 15, 56–61.

Hulbert, A. J. and Else, P. L. (1990) The cellular basis of endothermic metabolism – a role for leaky membranes. *News in Physiological Sciences* 5, 25–28.

Hulbert, A. J. and Else, P. L. (2000) Mechanisms underlying the cost of living in animals. *Annual Review of Physiology* 62, 207–235.

Hulbert, A. and Else, P. (2004) Basal metabolic rate: history, composition, regulation, and usefulness. *Physiological and Biochemical Zoology* 77, 869–876.

Hulbert, A. J., Else, P. L., Manolis, S. C., and Brand, M. D. (2002) Proton leak in hepatocytes and liver mitochondria from archosaurs (crocodiles) and allometric relationships for ectotherms. *Journal of Comparative Physiology* 172, 387–397.

Humboldt, A. (1850) *Aspects of nature in different lands and different climates, with scientific elucidations.* Philadelphia: Lea & Blanchard, translated by Mrs. Sabine.

Hume, I. D., Morgan, K. R., and Kenagy, G. J. (1993) Digesta retention and digestive performance in sciurid and microtine rodents: effects of hindgut morphology and body size. *Physiological Zoology* 66, 396–411.

Hummel, J., Sudekum, K. H., Streich, W. J., and Clauss, M. (2006) Forage fermentation patterns and their implications for herbivore ingesta retention times. *Functional Ecology* 20, 989–1002.

Humphreys, W. F. (1979) Production and respiration in animal populations. *Journal of Animal Ecology* 48, 427–453.

Hungate, B., Stiling, P., Dijkstra, P., et al. (2004) CO_2 elicits long-term decline in nitrogen fixation. *Science* 304, 1291.

Hunt, G. and Roy, K. (2006) Climate change, body size evolution, and Cope's Rule in deep-sea ostracodes. *Proceedings of the National Academy Of Sciences of the USA* 103, 1347–1352.

Hunt, G., Cronin, T. M., and Roy, K. (2005) Species-energy relationship in the deep sea: a test using the Quaternary fossil record. *Ecology Letters* 8, 1218–1234.

Hunt, R. (1978) *Plant Growth Analysis.* London: Edward Arnold.

Hurlbert, A. H. (2004) Species-energy relationships and habitat complexity in bird communities. *Ecology Letters* 7, 714–720.

Hurlbert, S. H. and Lombardi, C. M. (2009) Final collapse of the Neyman–Pearson decision theoretic framework and rise of the neoFisherian. *Annales Zoologici Fennici* 46, 311–349.

Hurlbert, A. H. and White, E. P. (2007) Ecological correlates of geographical range occupancy in North American birds. *Global Ecology and Biogeography* 16, 764–773.

Huxley J.S. (1932) *Problems of Relative Growth.* London: Methuen.

Huxley, J. S. and Tessier, G. (1936) Terminology in relative growth. *Nature* 137, 780–781.

Huxman, T. E., Smith, M. D., Fay, P. A., et al. (2004a) Convergence across biomes to a common rain-use efficiency. *Nature* 429, 651–654.

Huxman, T. E., Snyder, K. A., Tissue, D., et al. (2004b) Precipitation pulses and carbon fluxes in semiarid and arid ecosystems. *Oecologia* 141, 254–268.

Ibelings, B. W., Gsell, A. S., Mooij, W. M., Van Donk, E., Van Den Wyngaert, S., and De Senerpont Domis, L. N. (2011) Chytrid infections and diatom spring blooms: paradoxical effects of climate warming on fungal epidemics in lakes. *Freshwater Biology* 56, 754–766.

Iglesias-Rodriguez, M. D., Halloran, P. R., Rickaby, et al. (2008) Phytoplankton calcification in a high-CO_2 world. *Science* 320, 336–340.

Ingledew, W. J. (1990) Acidophiles. In: Edwards, C. (Ed.), *Microbiology of Extreme Environments.* Milton Keynes, UK: Open University Press, pp. 33–53.

Inglima, I., Alberti, G., Bertolini, T., et al. (2009) Precipitation pulses enhance respiration of Mediterranean ecosystems: the balance between organic and inorganic components of increased soil CO_2 efflux. *Global Change Biology* 15, 1289–1301.

Ingraham, J. L. (1958) Growth of psychrophilic bacteria. *Journal of Bacteriology* 76, 75–80.

Ings, T. C., Montoya, J. M., Bascompte, J., et al. (2009) Ecological networks – beyond food webs. *Journal of Animal Ecology* 78, 253–269.

Inskeep, W. P., Ackerman, G. G., Taylor, W. P., Kozubal, M., Korf, S., and Macur, R. E. (2005) On the energetics of chemolithotrophy in nonequilibrium systems: case studies of geothermal springs in Yellowstone National Park. *Geobiology* 3, 297–317.

Intergovernmental Panel on Climate Change (IPCC) (2007a) Summary for Policymakers. In: Solomon, S., Qin, D., Manning, M., et al. (Eds), *Climate Change 2007: The Physical Science Basis.* Contribution of Working Group I to the Fourth Assessment Report of the Intergovernmental Panel on Climate Change. Cambridge, UK; New York: Cambridge University Press.

Intergovernmental Panel on Climate (IPCC) (2007b) *Climate Change 2007: The Physical Science Basis.* Contribution of Working Group I to the Fourth Assessment Report of the Intergovernmental Panel on Climate Change. Cambridge, UK; New York: Cambridge University Press.

Irigoien, X., Huisman, J., and Harris, R. P. (2004) Global biodiversity patterns of marine phytoplankton and zooplankton. *Nature* 429, 863–867.

Irlich, U. M., Terblanche, J. S., Blackburn, T. M., and Chown, S. L. (2009) Insect rate–temperature relationships: environmental variation and the metabolic theory of ecology. *American Naturalist* 174, 819–835.

Isaac, N. J. B. and Carbone, C. (2010) Why are metabolic scaling exponents so controversial? Quantifying variance and testing hypotheses. *Ecology Letters* 13, 728–735.

Isaac, N. J. B., Jones, K. E., Gittleman, J. L., and Purvis, A. (2005) Correlates of species richness in mammals: body size, life-history and ecology. *American Naturalist* 165, 600–607.

Isaac, N. J. B., Girardello, M., Brereton, T. M., and Roy, D. B. (2011a) Butterfly abundance in a warming climate: patterns in space and time are not congruent. *Journal of Insect Conservation* 15, 233–240.

Isaac, N. J. B., Storch, D., and Carbone, C. (2011b) Variation in the size-density relationship challenges the notion of energy equivalence. *Biology Letters* 7, 615–618.

Isaac, N. J. B., Storch, D., and Carbone, C. (in press) The paradox of energy equivalence. *Global Ecology and Biogeography*.

Isobe, T., Feigelson, E. D., Akritas, M. G., and Babu, G. J. (1990) Linear regression in astronomy. I. *Astrophysical Journal* 364, 104–113.

Iverson, L. R., Schwartz, M. W., and Prasad, A. M. (2004) How fast and far might tree species migrate in the eastern United States due to climate change? *Global Ecology and Biogeography* 13, 209–219.

Jablonski, D. (2005) Mass extinctions and macroevolution. *Paleobiology* 31 (Suppl. to No. 2), 192–210.

Jablonski, D. and Hunt, G. (2006) Larval ecology, geographic range, and species survivorship in Cretaceous mollusks: organismic versus species-level explanations. *American Naturalist* 168, 556–564.

Jablonski, D., Roy, K., and Valentine, J. W. (2006) Out of the tropics: Evolutionary dynamics of the latitudinal diversity gradient. *Science* 314, 102–106.

Jansson, R. and Davies, T. J. (2008) Global variation in diversification rates of flowering plants: energy vs. climate change. *Ecology Letters* 11, 173–183.

Janzen, D. H. (1987) Insect diversity of a Costa Rican dry forest: why keep it, and how? *Biological Journal of the Linnean Society* 30, 343–356.

Jenerette, G., Scott, R. L., and Huxman, T. E. (2008) Whole ecosystem metabolic pulses following precipitation events. *Functional Ecology* 22, 924–930.

Jennings, S. and Blanchard, J. L. (2004) Fish abundance with no fishing: predictions based on macroecological theory. *Journal of Animal Ecology* 73, 632–642.

Jennings, S. and Mackinson, S. (2003) Abundance-body mass relationships in size-structured food webs. *Ecology Letters* 6, 971–974.

Jennings, S., Pinnegar, J. K., Polunin, N. V. C., and Boon, T. (2001) Weak cross-species relationships between body size and trophic level belie powerful size-based trophic structuring in fish communities. *Journal of Animal Ecology* 70, 934–944.

Jennings, S., D'oliveira, J. A. A., and Warr, K. J. (2007) Measurement of body size and abundance in tests of macroecological and food web theory. *Journal of Animal Ecology* 76, 72–82.

Jennings, S., Melin, F., Blanchard, J. L., Forster, R. M., Dulvy, N. K., and Wilson, R. W. (2008) Global-scale predictions of community and ecosystem properties from simple ecological theory. *Proceedings of the Royal Society B: Biological Sciences* 275, 1375–1383.

Jetz, W., Carbone, C., Fulford, J., and Brown, J. H. (2004) The scaling of animal space use. *Science* 306, 266–268.

Jetz, W., Sekercioglu, C. H., and Watson, J. E. (2008) Ecological correlates and conservation implications of overestimating species geographic ranges. *Conservation Biology* 22, 110–119.

Jeyasingh, P. D. (2007) Plasticity in metabolic allometry: the role of dietary stoichiometry. *Ecology Letters* 10, 282–289.

Jeyasingh, P., Ragavendran, A., Paland, S., Lopez, J., Sterner, R., and Colbourne, J. (2011) How do consumers deal with stoichiometric constraints? Lessons from functional genomics using *Daphnia pulex*. *Molecular Ecology* 20, 2341–2352.

Jiang, L. and Morin, P. J. (2004) Temperature-dependent interactions explain unexpected responses to environmental warming in communities of competitors. *Journal of Animal Ecology* 73, 569–576.

Jiguet, F., Julliard, R., Thomas, C. D., Dehorter, O., Newson, S. E., and Couvet, D. (2006) Thermal range predicts bird population resilience to extreme high temperatures. *Ecology Letters* 9, 1321–1330.

Jin, Q. and Bethke, C. M. (2003) A new rate law describing microbial respiration. *Applied and Environmental Microbiology* 69, 2340–2348.

Jin, Q. and Bethke, C. M. (2007) The thermodynamics and kinetics of microbial metabolism. *American Journal of Science* 307, 643–647.

Jobbágy, E. G. and Jackson, R. B. (2000a) Global controls of forest line elevation in the northern and southern hemispheres. *Global Ecology and Biogeography* 9, 253–268.

Jobbágy, E. G. and Jackson, R. B. (2000b) The vertical distribution of soil organic carbon and its relation to climate and vegetation. *Ecological Applications* 10, 423–436.

John, R., Dalling, J. W., Harms, K. E., et al. (2007) Soil nutrients influence spatial distributions of tropical tree species. *Proceedings of the National Academy of Sciences of the USA* 104, 864–869.

Johnson, C. N. (2002) Determinants of loss of mammal species during the Late Quaternary "megafauna" extinctions: life history and ecology, but not body size. *Proceedings of the Royal Society B: Biological Sciences* 269, 2221–2227.

Johnson, F. H. and Lewin, I. (1946) The growth rate of *E. coli* in relation to temperature, quinine and coenzyme. *Journal of Cellular and Comparative Physiology* 28, 47–75.

Johnson, F. H., Eyring, H., and Stover, B. J. (1974) *The Theory of Rate Processes in Biology and Medicine*. New York: John Wiley.

Johnson, M. D., Völker, J., Moeller, H. V., Laws, E., Breslauer, K. J., and Falkowski, P. G. (2009) Universal constant for heat production in protists. *Proceedings of the National Academy of Sciences of the USA* 106, 6696–6699.

Johnson, N. L., Kemp, A. W., and Kotz, S. (2005) *Univariate Discrete Distributions* (third edition). New York: Wiley-Interscience.

Johnson, N. L., Kotz, S., and Balakrishnan, N. (1994) *Continuous Univariate Distributions*, Volume 1 (second edition). New York: Wiley-Interscience.

Jones, H. G. (1978) Modeling diurnal trends of leaf water potential in transpiring wheat. *Journal of Applied Ecology* 15, 613–626.

Jones, H. G. (1992) *Plants and Microclimate: a Quantitative Approach to Environmental Plant Physiology*. Cambridge: Cambridge University Press.

Jones, K. E., Purvis, A., and Gittleman, J. L. (2003) Biological correlates of extinction risk in bats. *American Naturalist* 161, 601–614.

Jones, K. E., Bielby, J., Cardillo, M., et al. (2009) PanTHERIA: a species-level database of life history, ecology, and geography of extant and recently extinct mammals. *Ecology* 90, 2648.

Jones, K. M., Kobayashi, H., Davies, B. W., Taga, M. E., and Walker, G. C. (2007) How rhizobial symbionts invade plants: the Sinorhizobium–Medicago model. *Nature Reviews Microbiology* 5, 619–633.

Jones, R. L. and Hanson, H. C. (1985) *Mineral Licks, Geophagy, and Biogeochemistry of North American Ungulates*. Ames, IA: Iowa State University Press.

Joos, B., Lighton, J. R. B., Harrison, J. F., Suarez, R. K., and Roberts, S. P. (1997) Effects of ambient oxygen tension on flight performance, metabolism and water loss of the honeybee. *Physiological Zoology* 70, 167–174.

Juliano, S. A. (1986) Resistance to desiccation and starvation of two species of *Brachinus* (Coleoptera: Carabidae) from southeastern Arizona. *Canadian Journal of Zoology* 64, 73–80.

Jun, J., Pepper, J., Savage, V., Gillooly, J., and Brown, J. (2003) Allometric scaling of ant foraging trail networks. *Evolutionary Ecology Research* 5, 297–303.

Kaiser, A., Klok, C. J., Socha, J. J., Lee, W. K., Quinlan, M. C., and Harrison, J. F. (2007) Increase in tracheal investment with beetle size supports hypothesis of oxygen limitation on insect gigantism. *Proceedings of the National Academy of Sciences of the USA* 104, 13198–13203.

Kaitaniemi, P. and Lintunen, A. (2008) Precision of allometric scaling equations for trees can be improved by including the effect of ecological interactions. *Trees – Structure and Function* 22, 579–584.

Kangatharalingam, N. and Amy, P. S. (1994) *Helicobacter pylori* comb. nov. exhibits facultative acidophilism and obligate microaerophilism. *Applied and Environmental Microbiology* 60, 2176–2179.

Kaplan, H. S. and Robson, A. J. (2002) The emergence of humans: the coevolution of intelligence and longevity with intergenerational transfers. *Proceedings of the National Academy of Sciences of the USA* 99, 10221–10226.

Kaplan, H., Hill, K., Lancaster, J., and Hurtado, A. M. (2000) A theory of human life history evolution: diet, intelligence, and longevity. *Evolutionary Anthropology* 9, 156–185.

Kappes, M. A., Shaffer, S. A., Tremblay, Y., et al. (2010) Hawaiian albatrosses track interannual variability of marine habitats in the North Pacific. *Progress in Oceanography* 86, 246–260.

Karasov, W. H. and Diamond, J. (1985) Digestive adaptations for fueling the cost of endothermy. *Science* 228, 202–204.

Karasov, W. H. and Hume, I. D. (1997) Vertebrate gastrointestinal system. In: Dantzler, W. (Ed.), *Handbook of Comparative Physiology*. Bethesda, MD: American Physiological Society.

Karasov, W. H. and Martínez Del Rio, C. (2007) *Physiological Ecology: How Animals Process Energy, Nutrients, and Toxins*. Princeton, NJ: Princeton University Press.

Kartascheff, B., Heckmann, L., Drossel, B., and Guill, C. (2010) Why allometric scaling enhances stability in food web models. *Theoretical Ecology* 3, 195–208.

Kaspari, M. and Stevenson, B. S. (2008) Evolutionary ecology, antibiosis, and all that rot. *Proceedings of the National Academy of Sciences of the USA* 105, 19027–19028.

Kaspari, M. and Vargo, E. (1995) Colony size as a buffer against seasonality: Bergmann's rule in social insects. *American Naturalist* 145, 610–632.

Kaspari, M. and Yanoviak, S. P. (2009) Biogeochemistry and the structure of tropical brown food webs. *Ecology* 90, 3342–3351.

Kaspari, M., Yuan, M., and Alonso, L. (2003) Spatial grain and the causes of regional diversity gradients in ants. *American Naturalist* 161, 459–477.

Kaspari, M., Wright, J., Yavitt, J., Harms, K., Garcia, M., and Santana, M. (2008a) Multiple nutrients regulate litterfall and decomposition in a tropical forest. *Ecology Letters* 11, 35–43.

Kaspari, M., Yanoviak, S., and Dudley, R. (2008b) On the biogeography of salt limitation: a study of ant communities. *Proceedings of the National Academy of Sciences of the USA* 105, 17848–17851.

Kaspari, M., Yanoviak, S. P., Dudley, R., Yuan, M., and Clay, N. A. (2009) Sodium shortage as a constraint on the carbon cycle in an inland tropical forest. *Proceedings of the National Academy of Sciences of the USA* 106, 19405–19409.

Kay, A. 2002. Applying optimal foraging theory to assess nutrient availability ratios for ants. *Ecology* 83, 1935–1964.

Kay, A., Zumbusch, T., Heinen, J., Marsh, T., and Holway, D. (2010) Nutrition and interference competition have interactive effects on the behavior and performance of Argentine ants. *Ecology* 91, 57–64.

Kearney, M., Shine, R., and Porter, W. P. (2009) The potential for behavioral thermoregulation to buffer "cold-blooded" animals against climate warming. *Proceedings of the National Academy of Sciences of the USA* 106, 3835–3840.

Keil, P., Šímová, I., and Hawkins, B. (2008) Water-energy and the geographical species richness patterns of European and North African dragonflies. *Insect Conservation and Diversity* 1, 142–150.

Keim, C., Lopes Martins, J., Lins De Barros, H., Lins, U., and Farina, M. (2007) Structure, behavior, ecology and diversity of multicellular magnetotactic prokaryotes. *Magnetoreception and Magnetosomes in Bacteria*, pp. 103–132.

Keller, L. (1998) Queen lifespan and colony characteristics in ants and termites. *Insectes Sociaux* 45, 235–246.

Kellner, J. R. and Asner, G. P. (2009) Convergent structural responses of tropical forests to diverse disturbance regimes. *Ecology Letters* 12, 887–897.

Kelly, E., Chadwick, O., and Hilinski, T. (1998) The effect of plants on mineral weathering. *Biogeochemistry* 42, 21–53.

Kenrick, P. and Crane, P. R. (1997) The origin and early evolution of plants on land. *Nature* 389, 33–39.

Kerkhoff, A. J. and Enquist, B. J. (2006) Ecosystem allometry: the scaling of nutrient stocks and primary productivity across plant communities. *Ecology Letters* 9, 419–427.

Kerkhoff, A. J. and Enquist, B. J. (2007) The implications of scaling approaches for understanding resilience and reorganization in ecosystems. *Bioscience* 57, 489–499.

Kerkhoff, A. J. and Enquist, B. J. (2009) Multiplicative by nature: Why logarithmic transformation is necessary in allometry. *Journal of Theoretical Biology* 257, 519–521.

Kerkhoff, A. J., Enquist, B. J., Elser, J. J., and Fagan, W. F. (2005) Plant allometry, stoichiometry and the temperature-dependence of primary productivity. *Global Ecology and Biogeography* 14, 585–598.

Kerkhoff, A., Enquist, B., Elser, J., and Fagan, W. (2005) Plant allometry, stoichiometry and the temperature dependence of primary productivity. *Global Ecology and Biogeography* 14, 585–598.

Kerney, R., Kim, E., Hangarter, R. P., Heiss, A. A., Bishop, C. D., and Hall, B. K. (2011) Intracellular invasion of green algae in a salamander host. *Proceedings of the National Academy of Sciences of the USA* 108, 6497–6502.

Kerr, S. R. and Dickie, L. M. (2001) *The Biomass Spectrum: a Predator–Prey Theory of Aquatic Production*. New York: Columbia University Press.

Kiflawi, M. (2006) On optimal propagule size and developmental time. *Oikos* 113, 168–173.

Kiltie, R. A. (2000) Scaling of visual acuity with body size in mammals and birds. *Functional Ecology* 14, 226–234.

Kim, B. H. and Gadd, G. M. (2008) *Bacterial Physiology and Metabolism*. Cambridge: Cambridge University Press.

King, A. W., Gunderson, C. A., Post, W. M., Weston, D. J., and Wullschleger, S. D. (2006) Plant respiration in a warmer world. *Science* 312, 536–537.

King, D. A. and Louks, O. L. (1978) The theory of tree bole and branch form. *Radiation and Environmental Biophysics* 15, 141–165.

King, J. S., Hanson, P. J., Bernhardt, E., Deangelis, P., Norby, R. J., and Pregitzer, K. S. (2004) A multiyear synthesis of soil respiration responses to elevated atmospheric CO_2 from four forest FACE experiments. *Global Change Biology* 10, 1027.

Kingsolver, J. G. and Huey, R. B. (2008) Size, temperature, and fitness: three rules. *Evolutionary Ecology Research* 10, 251–268.

Kinlan, B. P. and Gaines, S. D. (2003) Propagule dispersal in marine and terrestrial environments: a community perspective. *Ecology* 84, 2007–2020.

Kinlan, B. P., Gaines, S. D., and Lester, S. E. (2005) Propagule dispersal and the scales of marine community process. *Diversity and Distributions* 11, 139–148.

Kiørboe, T. (1993) Turbulence, phytoplankton cell size, and the structure of pelagic food webs. *Advances in Marine Biology* 29, 1–72.

Kiørboe, T. (1996) Material flux in the water column. In: Jørgensen, B. B. and Richardson, K. (Eds), *Eutrophication in Coastal Marine Ecosystems*. Washington, DC: American Geophysical Union.

Kiørboe, T. (2008) *A Mechanistic Approach to Plankton Ecology*. Princeton, NJ: Princeton University Press.

Kirkton, S. D., Niska, J. A., and Harrison, J. F. (2005) Ontogenetic effects on aerobic and anaerobic metabolism during jumping in the American locust, *Schistocerca americana*. *Journal of Experimental Biology* 208, 3003–3012.

Kirschbaum, M. U. F. (1995) The temperature dependence of soil organic matter decomposition, and the effect of global warming on soil organic C storage. *Soil Biology and Biochemistry* 27, 753–760.

Klasing, K. C., Leschinsky, T. V., Adams, N. J. S., and Slotow, R. H. (1999) Functions, costs, and benefits of the immune system during development and growth. In: *Proceedings of the 22nd International Ornithological Congress*. Johannesburg, Birdlife South Africa.

Kleiber, M. (1932) Body size and metabolism. *Hilgardia* 6, 315–353.

Klein, R. G. (2009) *The Human Career: Human Biological and Cultural Origins*. Chicago: University of Chicago Press.

Klok, C. J. and Chown, S. L. (2005) Temperature- and body mass-related variation in cyclic gas exchange characteristics and metabolic rate of seven weevil species: broader implications. *Journal of Insect Physiology* 51, 789–801.

Klok, C. J. and Harrison, J. F. (2009) Atmospheric hypoxia limits selection for large body size in insects. *PLoS ONE* 4, e3876.

Klok, C. J., Kaiser, A., Lighton, J. R. B., and Harrison, J. F. (2010) Critical oxygen partial pressures and maximal tracheal conductances for *Drosophila melanogaster* reared for multiple generations in hypoxia or hyperoxia. *Journal of Insect Physiology* 56, 461–469.

Knies, J. L. and Kingsolver, J. G. (2010) Erroneous Arrhenius: modified Arrhenius model best explains the temperature dependence of ectotherm fitness. *American Naturalist* 176, 227–233.

Knoll, A. H. and Niklas, K. J. (1985) Adaptation and the fossil record of plants. *American Journal of Botany* 72, 886–887.

Kodric-Brown, A., Sibly, R. M., and Brown, J. H. (2006) The allometry of ornaments and weapons. *Proceedings of the National Academy of Sciences of the USA* 103, 8733–8738.

Koh, L. P., Sodhi, N. S., and Brook, B. W. (2004) Ecological correlates of extinction proneness in tropical butterflies. *Conservation Biology* 18, 1571–1578.

Kohler, P. (1985) The strategies of energy-conservation in helminths. *Molecular and Biochemical Parasitology* 17, 1–18.

Kohyama, T. (1993). Size-structured tree populations in the gap-dynamic forest: the forest architecture hypothesis for the stable coexistence of species. *Journal of Ecology* 81, 131–143.

Kolding, S. and Fenchel, T. M. (1981) Patterns of reproduction in different populations of five species of the amphipod genus Gammarus. *Oikos* 37, 167–172.

Kolokotrones, T., Savage, V., Deeds, E. J., and Fontana, W. (2010) Curvature in metabolic scaling. *Nature* 464, 753–756.

Kooijman, S. A. L. M. (1986) Energy budgets can explain body size relations. *Journal of Theoretical Biology* 121, 269–282.

Kooijman, S. A. L. M. (2000) *Dynamic Energy and Mass Budgets in Biological Systems*. Cambridge: Cambridge University Press.

Koontz, T. L., Petroff, A., West, G. B., and Brown, J. H. (2009) Scaling relations for a functionally two dimensional plant: *Chamaesyce setiloba* (Euphorbiaceae). *American Journal of Botany* 96, 877–884.

Kordas, R., Harley, C. D. G., and O'Connor, M. I. (2011) Community ecology in a warming world: thermal influence on interspecific interactions. *Journal of Experimental Marine Biology and Ecology* 400, 218–226.

Koyama, K. and Kikuzawa, K. (2009) Is whole-plant photosynthetic rate proportional to leaf area? A test of scalings and a logistic equation by leaf demography census. *American Naturalist* 173, 640–649.

Kozlowski, J. and Konarzewski, M. (2004) Is West, Brown and Enquist's model of allometric scaling mathematically correct and biologically relevant? *Functional Ecology* 18, 283–289.

Kozlowski, J. and Konarzewski, M. (2005) West, Brown and Enquist's model of allometric scaling again: the same questions remain. *Functional Ecology* 19, 739–743.

Kozlowski, J. and Weiner, J. (1997) Interspecific allometries are by-products of body size optimization. *American Naturalist* 149, 352–380.

Kozlowski, J., Konarzewski, M., and Gawelczyk, A. T. (2003) Cell size as a link between noncoding DNA and metabolic rate scaling. *Proceedings of the National Academy of Sciences of the USA* 100, 14080–14085.

Kozlowski, J., Czarnoleski, M., and Danko, M. (2004) Can optimal resource allocation models explain why ectotherms grow larger in cold? *Integrative and Comparative Biology* 44, 480–493.

Kraft, N. J. B., Valencia, R., and Ackerly, D. D. (2008) Functional traits and niche-based tree community assembly in an Amazonian forest. *Science* 322, 580.

Krebs, J. R. and Davies, N. B. (1993) *An Introduction to Behavioural Ecology*. Oxford: Blackwell.

Kreft, H. and Jetz, W. (2007) Global patterns and determinants of vascular plant diversity. *Proceedings of the National Academy of Sciences of the USA* 104, 5925–5930.

Krogh, A. (1916) *Respiratory Exchange of Animals and Man*. London: Longmans, Green.

Kroll, R. G. (1990) Alkalophiles. In: *Microbiology of Extreme Environments*. New York: McGraw-Hill, pp. 55–92.

Küppers, M. (1989) Ecological significance of above-ground architectural patterns in woody plants: a question of cost–benefit relationships. *TREE* 4, 375–379.

Kuris, A. M. and Lafferty, K. D. (2000) Parasite-host modeling meets reality: adaptive peaks and their ecological attributes. In: Poulin, R., Morand, S., and Skorping, A. (Eds), *Evolutionary Biology of Host–Parasite Relationships: Theory Meets Reality*. Amsterdam: Elsevier, pp. 9–26.

La Sorte, F. A. and Jetz, W. (2010) Avian distributions under climate change: towards improved projections. *Journal of Experimental Biology* 213, 862–869.

Lachenicht, M. W., Clusella-Trullas, S., Boardman, L., Le Roux, C., and Terblanche, J. S. (2010) Effects of acclimation temperature on thermal tolerance, locomotion performance and respiratory metabolism in *Acheta domesticus* L. (Orthoptera: Gryllidae). *Journal of Insect Physiology* 56, 822–830.

Lack, D. (1954). *The Natural Regulation of Animal Numbers*. Oxford: Clarendon Press.

Lafferty, K. D. (1993) Effects of parasitic castration on growth, reproduction and population dynamics of the marine snail *Cerithidea californica*. *Marine Ecology: Progress Series* 96, 229–237.

Lafferty, K. D. and Kuris, A. M. (2002) Trophic strategies, animal diversity and body size. *Trends in Ecology and Evolution* 17, 507–513.

Lalouette, L., Williams, C., Hervant, F., Sinclair, B., and Renault, D. (2011) Metabolic rate and oxidative stress in insects exposed to low temperature thermal fluctuations. *Comparative Biochemistry and Physiology A: Molecular and Integrative Physiology* 158, 229–234.

Lambers, H., Freijsen, N., Poorter, H., Hirose, T., and van der Werff. H. (1989) Analyses of growth based on net assimilation rate and nitrogen productivity: Their physiological background. In: Lambers, H., Cambridge, M. L., H. Konings, H., and Pons, T. L. (Eds), *Variation in Growth Rate and Productivity of Higher Plants*. The Hague, Netherlands: SPB Academic Publishing, pp. 1–17.

Langdon, C. (1988) On the causes of interspecific differences in the growth-irradiance relationship for phytoplankton.

Part II. A general review. *Journal of Plankton Research* 10, 1291–1312.

Langley, J. A., McKinley, D. C., Wolf, A. A., Hungate, B. A., Drake, B. G., and Megonigal, J. P. (2009) Priming depletes soil carbon and releases nitrogen in a scrub-oak ecosystem exposed to elevated CO_2. *Soil Biology and Biochemistry* 41, 54–60.

Laporte, L. F. (2000) *George Gaylord Simpson: Paleontologist and Evolutionist.* New York: Columbia University Press.

Laptikhovsky, V. (2006) Latitudinal and bathymetric trends in egg size variation: a new look at Thorson's and Rass's rules. *Marine Ecology: An Evolutionary Perspective* 27, 7–14.

Larowe, D. E. and Helgeson, H. C. (2007) Quantifying the energetics of metabolic reactions in diverse biogeochemical systems: electron flow and ATP synthesis. *Geobiology* 5, 153–168.

Larson, E. R. and Olden, J. D. (2010) Latent extinction and invasion risk of crayfishes in the southeastern United States. *Conservation Biology* 24, 1099–1110.

Latham, R. E. and Ricklefs, R. E. (1993) Global patterns of tree species richness in moist forests – energy-diversity theory does not account for variation in species richness. *Oikos* 67, 325–333.

Latimer, A. M. (2007) Geography and resource limitation complicate metabolism-based predictions of species richness. *Ecology* 88, 1895–1898.

Laurance, W. F. (1991) Ecological correlates of extinction proneness in Australian tropical rain forest mammals. *Conservation Biology* 5, 79–89.

Laurel, B. J. and Bradbury, I. R. (2006) "Big" concerns with high latitude marine protected areas (MPAs): trends in connectivity and MPA size. *Canadian Journal of Fisheries and Aquatic Sciences* 63, 2603–2607.

Lavorel, S., Grigulis, K., Lamarque, P., et al. (2011) Using plant functional traits to understand the landscape distribution of multiple ecosystem services. *Journal of Ecology* 99, 135–147.

Law, R. (1979) Ecological determinants in the evolution of life histories. In: Anderson, R. M., Turner, B. D., and Taylor, L. R. (Eds), *Population Dynamics.* Oxford: Blackwell Scientific Publications, pp. 81–103.

Layman, C. A., Winemiller, K. O., Arrington, A., and Jepsen, D. B. (2005) Body size and trophic position in a diverse tropical food web. *Ecology* 86, 2530–2535.

Leakey, A. D. B., Ainsworth, E. A., Bernacchi, C. J., Rogers, A., Long, S. P., and Ort, D. R. (2009) Elevated CO_2 effects on plant carbon, nitrogen, and water relations: six important lessons from FACE. *Journal of Experimental Botany* 60, 2859–2876.

Lebauer, D. S. and Treseder, K. K. (2008) Nitrogen limitation of net primary productivity in terrestrial ecosystems is globally distributed. *Ecology* 89, 371–379.

Lee, J. J., Soldo, A. T., Reisser, W., Lee, M. J., Jeon, K. W., and Görtz, H. D. (1985) The extent of algal and bacterial endo-symbioses in protozoa. *Journal of Eukaryotic Microbiology* 32, 391–403.

Lee, T. M. and Jetz, W. (2011) Unravelling the structure of species extinction risk for predictive conservation science. *Proceedings of the Royal Society B: Biological Sciences* 278, 1329–1338.

Legendre, P. and Legendre, L. (1998) *Numerical Ecology* (second edition). Amsterdam: Elsevier Science BV.

Lehman, T. M. and Woodward, H. N. (2008) Modeling growth rates for sauropod dinosaurs. *Paleobiology* 34, 264–281.

Leibold, M. A. (1996) A graphical model of keystone predators in food webs: trophic regulation of abundance, incidence, and diversity patterns in communities. *American Naturalist* 147, 784–812.

Leonard, W. R., Snodgrass, J. J., and Robertson, M. L. (2007) Effects of brain evolution on human nutrition and metabolism. *Nutrition* 27, 311–327.

Leroi, A. M., Bennett, A. F., and Lenski, R. E. (1994) Temperature acclimation and competitive fitness: an experimental test of the beneficial acclimation assumption. *Proceedings of the National Academy of Sciences of the USA* 91, 1917–1921.

Leroux, S. J., Schmiegelow, F. K. A., Lessard, R. B., and Cumming, S. G. (2007) Minimum dynamic reserves: a framework for determining reserve size in ecosystems structured by large disturbances. *Biological Conservation* 138, 464–473.

Lester, S. E. and Ruttenberg, B. I. (2005) The relationship between pelagic larval duration and range size in tropical reef fishes: a synthetic analysis. *Proceedings of the Royal Society B: Biological Sciences* 272, 585–591.

Lettini, S. E. and Sukhdeo, M. V. K. (2010) The energetic cost of parasitism in isopods. *Ecoscience* 17, 1–8.

Levin, L. A. and Bridges, T. S. (1995) Pattern and diversity in reproduction and development. In: McEdward, L. R. (Ed.), *Ecology of Marine Invertebrate Larvae.* Boca Raton, FL: CRC Press, pp. 1–48.

Levin, S. A. (1992) The problem of pattern and scale in ecology. *Ecology* 73, 1943–1967.

Levitan, D. R. (2000) Optimal egg size in marine invertebrates: theory and phylogenetic analysis of the critical relationship between egg size and development time in Echinoids. *American Naturalist* 156, 175–192.

Lewis, W. M. (1989) Further evidence of anomalous size scaling of respiration in phytoplankton. *Journal of Phycology* 25, 395–397.

Ley, R. E., Turnbaugh, P. J., Klein, S., and Gordon, J. I. (2006) Microbial ecology: human gut microbes associated with obesity. *Nature* 444, 1022–1023.

Li, W.-H., Ellsworth, D. L., Krushkal, J., Chang, B. H. J., and Hewett-Emmett, D. (1996) Rates of nucleotide substitution in primates and rodents and the generation-time effect hypothesis. *Molecular Phylogenetics and Evolution* 5, 182–187.

Li, W. K. W. (1985) Photosynthetic response to temperature of marine phytoplankton along a latitudinal gradient (16N to 74N). *Deep Sea Research A: Oceanographic Research Papers* 32, 1381–1385, 1387–1391.

Li, W. K. W. (2002) Macroecological patterns of phytoplankton in the northwestern North Atlantic Ocean. *Nature* 419, 154–157.

Li, W. K. W., McLaughlin, F. A., Lovejoy, C., and Carmack, E. C. (2009) Smallest algae thrive as the Arctic Ocean freshens. *Science* 326, 539–539.

Lieberman, D., Hartshorn, G. S., Lieberman, M., and Peralta, R. (1990) Forest dynamics at La Selva Biological Station, 1969–1985. In: Gentry, A. H. (Ed.), *Four Neotropical Rainforests*. New haven, CT: Yale University Press, pp. 509–521.

Liebig, J. V. (1855) *Principles of Agricultural Chemistry with Special Reference to the Late Researches made in England*. London: Dowden, Hutchinson & Ross.

Lieth, H. (1973) Primary production: terrestrial ecosystems. *Human Ecology* 1, 303–332.

Lifson, N. and McClintock, R. (1966) Theory of use of the turnover rates of body water for measuring energy and material balance. *Journal of Theoretical Biology* 12, 46–74.

Lighton, J. R. B. (1989) Individual and whole-colony respiration in an African formicine ant. *Functional Ecology* 3, 523–530.

Lighton, J. R. B. (1991) Ventilation in Namib Desert tenebrionid beetles: mass scaling and evidence of a novel quantized flutter-phase. *Journal of Experimental Biology* 159, 249–268.

Lighton, J. R. B. (2008) *Measuring Metabolic Rates : A Manual for Scientists*. Oxford; New York: Oxford University Press.

Lighton, J. R. B. and Berrigan, D. (1995) Questioning paradigms: caste-specific ventilation in harvester ants, *Messor pergandei* and *M. julianus* (Hymenoptera: Formicidae). *Journal of Experimental Biology* 198, 521–530.

Lighton, J. R. B. and Lovegrove, B. G. (1990) A temperature–induced switch from diffusive to conective ventilation in the honeybee. *Journal of Experimental Biology* 154, 509–516.

Lighton, J. R. B., Fukushi, T., and Wehner, R. (1993) Ventilation in *Cataglyphis bicolor*: regulation of carbon dioxide release from the thoracic and abdominal spiracles. *Journal of Insect Physiology* 39, 687–699.

Lighton, J. R. B., Brownell, P. H., Joos, B., and Turner, R. J. (2001) Low metabolic rate in scorpions: implication for population biomass and cannibalism. *Journal of Experimental Biology* 204, 607–613.

Lindeman, R. L. (1942) The trophic-dynamic aspect of ecology. *Ecology* 23, 399–418.

Lindstedt, S. L. and Boyce, M. 1985. Seasonality, fasting endurance, and body size in mammals. *American Naturalist* 125, 873–878.

Liow, L. H., Fortelius, M., Bingham, E., et al. (2008) Higher origination and extinction rates in larger mammals. *Proceedings of the National Academy of Sciences of the USA* 105, 6097–6102.

Liow, L. H., Quental, T. B., and Marshall, C. R. (2010) When can decreasing diversification rates be detected with molecular phylogenies and the fossil record? *Systematic Biology* 59, 646–659.

Lipp, A., Wolf, H., and Lehmann, F.-O. (2005) Walking on inclines: energetics of locomotion in the ant Camponotus. *Journal of Experimental Biology* 208, 707–719.

Lips, K. R., Reeve, J. D., and Witters, L. R. (2003) Ecological traits predicting amphibian population declines in Central America. *Conservation Biology* 17, 1078–1088.

Litchman, E. and Klausmeier, C. A. (2008) Trait-based community ecology of phytoplankton. *Annual Review of Ecology Evolution and Systematics* 39, 615–639.

Litchman, E., Klausmeier, C. A., Schofield, O. M., and Falkowski, P. G. (2007) The role of functional traits and trade-offs in structuring phytoplankton communities: scaling from cellular to ecosystem level. *Ecology Letters* 10, 1170–1181.

Litchman, E., Klausmeier, C. A., and Yoshiyama, K. (2009) Contrasting size evolution in marine and freshwater diatoms. *Proceedings of the National Academy of Sciences of the USA* 106, 2665–2670.

Liu, Y. (2007) Overview of some theoretical approaches for derivation of the Monod equation. *Applied Microbiology and Biotechnology* 73, 1241–1250.

Loarie, S. R., Duffy, P. B., Hamilton, H., Asner, G. P., Field, C. B., and Ackerly, D. D. (2009) The velocity of climate change. *Nature* 462, 1052–1055.

Loeuille, N. and Loreau, M. (2005) Evolutionary emergence of size-structured food webs. *Proceedings of the National Academy of Sciences of the USA* 102, 5761–5766.

Loeuille, N. and Loreau, M. (2006) Evolution of body size in food webs: does the energy equivalence rule hold? *Ecology Letters* 9, 171–178.

Loker, E. S. (1983) A comparative-study of the life-histories of mammalian schistosomes. *Parasitology* 87, 343–369.

Long, S. P., Ainsworth, E. A., Rogers, A., and Ort, D. R. (2004) Rising atmospheric carbon dioxide: plants FACE the future. *Annual Review of Plant Biology* 55, 591–628.

Long, Z. T., Steiner, C. F., Krumins, J. A., and Morin, P. J. (2006) Species richness and allometric scaling jointly determine biomass in model aquatic food webs. *Journal of Animal Ecology* 75, 1014–1023.

Lopez-Urrutia, A. (2008) The metabolic theory of ecology and algal bloom formation. *Limnology and Oceanography* 53, 2046–2047.

López-Urrutia, Á. and Morán, X.A.G. (2007) Resource limitation of bacterial production distorts the temperature dependence of oceanic carbon cycling. *Ecology* 88, 817–822.

López-Urrutia, Á., San Martin, E., Harris, R. P., and Irigoien, X. (2006) Scaling the metabolic balance of the oceans. *Proceedings of the National Academy of Sciences of the USA* 103, 8739–8744.

Lotka, A. J. (1922) Contribution to the energetics of evolution. *Proceedings of the National Academy of Sciences of the USA* 8, 147–151.

Lotka, A. J. (1925) *Elements of Physical Biology*. Baltimore, MD: Williams & Wilkins.

Lovegrove, B. (2000) The zoogeography of mammalian basal metabolic rate. *American Naturalist* 156, 201–219.

Lovegrove, B. (2003) The influence of climate on the basal metabolic rate of small mammals: a slow-fast metabolic continuum. *Journal of Comparative Physiology B: Biochemical Systemic and Environmental Physiology* 173, 87–112.

Lovegrove, B. G. (2009) Age at first reproduction and growth rate are independent of basal metabolic rate in mammals. *Journal of Comparative Physiology B: Biochemical Systemic and Environmental Physiology* 179, 391–401.

Lovelock, C. E., Feller, I. C., Ball, M. C., Ellis, J., and Sorrell, B. (2007) Testing the growth rate vs. geochemical hypothesis for latitudinal variation in plant nutrients. *Ecology Letters* 10, 1154–1163.

Loveridge, J. P. and Bursell, E. (1975) Studies on the water relations of adult locusts (*Orthoptera: Acrididae*). I. Respiration and the production of metabolic water. *Bulletin of Entomological Research* 65, 13–20.

Lucas, J. (1985) Metabolic rates and pit-construction costs of two antlion species. *Journal of Animal Ecology* 54, 295–309.

Luckinbill, L. S. (1973) Coexistence in laboratory populations of *Paramecium aurelia* and its predator *Didinium nasutum*. *Ecology* 54, 1320–13.27.

Luo, Y. (2007) Terrestrial carbon-cycle feedback to climate Warming. *Annual Review of Ecology Evolution and Systematics* 38, 683–712.

Luoma, S. N. and Rainbow, P. S. (2005) Why is metal bioaccumulation so variable? Biodynamics as a unifying concept. *Environmental Science and Technology* 39, 1921–1931.

Lupez-Barea, J. and Gumez-Ariza, J. L. (2006) Environmental proteomics and metallomics. *Proteomics* 6, S51–S62.

Luyssaert, S., Inglima, I., Jung, M., et al. (2007) CO_2 balance of boreal, temperate, and tropical forests derived from a global database. *Global Change Biology* 13, 2509–2537.

Lyons, S. K., Smith, F. A., and Brown, J. H. (2004) Of mice, mastodons and men: human-mediated extinctions on four continents. *Evolutionary Ecology Research* 6, 339–358.

Macarthur, R. H. and Pianka, E. R. (1966) On optimal use of a patchy environment. *American Naturalist* 100, 603–609.

Macarthur, R. H. and Wilson, E. O. (1967) *The Theory of Island Biogeography*. Princeton, NJ: Princeton University Press.

Mace, G. M., Collar, N. J., Gaston, K. J., et al. (2008) Quantification of extinction risk: IUCN's system for classifying threatened species. *Conservation Biology* 22, 1424–1442.

Machac, A., Janda, M., Dunn, R. R., and Sanders, N. J. (2011) Elevation gradients in phylogenetic structure of ant communities reveal the interplay of biotic and abiotic constraints on diversity. *Ecography* 34, 364–371.

Mackay, M. D., Neale, P. J., Arp, C. D., et al. (2009) Modeling lakes and reservoirs in the climate system. *Limnology and Oceanography* 54, 2315–2329.

Magnhagen, C. (1991) Predation risk as a cost of reproduction. *Trends in Ecology and Evolution* 6, 183–186.

Mahecha, M. D., Reichstein, M., Carvalhais, N., et al. (2010) Global convergence in the temperature sensitivity of respiration at ecosystem level. *Science* 329, 838–840.

Maier, R. M., Pepper, I. L., and Gerba, C. P. (2009) *Environmental Microbiology* (second edition). Burlington, MA: Academic Press.

Makarieva, A. M., Gorshkov, V. G., and Li, B. L. (2004) Ontogenetic growth: models and theory. *Ecological Modelling* 176, 15–26.

Makarieva, A. M., Gorshkov, V. G., and Li, B. L. (2005a) Energetics of the smallest: do bacteria breathe at the same rate as whales? *Proceedings of the Royal Society B: Biological Sciences* 272, 2219–2224.

Makarieva, A. M., Gorshkov, V. G., and Li, B. L. (2005b) Revising the distributive networks models of West, Brown and Enquist (1997) and Banavar, Maritan and Rinaldo (1999): metabolic inequity of living tissues provides clues for the observed allometric scaling rules. *Journal of Theoretical Biology* 237, 291–301.

Makarieva, A. M., Gorshkov, V. G., Li, B.-L., Chown, S. L., Reich, P. B., and Gavrilov, V. M. (2008) Mean mass-specific metabolic rates are strikingly similar across life's major domains: evidence for life's metabolic optimum. *Proceedings of the National Academy of Sciences of the USA* 105, 16994–16999.

Mäkelä, A. and Valentine, H. T. (2006) Crown ratio influences allometric scaling in trees. *Ecology* 87, 2967–2972.

Marañón, E. (2008) Inter-specific scaling of phytoplankton production and cell size in the field. *Journal of Plankton Research* 30, 157–163.

Marba, N., Duarte, C. M., and Agusti, S. (2007) Allometric scaling of plant life history. *Proceedings of the National Academy of Sciences of the USA* 104, 15777–15780.

Margules, C. R. and Pressey, R. L. (2000) Systematic conservation planning. *Nature* 405, 243–253.

Marquet, P. A., Navarrete, S. A., and Castilla, J. C. (1990) Scaling population density to body size in rocky intertidal communities. *Science* 250, 1125–1127.

Marquet, P. A., Navarrete, S. A., and Castilla, J. C. (1995) Body-size, population-density, and the energetic equivalence rule. *Journal of Animal Ecology* 64, 325–332.

Marquet, P. A., Quinones, R. A., Abades, S., et al. (2005) Scaling and power-laws in ecological systems. *Journal of Experimental Biology* 208, 1749–1769.

Marron, M., Markow, T., Kain, K., and Gibbs, A. (2003) Effects of starvation and desiccation on energy metabolism in desert and mesic Drosophila. *Journal of Insect Physiology* 49, 261–270.

Martin, A. P. (1999) Substitution rates of organelle and nuclear genes in sharks: implicating metabolic rate (again). *Molecular Biology and Evolution* 16, 996–1002.

Martin, A. P., and Palumbi, S. R. (1993) Body size, metabolic rate, generation time, and the molecular clock. *Proceedings of the National Academy of Sciences of the USA* 90, 4087–4091.

Martin, H. G. and Goldenfeld, N. (2006) On the origin and robustness of power-law species-area relationships in ecology. *Proceedings of the National Academy of Sciences of the USA* 103, 10310–10315.

Martin, J. (1990) Glacial-interglacial CO_2 change: the iron hypothesis. *Paleoceanography* 5, 1–13.

Martin, M. M. and Van't Hof, H. M. (1988) The cause of reduced growth of *Manduca sexta* larvae on a low-water diet: increased metabolic processing costs or nutrient limitation? *Journal of Insect Physiology* 34, 515–525.

Martin, P. S. (1984) Prehistoric over kill: the global model. In: Martin, P.S. and Klein, R. (Eds), *Quaternary Extinctions*. Tucson, AZ: University of Arizona Press, pp. 354–403.

Martin, T. E., Auer, S. K., Bassar, R. D., Niklison, A. M., and Lloyd, P. (2007) Geographic variation in avian incubation periods and parental influences on embryonic temperature. *Evolution* 61, 2558–2569.

Martinez, N. D., Williams, R. J., and Dunne, J. A. (2006) Diversity, complexity, and persistence in large model ecosystems. In: Pascual, M. and Dunne, J. A. (Eds), *Ecological Networks: Linking Structure to Dynamics in Food Webs*. New York: Oxford University Press.

Martinez Del Rio, C. (2008) Metabolic theory or metabolic models? *Trends in Ecology and Evolution* 23, 256–260.

Martins, E. P. (1994) Estimating the rate of phenotypic evolution from comparative data. *American Naturalist* 144, 193–209.

Matsura, T. (1981) Responses to starvation in a mantis, *Paratenodera* angustipennis (S.). *Oecologia* 50, 291–295.

Mattila, T. M. and Bokma, F. (2008) Extant mammal body masses suggest punctuated equilibrium. *Proceedings of the Royal Society B: Biological Sciences* 275, 2195–2199.

Maurer, B. A. and Brown, J. H. (1988) Distribution of energy use and biomass among species of North American terrestrial birds. *Ecology* 69, 1923–1932.

May, R. M. (1972) Will a large complex system be stable? *Nature* 238, 413–414.

Maynard Smith, J. and Szathmary, E. (1995) *The Major Transitions in Evolution*. Oxford: Oxford University Press.

May, R. M. (1978) The dynamics and diversity of insect faunas. In: Mound, L. A. and Waloff, N. (Eds), *Diversity of Insect Faunas*. Oxford: Blackwell Scientific, pp. 188–204.

May, R. M. (1986) The search for patterns in the balance of nature: advances and retreats. *Ecology* 67, 1115–1126.

Mayr, E. (1956) Geographical character gradients and climatic adaptation. *Evolution* 10, 105–108.

McArdle, B. H. (2003) Lines, models and errors: regression in the field. *Limnology and Oceanography* 48, 1363–1366.

McCain, C. M., and Sanders, N. J. (2010) Metabolic theory and elevational diversity of vertebrate ectotherms. *Ecology* 91, 601–609.

McCann, K. S. (2000) The diversity-stability debate. *Nature* 405, 228–233.

McCann, K., Hastings, A., and Huxel, G. R. (1998) Weak trophic interactions and the balance of nature. *Nature* 395, 794–798.

McCarthy, H. R., Oren, R., Johnsen, K. H., et al. (2010) Reassessment of plant carbon dynamics at the Duke free-air CO_2 enrichment site: interactions of atmospheric $[CO_2]$ with nitrogen and water availability over stand development. *New Phytologist* 185, 514–528.

McCarthy, M. A., Thompson, C. J., and Possingham, H. P. (2009) Designing nature reserves in the face of uncertainty. *Nature Proceedings*, available from: http://hdl.handle.net/10101/npre.2009.3387.1.

McCarty, J. P. (2001) Ecological consequences of recent climate change. *Conservation Biology* 15, 320–331.

McCoy, M. W. and Gillooly, J. F. (2008) Predicting natural mortality rates of plants and animals. *Ecology Letters* 11, 710–716.

McCoy, M. W. and Gillooly, J. F. (2009) Corrigendum. *Ecology Letters* 12, 731–733.

McCulloh, K. A. and Sperry, J. S. (2005) Patterns in hydraulic architecture and their implications for transport efficiency. *Tree Physiology* 25, 257–267.

McCulloh, K. A., Sperry, J. S., and Adler, F. R. (2003) Water transport in plants obeys Murray's law. *Nature* 421, 939–942.

McCulloh, K. A., Sperry, J. S., and Adler, F. R. (2004) Murray's law and the hydraulic vs. mechanic functioning of wood. *Functional Ecology* 18, 931–938.

McCulloh, K., Sperry, J. S., Lachenbruch, B., Meinzer, F. C., Reich, P. B., and Voelker, S. (2010) Moving water well: comparing hydraulic efficiency in twigs and trunks of coniferous, ring-porous, and diffuse-porous saplings from temperate and tropical forests. *New Phytologist* 186, 439–450.

McGaughran, A., Redding, G. P., Stevens, M. I., and Convey, P. (2009) Temporal metabolic rate variation in a continental Antarctic springtail. *Journal of Insect Physiology* 55, 129–134.

McGill, B. J., Maurer, B. A., and Weiser, M. D. (2006) Empirical evaluation of neutral theory. *Ecology* 87, 1411–1423.

McGill, B. (2003) Strong and weak tests of macroecological theory. *Oikos* 102, 679–685.

McGill, B. J. (2008) Exploring predictions of abundance from body mass using hierarchical comparative approaches. *American Naturalist* 172, 88–101.

McGill, B. (2010) Towards a unification of unified theories of biodiversity. *Ecology Letters* 13, 627–642.

McGill, B. J. and Mittelbach, G. G. (2006) An allometric vision and motion model to predict prey encounter rates. *Evolutionary Ecology Research* 8, 691–701.

McGill, B. J., Enquist, B. J., Weiher, E., and Westoby, M. (2006) Rebuilding community ecology using functional traits. *Trends in Ecology and Evolution* 21, 175–185.

McGill, B. J., Etienne, R. S., Gray, J. S., et al. (2007) Species abundance distributions: moving beyond single prediction theories to integration within an ecological framework. *Ecology Letters* 10, 995–1015.

McGlynn, T. P., Weiser, M. D., and Dunn, R. R. (2010) More individuals but fewer species: testing the "more individuals hypothesis" in a diverse tropical fauna. *Biology Letters* 6, 490–493.

McKechnie, A. E. and Wolf, B. O. (2009) Climate change increases the likelihood of catastrophic avian mortality events during extreme heat waves. *Biology Letters* 6, 253–256.

McKinney, M.L. (1997) Extinction vulnerability and selectivity: combining ecological and paleontological views. *Annual Review of Ecology and Systematics* 28, 495–516.

McMahon, T. (1973) Size and shape in biology. *Science* 179, 1201–1204.

McMahon, T.A. (1975) Allometry and biomechanics: limb bones in adult ungulates. *American Naturalist* 109, 547–563.

McMahon, T. A. and Bonner, J. T. (1983) *On Size and Life*. New York: Scientific American Books.

McMahon, T. A. and Kronauer, R. E. (1976) Tree structures: deducing the principle of mechanical design. *Journal of Theoretical Biology* 59, 443–466.

McNab, B. K. (1986) The influence of food habits on the energetics of eutherian mammals. *Ecological Monographs* 56, 1–19.

McNab, B. K. (1988) Complications inherent in scaling basal rate of metabolism in mammals. *Quarterly Review of Biology* 63, 25–54.

McNab, B. K. (1997) On the utility of uniformity in the definition of basal rate of metabolism. *Physiological Zoology* 70, 718–720.

McNab, B. K. (2002) *The physiological ecology of vertebrates: a view from energetics*, Ithaca, NY: Comstock Publishing Assoc.

McNab, B. K. (2003) Ecology shapes bird bioenergetics. *Nature* 426, 620–621.

McNab, B. K. (2008) An analysis of the factors that influence the level and scaling of mammalian BMR. *Comparative Biochemistry and Physiology A: Molecular and Integrative Physiology* 151, 5–28.

McNab, B. K. (2009) Resources and energetics determined dinosaur maximal size. *Proceedings of the National Academy of Sciences of the USA* 106, 12184–12188.

McNamara, J. and Houston, A. (1992) Risk-sensitive foraging: a review of the theory. *Bulletin of Mathematical Biology* 54, 355–378.

McPeek, M. A. and Brown, J. M. (2007) Clade age and not diversification rate explains species richness among animal taxa. *American Naturalist* 169, E97–E106.

McWilliams, S. R., Guglielmo, C., Pierce, B., and Klaassen, M. (2004) Flying, fasting, and feeding in birds during migration: a nutritional and physiological ecology perspective. *Journal of Avian Biology* 35, 377–393.

Medrano, J. F. and Gall, G. A. E. (1976) Food consumption, feed efficiency, metabolic rate and utilization of glucose in lines of *Tribolium castaneum* selected for 21-day pupa weight. *Genetics* 83, 393–407.

Medvigy, D., Wofsy, S. C., Munger, J. W., and Moorcroft, P. R. (2010) Responses of terrestrial ecosystems and carbon budgets to current and future environmental variability. *Proceedings of the National Academy of Sciences of the USA* 107, 8275–8280.

Meehan, T. D. (2006) Energy use and animal abundance in litter and soil communities. *Ecology* 87, 1650–1658.

Meehan, T. and Lindroth, R. (2007) Modeling nitrogen flux by larval insect herbivores from a temperate hardwood forest. *Oecologia* 153, 833–843.

Meehan, T. D., Jetz, W., and Brown, J. H. (2004) Energetic determinants of abundance in winter landbird communities. *Ecology Letters* 7, 532–537.

Meerhoff, M., Clemente, J. M., De Mello, F. T., Iglesias, C., Pedersen, A. R., and Jeppesen, E. (2007) Can warm climate-related structure of littoral predator assemblies weaken the clear water state in shallow lakes? *Global Change Biology* 13, 1888–1897.

Mei, Z. P., Finkel, Z. V., and Irwin, A. J. (2009) Light and nutrient availability affect the size-scaling of growth in phytoplankton. *Journal of Theoretical Biology* 259, 582–588.

Meinzer, F. (2003) Functional convergence in plant responses to the environment. *Oecologia* 134, 1–11.

Meinzer, F., Bond, B., Warren, J., and Woodruff, D. (2005) Does water transport scale universally with tree size? *Functional Ecology* 19, 558–565.

Memmott, J., Martinez, N. D., and Cohen, J. E. (2000) Predators, parasitoids and pathogens: species richness, trophic generality and body sizes in a natural food web. *Journal of Animal Ecology* 69, 1–15.

Mencuccini, M. (2002) Hydraulic constraints in the functional scaling of trees. *Tree Physiology* 22, 553–565.

Mencuccini, M., Holtta, T., Petit, G., and Magnani, F. (2007) Sanio's law revisited. Size dependent changes in the xylem architecture of trees. *Ecology Letters* 10, 1084–1093.

Menge, D. N. L. and Field, C. B. (2007) Simulated global changes alter phosphorus demand in annual grassland. *Global Change Biology* 13, 2582–2591.

Mentis, M. T. (1988) Hypothetico-deductive and inductive approaches in ecology. *Functional Ecology* 2, 5–14.

Metzger, R. J., Klein, O. D., Martin, G. R., and Krasnow, M. A. (2008) The branching programme of mouse lung development. *Nature* 453, 745–750.

Michener, C. D. (1964) Reproductive efficiency in relation to colony size in hymenopterous societies. *Insectes Sociaux* 4, 317–342.

Michener, C. D. (1974) *The Social Behavior of Bees*. Harvard, MA: Harvard University Press.

Michod, R.E. (2000) *Darwinian Dynamics: Evolutionary Transitions in Fitness and Individuality*. Princeton, NJ: Princeton University Press.

Mileikovsky, S.A. (1971) Types of larval development in marine bottom invertebrates, their distribution and ecological significance: a reevaluation. *Marine Biology* 10, 193–213.

Milewski, A. V. and Diamond, R. E. (2000) Why are very large herbivores absent from Australia? A new theory of micronutrients. *Journal of Biogeography* 27, 957–978.

Miller, M. B. and Bassler, B. L. (2001) Quorum sensing in bacteria. *Annual Reviews in Microbiology* 55, 165–199.

Miller, P. L. (1966) The supply of oxygen to the active flight muscles of some large beetles. *Journal of Experimental Biology* 45, 285–304.

Miller, P. L. (1981) Ventilation in active and in inactive insects. In: Herreid, C. F. and Fourtner, C. F. (Eds), *Locomotion and Energetics in Arthropods*. New York: Plenum Press.

Millien, V., Kathleen Lyons, S., Olson, L., Smith, F. A., Wilson, A. B., and Yom-Tov, Y. (2006) Ecotypic variation in the context of global climate change: revisiting the rules. *Ecology Letters* 9, 853–869.

Minelli, A., Maruzzo, D., and Fusco, G. (2010) Multi-scale relationships between numbers and size in the evolution of arthropod body features. *Arthropod Structure and Development* 39, 468–477.

Mittermeier, R. A., Gil, P. R., Hoffman, M., et al. (2005) *Hotspots Revisited: Earth's Biologically Richest and Most Endangered Terrestrial Ecoregions*. Washington, DC: Conservation International.

Molina, N. and Van Nimwegen, E. (2008) The evolution of domain-content in bacterial genomes. *Biology Direct* 3, 51.

Moller, A. P. and Birkhead, T. R. (1991) Frequent copulations and mate guarding as alternative paternity guards in birds: a comparative study. *Behaviour* 118, 170–186.

Moloney, C. L. and Field, J. G. (1991) The size-based dynamics of plankton food webs. 1. A simulation model of carbon and nitrogen flows. *Journal of Plankton Research* 13, 1003–1038.

Montoya, J. M., Pimm, S., and Sole, R. V. (2006) Ecological networks and their fragility. *Nature* 442, 259–264.

Moorcroft, P. R. (2006) How close are we to a predictive science of the biosphere? *Trends in Ecology and Evolution (personal edition)* 21, 400–407.

Moorcroft, P. R., Hurtt, G. C., and Pacala, S. W. (2001) A method for scaling vegetation dynamics: the ecosystem demography model (ED). *Ecological Monographs* 71, 557–585.

Moore, J. (2002) *Parasites and the Behavior of Animals*. Oxford: Oxford University Press.

Morán, X. A. G., López-Urrutia, Á., Calvo-Díaz, A., and Li, W. K. W. (2010) Increasing importance of small phyto-plankton in a warmer ocean. *Global Change Biology* 16, 1137–1144.

Morand, S. (1996) Life-history traits in parasitic nematodes: a comparative approach for the search for invariants. *Functional Ecology* 10, 210–218.

Morand, S. and Poulin, R. (2002) Body size-density relationships and species diversity in parasitic nematodes: patterns and likely processes. *Evolutionary Ecology Research* 4, 951–961.

Morgan, K. R., Shelley, T. F., and Kimsey, L. S. (1985) Body temperature regulation, energy metabolism and foraging in light-seeking and shade-seeking robber flies. *Journal of Comparative Physiology B* 155, 561–570.

Mori, S., Yamaji, K., Ishida, A., et al. (2010). Mixed-power scaling of whole-plant respiration from seedlings to giant trees. *Proceedings of the National Academy of Sciences of the USA* 107, 1447–1451.

Morita, R. Y. (1980) Biological limits of temperature and pressure. *Origins of Life and Evolution of Biospheres* 10, 215–222.

Morlon, H., White, E. P., Etienne, R. S., et al. (2009) Taking species abundance distributions beyond individuals. *Ecology Letters* 12, 488–501.

Moses, M. E. (2009) Engineering: world wide ebb. *Nature* 457, 660–661.

Moses, M. E., Hou, C., Woodruff, W. H., et al. (2008a) Revisiting a model of ontogenetic growth: estimating model parameters from theory and data. *American Naturalist* 171, 632–645.

Moses, M. E., Forrest, S., Davis, A. L., Loder, M., and Brown, J. H. (2008b) Scaling theory for information networks. *Journal of the Royal Society's Interface* 5, 1469–1480.

Mouritsen, K. N. and Poulin, R. (2005) Parasite boosts biodiversity and changes animal community structure by trait-mediated indirect effects. *Oikos* 108, 344–350.

Muldavin, E., Moore, D., Collins, S., Wetherill, K., and Lightfoot, D. (2008) Aboveground net primary production dynamics in a northern Chihuahuan Desert ecosystem. *Oecologia* 155, 123–132.

Muller-Landau, H. C., Condit, R. S., Harms, K. E., et al. (2006a) Comparing tropical forest tree size distributions with the predictions of metabolic ecology and equilibrium models. *Ecology Letters* 9, 589–602.

Muller-Landau, H. C., Condit, R. S., Chave, J., et al. (2006b) Testing metabolic ecology theory for allometric scaling of tree size, growth, and mortality in tropical forests. *Ecology Letters* 9, 575–588.

Munch, S. B. and Salinas, S. (2009) Latitudinal variation in lifespan within species is explained by the metabolic theory of ecology. *Proceedings of the National Academy of Sciences of the USA* 106, 13860–13864.

Munday, P. L., Leis, J. M., Lough, J. M., et al. (2009) Climate change and coral reef connectivity. *Coral Reefs* 28, 379–395.

Muñoz, G. and Cribb, T. H. (2005) Infracommunity structure of parasites of *Hemigymnus melapterus* (Pisces: Labridae) from Lizard Island, Australia: the importance of habitat and parasite body size. *Journal of Parasitology* 91, 38–44.

Munshi-South, J. and Wilkinson, G. S. (2010) Bats and birds: exceptional longevity despite high metabolic rates. *Ageing Research Reviews* 9, 12–19.

Murray, C. D. (1927) A relationship between circumference and weight in trees and its bearing on branching angles. *Journal of General Physiology* 10, 725–739.

Murren, C. J. (2002) Phenotypic integration in plants. *Plant Species Biology* 17, 89–99.

Myers, N., Mittermeier, R. A., Mittermeier, C. G., Da Fonseca, G. A. B., and Kent, J. (2000) Biodiversity hotspots for conservation priorities. *Nature* 403, 853–858.

Myers, R. M., Bowen, K. G., and Barrowman, N. J. (1999) Maximum reproductive rate of fish at low population sizes. *Canadian Journal of Fisheries and Aquatic Science* 56, 2404–2419.

Nabholz, B., Glemin, S., and Galtier, N. (2008) Strong variations of mitochondrial mutation rate across mammals: the longevity hypothesis. *Molecular Biology and Evolution* 25, 120–130.

Nadell, C. D., Xavier, J. B., and Foster, K. R. (2009) The sociobiology of biofilms. *FEMS Microbiology Reviews* 33, 206–224.

Nagy, K. A. (2005) Field metabolic rate and body size. *Journal of Experimental Biology* 208, 1621–1625.

Nagy, K. A. and Bradshaw, S. D. (2000) Scaling of energy and water fluxes in free-living arid-zone Australian marsupials. *Journal of Mammalogy* 81, 962–970.

Nagy, K. A., Girard, I. A., and Brown, T. K. (1999) Energetics of free-ranging mammals, reptiles, and birds. *Annual Review of Nutrition* 19, 247–277.

Nee, S., Read, A. F., Greenwood, J. J. D., and Harvey, P. H. (1991) The relationship between abundance and body size in British birds. *Nature* 351, 312–313.

Nee, S., Mooers, A. O., and Harvey, P. H. (1992) Tempo and mode of evolution revealed from molecular phylogenies. *Proceedings of the National Academy of Sciences of the USA* 89, 8322–8326.

Nelson, J. B. T. (1978) *The Sulidae: Gannets and Boobies*. Oxford: Oxford University Press.

Nemani, R. R., Keeling, C. D., Hashimoto, H., et al. (2003) Climate-driven increases in global terrestrial net primary production from 1982 to 1999. *Science* 300, 1560–1563.

Nespolo, R., Lardies, M., and Bozinovic, F. (2003) Intrapopulational variation in the standard metabolic rate of insects: repeatability, thermal dependence and sensitivity (Q10) of oxygen consumption in a cricket. *Journal of Experimental Biology* 206, 4309–4315.

Neutel, A.-M., Heesterbeek, J. A. P., and De Ruiter, P. C. (2002) Stability in real food webs: weak links in long loops. *Science* 296, 1120–1123.

Newman, M. E. J. (2005) Power laws, Pareto distributions and Zipf's law. *Comtemporary Physics* 46, 323–351.

Nielsen, M., Elmes, G., and Kipyatkov, V. (1999) Respiratory Q10 varies between populations of two species of Myrmica ants according to the latitude of their sites. *Journal of Insect Physiology* 45, 559–564.

Nielsen, S. L. (2006) Size-dependent growth rates in eukaryotic and prokaryotic algae exemplified by green algae and cyanobacteria: comparisons between unicells and colonial growth forms. *Journal of Plankton Research* 28, 489–498.

Nijhout, H. F. and Emlen, D. J. (1998) Competition among body parts in the development and evolution of insect morphology. *Proceedings of the National Academy of Sciences of the USA* 95, 3685–3689.

Niklas, K. J. (1982) Computer-simulations of early land plant branching morphologies : canalization of patterns during evolution. *Paleobiology* 8, 196–210.

Niklas, K. J. (1994a) Morphological evolution through complex domains of fitness. *Proceedings of the National Academy of Sciences of the USA* 91, 6772–6779.

Niklas, K. J. (1994b) *Plant Allometry: The Scaling of Form and Process*. Chicago: University of Chicago Press.

Niklas, K. J. (1994c) Size-dependent variations in plant growth rates and the "3/4-power rule." *American Journal of Botany* 81, 134–145.

Niklas, K. J. (1997) *The Evolutionary Biology of Plants*. Chicago: University of Chicago Press.

Niklas, K. J. (2004) Plant allometry: Is there a grand unifying theory? *Biological Reviews* 79, 871–889.

Niklas, K. J. (2008) Carbon/nitrogen/ phosphorus allometric relations across species. In: White, P. J. and Hammond, J. P. (Eds), *The Ecophysiology of Plant-Phosphorus Interactions*. Springer.

Niklas, K. J. and Enquist, B. J. (2001) Invariant scaling relationships for interspecific plant biomass production rates and body size. *Proceedings of the National Academy of Sciences of the USA* 98, 2922–2927.

Niklas, K. J. and Enquist, B. J. (2002a) Canonical rules for plant organ biomass partitioning and annual allocation. *American Journal of Botany* 89, 812–819.

Niklas, K. J. and Enquist, B. J. (2002b) On the vegetative biomass partitioning of seed plant leaves, stems, and roots. *American Naturalist* 159, 482–497.

Niswander, R. E. (1951) Life history and respiration of the milkweed bug *Oncopeltus fasciatus* (Dallas). *Ohio Journal of Science* 51, 27–33.

Niven, J. E. and Scharlemann, J. P. (2005) Do insect metabolic rates at rest and during flight scale with body mass? *Biology Letters* 1, 346–349.

Norberg, J., Swaney, D. P., Dushoff, J., Lin, J., Casagrandi, R., and Levin, S. A. (2001) Phenotypic diversity and ecosystem functioning in changing environments: a theoretical framework. *Proceedings of the National Academy of Sciences of the USA* 98, 11376–11381.

Novotny, V., Drozd, P., Miller, S. E., et al. (2006) Why are there so many species of herbivorous insects in tropical rainforests? *Science* 313, 1115–1118.

Nowak, M. A. (2006) Five rules for the evolution of cooperation. *Science* 314, 1560–1563.

NRC (2000) *Nutrient Requirements of Beef Cattle* (seventh revised edition). Washington, DC: National Academies Press.

NRC (2005) *Mineral Tolerance of Animals*. Washington, DC: National Academies Press.

Nunn, G. B. and Stanley, S. E. (1998) Body size effects and rates of cytochrome b evolution in tube-nosed seabirds. *Molecular Biology and Evolution* 15, 1360–1371.

O'Connor, M. I. (2009) Warming strengthens an herbivore-plant interaction. *Ecology* 90, 388–398.

O'Connor, M. I., Bruno, J. F., Gaines, S. D., et al. (2007). Temperature control of larval dispersal and implications for marine ecology, evolution, and conservation. *Proceedings of the National Academy of Sciences of the USA* 104, 1266–1271.

O'Connor, M. I., Piehler, M. F., Leech, D. M., Anton, A., and Bruno, J. F. (2009) Warming and resource availability shift food web structure and metabolism. *PLoS Biology* 7, e1000178.

O'Connor, M. P., Agosta, S. J., Hansen, F., et al. (2007a) Phylogeny, regression, and the allometry of physiological traits. *American Naturalist* 170, 431–442.

O'Connor, M. P., Kemp, S. J., Agosta, S. J., et al. (2007b) Reconsidering the mechanistic basis of the metabolic theory of ecology. *Oikos* 116, 1058–1072.

O'Connor, P. M. and Claessens, L. (2005) Basic avian pulmonary design and flow-through ventilation in non-avian theropod dinosaurs. *Nature* 436, 253–256.

Odum, E. P. (1969) The strategy of ecosystem development. *Science* 164, 262–270.

Odum, E. P. (1971) *Fundamentals of Ecology*. Philadelphia: Saunders.

Odum, H. T. (1971) *Environment, Power, and Society*. New York: John Wiley.

Odum, H. T. (2007) *Environment, Power, and Society for the Twenty-First Century: The Hierarchy of Energy*. New York: Columbia University Press.

O'Flaherty, V., Mahony, T., O'Kennedy, R., and Colleran, E. (1998) Effect of pH on growth kinetics and sulphide toxicity thresholds of a range of methanogenic, syntrophic and sulphate-reducing bacteria. *Process Biochemistry* 33, 555–569.

O'Gorman, E. J., Jacob, U., Jonsson, T., and Emmerson, M. C. (2009) Interaction strength, food web topology and the relative importance of species in food webs. *Journal of Animal Ecology* 79, 682–692.

O'Grady, J. J., Reed, D. H., Brook, B. W., and Frankham, R. (2004) What are the best correlates of predicted extinction risk? *Biological Conservation* 118, 513–520.

Oksanen, L., Fretwell, S. D., Arruda, J., and Niemelä, P. (1981) Exploitation ecosystems in gradients of primary productivity. *American Naturalist* 118, 240–261.

Olden, J. D., Hogan, Z. S., and Vander Zanden, M. J. (2007) Small fish, big fish, red fish, blue fish: size-biased extinction risk of the world's freshwater and marine fishes. *Global Ecology and Biogeography* 16, 694–701.

Olden, J. D., Poff, N. L., and Bestgen, K. R. (2008) Trait synergisms and the rarity, extirpation, and extinction risk of desert fishes. *Ecology* 89, 847–856.

Oliver, D. (1971) Life history of the Chironomidae. *Annual Review of Entomology* 16, 211–230.

Olson, D. M. and Dinerstein, E. (2002) The Global 200: priority ecoregions for global conservation. *Annals of the Missouri Botanical Garden* 89, 199–224.

O'Meara, B. C., Ané, C., Sanderson, M. J., Wainwright, P. C., and Hansen, T. (2006) Testing for different rates of continuous trait evolution using likelihood. *Evolution* 60, 922–933.

Ophir, A. G., Schrader, S. B., and Gillooly, J. F. (2010) Energetic cost of calling: general constraints and species-specific differences. *Journal of Evolutionary Biology* 23, 1564–1569.

Orchard, V. A. and Cook, F. (1983) Relationship between soil respiration and soil moisture. *Soil Biology and Biochemistry* 15, 447–453.

Orcutt, B. and Meile, C. (2008) Constraints on mechanisms and rates of anaerobic oxidation of methane by microbial consortia: process-based modeling of ANME-2 archaea and sulfate reducing bacteria interactions. *Biogeosciences Discussions* 5, 1933–1967.

Oren, R., Ellsworth, D. S., Johnsen, K. H., et al. (2001) Soil fertility limits carbon sequestration by forest ecosystems in a CO_2-enriched atmosphere. *Nature* 411, 469–472.

Orme, D., Freckleton, R., Thomas, G., Petzoldt, T., Fritz, S. A., and Isaac, N. J. B. (2011) CAPER: Comparative Analyses of Phylogenetics and Evolution in R. R package version 0.4. http://CRAN.R-project.org/package=caper.

Otto, S. B., Rall, B. C., and Brose, U. (2007) Allometric degree distributions facilitate food web stability. *Nature* 450, 1226–1230.

Owens, I. P. F. and Bennett, P. M. (2000) Ecological basis of extinction risk in birds: habitat loss versus human persecution and introduced predators. *Proceedings of the National Academy of Sciences of the USA* 97, 144–148.

Owens, I. P. F., Bennett, P. M., and Harvey, P. H. (1999) Species richness among birds: body size, life history, sexual selection or ecology. *Proceedings of the Royal Society B: Biological Sciences* 266, 933–939.

Packard, G. C. (2009) On the use of logarithmic transformations in allometric analyses. *Journal of Theoretical Biology* 257, 515–518.

Packard, G. C. and Birchard, G. F. (2008) Traditional allometric analysis fails to provide a valid predictive model for mammalian metabolic rates. *Journal of Experimental Biology* 211, 3581–3587.

Packard, G. C. and Boardman, T. J. (2008) Model selection and logarithmic transformation in allometric analysis. *Physiological and Biochemical Zoology* 81, 496–507.

Packard, G. C. and Boardman, T. J. (2009a) Bias in interspecific allometry: examples from morphological scaling in varanid lizards. *Biological Journal of the Linnean Society* 96, 296–305.

Packard, G. C. and Boardman, T. J. (2009b) A comparison of methods for fitting allometric equations to filed metabolic rates of animals. *Journal of Comparative Physiology B* 179, 175–182.

Packard, G. C., Boardman, T. J., and Birchard, G. F. (2009) Allometric equations for predicting body mass of dinosaurs. *Journal of Zoology* 279, 102–110.

Packard, G. C., Birchard, G. F., and Boardman, T. J. (2010) Fitting statistical models in bivariate allometry. *Biological Reviews* doi: 10.1111/j.1469-185X.2010.00160.x.

Pagel, M. (1997) Inferring evolutionary processes from phylogenies. *Zoologica Scripta (Journal of the Royal Swedish Academy)* 25th Anniversary Special Issue on Phylogenetics and Systematics 26, 331–348.

Pagel, M. (1999) Inferring the historical patterns of biological evolution. *Nature* 401, 877–884.

Paine, R. T. (1988) Food webs: road maps of interactions or grist for theoretical development? *Ecology* 69, 1648–1654.

Paine, R. T. (2002) Trophic control of production in a rocky intertidal community. *Science* 296, 736–739.

Paine, R. T. (2010) Macroecology: does it ignore or can it encourage further ecological syntheses based on spatially local experimental manipulations? *American Naturalist* 176, 385–393.

Panikov, N. S. (1995) *Microbial Growth Kinetics*. London: Chapman & Hall.

Park, M. G., Yih, W., and Coats, D. W. (2004) Parasites and phytoplankton, with special emphasis on dinoflagellate infections. *Journal of Eukaryotic Microbiology* 51, 145–155.

Parmesan, C. and Yohe, G. (2003) A globally coherent fingerprint of climate change impacts across natural systems. *Nature* 421, 37–42.

Parra, R. and Montgomery, G. G. (1978) Comparisons of foregut and hindgut fermentation in herbivores. In: *The Ecology of Arboreal Folivores*. Washington, DC: Smithsonian Institution.

Partridge, L., Barrie, B., Fowler, K., and Vernon, F. (1994) Evolution and development of body size and cell size in *Drosophila melanogaster* in response to temperature. *Evolution* 48, 1269–1276.

Pasciak, W. J. and Gavis, J. (1974) Transport limitation of nutrient uptake in phytoplankton. *Limnology and Oceanography* 19, 881–889.

Pascoe, D., Richards, R. J., and James, B. L. (1968) Oxygen uptake metabolic rate reduced weight length and number of cercariae in starving sporocysts of *Cercaria dichotoma*. *Experimental Parasitology* 23, 171.

Patterson, M. R. (1992) A mass transfer explanation of metabolic scaling relations in some aquatic invertebrates and algae. *Science* 255, 1421–1423.

Pauly, D. (1980) On the interrelationships between natural mortality, growth parameters, and mean environmental

temperature in 175 fish stocks. *Journal du Conseil, Conseil International pour l'Exploration de la Mer* 39, 175–192.

Pauly, D. (2007) *Gasping Fish and Panting Squids: Oxygen, Temperature and the Growth of Water-Breathing Animals*. Oldendorf/Luhe, Germany.

Pauly, D. and Christensen, V. (1995) Primary production required to sustain global fisheries. *Nature* 374, 255–257.

Pautasso, M. and Gaston, K. J. (2005) Resources and global avian assemblage structure in forests. *Ecology Letters* 8, 282–289.

Payne, J. L. and Finnegan, S. (2007) The effect of geographic range on extinction risk during background and mass extinction. *Proceedings of the National Academy of Sciences of the USA* 104, 10506–10511.

Payne, J. L., Boyer, A. G., Brown, J. H., et al. (2009) Two-phase increase in the maximum size of life over 3.5 billion years reflects biological innovation and environmental opportunity. *Proceedings of the National Academy of Sciences* 106(1), 24–27.

Pearsall, W. H. (1927) Growth studies. VI. On the relative sizes of growing plant organs. *Annals of Botany* 41, 549.

Pearse, J. S. and Lockhart, S. J. (2004) Reproduction in cold water: paradigm changes in the 20th century and a role for cidaroid sea urchins. *Deep-Sea Research II: Topical Studies in Oceanography* 51, 1533–1549.

Pearse, J. S., McClintock, J. B., and Bosch, I. (1991) Reproduction of Antarctic benthic marine invertebrates: tempos, modes and timing. *American Zoologist* 31, 65–80.

Pennisi, E. (2010) Conservation: filling gaps in global biodiversity estimates. *Science* 330, 24.

Pennycuick, C. J. (1982) The flight of petrels and albatrosses (Procellariiformes), observed in South Georgia and its vicinity. *Philosophical Transactions of the Royal Society of London B: Biological Sciences* 300, 75–106.

Pennycuick, C. J. (1989) *Bird Flight Performance: A practical calculation manual* (first edition). Oxford: Oxford University Press.

Pennycuick, C. J., Croxall, J. P., and Prince, P. A. (1984) Scaling of foraging radius and growth rate in petrels and albatrosses (Procellariiformes). *Ornis Scandinavica* 15, 145–154.

Pereira, H. M., Daily, G. C., and Roughgarden, J. (2004) A framework for assessing the relative vulnerability of species to land-use change. *Ecological Applications* 14, 730–742.

Perry, G. and Pianka, E. R. (1997) Animal foraging: past, present and future. *Trends in Ecology and Evolution* 12, 360–364.

Persson, L., Leonardsson, K., De Roos, A. M., Gyllenberg, M., and Christensen, B. (1998) Ontogenetic scaling of foraging rates and the dynamics of a size-structured consumer-resource model. *Theoretical Population Biology* 54, 270–293.

Petchey, O. L., McPhearson, P. T., Casey, T. M., and Morin, P. J. (1999) Environmental warming alters food-web structure and ecosystem function. *Nature* 402, 69–72.

Petchey, O. L., Beckerman, A. P., Riede, J. O., and Warren, P. H. (2008a) Size, foraging, and food web structure. *Proceedings of the National Academy of Sciences of the USA* 105, 4191–4196.

Petchey, O. L., Eklöf, A., Borrvall, C., and Ebenman, B. (2008b) Trophically unique species are vulnerable to cascading extinction. *American Naturalist* 171, 568–579.

Petchey, O. L., Brose, U., and Rall, B. C. (2010a) Predicting the effects of temperature on food web connectance. *Philosophical Transactions of the Royal Society B: Biological Sciences* 365, 2081–2091.

Petchey, O. L., Morin, P. J., and Olff, H. (2010b) The topology of ecological interaction networks: the state of the art. In: Verhoef, H. A. and Morin, P. J. (Eds), *Community Ecology: Processes, Models, and Applications*. Oxford: Oxford University Press.

Petchey, O. L., Beckerman, A. P., Riede, J. O., and Warren, P. H. (2011) Fit, efficiency, and biology: some thoughts on judging food web models. *Journal of Theoretical Biology* 279, 169–171.

Peters, R. H. (1983) *The Ecological Implications of Body Size*. Cambridge: Cambridge University Press.

Peters, R. H. and Downing, J.A. (1984) Empirical analysis of zooplankton filtering and feeding rates. *Limnology and Oceanography* 29, 763–778.

Peters, R. H. and Wassenberg, K. (1983) The effect of body size on animal abundance. *Oecologia* 60, 89–96.

Peterson, A. T., Ball, L. G., and Cohoon, K. P. (2002) Predicting distributions of Mexican birds using ecological niche modelling methods. *IBIS* 144, E27–E32.

Petit, G. and Anfodillo, T. (2009) Plant physiology in theory and practice: an analysis of the WBE model for vascular plants. *Journal of Theoretical Biology* 259, 1–4.

Petit, G., Anfodillo, T., and De Zan, C. (2009) Degree of tapering of xylem conduits in stems and roots of small *Pinus cembra* and *Larix decidua* trees. *Botany–Botanique* 87, 501–508.

Petroff, A. P., Sim, M. S., Maslov, A., Krupenin, M., Rothman, D. H., and Bosak, T. (2010) Biophysical basis for the geometry of conical stromatolites. *Proceedings of the National Academy of Sciences of the USA* 107, 9956–9961.

Petz, M., Stabentheiner, A., and Crailsheim, K. (2004) Respiration of individual honeybee larvae in relation to age and ambient temperature. *Journal of Comparative Physiology B: Biochemical Systemic and Environmental Physiology* 174, 511–518.

Pfeiffer, T., Schuster, S., and Bonhoeffer, S. (2001) Cooperation and competition in the evolution of ATP-producing pathways. *Science* 292, 504–507.

Phillimore, A. B. and Price, T. D. (2008) Density-dependent cladogenesis in birds. *PLoS Biology* 6, e71.

Phillips, O. and Miller, J. S. (2002) Global patterns of plant diversity: Alwyn Gentry's forest transect dataset. *Monographs in Systematic Botany. Missouri Botanical Garden* 89, 1–319.

Piao, S., Ciais, P., Friedlingstein, P., et al. (2008) Net carbon dioxide losses of northern ecosystems in response to autumn warming. *Nature* 451, 49–52.

Pickett, S. T. A. and Thompson, J. N. (1978) Patch dynamics and the design of nature reserves. *Biological Conservation* 13, 27–37.

Pilli, R., Anfodillo, T., and Carrer, M. (2006) Towards a functional and simplified allometry for estimating forest biomass. *Forest Ecology and Management* 237, 583–593.

Pimm, S. L. and Lawton, J. H. (1977) Number of trophic levels in ecological communities. *Nature* 268, 329–331.

Pimm, S. L. and Raven, P. (2000) Biodiversity: extinction by numbers. *Nature* 403, 843–845.

Pimm, S. L., Jones, H. L., and Diamond, J. (1988) On the risk of extinction. *American Naturalist* 132, 757–785.

Pirt, S. J. (1965) The maintenance energy of bacteria in growing cultures. *Proceedings of the Royal Society of London B: Biological Sciences* 163, 224–231.

Platt, J. R. (1964) Strong inference. *Science* 146, 347–353.

Pol, A., Heijmans, K., Harhangi, H. R., Tedesco, D., Jetten, M. S. M., and Den Camp, H. J. M. O. (2007) Methanotrophy below pH 1 by a new Verrucomicrobia species. *Nature* 450, 874–878.

Polidoro, B. A., Carpenter, K. E., Collins, L., et al. (2010) The loss of species: mangrove extinction risk and geographic areas of global concern. *PloS One* 5, e10095.

Polis, G. A. (1991) Complex trophic interactions in deserts: an empirical critique of food web ecology. *American Naturalist* 138, 123–155.

Pontzer, H., Allen, V., and Hutchinson, J. R. (2009) Biomechanics of running indicates endothermy in bipedal dinosaurs. *Plos One* 4, A127–A135.

Poorter, H. (1989) Interspecific variation in relative growth rate: on ecological causes and physiological consequences. In: Lambers, H., Cambridge, M. L., Konings, H., and Pons, T. L. (Eds), *Causes and Consequences of Variation in Growth Rate and Productivity in Higher Plants*. The Hague, Netherlands: SPB Academic Publishing.

Pope, J. G. (1979) A modified cohort analysis in which constant natural mortality is replaced by estimates of predation levels. *International Council for the Exploration of the Sea, Committee Meeting* 1979/ H, 16.

Pope, J. G., Rice, J. C., Daan, N., Jennings, S., and Gislason, H. (2006) Modelling an exploited marine fish community with 15 parameters: results from a simple size-based model. *ICES Journal of Marine Science* 63, 1029–1044.

Porter, R. K. (2001) Allometry of mammalian cellular oxygen consumption. *Cellular and Molecular Life Sciences* 58, 815–822.

Portner, H. O. (2002) Physiological basis of temperature-dependent biogeography: trade-offs in muscle design and performance in polar ectotherms. *Journal of Experimental Biology* 205, 2217–2230.

Posada, D. and Buckley, T. R. (2004) Model selection and model averaging in phylogenetics: advantages of Akaike information criterion and Bayesian approaches over likelihood ratio tests. *Systematic Biology* 53, 793–808.

Posada, D. and Crandall, K. A. (2001) Selecting the best-fit model of nucleotide substitution. *Systematic Biology* 50, 580–601.

Potts, S. G., Biesmeijer, J. C., Kremen, C., Neumann, P., Schweiger, O., and Kunin, W. E. (2010) Global pollinator declines: trends, impacts and drivers. *Trends in Ecology and Evolution* 25, 345–353.

Poulin, F. J. and Franks, P. J. S. (2010) Size-structured planktonic ecosystems: constraints, controls and assembly instructions. *Journal of Plankton Research* 32, 1121–1130.

Poulin, R. (1995) Clutch size and egg size in free-living and parasitic copepods: a comparative analysis. *Evolution* 49, 325–336.

Poulin, R. (1997) Egg production in adult trematodes: adaptation or constraint? *Parasitology* 114, 195–204.

Poulin, R. (1999) Body size vs abundance among parasite species: positive relationships? *Ecography* 22, 246–250.

Poulin, R. and George-Nascimento, M. (2007) The scaling of total parasite biomass with host body mass. *International Journal for Parasitology* 37, 359–364.

Prestwich, K. N. (1994) The energetics of acoustic signaling in anurans and insects. *American Zoologist* 34, 625–643.

Prestwich, K. N., Brugger, K. E., and Topping, M. (1989) Energy and communication in three species of hylid frogs: power input, power output and efficiency. *Journal of Experimental Biology* 144, 53–80.

Price, C. A. and Enquist, B. J. (2006) Scaling mass and morphology in plants with minimal branching: an extension of the WBE model. *Functional Ecology* 20, 11–20.

Price, C. A. and Enquist, B. J. (2007) Scaling mass and morphology in leaves: an extension of the WBE model. *Ecology* 88, 1132–1141.

Price, C. A., Enquist, B. J., and Savage, V. M. (2007) A general model for allometric covariation in botanical form and function. *Proceedings of the National Academy of Sciences of the USA* 104, 13204–13209.

Price, C. A., Ogle, K., White, E. P., and Weitz, J. S. (2009) Evaluating scaling models in biology using hierarchical Bayesian approaches. *Ecology Letters* 12, 641–651.

Price, C. A., Gilooly, J. F., Allen, A. P., Weitz, J. S., and Niklas, K. J. (2010) The metabolic theory of ecology: prospects and challenges for plant biology. *New Phytologist* 188, 696–710.

Price, N. D., Reed, J. L., and Palsson, B. Ø. (2004) Genome-scale models of microbial cells: evaluating the consequences of constraints. *Nature Reviews Microbiology* 2, 886–897.

Price, P. B. and Sowers, T. (2004) Temperature dependence of metabolic rates for microbial growth, maintenance, and survival. *Proceedings of the National Academy of Sciences of the USA* 101, 4631–4636.

Price, P. W. (1980) *Evolutionary Biology of Parasites*. Princeton, NJ: Princeton University Press.

Price, S. A. and Gittleman, J. L. (2007) Hunting to extinction: biology and regional economy influence extinction risk and the impact of hunting in artiodactyls. *Proceedings of the Royal Society B: Biological Sciences* 274, 1845–1851.

Promislow, D. E. L. and Harvey, P. H. (1990) Living fast and dying young: a comparative analysis of life-history variation among mammals. *Journal of Zoology* 220, 417–437.

Pulliam, H. R. (1985) Foraging efficiency, resource partitioning, and the coexistence of sparrow species. *Ecology* 66, 1829–1836.

Purugganan, M. D. and Fuller, D. Q. (2009) The nature of selection during plant domestication. *Nature* 457, 843–848.

Promislow, D. E. L. and Harvey, P. H. (1990) Living fast and dying young – a comparative-analysis of life-history variation among mammals. *Journal of Zoology*, 220, 417–437.

Purvis, A. and Harvey, P. H. (1995) Mammal life-history evolution: a comparative test of Charnov's model. *Journal of Zoology* 237, 259–283.

Purvis, A., Agapow, P.-M., Gittleman, J. L., and Mace, G. M. (2000a) Nonrandom extinction and the loss of evolutionary history. *Science* 288, 328–330.

Purvis, A., Jones, K. E., and Mace, G. M. (2000b) Extinction. *BioEssays* 22, 1123–1133.

Purvis, A., Gittleman, J. L., Cowlishaw, G., and Mace, G. M. (2000c) Predicting extinction risk in declining species. *Proceedings of the Royal Society B: Biological Sciences* 267, 1947–1952.

Pyke, G. H. (1978) Optimal foraging in hummingbirds: testing the marginal value theorem. *Integrative and Comparative Biology* 18, 739.

Pyke, G. H. (1984) Optimal foraging theory: a critical review. *Annual Review of Ecology and Systematics* 15, 523–575.

Pyke, G. H., Pulliam, H. R., and Charnov, E. L. (1977) Optimal foraging: a selective review of theory and tests. *Quarterly Review of Biology* 52, 137–154.

Queller, D. C. and Strassmann, J. E. (2009) Beyond society: the evolution of organismality. *Philosophical Transactions of the Royal Society B: Biological Sciences* 364, 3143–3155.

Rabosky, D. L. (2009) Ecological limits and diversification rate: alternative paradigms to explain the variation in species richness among clades and regions. *Ecology Letters* 12, 735–743.

Rabosky, D. L. (2010) Primary controls on species richness in higher taxa. *Systematic Biology* 59, 634–643.

Raes, J. and Bork, P. (2008) Molecular eco-systems biology: towards an understanding of community function. *Nature Reviews Microbiology* 6, 693–699.

Raffaelli, D. (2007) Food webs, body size, and the curse of the Latin binomial. In: Rooney, N., McCann, K. S., and Noakes, D. L. G. (Eds), *From Energetics to Ecosystems: The Dynamics and Structure of Ecological Systems*. Dordrecht: Springer.

Raich, J., Russell, A., Kitayama, K., Parton, W., and Vitousek, P. (2006) Temperature influences carbon accumulation in moist tropical forests. *Ecology* 87, 76–87.

Raichlen, D. A., Gordon, A. D., Muchlinski, M. N., and Snodgrass, J. J. (2010) Causes and significance of variation in mammalian basal metabolism. *Journal of Comparative Physiology B: Biochemical Systemic and Environmental Physiology* 180, 301–311.

Rall, B. C., Guill, C., and Brose, U. (2008) Food-web connectance and predator interference dampen the paradox of enrichment. *Oikos* 117, 202–213.

Rall, B. C., Kalinkat, G., Ott, D., Vucic-Pestic, O., and Brose, U. (2011) Taxonomic versus allometric constraints on nonlinear interaction strengths. *Oikos* 120, 483–492.

Rand, D. M. (1994) Thermal habit, metabolic rate and the evolution of mitochondrial DNA. *Trends in Ecology and Evolution* 9, 125–131.

Rands, M. R. W., Adams, W. M., Bennun, L., et al. (2010) Biodiversity conservation: challenges beyond 2010. *Science* 329, 1298–1303.

Rascón, B. and Harrison, J. F. (2005) Oxygen partial pressure effects on metabolic rate and behavior of tethered flying locusts. *Journal of Insect Physiology* 51, 1193–1199.

Ratkowsky, D. A., Olley, J., McMeekin, T. A., and Ball, A. (1982) Relationship between temperature and growth rate of bacterial cultures. *Journal of Bacteriology* 149, 1–5.

Ratkowsky, D. A., Lowry, R. K., McMeekin, T. A., Stokes, A. N., and Chandler, R. E. (1983) Model for bacterial culture growth rate throughout the entire biokinetic temperature range. *Journal of Bacteriology* 154, 1222–1226.

Ratkowsky, D. A., Olley, J., and Ross, T. (2005) Unifying temperature effects on the growth rate of bacteria and the stability of globular proteins. *Journal of Theoretical Biology* 233, 351–362.

Raven, J. A. (1984) A cost-benefit analysis of photon absorption by photosynthetic unicells. *New Phytologist* 98, 593–625.

Raven, J. A. and Geider, R. J. (1988) Temperature and algal growth. *New Phytologist* 110, 441–461.

Raven, J. A. and Handley, L. L. (1987) Transport processes and water relations. *New Phytologist (Suppl.)* 106, 217–233.

Raymond, J. and Segre, D. (2006) The effect of oxygen on biochemical networks and the evolution of complex life. *Science* 311, 1764–1767.

Read, A. F. and Harvey, P. H. (1989) Life-history differences among the eutherian radiations. *Journal of Zoology* 219, 329–353.

Redfield, A. C. (1958) The biological control of chemical factors in the environment. *American Scientist* 46, 205–221.

Reed, R. N. and Shine, R. (2002) Lying in wait for extinction: ecological correlates of conservation status among Australian elapid snakes. *Conservation Biology* 16, 451–461.

Reich, P. B. (2010) The carbon dioxide exchange. *Science* 329, 774–775.

Reich, P. B., Tjoelker, M. G., Machado, J.-L., and Oleksyn, J. (2006) Universal scaling of respiratory metabolism, size and nitrogen in plants. *Nature* 439, 457–461.

Reichstein, M., Tenhunen, J. D., Roupsard, O., et al. (2002) Severe drought effects on ecosystem CO_2 and H_2O fluxes at three Mediterranean evergreen sites: revision of current hypotheses? *Global Change Biology* 8, 999–1017.

Reichstein, M., Rey, A., Freibauer, A., et al. (2003) Modeling temporal and large-scale spatial variability of soil respiration from soil water availability, temperature and vegetation productivity indices. *Global Biogeochemical Cycles* 17, 15-1–15-15.

Reichstein, M., Falge, E., Baldocchi, D., et al. (2005a) On the separation of net ecosystem exchange into assimilation and ecosystem respiration: review and improved algorithm. *Global Change Biology* 11, 1424–1439.

Reichstein, M., Subke, J.-A., Angeli, A. C., and Tenhunen, J. D. (2005b) Does the temperature sensitivity of decomposition of soil organic matter depend upon water content, soil horizon, or incubation time? *Global Change Biology* 11, 1754–1767.

Reiners, W. A. (1986) Complementary models for ecosystems. *American Naturalist* 127, 59–73.

Reinhold, K. (1999) Energetically costly behaviour and the evolution of resting metabolic rate in insects. *Functional Ecology* 13, 217–224.

Reis, P. M., Jung, S., Aristoff, J. M., and Stocker, R. (2010) How cats lap: water uptake by *Felis catus*. *Science* 330, 1231–1234.

Reiss, M. J. (1989) *The Allometry of Growth and Reproduction.* Cambridge: Cambridge University Press.

Renault, D., Hervant, F.. and Vernon, P. (2002) Comparative study of the metabolic responses during food shortage and subsequent recovery at different temperatures in the adult lesser mealworm, *Alphitobius diaperinus* (Coleoptera: Tenebrionidae). *Physiological Entomology* 27, 291–301.

Reuman, D. C., Mulder, C., Raffaelli, D., and Cohen, J. (2008) Three allometric relations of population density to body mass: theoretical integration and empirical tests in 149 food webs. *Ecology Letters* 11, 1216–1228.

Reuman, D. C., Mulder, C., Banasek-Richter, C., et al. (2009) Allometry of body size and abundance in 166 food webs. *Advances in Ecological Research* 41, 1–44.

Revell, L. J., Harmon, L. J., and Collar, D. C. (2008) Phylogenetic signal, evolutionary process, and rate. *Systematic Biology* 57, 591–601.

Rex, M. A., Etter, R. J., Morris, J. S., et al. (2006) Global bathymetric patterns of standing stock and body size in the deep-sea benthos. *Marine Ecology: Progress Series* 217, 1–8.

Reynolds, C. S. (1984) *The Ecology of Freshwater Phytoplankton.* Cambridge: Cambridge University Press.

Reynolds, J. D. (2003) Life histories and extinction risk. In: Blackburn, T. M. and Gaston, K. J. (Eds), *Macroecology*. Oxford: Blackwell Publishing, pp. 195–217.

Reynolds, J. D., Dulvy, N. K., Goodwin, N. B., and Hutchings, J. A. (2005) Biology of extinction risk in marine fishes. *Proceedings of the Royal Society B: Biological Sciences* 272, 2337–2344.

Rhee, G. Y. (1978) Effects of N:P atomic ratios and nitrate limitation on algal growth, cell composition and nitrate uptake. *Limnological Oceanography* 23, 10–25.

Rheindt, F. E., Grafe, T. U., and Abouheif, E. (2004) Rapidly evolving traits and the comparative method: how important is testing for phylogenetic signal? *Evolutionary Ecology Research* 6, 377–396.

Rice, J. and Gislason, H. (1996) Patterns of change in the size spectra of numbers and diversity of the North Sea fish assemblage, as reflected in surveys and models. *ICES Journal of Marine Science* 53, 1214–1225.

Richter, J. P. (1970) *The Notebooks of Leonardo da Vinci 1452–1519*. New York: Dover.

Ricklefs, R. E. (1987) Community diversty: relative roles of local and regional processes. *Science* 235, 167–171.

Ricklefs, R. E. (2003) Is rate of ontogenetic growth constrained by resource supply or tissue growth potential? A comment on West et al.'s model. *Functional Ecology* 17, 384–393.

Ricklefs, R. E. (2006a) Evolutionary diversification and the origin of the diversity-environment relationship. *Ecology* 87 (Supplement), S3–S13.

Ricklefs, R. E. (2006b) Global variation in the diversification rate of passerine birds. *Ecology* 87, 2468–2478.

Ricklefs, R. E. (2007) Estimating diversification rates from phylogenetic information. *Trends in Ecology and Evolution* 22, 601–610.

Ricklefs, R. E. (2008) Disintegration of the ecological community. *American Naturalist* 172, 741–750.

Ricklefs, R. E. (2010a) Insights from comparative analyses of aging in birds and mammals. *Aging Cell* 9, 273–284.

Ricklefs, R. E. (2010b) Life-history connections to rates of aging in terrestrial vertebrates. *Proceedings of the National Academy of Sciences of the USA* 107, 10314–10319.

Ricklefs, R. E. and Schluter, D. (Eds) (1993) *Species Diversity in Ecological Communities: Historical and Geographical Perspectives*. Chicago: University of Chicago Press.

Riede, J. O., Brose, U., Ebenman, B., et al. (2011) Stepping in Elton's footprints: a general scaling model for body masses and trophic levels across ecosystems. *Ecology Letters* 14, 169–178.

Rigby, M. C., Hechinger, R. F., and Stevens, L. (2002) Why should parasite resistance be costly? *Trends in Parasitology* 18, 116–120.

Riviere, J. E. (1999) *Comparative Pharmacokinetics. Principles, Techniques and Applications*. Ames, IA: Iowa State University Press.

Roberts, M. F., Lightfoot, E. N., and Porter, W. P. (2010) A new model for the body size–metabolism relationship. *Physiological and Biochemical Zoology* 83, 395–405.

Roberts, S. P., Harrison, J. F., and Hadley, N. F. (1998) Mechanisms of thermal balance in flying *Centris pallida* (Hymenoptera: Anthophoridae). *Journal of Experimental Biology* 201, 2321–2331.

Roberts, S. P., Harrison, J. F., and Dudley, R. (2004) Allometry of kinematics and energetics in carpenter bees (*Xylocopa varipuncta*) hovering in variable-density gases. *Journal of Experimental Biology* 207, 993–1004.

Robinson, D. (2004) Scaling the depths: below-ground allocation in plants, forests and biomes. *Functional Ecology* 18, 290–295.

Robinson, D. (2007) Implications of a large global root biomass for carbon sink estimates and for soil carbon dynamics. *Proceedings of the Royal Society B: Biological Sciences* 274, 2753–2759.

Robinson, P. W., Simmons, S. E., Crocker, D. E., and Costa, D. P. (2010) Measurements of foraging success in a highly pelagic marine predator, the northern elephant seal. *Journal of Animal Ecology* 79, 1146–1156.

Robinson, W. R., Peters, R. H., and Zimmerman, J. (1983) The effects of body size and temperature on metabolic rate of organisms. *Canadian Journal of Zoology* 61, 281–288.

Roces, F. and Lighton, J. R. B. (1995) Larger bites of leaf-cutting ants. *Nature* 373, 392.

Roff, D. A. (1992) *The Evolution of Life Histories: Theory and Analysis*. New York: Chapman & Hall.

Roff, D. A. (2002) *Life History Evolution*. Sunderland, MA: Sinauer.

Rogers, D. and Randolph, S. (2000) The global spread of malaria in a future, warmer world. *Science* 289, 1763.

Rohde, K. (1992) Latitudinal gradients in species diversity: the search for the primary cause. *Oikos* 65, 514–527.

Rohr, R. P., Scherer, H., Kehrli, P., Mazza, C., and Bersier, L. F. (2010) Modeling food webs: exploring unexplained structure using latent traits. *American Naturalist* 176, 170–177.

Rojstaczer, S., Sterling, S. M., and Moore, N. J. (2001) Human appropriation of photosynthesis products. *Science* 294, 2549–2552.

Romanuk, T. N., Zhou, Y., Brose, U., Berlow, E. L., Williams, R. J., and Martinez, N. D. (2009) Predicting invasion success in complex ecological networks. *Philosophical Transactions of the Royal Society B: Biological Sciences* 364, 1743–1754.

Romanuk, T. N., Hayward, A., and Hutchings, J. A. (2011) Trophic level scales positively with body sizes in fishes. *Global Ecology and Biogeography* 20, 231–240.

Root, R. B. (1967) The niche exploitation pattern of the Blue-gray Gnatcatcher. *Ecological Monographs* 37, 317–350.

Root, T. L., Price, J. T., Hall, K. R., Schneider, S. H., Rosenzweig, C., and Pounds, J. A. (2003) Fingerprints of global warming on wild animals and plants. *Nature* 421, 57–60.

Rosa, R. and Seibel, B. A. (2010) Slow pace of life of the Antarctic colossal squid. *Journal of the Marine Biological Association of the UK* 90, 1375–1378.

Rose, J. M. and Caron, D. A. (2007) Does low temperature constrain the growth rates of heterotrophic protists? Evidence and implications for algal blooms in cold waters. *Limnology and Oceanography* 52, 886–895.

Rosensweig, M. (1968) Net primary productivity of terrestrial environments: predictions from climatological data. *American Naturalist* 102, 67–84.

Rossberg, A. G., Ishii, R., Amemiya, T., and Itoh, K. (2008) The top-down mechanism for body-mass–abundance scaling. *Ecology* 89, 567–580.

Rosso, L., Lobry, J. R., Bajard, S., and Flandrois, J. P. (1995) Convenient model to describe the combined effects of temperature and pH on microbial growth. *Applied and Environmental Microbiology* 61, 610–616.

Rothschild, L. J. and Mancinelli, R. L. (2001) Life in extreme environments. *Nature* 409, 1092–1101.

Rubenstein, D. and Kealey, J. (2010) Cooperation, conflict, and the evolution of complex animal societies. *Nature Education Knowledge* 1, 47.

Rüber, L. and Zardoya, R. (2005) Rapid cladogenesis in marine fishes revisited. *Evolution* 59, 1119–1127.

Rubner, M. 1883. Über den einfluss der körpergrösse auf stoff- und kraftwechsel. *Zeitschrift Biol* 19, 536–562.

Rudolf, V. and Lafferty, K. D. (2011) Stage structure alters how complexity affects stability of ecological networks. *Ecology Letters* 14, 75–79.

Rueppell, O. and Kirkman, R. W. (2005) Extraordinary starvation resistance in *Temnothorax rugatulus* (Hymenoptera, Formicidae) colonies: demography and adaptive behavior. *Insectes Sociaux* 52, 282–290.

Ruf, C. and Fiedler, K. (2002) Tent-based thermoregulation in social caterpillars of *Eriogaster lanestris* (Lepidoptera: Lasiocampidae): behavioral mechanisms and physical features of the tent. *Journal of Thermal Biology* 27, 493–501.

Ruppert, E. E. and Barnes, R. D. (1994) *Invertebrate Zoology* (sixth edition). Fort Worth, TX; London: Saunders College Publishing.

Russell, J. B. and Cook, G. M. (1995) Energetics of bacterial growth: balance of anabolic and catabolic reactions. *Microbiology and Molecular Biology Reviews* 59, 48–62.

Russo, S. E., Wiser, S. K., and Coomes, D. A. (2007) Growth-size scaling relationships of woody plant species differ from predictions of the Metabolic Ecology Model. *Ecology Letters* 10, 889–901.

Russo, S. E., Robinson, S. K., and Terborgh, J. (2003) Size-abundance relationships in an Amazonian bird community: implications for the energetic equivalence rule. *American Naturalist* 161, 267–283.

Rustad, L., Campbell, J., Marion, G., et al. (2001) A meta-analysis of the response of soil respiration, net nitrogen mineralization, and aboveground plant growth to experimental ecosystem warming. *Oecologia* 126, 543–562.

Ryan, M. G. and Yoder, B. J. (1997) Hydraulic limits to tree height and tree growth. *Bioscience* 47, 235–242.

Ryan, M. J. (1986) Factors influencing the evolution of acoustic communication: biological constraints. *Brain Behavior and Evolution* 28, 70–82.

Ryan, M. J. (1988) Energy, calling, and selection. *American Zoologist* 28, 885–898.

Sack, L., Maranon, T., and Grubb, P. J. (2002) Global allocation rules for patterns of biomass partitioning. *Science* 296, A1923–A1923.

Sage, R. F. and Kubien, D. S. (2007) The temperature response of C3 and C4 photosynthesis. *Plant Cell and Environment* 30, 1086–1106.

Saito, M., Sigman, D., and Morel, F. (2003) The bioinorganic chemistry of the ancient ocean: the co-evolution of cyanobacterial metal requirements and biogeochemical cycles at the Archean-Proterozoic boundary? *Inorganica Chimica Acta* 356, 308–318.

Saito, M., Goepfert, T., and Ritt, J. (2008) Some thoughts on the concept of colimitation: Three definitions and the importance of bioavailability. *Limnology and Oceanography* 53, 276–290.

Saito, R., Yamaguchi, A., Saitoh, S.-I., Kuma, K., and Imai, I. (2011) East-west comparison of the zooplankton community in the subarctic Pacific during summers of 2003–2006. *Journal of Plankton Research* 33, 145–160.

Salt, D., Baxter, I., and Lahner, B. (2008) Ionomics and the study of the plant ionome. *Annual Review of Plant Biology* 59, 709.

Salvucci, M. E. and Crafts-Brandner, S. J. (2000) Effects of temperature and dietary sucrose concentration on respiration in the silverleaf whitefly, *Bemisia argentifolii*. *Journal of Insect Physiology* 46, 1461–1467.

Samaniego, H. and Moses, M. E. (2008) Cities as organisms: allometric scaling of urban road networks in the USA. *Journal of Transport and Land Use* 1, 1.

Sampson, D. A., Janssens, I. A., Yuste, J. C., and Ceulemans, R. (2007) Basal rates of soil respiration are correlated with photosynthesis in a mixed temperate forest. *Global Change Biology* 13, 2008–2017.

Sander, P. M. and Clauss, M. (2008) Sauropod gigantism. *Science* 322, 200–201.

Sander, P. M., Klein, N., Buffetaut, E., Cuny, G., Suteethorn, V., and Le Loeuff, J. (2004) Adaptive radiation in sauropod dinosaurs: bone histology indicates rapid evolution of giant body size through acceleration. *Organisms Diversity and Evolution* 4, 165–173.

Sander, P. M., Christian, A., Clauss, M., et al. (2011) Biology of the sauropod dinosaurs: the evolution of gigantism. *Biological Reviews* 86, 117–155.

Sanders, N. J., Lessard, J. P., Fitzpatrick, M. C., and Dunn, R. R. (2007) Temperature, but not productivity or geometry, predicts elevational diversity gradients in ants across spatial grains. *Global Ecology and Biogeography* 16, 640–649.

Sanderson, M. J. (2003) r8s: inferring absolute rates of molecular evolution and divergence times in the absence of a molecular clock. *Bioinformatics* 19, 301–302.

Sarmento, H., Montoya, J. M., Vázquez-Domínguez, E., Vaqué, D., and Gasol, J. M. (2010) Warming effects on marine microbial food web processes: how far can we go when it comes to predictions? *Philosophical Transactions of the Royal Society B: Biological Sciences* 365, 2137–2149.

Sato, K., Sakamoto, K. Q., Watanuki, Y., et al. (2009) Scaling of soaring seabirds and implications for flight abilities of giant pterosaurs. *Plos One* 4.

Sauer, J. R., Hines, J. E., Fallon, J. E., Pardieck, K. L., Ziolkowski, J., and Link, W. A. (2011) *The North American Breeding Bird Survey, Results and Analysis 1966 – 2009* [Online]. Laurel, MD: USGS Patuxent Wildlife Research Center. Available at http://www.mbr-pwrc.usgs.gov/bbs/bbs.html.

Savage, M. and Swetnam T. W. (1990) Early 19th-century fire decline following sheep pasturing in a Navajo Ponderosa pine forest. *Ecology* 71, 2374–2378.

Savage, V. M. (2004) Improved approximations to scaling relationships for species, populations, and ecosystems across latitudinal and elevational gradients. *Journal of Theoretical Biology* 227, 525–534.

Savage, V. M. and West, G. B. (2007) A quantitative, theoretical framework for understanding mammalian sleep. *Proceedings of the National Academy of Sciences of the USA* 104, 1051–1056.

Savage, V. M., Gillooly, J. F., Brown, J. H., West, G. B., and Charnov, E. L. (2004a) Effects of body size and temperature on population growth. *American Naturalist* 163, 429–441.

Savage, V. M., Gillooly, J. F., Woodruff, W.H., et al. (2004b) The predominance of quarter-power scaling in biology. *Functional Ecology* 18, 257–282.

Savage, V. M., Allen, A. P., Brown, J. H., et al. (2007) Scaling of number, size, and metabolic rate of cells with body size in mammals. *Proceedings of the National Academy of Sciences of the USA* 104, 4718–4723.

Savage, V. M., Deeds, E. J., and Fontana, W. (2008) Sizing up allometric scaling theory. *Plos Computational Biology* 4, e1000171. doi:10.1371/journal.pcbi.1000171.

Savage, V.M., Bentley, L. P., Enquist, B. J., et al. (2010) Hydraulic tradeoffs and space filling enable better predictions of vascular structure and function in plants. *Proceedings of the National Academy of Sciences of the USA* 107, 22722–22727.

Savageau, M. A. (1995) Michaelis-Menten mechanism reconsidered: implications of fractal kinetics. *Journal of Theoretical Biology* 176, 115–124.

Saz, H. J. (1981) Energy metabolisms of parasitic helminths – adaptations to parasitism. *Annual Review of Physiology* 43, 323–341.

Scheffer, M., Carpenter, S., Foley, J. A., Folke, C., and Walker, B. (2001) Catastrophic shifts in ecosystems. *Nature* 413, 591–596.

Schemske, D. W. (2009) Biotic interactions and speciation in the tropics. In: Butlin, R. K., Bridle, J. R., and Schluter, D. (Eds), *Speciation and Patterns of Diversity*. Cambridge: Cambridge University Press, pp. 219–239.

Schink, B. (1997) Energetics of syntrophic cooperation in methanogenic degradation. *Microbiology and Molecular Biology Reviews* 61, 262–280.

Schleper, C., Puehler, G., Holz, I. et al. (1995) Picrophilus gen. nov., fam. nov.: a novel aerobic, heterotrophic, thermoacidophilic genus and family comprising archaea capable of growth around pH 0. *Journal of Bacteriology* 177, 7050–7059.

Schmickl, T. and Crailsheim, K. (2001) Cannibalism and early capping: strategy of honeybee colonies in times of experimental pollen shortages. *Journal of Comparative Physiology A: Neuroethology Sensory Neural and Behavioral Physiology* 187, 541–547.

Schmidt-Nielsen, K. (1984) *Scaling: Why is Animal Size so Important?* Cambridge: Cambridge University Press.

Schmidt-Nielsen, K. (1990) *Animal Physiology: Adaptation and Environment*. Cambridge: Cambridge University Press.

Schmitz, O. J., Post, E., Burns, C. E., and Johnston, K. M. (2003) Ecosystem responses to global climate change: moving beyond color mapping. *Bioscience* 53, 1199–1205.

Schmolz, E., Hoffmeister, D., and Lamprecht, I. (2002) Calorimetric investigations on metabolic rates and thermoregulation of sleeping honeybees (*Apis mellifer carnica*). *Thermochimica Acta* 382, 221–227.

Schneider, D. C. (1994) *Quantitative Ecology: Spatial and Temporal Scaling*. San Diego, CA: Academic Press.

Schneiderman, H. A. and Williams, C. M. (1953) The physiology of insect diapause. VII. The respiratory metabolism of the Cecropia silkworm during diapause and development. *Biological Bulletin* 105, 320.

Schoenemann, P. T. (2006) Evolution of the size and functional areas of the human brain. *Annual Review of Anthropology* 35, 379–406.

Schoenemann, P. T. (2009) Evolution of brain and language. *Language Learning* 59, 162–186.

Schoener, T. W. (1971) A theory of feeding strategies. *Annual Review of Ecology and Systematics* 2, 369–404.

Schoener, T. W. (1983) Rate of species turnover decreases from lower to higher organisms: a review of the data. *Oikos* 41, 372–377.

Scholander, P. F. (1955) Evolution of climatic adaptation in homeotherms. *Evolution* 9, 15–26.

Schoolfield, R., Sharpe, P., and Magnuson, C. (1981) Nonlinear regression of biological temperature-dependent rate models based on absolute reaction-rate theory. *Journal of Theoretical Biology* 88, 719–731.

Schreer, J. F. and Kovacs, K. M. (1997) *Allometry of Diving Capacity in Air-Breathing Vertebrates*. Ottawa: National Research Council of Canada.

Schreiber, E. A. and Burger, J. (2001) *Biology of Marine Birds*. Oca Raon, FL: CRC Press.

Schrempf, A. and Heinze, J. (2007) Back to one: consequences of derived monogyny in an ant with polygynous ancestors. *Journal of Evolutionary Biology* 20, 792–799.

Schulte, P. J. and Costa, D. G. (1996) A mathematical description of water flow through plant tissues. *Journal of Theoretical Biology* 180, 61–70.

Schulz, H. N. and Jørgensen, B. B. (2001) Big bacteria. *Annual Reviews in Microbiology* 55, 105–137.

Schulz, H. N., Brinkhoff, T., Ferdelman, T. G., Mariné, M. H., Teske, A., and Jørgensen, B. B. (1999) Dense populations of a giant sulfur bacterium in Namibian shelf sediments. *Science* 284, 493–495.

Schultz, T. D., Quinlan, M. C., and Hadley, N. F. (1992) Preferrered body temperature, metabolic physiology, and water balance of adult *Cicindela longilabris* : a comparison of populations from boreal habitats and climatic refugia. *Physiological Zoology* 65, 226–242.

Schulz, T. M. and Bowen, W. D. (2004) Pinniped lactation strategies: evaluation of data on maternal and offspring life history traits. *Marine Mammal Science* 20, 86–114.

Schwaderer, A. S., Yoshiyama, K., Pinto, P. D. T., Swenson, N. G., Klausmeier, C. A., and Litchman, E. (2011) Eco-evolutionary patterns in light utilization traits and distributions of freshwater phytoplankton. *Limnology and Oceanography* 56, 589–598.

Schwalm, C. R., Williams, C. A., Schaefer, K., et al. (2009) Assimilation exceeds respiration sensitivity to drought: a FLUXNET synthesis. *Global Change Biology* 16, 657–670.

Sears, R. and Perrin, W. F. (2009) Blue Whale. In: Perrin, W. F., Würsig, B., and Thewissen, J. G. M. (Eds), *Encyclopedia of Marine Mammals*. San Francisco; San Diego, CA: Academic Press, pp. 120–124.

Secor, S. (2009) Specific dynamic action: a review of the postprandial metabolic response. *Journal of Comparative Physiology B: Biochemical Systemic and Environmental Physiology* 179, 1–56.

Seddon, J. M., Baverstock, P. R., and Georges, A. (1998) The rate of mitochondrial 12S rRNA gene evolution is similar in freshwater turtles and marsupials. *Journal of Molecular Evolution* 46, 460–464.

Seibel, B. A. and Drazen, J. C. (2007) The rate of metabolism in marine animals: environmental constraints, ecological demands and energetic opportunities. *Philosophical Transactions of the Royal Society B: Biological Sciences* 362, 2061–2078.

Shaffer, S. A. (2011) A review of seabird energetics using the doubly labeled water method. *Comparative Biochemistry and Physiology* 158, 315–322.

Shaffer, S. A., Costa, D. P., and Weimerskirch, H. (2001) Behavioural factors affecting foraging effort of breeding wandering albatrosses. *Journal of Animal Ecology* 70, 864–874.

Shaffer, S. A., Costa, D. P., and Weimerskirch, H. (2004) Field metabolic rates of black-browed albatrosses *Thalassarche melanophrys* during the incubation stage. *Journal of Avian Biology* 35, 551–558.

Shanks, A. L., Grantham, B. A., and Carr, M. H. (2003) Propagule dispersal distance and the size and spacing of marine reserves. *Ecological Applications* 13, S159–S169.

Shapiro, J. A. (1998) Thinking about bacterial populations as multicellular organisms. *Annual Reviews in Microbiology* 52, 81–104.

Shaw, M. R., Zavaleta, E. S., Chiariello, N. R., Cleland, E. E., Mooney, H. A., and Field, C. B. (2002) Grassland responses to global environmental changes suppressed by elevated CO_2. *Science* 298, 1987–1990.

Sheldon, R. W., Prakash, A., and Sutcliffe Jr, W. H. (1972) The size distribution of particles in the ocean. *Limnology and Oceanography* 17, 327–340.

Sheldon, R. W., Sutcliffe Jr, W. H., and Paranjape, M. A. (1977) Structure of pelagic food chain and relationship between plankton and fish production. *Journal of the Fisheries Research Board of Canada* 34, 2344–2355.

Shigenaga, M. K., Gimeno, C. J., and Ames, B. N. (1989) Urinary 8-hydroxy-2′-deoxyguanosine as a biological marker of in vivo oxidative DNA damage. *Proceedings of the National Academy of Sciences of the USA* 86, 9697–9701.

Shik, J. (2010) The metabolic costs of building ant colonies from variably sized subunits. *Behavioral Ecology and Sociobiology*, 1–10.

Shinozaki, K., Yoda, K., Hozumi, K., and Kira, T. (1964a) A quantitative analysis of plant form – the pipe model theory I. Basic analysis. *Japanese Journal of Ecology* 14, 97–105.

Shinozaki, K., Yoda, K., Hozumi, K., and Kira, T. (1964b) A quantitative analysis of plant form – the pipe model theory II. Further evidence of the thoery and its application in forest ecology. *Japanese Journal of Ecology* 14, 133–139.

Shipley, B. (2010) *From Plant Traits to Vegetation Structure*. Cambridge: Cambridge University Press.

Shock, E. L. and Holland, M. E. (2007) Quantitative habitability. *Astrobiology* 7, 839–851.

Shock, E. L., M. Holland, M., Meyer-Dombard, D. A., Amend, J. P., Osburn, G. R., and Fischer, T. P. (2010) Quantifying inorganic sources of geochemical energy in hydrothermal ecosystems, Yellowstone National Park, USA. *Geochimica et Cosmochimica Acta* 74, 4005–4043.

Sibly, R. M. (1983) Optimal group size is unstable. *Animal Behaviour* 31, 947–948.

Sibly, R. M. (2002) Life history theory. In: Pagel, M. (Ed.), *Encyclopedia of Evolution*. Oxford: Oxford University Press.

Sibly, R. M. and Atkinson, D. (1994) How rearing temperature affects optimal adult Size in ectotherms. *Functional Ecology* 8, 486–493.

Sibly, R. M. and Brown, J. H. (2007) Effects of body size and lifestyle on evolution of mammal life histories. *Proceedings of the National Academy of Sciences of the USA* 104, 17707–17712.

Sibly, R. M. and Brown, J. H. (2009) Mammal reproductive strategies driven by offspring mortality-size relationships. *American Naturalist* 173, E185–E199.

Sibly, R. M. and Calow, P. (1986) *Physiological Ecology of Animals*. Oxford: Blackwell Scientific Publications.

Sibly, R. and Calow, P. (1987) Ecological compensation – a complication for testing life-history theory. *Journal of Theoretical Biology*, 125, 177–186.

Sicko-Goad, L. M., Schelske, C. L., and Stoermer, E. F. (1984) Estimation of intracellular carbon and silica content of diatoms from natural assemblages using morphometric techniques. *Limnology and Oceanography* 29, 1170–1178.

Sieg, A. E., O'Connor, M. P., McNair, J. N., Grant, B. W., Agosta, S. J., and Dunham, A. E. (2009) Mammalian metabolic allometry: Do intraspecific variation, phylogeny, and regression models matter? *American Naturalist* 174, 720–733.

Siegel, D. A., Kinlan, B. P., Gaylord, B., and Gaines, S. D. (2003) Lagrangian descriptions of marine larval dispersal. *Marine Ecology: Progress Series* 260, 83–96.

Siegrist, H. and Gujer, W. (1985) Mass transfer mechanisms in a heterotrophic biofilm. *Water Research* 19, 1369–1378.

Silva, M. and Downing, J. A. (1994) Allometric scaling of minimal mammal densities. *Conservation Biology* 8, 732–743.

Silver, S. (1998) Genes for all metals – a bacterial view of the Periodic Table. *Journal of Industrial Microbiology and Biotechnology* 20, 1–12.

Silvertown, J., McConway, K., Gowing, D., et al. (2006) Absence of phylogenetic signal in the niche structure of meadow plant communities. *Proceedings of the Royal Society B: Biological Sciences* 273, 39–44.

Simberloff, D. (2004) Community ecology: is it time to move on? *American Naturalist* 163, 787–799.

Simberloff, D. S. and Abele, L. G. (1976) Island biogeography theory and conservation practice. *Science* 191, 285–286.

Šímová, I., Storch, D., Keil, P., Boyle, B., Phillips, O. L., and Enquist, B. J. (2011) Global species-energy relationship in forest plots: role of abundance, temperature and species' climatic tolerances. *Global Ecology and Biogeography* 20, 842–856.

Simpson, G. G. (1944) *Tempo and Mode in Evolution*. New York: Columbia University Press.

Simpson, G. G. (1953) *The Major Features of Evolution*. New York: Columbia University Press.

Simpson, S. J. and Raubenheimer, D. (2001) The geometric analysis of nutrient-allelochemical interactions: a case study using locusts. *Ecology* 82, 422–439.

Sinclair, B. J., Bretman, A., Tregenza, T. O. M., Tomkins, J. L., and Hosken, D. J. (2011) Metabolic rate does not decrease with starvation in *Gryllus bimaculatus* when changing fuel use is taken into account. *Physiological Entomology* 36, 84–89.

Sinervo, B., Mendez-De-La-Cruz, F., Miles, D. B., et al. (2010) Erosion of lizard diversity by climate change and altered thermal niches. *Science* 328, 894–899.

Sinsabaugh, R. L. and Follstad Shah, J. J. (2011) Ecoenzymatic stoichiometry of recalcitrant organic matter decomposition: the growth rate hypothesis in reverse. *Biogeochemistry* 102, 1–13.

Sinsabaugh, R. L. and Shah, J. J. (2010) Integrating resource utilization and temperature in metabolic scaling of riverine bacterial production. *Ecology* 91, 1455–1465.

Sinsabaugh, R. L., Hill, B. H., and Shah, J. J. F. (2009) Ecoenzymatic stoichiometry of microbial organic nutrient acquisition in soil and sediment. *Nature* 462, 795–798.

Sinsabaugh, R. L., Van Horn, D. J., Follstad Shah, J. J., and Findlay, S. (2010) Ecoenzymatic stoichiometry in relation to productivity for freshwater biofilm and plankton communities. *Microbial Ecology* 60, 885–893.

Sinsabaugh, R. L., Follstad Shah, J. J., Hill, B. H., and Elonen, C. M. (2011) Ecoenzymatic stoichiometry of stream sediments with comparison to terrestrial soils. *Biogeochemistry* DOI: 10.1007/s10533-011-9676-x.

Skorping, A., Read, A. F., and Keymer, A. E. (1991) Life-history covariation in intestinal nematodes of mammals. *Oikos* 60, 365–372.

Smayda, T. J. (1970) The suspension and sinking of phytoplankton in the sea. *Oceanography and Marine Biology Annual Review* 8, 353–414.

Smetacek, V. (1999) Diatoms and the ocean carbon cycle. *Protist* 150, 25–32.

Smetacek, V. (2001) A watery arms race. *Nature* 411, 745.

Smil, V. (2000) *Cycles of Life: Civilization and the Biosphere*. New York: Scientific American Library.

Smil, V. (2008) *Energy in Nature and Society: General Energetics of Complex Systems*. Cambridge, MA: MIT.

Smith, C. C. and Fretwell, S. D. (1974) Optimal balance between size and number of offspring. *American Naturalist* 108, 499–506.

Smith, F. A. and Betancourt, J. L. (1998) Response of bushy-tailed woodrats (*Neotoma cinerea*) to Late Quaternary climatic change in the Colorado Plateau. *Quaternary Research* 50, 1–11.

Smith, F. A. and Betancourt, J. L. (2003) The effect of Holocene temperature fluctuations on the evolution and ecology of Neotoma (woodrats) in Idaho and northwestern Utah. *Quaternary Research* 59, 160–171.

Smith, F. A. and Betancourt, J. L. (2006) Predicting woodrat (Neotoma) responses to anthropogenic warming from studies of the palaeomidden record. *Journal of Biogeography* 33, 2061–2076.

Smith, F. A., Betancourt, J. L., and Brown, J. H. (1995) Evolution of body size in the woodrat over the past 25 000 years of climate change. *Science* 270, 2012–2014.

Smith, F. A., Browning, H., and Shepherd, U. L. (1998) The influence of climate change on the body mass of woodrats Neotoma in an arid region of New Mexico, USA. *Ecography* 21, 140–148.

Smith, F. A., Crawford, D. L., Harding, L. E., et al. (2009) A tale of two species: extirpation and range expansion during the late Quaternary in an extreme environment. *Global and Planetary Change* 65, 122–133.

Smith, F. A., Boyer, A. G., Brown, J. H., et al. (2010a) The evolution of maximum body size of terrestrial mammals. *Science* 330, 1216–1219.

Smith, F. A., Elliott, S. M., and Lyons, S. K. (2010b) Methane emissions from extinct megafauna. *Geoscience* 3, 374–375.

Smith, J. A. C., Schulte, P. J., and Nobel, P. S. (1987) Water flow and water storage in Agave deserti: osmotic implications of crassulacean acid metabolism. *Plant Cell and Environment* 10, 639–648.

Smith, J. M. (1964) Group selection and kin selection. *Nature* 201, 1145–1147.

Smith, R. J. (2009) Use and misuse of the reduced major axis for line-fitting. *American Journal of Physical Anthropology* 140, 476–486.

Smith, V. H., Foster, B. L., Grover, J. P., Holt, R. D., Leibold, M. A., and Denoyelles, F. (2005) Phytoplankton species richness scales consistently from laboratory microcosms to the world's oceans. *Proceedings of the National Academy of Sciences of the USA* 102, 4393–4396.

Socha, J. J., Westneat, M. W., Harrison, J. F., Waters, J. S., and Lee, W. K. (2007) Real-time phase-contrast x-ray imaging: a new technique for the study of animal form and function. *BMC Biology* 5, 6.

Socha, J. J., Lee, W.-K., Harrison, J. F., Waters, J. S., Fezzaa, K., and Westneat, M. W. (2008) Correlated patterns of tracheal compression and convective gas exchange in a carabid beetle. *Journal of Experimental Biology* 211, 3409–3420.

Socha, J. J., Förster, T. D., and Greenlee, K. J. (2010) Issues of convection in insect respiration: Insights from synchrotron X-ray imaging and beyond. *Respiratory Physiology and Neurobiology* 173, S65–S73.

Sodhi, N. S., Bickford, D., Diesmos, A. C., et al. (2008) Measuring the meltdown: drivers of global amphibian extinction and decline. *PLoS ONE* 3, e1636.

Sokal, R. R. and Rohlf, F. J. (1995) *Biometry: The Principles and Practice of Statistics in Biological Research* (third edition). New York: W. H. Freeman & Co.

Sommer, U. (1984) The paradox of the plankton: fluctuations of phosphorus availability maintain diversity of phytoplankton in flow-through cultures. *Limnology and Oceanography* 29, 633–636.

Sommer, U. and Lewandowska, A. (2011) Climate change and the phytoplankton spring bloom: warming and overwintering zooplankton have similar effects on phytoplankton. *Global Change Biology* 17, 154–162.

Sorensen, A. A., Busch, T. M., and Vinson, S. B. (1983) Factors affecting brood cannibalism in laboratory colonies of the imported fire ant, *Solenopsis invicta* Buren (Hymenoptera: Formicidae). *Journal of the Kansas Entomological Society* 56, 140–150.

Sorensen, A. A., Busch, T. M., and Vinson, S. B. (1985) Control of food influx by temporal subcastes in the fire ant, *Solenopsis invicta*. *Behavioral Ecology and Sociobiology* 17, 191–198.

Sousa, T., Marques, G. M., and Domingos, T. (2009) Comment on "Energy Uptake and Allocation During Ontogeny." *Science* 325, 1206b.

Sousa, W. P. (1983) Host life history and the effect of parasitic castration on growth a field study of *Cerithidea californica* (Gastropoda: Prosobranchia) and its trematode parasites. *Journal of Experimental Marine Biology and Ecology* 73, 273–296.

Southwick, E. E. (1985) Allometric relations, metabolism and heat conductance in clusters of honeybees at cool temperatures. *Journal of Comparative Physiology* 156, 143–149.

Southwick, E. E., Roubik, D. W., and Williams, J. M. (1990) Comparative energy balance in groups of Africanized and European honey bees: ecological implications. *Comparative Biochemistry and Physiology* 97A, 1–7.

Speakman, J. R. (2005a) Body size, energy metabolism and lifespan. *Journal of Experimental Biology* 208, 1717–1730.

Speakman, J. R. (2005b) Correlations between physiology and lifespan – two widely ignored problems with comparative studies. *Aging Cell* 4, 167–175.

Spear, J. R., Walker, J. J., McCollom, T. M., and Pace, N. R. (2005) Hydrogen and bioenergetics in the Yellowstone geothermal ecosystem. *Proceedings of the National Academy of Sciences of the USA* 102, 2555–2560.

Sperry, J. S., Alder, N. N., and Eastlack, S. E. (1993) The effect of reduced hydraulic conductance on stomatal conductance and xylem cavitation. *Journal of Experimental Botany* 44, 1075–1082.

Sperry, J. S., Hacke, U. G., and Pittermann, J. (2006) Size and function in conifer tracheids and angiosperm vessels. *American Journal of Botany* 93, 1490–1500.

Sperry, J. S., Meinzer, F. C., and McCulloh, K. A. (2008) Safety and efficiency conflicts in hydraulic architecture: scaling from tissues to trees. *Plant Cell and Environment* 31, 632–645.

Spotila, J. R., Lommen, P. W., Bakken, G. S., and Gates, D. M. (1973) Mathematical-model for body temperatures of large reptiles – implications for dinosaur ecology. *American Naturalist* 107, 391–404.

Staley, J. T. and Reysenbach, A. L. (2002) *Biodiversity of Microbial Life: Foundation of Earth's Biosphere*. New York: Wiley-Liss.

Stallmann, R. R., and Harcourt, A. H. (2006) Size matters: the (negative) allometry of copulatory duration in mammals. *Biological Journal of the Linnean Society* 87, 185–193.

Stang, M., Klinkhamer, P. G. L., and Van Der Meijden, E. (2006) Size constraints and flower abundance determine the number of interactions in a plant-flower visitor web. *Oikos* 112, 111–121.

Stanley, S. M. (1973) Effects of competition on rates of evolution, with special reference to bivalve mollusks and mammals. *Systematic Zoology* 22, 486–506.

Stanley, S. M. (1979) *Macroevolution: Pattern and Process*. San Francisco, CA: W. B. Freeman & Co.

Staples, J. F. and Brown, J. C. L. (2008) Mitochondrial metabolism in hibernation and daily torpor: a review. *Journal of Comparative Physiology B: Biochemical Systemic and Environmental Physiology* 178, 811–827.

Stearns, S. C. (1992) *The Evolution of Life Histories*. Oxford: Oxford University Press.

Stegelmann, C., Andreasen, A., and Campbell, C. T. (2009) Degree of rate control: how much the energies of intermediates and transition states control rates. *Journal of the American Chemical Society* 131, 8077–8082.

Stegen, J. C., Enquist, B. J., and Ferriere, R. (2009) Advancing the metabolic theory of biodiversity. *Ecology Letters* 12, 1001–1015.

Stegen, J. C., Swenson, N. G., Enquist, B. J., et al. (2011) Variation in above-ground forest biomass across broad climatic gradients. *Global Ecology and Biogeography* 20, 744–754.

Stemberger, R. S. and Gilbert, J. J. (1985) Body size, food concentration, and population growth in planktonic rotifers. *Ecology* 66, 1151–1159.

Stenseth, N. C., Mysterud, A., Ottersen, G., Hurrell, J. W., Chan, K.-S., and Lima, M. (2002) Ecological effects of climate fluctuations. *Science* 297, 1292–1296.

Stephens, D. and Krebs, J. (1986) *Foraging Theory*. Princeton, NJ: Princeton University Press.

Sterner, R. (1997) Modelling interactions of food quality and quantity in homeostatic consumers. *Freshwater Biology* 38, 473–481.

Sterner, R. W. (2008) On the phosphorus limitation paradigm for lakes. *International Review of Hydrobiology* 93, 433–445.

Sterner, R. W. and Elser, J. J. (2002) *Ecological Stoichiometry: The Biology of Elements from Molecules to the Biosphere*. Princeton, NJ: Princeton University Press.

Stevens, E. D. and Josephson, R. K. (1977) Metabolic rate and body temperature in singing katydids. *Physiological Zoology* 50, 31–42.

Stevenson, B. S. and Schmidt, T. M. (2004) Life history implications of rRNA gene copy number in *Escherichia coli*. *Applied and Environmental Microbiology* 70, 6670–6677.

Stoddard, P. K., and Salazar, V. L. (2011) Energetic cost of communication. *Journal of Experimental Biology* 214, 200–205.

Stoks, R., Block, M. D., and McPeek, M. A. (2006) Physiological costs of compensatory growth in a damselfly. *Ecology* 87, 1566–1574.

Storch, D. (2003) Comment on "Global biodiversity, biochemical kinetics and the energetic-equivalence rule." *Science* 299, 346b.

Storch, D. and Šizling, A. L. (2008) The concept of taxon invariance in ecology: do diversity patterns vary with changes in taxonomic resolution? *Folia Geobotanica* 43, 329–344.

Storch, D., Davies, R. G., Zajíček, S., et al. (2006) Energy, range dynamics and global species richness patterns: reconciling mid-domain effects and environmental determinants of avian diversity. *Ecology Letters* 9, 1308–1320.

Stouffer, D. B. (2010) Scaling from individuals to networks in food webs. *Functional Ecology* 24, 44–51.

Stouffer, D. B., Camacho, J., Guimera, R., Ng, C. A., and Amaral, L. A. N. (2005) Quantitative patterns in the structure of model and empirical food webs. *Ecology* 86, 1301–1311.

Stouffer, D. B., Camacho, J., and Amaral, L. A. N. (2006) A robust measure of food web intervality. *Proceedings of the National Academy of Sciences of the USA* 103, 19015–19020.

Stouffer, D. B., Rezende, E. L., and Amaral, L. A. N. (2011) The role of body mass in diet contiguity and food-web structure. *Journal of Animal Ecology* 80, 632–639.

Strassmann, J. E. and Queller, D. C. (2010) The social organism: congresses, parties, and committees. *Evolution* 64, 605–616.

Suarez, R. K. and Darveau, C. A. (2005) Multi-level regulation and metabolic scaling. *Journal of Experimental Biology* 208, 1627–1634.

Suding, K. N. and Goldstein, L. J. (2008) Testing the Holy Grail framework: using functional traits to predict ecosystem change. *New Phytologist* 180, 559–562.

Sunda, W. and Huntsman, S. (1997) Interrelated influence of iron, light and cell size on marine phytoplankton growth. *Nature* 390, 389–392.

Surovell, T., Waguespack, N., and Brantingham, P. J. (2005) Global archaeological evidence for proboscidean overkill. *Proceedings of the National Academy of Sciences of the USA* 102, 6231–6236.

Suryan, R. M., Anderson, D. J., Shaffer, S. A., et al. (2008) Wind, waves, and wing loading: their relative importance to the at-sea distribution and movements of North and Central Pacific albatrosses. *PLoS One* 3(12), e4016, doi:10.1371/journal.pone.0004016.

Sutcliffe, W. (1970) Relationship between growth rate and ribonucleic acid concentration in some invertebrates. *Journal of the Fisheries Research Board of Canada* 27, 606–609.

Sutherland, W. J., Grafen, A., and Harvey, P. H. (1986) Life history correlations and demography. *Nature* 320, 88.

Sutton, F. M. and Morgan, J. W. (2009) Functional traits and prior abundance explain native plant extirpation in a fragmented woodland landscape. *Journal of Ecology* 97, 718–727.

Svenning, J.-C., Borchsenius, F., Bjorholm, S., and Balslev H. (2007) High tropical net diversification drives the New World latitudinal gradient in palm (Arecaceae) species richness. *Journal of Biogeography* 35, 394–406.

Swift, M., Heal, O., and Andersen, J. (1979) *Decomposition in Terrestrial Ecosystems*. Berkeley, CA: University of California Press.

Symonds, M. R. E. and Elgar, M. A. (2002) Phylogeny affects estimation of metabolic scaling in mammals. *Evolution* 56, 2330–2333.

Szathmary, E. and Smith, J. M. (1995) The major evolutionary transitions. *Nature* 374, 227–232.

Taguchi, S. (1976) Relationship between photosynthesis and cell size of marine diatoms. *Journal of Phycology* 12, 185–189.

Taiz, L. and Zeiger, E. (1998) *Plant Physiology* (second edition). Sunderland, MA: Sinauer.

Tamura, M., Shimada, S., and Horiguchi, T. (2005) *Galeidiniium rugatum* gen. et sp. nov. (Dinophyceae), a new coccoid dinoflagellate with a diatom endosymbiont. *Journal of Phycology* 41, 658–671.

Taper, M. L. and Marquet, P. A. (1996) How do species really divide resources? *American Naturalist* 147, 1072–1086.

Taylor, C. R. and Weibel, E. R. (1981) Design of the mammalian respiratory system. I. Problem and strategy. *Respiration Physiology* 44, 1–10.

Terblanche, J. S. and Chown, S. L. (2007) The effects of temperature, body mass and feeding on metabolic rate in the tsetse fly *Glossina morsitans centralis*. *Physiological Entomology* 32, 175–180.

Terborgh, J. (1976) Island biogeography and conservation: strategy and limitations. *Science* 193, 1029–1030.

Teske, A. and Stahl, D.A. (2002) Microbial mats and biofilms: evolution, structure, and function of fixed microbial communities. In: *Biodiversity of Microbial Life*. New York: Wiley-Liss, pp. 49–101.

Tewksbury, J. J., Huey, R. B., and Deutsch, C. A. (2008) Putting the heat on tropical animals. *Science* 320, 1296–1297.

Thauer, R. K., Jungermann, K., and Decker, K. (1977. Energy conservation in chemotrophic anaerobic bacteria. *Microbiology and Molecular Biology Reviews* 41, 100–180.

Thiel, H. (1975) The size structure of the deep-sea benthos. *Internationale Revue des Gesamten Hydrobiologie* 60, 576–606.

Thierry, A., Petchey, O. L., Beckerman, A. P., Warren, P. H., and Williams, R. J. (2011) The consequences of size dependent foraging for food web topology. *Oikos* 120, 493–502.

Thomey, M. L., Collins, S. L., Vargas, R., et al. (2011) Effect of precipitation variability on net primary production and soil respiration in a Chihuahuan Desert grassland. *Global Change Biology* 17, 1505.

Thompson, D. A. W. (1942) *On Growth and Form*. Cambridge: Cambridge University Press.

Thorson, G. (1936) The larval development, growth and metabolism of Arctic marine bottom invertebrates compared with those of other seas. *Meddelelser om Gronland* 100, 1–155.

Thorson, G. (1950) Reproductive and larval ecology of marine bottom invertebrates. *Biological Review* 25, 1–45.

Thuiller, W., Araújo, M. B., and Lavorel, S. (2004) Do we need land-cover data to model species distributions in Europe? *Journal of Biogeography* 31, 353–361.

Tienungoon, S., Ratkowsky, D. A., McMeekin, T. A., and Ross, T. (2000) Growth limits of *Listeria monocytogenes* as a function of temperature, pH, NaCl, and lactic acid. *Applied and Environmental Microbiology* 66, 4979–4987.

Tilman, D. (1982) *Resource Competition and Community Structure*. Princeton, NJ: Princeton University Press.

Tilman, D. (1987) Secondary succession and the pattern of plant dominance along experimental nitrogen gradients. *Ecological Monographs* 57, 189–214.

Tilman, D. (1988) *Plant Strategies and the Dynamics and Structure of Plant Communities*. Princeton, NJ: Princeton University Press.

Tilman, D., Hillerislambers, J., Harpole, S., et al. (2004). Does metabolic theory apply to community ecology? It's a matter of scale. *Ecology* 85, 1797–1799.

Tittensor, D. P., Mora, C., Jetz, W., et al. (2010) Global patterns and predictors of marine biodiversity across taxa. *Nature* 466, 1098–1101.

Tjoelker, M. G., Oleksyn, J., and Reich, P. B. (2001) Modelling respiration of vegetation: evidence for a general temperature-dependent Q10. *Global Change Biology* 7, 223–230.

Tokeshi, M. (1993) Species abundance patterns and community structure. *Advances in Ecological Research* 24, 111–186.

Tomkins, J. L., Lebas, N. R., Witton, M. P., Martill, D. M., and Humphries, S. (2010) Positive allometry and the prehistory of sexual selection. *American Naturalist* 176, 141–148.

Traill, L. W., Bradshaw, C. J. A., and Brook, B. W. (2007) Minimum viable population size : a meta-analysis of 30 years of published estimates. *Biological Conservation* 139, 157–166.

Traniello, J. F. A., Fujita, M. S., and Bowen, R. V. (1984) Ant foraging behavior: ambient temperature influences prey selection. *Behavioral Ecology and Sociobiology* 15, 65–68.

Trick, C. G., Bill, B. D., Cochlan, W. P., Wells, M. L., Trainer, V. L., and Pickell, L. D. (2010) Iron enrichment stimulates toxic diatom production in high-nitrate, low-chlorophyll areas. *Proceedings of the National Academy of Sciences of the USA* 107, 5887–5892.

Trillmich, F. and Trillmich, K. G. K. (1984) The mating systems of pinnipeds and marine iguanas: convergent evolution of polygyny. *Biological Journal of the Linnaean Society* 21, 209–216.

Trillmich, F. and Weissing, F. J. (2006) Lactation patterns of pinnipeds are not explained by optimization of maternal energy delivery rates. *Behavioral Ecology and Sociobiology* 59, 1–13.

Trivers, R. L. (1971) The evolution of reciprocal altruism. *Quarterly Review of Biology* 46, 35–57.

Trouvé, S., Sasal, P., Jourdane, J., Renaud, F., and Morand, S. (1998) The evolution of life-history traits in parasitic and free-living Platyhelminthes: a new perspective. *Oecologia* 115, 370–378.

Trut, M. L. (2001) Canid domestication: the farm fox experiment. *American Scientist* 87, 160–169.

Turcotte, D. L., Pelletier, J. D., and Newman, W. I. (1998) Networks with side branching in biology. *Journal of Theoretical Biology* 193, 577–592.

Turner, D. P., Ritts, W. D., Cohen, W. B., et al. (2006) Evaluation of MODIS NPP and GPP products across multiple biomes. *Remote Sensing of Environment* 102, 282–292.

Tyree, M. T. and Ewers, F. W. (1991) The hydraulic architecture of trees and other woody plants. *New Phytologist* 119, 345–360.

Tyree, M. T. and Sperry, J. S. (1988) Do woody plants operate near the point of catastrophic xylem disfunction caused by dynamic water stress? Answers from a model. *Plant Physiology* 88, 574–580.

Tyree, M. T., Graham, M. E. D., Cooper, K. E., and Bazos, L. J. (1983) The hydraulic architecture of *Thuja occidentalis*. *Canadean Journal of Botany* 61, 2105–2111.

Ulijaszek, S. J. (1995) *Human Energetics in Biological Anthropology*. Cambridge: Cambridge University Press.

Vallino, J. J. (2010) Ecosystem biogeochemistry considered as a distributed metabolic network ordered by maximum entropy production. *Philosophical Transactions of the Royal Society B: Biological Sciences* 365, 1417.

Van Den Honert, T. H. (1948) Water transport in plants as a catenary process. *Discussions of the Faraday Society* 3, 146–153.

Van Den Oever, L., Baas, P., and Zandee, M. (1981) Comparative wood anatomy of Symplocos and latitude and altitude of provenance. *IAWA Bulletin* 2, 3–24.

Van Der Have, T. M. and De Jong, G. (1996) Adult size in ectotherms: temperature effects on growth and differentiation. *Journal of Theoretical Biology* 183, 329–340.

Van Der Meer, J. (2006) Metabolic theories in ecology. *Trends in Ecology and Evolution* 21, 136–140.

Van Valen, L. (1973) A new evolutionary law. *Evolutionary Theory* 1, 1–30.

Van Valen, L. (1975) Group selection, sex, and fossils. *Evolution* 29, 87–93.

Van Valkenburgh, B. (1999) Major patterns in the history of carnivorous mammals. *Annual Reviews of Earth and Planetary Science* 27, 463–493.

Van Voorhies, W. A. (2009) Metabolic function in *Drosophila melanogaster* in response to hypoxia and pure oxygen. *Journal of Experimental Biology* 212, 3132–3141.

Van Voorhies, W. A., Khazaeli, A. A., and Curtsinger, J. W. (2004) Lack of correlation between body mass and metabolic rate in *Drosophila melanogaster*. *Journal of Insect Physiology* 50, 445–453.

Van Wassenbergh, S., Aerts, P., and Herrel, A. (2006) Scaling of suction feeding performance in the catfish *Clarias gariepinus*. *Physiological and Biochemical Zoology* 79, 43–56.

Van Zyl, A., Van Der Linde, T., and Grimbeek, R. (1997) Metabolic rates of pitbuilding and non-pitbuilding antlion larvae (Neuroptera: Myrmeleontidae) from southern Africa. *Journal of Arid Environments* 37, 355–365.

Vance, R. R. (1973) On reproductive strategies in marine bottom invertebrates. *American Naturalist* 107, 339–352.

Vander Zanden, M. J. and Fetzer, W. W. (2007) Global patterns of aquatic food chain length. *Oikos* 116, 1378–1388.

Vazquez, A., Liu, J. X., Zhou, Y., and Oltvai, Z. N. (2010) Catabolic efficiency of aerobic glycolysis: the Warburg effect revisited. *BMC Systems Biology* 4, 58.

Verdy, A., Follows, M. and Flierl, G. (2009) Optimal phytoplankton cell size in an allometric model. *Marine Ecology–Progress Series* 379, 1–12.

Vezina, A. F. (1985) Empirical relationships between predator and prey size among terrestrial vertebrate predators. *Oecologia* 67, 555–565.

Viswanathan, G. M., Afanasyev, V., Buldyrev, S. V., Murphy, E. J., Prince, P. A., and Stanley, H. E. (1996) Lévy flight search patterns of wandering albatrosses. *Nature* 381, 413–415.

Vitousek, P. (1994) Beyond global warming: ecology and global change. *Ecology* 75, 1861–1876.

Vitousek, P. (2004) *Nutrient Cycling and Limitation: Hawai'i as a Model System*. Princeton, NJ: Princeton University Press.

Vitousek, P. M., Mooney, H. A., Lubchenco, J., and Melillo, J. M. (1997) Human domination of Earth's ecosystems. *Science* 277, 494–499.

Vladimirova, I. G. and Zotin, A. I. (1985) *Respiration Rates of Protozoa* (in Russian). All-Russian Institute of Scientific and Technical Information (VINITI).

Voet, D. and Voet, J. G. (1995) *Biochemistry*. New York: J. Wiley & Sons.

Vogt, J. T. and Appel, A. G. (1999) Standard metabolic rate of the fire ant, *Solenopsis invicta* Buren: effects of temperature, mass, and caste. *Journal of Insect Physiology* 45, 655–666.

Voigt, W., Perner, J., Davis, A. J., et al. (2003) Trophic levels are differentially sensitive to climate. *Ecology* 84, 2444–2453.

Von Bertalanffy, L. (1951) Metabolic types and growth types. *American Naturalist* 85, 111–117.

Von Bertalanffy, L. (1957) Quantitative laws in metabolism and growth. *Quarterly Review of Biology* 32, 217–231.

Von Bertalanffy, L. and Nowinski, W. W. (1960) Principles and theory of growth. In: *Fundamental Aspects of Normal and Malignant Growth*. Amsterdam: Elsevier.

Von Brand, T. (1960) Influence of size, motility, starvation, and age on metabolic rate. In: Sasser, J. N. and Jenkins, W. R. (Eds), *Nematology: Fundamentals and Recent Advances with Emphasis on Plant Parasitic and Soil Forms*. Chapel Hill, NC: University of North Carolina Press, pp. 233–241.

Von Brand, T. (1973) *Biochemistry of Parasites* (second edition). New York: Academic Press.

Von Brand, T. (1979) *Biochemistry and Physiology of Endoparasites*. Amsterdam; New York: Elsevier/North-Holland Biomedical Press.

Von Brand, T. and Alling, D. W. (1962) Relations between size and metabolism in larval and adult *Taenia taeniaeformis*. *Comparative Biochemistry and Physiology* 5, 141–148.

Von Dohlen, C. D., Kohler, S., Alsop, S. T., and McManus, W. R. (2001) Mealybug proteobacterial endosymbionts contain proteobacterial symbionts. *Nature* 412, 433–436.

Vucic-Pestic, O., Rall, B. C., Kalinkat, G., and Brose, U. (2010) Allometric functional response model: body masses constrain interaction strengths. *Journal of Animal Ecology* 79, 249–256.

Vucic-Pestic, O., Ehnes, R. B., Rall, B. C., and Brose, U. (2011) Warming up the system: higher predator feeding rates but lower energetic efficiencies. *Global Change Biology* 17, 1301–1310.

Vuong, Q. H. (1989) Likelihood ratio tests for model selection and non-nested hypotheses. *Econometrica* 57, 307–333.

Waddell, T. G., Repovic, P., Melendez-Hevia, E., Heinrich, R., and Montero, F. (1997) Optimization of glycolysis: a new look at the efficiency of energy coupling. *Biochemical Education* 25, 204–205.

Waite, A., Fisher, A., Thompson, P. A., and Harrison, P. J. (1997) Sinking rate versus cell volume relationships illuminate sinking rate control mechanisms. *Marine Ecology: Progress Series* 157, 97–108.

Waldron, K. and Robinson, N. (2009) How do bacterial cells ensure that metalloproteins get the correct metal? *Nature Reviews Microbiology* 7, 25–35.

Walker, K., Skelton, H., and Smith, K. (2002) Cutaneous lesions showing giant yeast forms of *Blastomyces dermatitidis*. *Journal of Cutaneous Pathology* 29, 616–618.

Walker, R., Hill, K., Burger, O., and Hurtado, A. M. (2006) Life in the slow lane revisited: ontogenetic separation between Chimpanzees and humans. *American Journal of Physical Anthropology* 129, 577–583.

Walker, T. W. and Syers, J. K. (1976) The fate of phosphorus during pedogenesis. *Geoderma* 15, 1–19.

Walker, T. D. and Valentine, J. W. (1984) Equilibrium models of evolutionary species diversity and the number of empty niches. *American Naturalist* 124, 887–899.

Wallenstein, M., Allison, S. D., Ernakovich, J., Steinweg, J. M., and Sinsabaugh, R. (2011) Controls on the temperature sensitivity of soil enzymes: a key driver of *in situ* enzyme activity rates. In: Shukla, G. and Varma, A. (Eds), *Soil Enzymology*. Berlin; Heidelberg: Springer.

Walters, R. J. and Hassall, M. (2006) The temperature-size rule in ectotherms: May a general explanation exist after all? *American Naturalist* 167, 510–523.

Walther, G.-R. (2010) Community and ecosystem responses to recent climate change. *Philosophical Transactions of the Royal Society B: Biological Sciences* 365, 2019–2024.

Walther, G.-R., Post, E., Convey, P., et al. (2002) Ecological responses to recent climate change. *Nature* 416, 389–395.

Wang, Z., O'Connor, T. P., Heshka, S., and Heymsfield, S. B. (2001) The reconstrucction of Kleiber's law at the organ-tissue level. *Journal of Nutrition* 131, 2967–2970.

Wang, Z., Brown, J. H., Tang, Z., and Fang, J. (2009) Temperature dependence, spatial scale, and tree species diversity in eastern Asia and North America. *Proceedings of the National Academy of Sciences of the USA* 106, 13388–13392.

Wang, Z., Fang, J., Tang, Z., and Lin, X. (2011) Patterns, determinants and models of woody plant diversity in China. *Proceedings of the Royal Society of London B*. doi: 10.1098/rspb.2010.1897.

Wardle, D. A., Walker, L. R., and Bardgett, R. D. (2004) Ecosystem properties and forest decline in contrasting long-term chronosequences. *Science* 305, 509–513.

Ware, D. M. (2000) Aquatic ecosystems: properties and models. In: Harrison, P. J. and Parsons, T. R. (Eds), *Fisheries Oceanography: An Integrative Approach to Fisheries Ecology and Management*. Oxford: Blackwell Science, pp. 161–194.

Warham, J. (1990) *The Petrels: Their Ecology and Breeding Systems*. San Diego; London: Academic Press.

Warne, R. W. and Charnov, E. L. (2008) Reproductive allometry and the size-number trade-off for lizards. *American Naturalist* 172, E80–E98.

Warren, C. P., Pascual, M., Lafferty, K. D., and Kuris, A. M. (2010) The inverse niche model for food webs with parasites. *Theoretical Ecology* 3, 285–294.

Warren, P. H. (2005) Wearing Elton's Wellingtons: why body size still matters in food webs. In: de Ruiter, P., Wolters, V., and Moore, J. C. (Eds), *Dynamic Food Webs: Multispecies Assemblages, Ecosystem Development, and Environmental Change*. Amsterdam; Boston, MA: Academic Press.

Warren, P. H. and Lawton, J. H. (1987) Invertebrate predator-prey body size relationships: an explanation for upper triangular food webs and patterns in food web structure. *Oecologia* 74, 231–235.

Warton, D. I., Wright, I. J., Falster, D. S., and Westoby, M. (2006) Bivariate line-fitting methods for allometry. *Biological Reviews* 81, 259–291.

Wasserthal, L. T. (1996) Interaction of circulation and tracheal ventilation in holometabolous insects. *Advances in Insect Physiology* 26, 297–351.

Wasserthal, L. T. (2001) Flight-motor-driven respiratory air flow in the hawkmoth *Manduca sexta*. *Journal of Experimental Biology* 204, 2209–2220.

Waters, J. S., Holbrook, C. T., Fewell, J. H., and Harrison, J. F. (2010) Allometric scaling of metabolism, growth, and activity in whole colonies of the seed-harvester ant. *Pogonomyrmex californicus*. *American Naturalist* 176, 501–510.

Weaver, R. S., Kirchman, D. L., and Hutchins, D. A. (2003) Utilization of iron/organic ligand complexes by marine bacterioplankton. *Aquatic Microbial Ecology* 31, 227–239.

Webb, E., Moffett, J., and Waterbury, J. (2001) Iron stress in open-ocean cyanobacteria (*Synechococcus*, *Trichodesmium*, and *Crocosphaera* spp.): identification of the IdiA protein. *Applied and Environmental Microbiology* 67, 5444.

Weibel, E. R. (2000) *Symmorphosis, on Form and Function Shaping Life*. Cambridge, MA: Harvard University Press.

Weibel, E. R., Taylor, C.R., and Hoppeler, H. (1991) The concept of symmorphosis: a testable hypothesis of structure-function relationship. *Proceedings of the National Academy of Sciences of the USA* 88, 10357–10361.

Weibel, E. R., Taylor, C.R., and Bolis, L. (Eds) (1998) *Principles of Animal Design: The Optimization and Symmorphosis Debate.* Cambridge, MA: Cambridge University Press.

Weier, J., Feener, D., and Lighton, J. (1995) Inter-individual variation in energy cost of running and loading in the seed-harvester ant, *Pogonomyrmex maricopa. Journal of Insect Physiology* 41, 321–327.

Weimerskirch, H. (2007) Are seabirds foraging for unpredictable resources? *Deep Sea Research II: Topical Studies in Oceanography* 54, 211–223.

Weimerskirch, H., Guionnet, T., Martin, J., Shaffer, S. A., and Costa, D. P. (2000) Fast and fuel efficient? Optimal use of wind by flying albatrosses. *Proceedings of the Royal Society B: Biological Sciences* 267, 1869–1874.

Weir, J. T. and Schluter, D. (2007) The latitudinal gradient in recent speciation and extinction rates of birds and mammals. *Nature* 315, 1574–1576.

Weis-Fogh, T. (1967) Respiration and tracheal ventilation in locusts and other flying insects. *Journal of Experimental Biology* 47, 561–587.

Weitz, J. S., Ogle, K., and Horn, H. S. (2006) Ontogenetically stable hydraulic design in woody plants. *Functional Ecology* 20, 191–199.

Weitz, J. S. and Levin, S. A. (2006) Size and scaling of predator–prey dynamics. *Ecology Letters* 9, 548–557.

Welch, J. J., Bininda-Emonds, O. R. P., and Bromham, L. (2008) Correlates of substitution rate variation in mammalian protein-coding sequences. *BMC Evolutionary Biology* 8, 53.

Werner, E. E. and Gilliam, J. F. (1984) The ontogenetic niche and speices interactions in size-structured populations. *Annual Review of Ecology and Systematics* 15, 393–425.

West, G. B. and Brown, J. H. (2005) Constructing a theory for scaling and more – Reply. *Physics Today* 58, 20–21.

West, G. B., Brown, J. H., and Enquist, B. J. (1997) A general model for the origin of allometric scaling laws in biology. *Science* 276, 122–126.

West, G., Brown, J., and Enquist, B. (1999a) The fourth dimension of life: fractal geometry and allometric scaling of organisms. *Science* 284, 1677–1679.

West, G., Brown, J., and Enquist, B. (1999b) A general model for the structure and allometry of plant vascular systems. *Nature* 400, 664–667.

West, G. B., Brown, J. H., and Enquist, B. J. (2001) A general model for ontogenetic growth. *Nature* 413, 628–631.

West, G. B., Woodruff, W. H., and Brown, J. H. (2002) Allometric scaling of metabolic rate from molecules and mitochondria to cells and mammals. *Proceedings of the National Academy of Sciences of the USA* 99 (suppl 1), 2473–2478.

West, G. B., Brown, J. H., and Enquist, B. J. (2004) Growth models based on first principles or phenomenology? *Functional Ecology* 18, 188–196.

West, G. B., Enquist, B. J., and Brown, J. H. (2009) A general quantitative theory of forest structure and dynamics. *Proceedings of the National Academy of Sciences of the USA* 106, 7040–7045.

Westerhoff, H. V. and Palsson, B. O. (2004) The evolution of molecular biology into systems biology. *Nature Biotechnology* 22, 1249–1252.

Western, D. (1979) Size, life history and ecology in mammals. *African Journal of Ecology* 17, 185–204.

Westoby, M. (1984) The self-thinning rule. *Advances in Ecological Research* 14, 167–225.

Westoby, M. and Wright, I. J. (2006) Land-plant ecology on the basis of functional traits. *Trends in Ecology and Evolution* 21, 261–268.

White, E. P., Enquist, B. J., and Green, J. L. (2008) On estimating the exponent of power law frequency distributions. *Ecology* 89, 905–912.

White, E. P., Ernest, S. K. M., Kerkhoff, A. J., and Enquist, B. J. (2007) Relationships between body size and abundance in ecology. *Trends in Ecology and Evolution* 22, 323–330.

White, A. and Blum, A. (1995) Effects of climate on chemical weathering in watersheds. *Geochimica et Cosmochimica Acta* 59, 1729–1747.

White, A. F., Blum, A. E., Bullen, T. D., Vivit, D. V., Schulz, M., and Fitzpatrick, J. (1999) The effect of temperature on experimental and natural chemical weathering rates of granitoid rocks. *Geochimica et Cosmochimica Acta* 63, 3277–3291.

White, C. R. and Seymour, R. S. (2003) Mammalian basal metabolic rate is proportional to body mass$^{2/3}$. *Proceedings of the National Academy of Sciences of the USA* 100, 4046–4049.

White, C. R. and Seymour, R. S. (2004) Does basal metabolic rate contain a useful signal? Mammalian BMR allometry and correlations with a selection of physiological, ecological, and life-history variables. *Physiological and Biochemical Zoology* 77, 929–941.

White, C. R. and Seymour, R. S. (2005) Allometric scaling of mammalian metabolism. *Journal of Experimental Biology* 208, 1611–1619.

White, C. R., Phillips, N. F., and Seymour, R. S. (2006) The scaling and temperature dependence of vertebrate metabolism. *Biology Letters* 2, 125–127.

White, C. R., Blackburn, T.M., and Seymour, R. S. (2009) Phylogenetically informed analysis of the allometry of mammalian basal metabolic rate supports neither geometric nor quarter-power scaling. *Evolution* 63, 2658–2667.

White, E. P., Ernest, S. K. M., and Thibault, K. M. (2004) Trade-offs in community properties through time in a desert rodent community. *American Naturalist* 164, 670–676.

White, E. P., Ernest, S. K. M., Kerkhoff, A. J., and Enquist, B. J. (2007) Relationships between body size and abundance in ecology. *Trends in Ecology and Evolution* 22, 323–330.

White, E. P., Enquist, B. J., and Green, J. L. (2008) On estimating the exponent of power-law frequency distributions. *Ecology* 89, 905–912.

White, E. P., Ernest, S. K. M., Adler, P. B., Hurlbert, A. H., and Lyons, S. K. (2010) Integrating spatial and temporal approaches to understanding species richness. *Philosophical Transactions of the Royal Society B: Biological Sciences* 365, 3633–3643.

White, T. (1978) The importance of a relative shortage of food in animal ecology. *Oecologia* 33, 71–86.

White, T. (1993) *The Inadequate Environment: Nitrogen and the Abundance of Animals.* Berlin: Springer.

Whitehead, H. (2009) Sperm Whale. In: Perrin, W. F., Würsig, B., and Thewissen, J. G. M. (Eds), *Encyclopedia of Marine Mammals.* San Francisco, CA: Academic Press, pp. 1091–1097.

Whitfield, J. (2006) *In the Beat of a Heart: Life, Energy, and the Unity of Nature.* Washington, DC: National Academies Press.

Whitman, W. B., Coleman, D. C., and Wiebe, W. J. (1998) Prokaryotes: the unseen majority. *Proceedings of the National Academy of Sciences of the USA* 95, 6578.

Wiegel, F. W. and Perelson, A. S. (2004) Some scaling principles for the immune system. *Immunology and Cell Biology* 82, 127–131.

Wiens, J. J. (2007) Global patterns of diversification and species richness in amphibians. *American Naturalist* 170, S86–S106.

Wiens, J. J. and Donoghue, M. J. (2004) Historical biogeography, ecology and species richness. *Trends in Ecology and Evolution* 19, 639–644.

Wilcox, B. A. and Murphy, D. D. (1985) Conservation strategy: effects of fragmentation on extinction. *American Naturalist* 125, 879–887.

Williams, P., Colla, S., and Xie, Z. (2009) Bumblebee vulnerability: common correlates of winners and losers across three continents. *Conservation Biology* 23, 931–940.

Williams, R. J. and Martinez, N. D. (2000) Simple rules yield complex food webs. *Nature* 409, 180–183.

Williams, R. J. and Martinez, N. D. (2004) Stabilization of chaotic and non-permanent food-web dynamics. *European Physical Journal B* 38, 297–303.

Williams, R. J. (2008) Effects of network and dynamical model structure on species persistence in large model food webs. *Theoretical Ecology* 1, 141–151.

Williams, R. J. P. and Frausto Da Silva, J. J. P. (2006) *The Chemistry of Evolution: The Development of our Ecosystems.* New York; London: Elsevier.

Williams, R. J., Anandanadesan, A., and Purves, D. (2010) The probabilistic niche model reveals the niche structure and role of body size in a complex food web. *PLoS ONE* 5.

Williamson, K. and McCarty, P. L. (1976a) A model of substrate utilization by bacterial films. *Journal of the Water Pollution Control Federation* 48, 9–24.

Williamson, K. and McCarty, P. L. (1976b) Verification studies of the biofilm model for bacterial substrate utilization. *Journal of the Water Pollution Control Federation* 48, 281–296.

Willmer, P., Stone, G., and Johnston, I. (2000) *Environmental Physiology of Animals.* Oxford: Blackwell Science.

Wilson, E. O. (1968) The ergonomics of caste in the social insects. *American Naturalist* 102, 41–66.

Wilson, E. O. (1971) *The Insect Societies.* Cambridge, MA: Harvard University Press.

Wilson, E. O. (1985) The biological diversity crisis. *BioScience* 35, 700–706.

Wilson, E. O. (1992) *The Diversity of Life.* Cambridge, MA: Harvard University Press.

Wilson, R. W., Millero, F. J., Taylor, J. R., et al. (2009) Contribution of fish to the marine inorganic carbon cycle. *Science* 323, 359–362.

Winberg, G. G. (1956) Rate of metabolism and food requirements of fishes. *Journal of the Fisheries Research Board of Canada* 194, 1–253.

Winder, M., Reuter, J. E., and Schladow, S. G. (2009) Lake warming favors small-sized planktonic diatom species. *Proceedings of the Royal Society B: Biological Sciences* 276, 427–435.

Witman, J. D., Etter, R. J., and Smith, F. (2004) The relationship between regional and local species diversity in marine benthic communities: a global perspective. *Proceedings of the National Academy of Sciences of the USA* 101, 15664–15669.

Witton, M. P. and Habib, M. B. (2010) On the size and flight diversity of giant pterosaurs, the use of birds as pterosaur analogues and comments on pterosaur flightlessness. *Plos One* 5.

Wolf, A., Berry, J. A., and Asner, G. P. (2010) Allometric constraints on sources of variability in multi-angle reflectance measurements. *Remote Sensing of Environment* 114, 1205–1219.

Wolf, A., Ciais, P., Bellassen, V., Delbart, N., Field, C., and Berry, J. (2011) Forest biomass allometry in global land surface models. *Global Biogeochemical Cycles* 25, GB3015, doi:10.1029/2010GB003917.

Wolf, B. O. and Walsberg, G. E. (1996) Thermal effects of radiation and wind on a small bird and implications for microsite selection. *Ecology* 77, 2228–2236.

Wood, C. L., Byers, J. E., Cottingham, K. L., Altman, I., Donahue, M. J., and Blakeslee, A. M. H. (2007) Parasites alter community structure. *Proceedings of the National Academy of Sciences of the USA* 104, 9335–9339.

Woods, H. A., Makino, W., Cotner, J. B., et al. (2003) Temperature and the chemical composition of poikilothermic organisms. *Functional Ecology* 17, 237–245.

Woodward, G. and Warren, P. H. (2007) Body size and predatory interactions in freshwaters: scaling from individuals to communities. In: Hildrew, A. G., Raffaelli, D., and Edmonds-Brown, R. (Eds), *Body Size: The Structure and*

Function of Aquatic Ecosystems. Cambridge: Cambridge University Press.

Woodward, G., Ebenman, B., Emmerson, M., et al. (2005a) Body size in ecological networks. *Trends in Ecology and Evolution* 7, 402–409.

Woodward, G., Speirs, D. C., and Hildrew, A. G. (2005b) Quantification and resolution of a complex, size-structured food web. In: Caswell, H. (Ed.), *Food Webs: From Connectivity to Energetics.* London: Elsevier Academic Press.

Woodward, G., Benstead, J. P., Beveridge, O. S., et al. (2010a) Ecological networks in a changing climate. *Advances in Ecological Research* 42, 72–122.

Woodward, G., Blanchard, J., Lauridsen, R. B., et al. (2010b) Individual-based food webs: species identity, body size, and sampling effects. *Advances in Ecological Research* 43, 43.

Woodward, G., Perkins, D. M., and Brown, L. E. (2010c) Climate change and freshwater ecosystems: impacts across multiple levels of organization. *Philosophical Transactions of the Royal Society B: Biological Sciences* 365, 2093–2106.

Wootton, J. T. and Emmerson, M. (2005) Measurement of interaction strength in nature. *Annual Review of Ecology Evolution and Systematics* 36, 419–444.

World Wildlife Fund for Nature (WWF) (2004) *Living Planet Report.* Gland, Switzerland: WWF–World Wildlife Fund for Nature. Available from: http://www.panda.org/news_facts/publications/livingplanetreport/index.cfm.

Worm, B., Hilborn, R., Baum, J. K., et al. (2009) Rebuilding global fisheries. *Science* 325, 578–585.

Wright, D. H. (1983) Species-energy theory: an extension of species-area theory. *Oikos* 41, 496–506.

Wright, I. J., Reich, P. B., Westoby, M., et al. (2004) The worldwide leaf economics spectrum. *Nature* 428, 821–827.

Wright, S. (1932) The roles of mutation, inbreeding, crossbreeding and selection in evolution. *Proceedings of the Sixth International Congress on Genetics* 1, 356–366.

Wright, S., Keeling, J., and Gillman, L. (2006) The road from Santa Rosalia: a faster tempo of evolution in tropical climates. *Proceedings of the Natial Academy of Sciences of the USA* 103, 7718–7722.

Wright, S. D., Gillman, L. N., Ross, H. A., and Keeling, D. J. (2010) Energy and tempo of evolution in amphibians. *Global Ecology and Biogeography* 19, 733–740.

Wu, Z., Dijkstra, P., Koch, G. W., Peñuelas, J., and Hungate, B. A. (2011) Responses of terrestrial ecosystems to temperature and precipitation change: a meta-analysis of experimental manipulation. *Global Change Biology* 17, 927.

Xiao, X., White, E. P., Hooten, M. B., and Durham, S. L. (2011) On the use of log-transformation vs. nonlinear regression for analyzing biological power-laws. *Ecology* 92, 1887–1894.

Yagi, M., Kanda, T., Takeda, T., Ishimatsu, A., and Oikawa, S. (2010) Ontogenetic phase shifts in metabolism: links to development and anti-predator adaptation. *Proceedings*

of the Royal Society B: Biological Sciences 277, 2793–2801.

Yampolsky, L. Y. and Scheiner, S. M. (1996) Why larger offspring at lower temperatures? A demographic approach. *American Naturalist* 147, 86–100.

Yoda, K., Kira, T., Ogawa, H., and Hozumi, K. (1963) Self-thinning in overcrowded pure stands under cultivated and natural conditions. *Journal of Biology: Osaka City Univeristy* 14, 107–129.

Yodzis, P. and Innes, S. (1992) Body size and consumer-resource dynamics. *American Naturalist* 139, 115–1175.

Yom-Tov, Y. (2001) Global warming and body mass decline in Israeli passerine birds. *Proceedings of the Royal Society of London B: Biological Sciences* 268, 947–952.

Yom-Tov, Y. and Nix, H. (1986) Climatological correlates for body size of five species of Australian mammals. *Biological Journal of the Linnean Society* 29, 245–262.

Yom-Tov, Y. and Yom-Tov, S. (2004) Climatic change and body size in two species of Japanese rodents. *Biological Journal of the Linnean Society* 82, 263–267.

Yom-Tov, Y. and Yom-Tov, J. (2005) Global warming, Bergmann's rule and body size in the masked shrew *Sorex cinereus* Kerr in Alaska. *Journal of Animal Ecology* 74, 803–808.

Yom-Tov, Y., Yom-Tov, S., Wright, J., Thorne, C., and Du Feu, R. (2006) Recent changes in body weight and wing length among some British passerine birds. *Oikos* 112, 91–101.

Yoshiyama, K. and Klausmeier, C. A. (2008) Optimal cell size for resource uptake in fluids: a new facet of resource competition. *American Naturalist* 171, 59–70.

Yvon-Durocher, G., Jones, J. I., Trimmer, M., Woodward, G., and Montoya, J. M. (2010) Warming alters the metabolic balance of ecosystems. *Philosophical Transactions of the Royal Society B: Biological Sciences* 365, 2117–2126.

Yvon-Durocher, G., Reiss, J., Blanchard, J., et al. (2011a) Across ecosystem comparisons of size structure: methods, approaches and prospects. *Oikos* 120, 550–563.

Yvon-Durocher, G., Montoya, J. M., Trimmer, M., and Woodward, G. U. Y. (2011b) Warming alters the size spectrum and shifts the distribution of biomass in freshwater ecosystems. *Global Change Biology* 17, 1681–1694.

Zaehle, S. (2005) Effect of height on tree hydraulic conductance incompletely compensated by xylem tapering. *Functional Ecology* 19, 359–364.

Zanotto, F., Gouveia, S., Simpson, S., and Calder, D. (1997) Nutritional homeostasis in locusts: Is there a mechanism for increased energy expenditure during carbohydrate overfeeding? *Journal of Experimental Biology* 200, 2437.

Zarkesh-Ha, P., Bezerra, G. B. P., Forrest, S., and Moses, M.E. (2010) Hybrid network on chip (HNoC): local buses with a global mesh architecture. In *Proceedings of the 12th ACM/ IEEE International Workshop on System Level interconnect Prediction* (Anaheim, California, USA, June 13, 2010). SLIP '10. New York: ACM, pp. 9–14.

Zeuthen, E. (1953) Oxygen uptake as related to body size in organisms. *Quarterly Review of Biology* pp. 1–12.

Zhou, J., Kang, S., Schadt, C., and Garten, C. J. (2008) Spatial scaling of functional gene diversity across various microbial taxa. *Proceedings of the National Academy of Sciences of the USA* 105, 7768–7773.

Zhou, S., Criddle, R., and Mitcham, E. (2000) Metabolic response of *Platynota stultana* pupae to controlled atmospheres and its relation to insect mortality response. *Journal of Insect Physiology* 46, 1375–1385.

Zimmermann, M. H. (1978) Structural requirements for optimal water conduction in tree stems. In: Tomlinson, P. B. and Zimmerman, M. H. (Eds), *Tropical Trees as Living Systems*. New York: Cambridge University Press.

Zimmermann, M. H. (1983) *Xylem Structure and the Ascent of Sap*. Berlin: Springer-Verlag.

Zimmermann, M. H. and Brown, C. L. (1971) *Trees: Structure and Function*. New York: Springer.

Zook, A. E., Eklof, A., and Allesina, S. (2011) Food webs: ordering species according to body size yields high degree of intervality. *Journal of Theoretical Biology* 271, 106–113.

Zuo, W. Y., Moses, M. E., Hou, C., Woodruff, W. H., West, G. B., and Brown, J. H. (2009) Response to Comments on "Energy Uptake and Allocation During Ontogeny." *Science* 325.

INDEX

abundance
    definition, 306
    see also community abundance; species abundance
        distributions
acclimation, 176, 288–9, 291
acorn ant, 206
activation energies, 22, 23, 31, 45
    central tendency, 45
    inherent, 103
    insects, 201, 202
    for photosynthesis, 30, 176
    for respiration, 30
adaptation, 176, 194
adaptive dynamics models, 163
aerobic respiration, 236, 238
affinity specialists, 157–8
African fruit beetle, 206
age at first reproduction, metabolism and, 220–1
aggregation effects, 42
aging, 220, 224
agriculturalists, 255, 256
albatrosses, 228–9, 230, 303
    field metabolic rates, 231, 232
    prey processing, 229, 230
    reproduction, 226
    soaring flight, 228–9, 232
allocation, principle of, 58, 65, 217
allocation rule, 81
allometric constant, 176
allometric food web dynamics models, 95
allometric food web structure models, 93

allometric similarity, across taxa, 184
allometry
    comparative (evolutionary/interspecific), 49, 50
    definition, 146
    early studies, 48–9
    intraspecific, 49, 50
    ontogenetic, 49, 50
    research history, 165–6
alternative models, comparison to, 18–19
alternative reproductive tactics (ARTs), 72–3
altitudinal diversity gradient, 303
amphibians, diversification rates,
        119
amplification, 299
anabolic processes, 51
angiosperms, 170–3, 176
    diversification rates, 119
animal movement patterns, 76
ant-lion larvae, 211
anthropology, 249
ants
    colonies, 199, 301
    foraging networks, 301
    lifespan of queens, 75
    local diversity, 130
    metabolic rates, 205, 207
ape package, 17
apical dominance, 182
aquatic food webs, 163
archaea, 137, 146
area-filling branch lengths, 298

Metabolic Ecology: A Scaling Approach, First Edition. Edited by Richard M. Sibly, James H. Brown, Astrid Kodric-Brown.
© 2012 John Wiley & Sons, Ltd. Published 2012 by John Wiley & Sons, Ltd.

visualization
    frequency distributions, 13
    functional relationships, 11–13
volume flow rate, scaling exponents, 167
vulnerability, 91

wandering albatross, 226
water
    availability, and species richness, 130
    and ecosystem energetics, 108–10
    as elemental availability driver, 38
    limitation, 126, 153, 278
        and insect metabolic rates, 206
    most living mass consists of, 36
water column, stratification, 162
WBE model, 24–6, 53–4, 170–5, 177, 308
    applications beyond biology, 293–301
        cities, 294, 295–8
        computers, 294, 298–301
    core assumptions, 24, 171
    core predictions, 172–4, 178
    elaborations, 176, 177
    empirical issues with, 27–9, 177
    extension to ecology, 182
    predicted scaling exponents, 167, 172–4, 178
    secondary assumptions, 171–2, 174–5, 178
    theoretical criticisms, 26–7, 177
    usefulness, 32–3
WBE2 model, 165, 175, 177, 308
    criticisms, 177

as intraspecific, 175
    merging with trait-based plant ecology, 175–7
wealth creation, 297, 298
weaning mass, 61
weapons, 73
West, Brown, and Endquist model *see* WBE model
whales, 226, 227, 267
whiteflies, 199, 201, 202
whole-organism metabolic rate (*I*), 22
whole-plant respiration rate, 174
whole-plant scaling relationships, 169
wildlife reserves, 277
Wilson's storm petrel, 226
wind, as elemental availability driver, 38
wire scaling, 300
wire width preserving, 298
within-plant scaling relationships, 169
woodrats, 284

xylem flow rate, 174
xylem flow resistance, 168
xylem flux, 25, 100

yeasts, 146, 147, 151

zero-order frameworks, 185, 186
zero-sum dynamics, 84
zooplankton
    development rate, 59–61
    foraging rates, 69
    grazing by, 159